Biomarkers in Drug Discovery and Development

A Handbook of Practice, Application, and Strategy

Edited by

Ramin Rahbari, MS, MBA
Innovative Scientific Management
New York, New York

Jonathan Van Niewaal, MBA
Innovative Scientific Management
Woodbury, Minnesota

Michael R. Bleavins, PhD, DABT
White Crow Innovation
Dexter, Michigan

Second Edition

WILEY

This edition first published 2020

Edition History
John Wiley & Sons Inc. (1e, 2010)

Registered Office
John Wiley & Sons, Inc., 111 River Street, Hoboken, NJ 07030, USA

Editorial Office
111 River Street, Hoboken, NJ 07030, USA

For details of our global editorial offices, customer services, and more information about Wiley products visit us at www.wiley.com.

Wiley also publishes its books in a variety of electronic formats and by print-on-demand. Some content that appears in standard print versions of this book may not be available in other formats.

Library of Congress Cataloging-in-Publication Data is available for this title

Hardback ISBN: 9781119187509

Cover image: Courtesy of Ramin Rahbari
Cover design by Wiley

Set in 10/12pt WarnockPro by SPi Global, Chennai, India

10 9 8 7 6 5 4 3 2 1

Contents

List of Contributors

Salvatore Alesci
Takeda Pharmaceuticals
Cambridge, MA
USA

Sara Assadian
PROOF Centre of Excellence
Vancouver, BC
Canada

and

University of British Columbia
Vancouver, British Columbia
Canada

Robert Balshaw
PROOF Centre of Excellence and
Biomarkers in Transplantation Team
Vancouver, BC
Canada

Mats Bergstrom
Independent Consultant
Uppsala
Sweden

Sven A. Beushausen
Zoetic Pharmaceuticals
Amherst
New York, NY
USA

Michael R. Bleavins
White Crow Innovation
Dexter, MI
USA

Alan P. Brown
Novartis Institutes for Biomedical
Research
Cambridge, MA
USA

Michael Burgess
University of British Columbia
Vancouver, BC
Canada

Bruce D. Car
Bristol-Myers Squibb Co.
Princeton, NJ
USA

Stephen Y. Chan
University of Pittsburgh Medical
Center
Pittsburgh, PA
USA

Kay A. Criswell
Westbrook Biomarker & Pharmaceutical Consulting, LLC
Westbrook, CT
USA

Breanne Crouch
PROOF Centre of Excellence
Vancouver, BC
Canada

and

University of British Columbia
Vancouver, British Columbia
Canada

Miranda K. Culley
Center for Pulmonary Vascular Biology and Medicine, Pittsburgh Heart, Lung, Blood, and Vascular Medicine Institute
University of Pittsburgh School of Medicine
Pittsburgh, PA
USA

Ian Dews
Envestia Ltd.
Thame, Oxfordshire
UK

Andrew Dorner
Takeda Pharmaceuticals International Co.
Cambridge, MA
USA

Gregory J. Downing
Innovation Horizons, LLC
Washington, DC
USA

Robert W. Dunstan
Abbvie
Worcester, MA
USA

Giora Z. Feuerstein
United States Department of Defense
Defense Threat Reduction Agency
Fort Belvoir, VA
USA

Edward P. Ficaro
INVIA Medical Imaging Solutions
Ann Arbor, MI
USA

William R. Foster
Bristol-Myers Squibb Co.
Princeton, NJ
USA

Ross A. Fredenburg
Amathus Therapeutics, Inc.
Cambridge, MA
USA

Gregory P. Fusco
Epividian, Inc.
Chicago, IL
USA

Brian Gemzik
Bristol-Myers Squibb Co.
Princeton, NJ
USA

Federico Goodsaid
Regulatory Pathfinders
San Juan
PR USA

Michael Hehenberger
HM NanoMed
Westport, CT
USA

Nita Ichhpurani
Innovative Scientific Management
Toronto, ON
Canada

Malle Jurima-Romet
Celerion
Montreal, QC
Canada

Paul Keown
PROOF Centre of Excellence and Biomarkers in Transplantation Team
Vancouver, BC
Canada

Anthony A. Killeen
University of Minnesota
Minneapolis, MN
USA

Ji-Young V. Kim
PROOF Centre of Excellence and Biomarkers in Transplantation Team
Vancouver, BC
Canada

Lena King
Innovative Scientific Management
Guelph, ON
Canada

Lars Klareskog
Karolinska Institute
Stockholm
Sweden

Frank Kramer
Bayer AG
Wuppertal
Germany

Karen Lam
PROOF Centre of Excellence
Vancouver, BC
Canada

and

University of British Columbia
Vancouver, British Columbia
Canada

Jean W. Lee
BioQualQuan
Camarillo, CA
USA

Deanne Lister
Invicro, a KonicaMinolta Company, San Diego, CA and Department of Radiology
University of California, San Diego, Molecular Imaging Center, Sanford Consortium for Regenerative Medicine

Bin Li
Takeda Pharmaceuticals International Co.
Cambridge, MA
USA

Xiaowu Liang
ImmPORT Therapeutics
Irvine
California

Calvert Louden
Johnson & Johnson Pharmaceuticals
Raritan, NJ
USA

William B. Mattes
National Center for Toxicological Research
US FDA
Jefferson, AR
USA

Patrick McConville
Invicro, a KonicaMinolta Company, San Diego, CA and Department of Radiology
University of California, San Diego, Molecular Imaging Center, Sanford Consortium for Regenerative Medicine

Bruce McManus
PROOF Centre of Excellence
Vancouver, BC
Canada

and

University of British Columbia
Vancouver, British Columbia
Canada

Robert McMaster
PROOF Centre of Excellence and
Biomarkers in Transplantation Team
Vancouver, BC
Canada

Jonathan B. Moody
INVIA Medical Imaging Solutions
Ann Arbor, MI
USA

Philip S. Murphy
GlaxoSmithKline Research and
Development
Stevenage
UK

Raymond T. Ng
PROOF Centre of Excellence and
Biomarkers in Transplantation Team
Vancouver, BC
Canada

Matthias Ocker
Bayer AG
Germany
Berlin

and

Charite University Medicine
Berlin
Germany

Damian O'Connell
Experimental Drug Development
Centre A*STAR
Singapore

J. Fred Pritchard
Celerion
Lincoln Nebraska, NE
USA

M. Lynn Pritchard
Branta Bioscience, LLC
Littleton, NC
USA

Ramin Rahbari
Innovative Scientific Management
New York, NY
USA

Ruth A. Roberts
Apconix
Alderley Edge, Cheshire
UK

Robert R. Ruffolo
Ruffolo Consulting
Spring City, PA
USA

J. Lynn Rutkowski
Ossianix
Philadelphia, PA
USA

Zaki Shaikhibrahim
Bayer AG
Germany
Berlin

Hyunjin Shin
Takeda Pharmaceuticals
International Co.
Cambridge, MA
USA

Frank D. Sistare
Merck Research Laboratories
West Point, PA
USA

Scott J. Tebbutt
PROOF Centre of Excellence and
Biomarkers in Transplantation Team
Vancouver, BC
Canada

William L. Trepicchio
Takeda Pharmaceuticals
International Co.
Cambridge, MA
USA

Christina Trollmo
Roche Pharmaceuticals
Stockholm
Sweden

Jin Wang
Amgen, Inc.
Thousand Oaks, CA
USA

Frank L. Walsh
Wyeth Research
Collegeville, PA
USA

Yuling Wu
MedImmune
Gaithersburg, MD
USA

Mary Zacour
BioZac Consulting
Montreal, QC
Canada

Preface

Since the first edition of *Biomarkers in Drug Development: A Handbook of Practice, Application, and Strategy* was published in 2010, biomarkers have become even more significant, valuable, and important in the decision-making multiple criteria for the development of new drugs. In particular, previously novel biomarkers in nonclinical studies have transitioned into clinical trials. Companies and regulatory agencies have become more comfortable with the inclusion of biomarkers in *ex vivo* experiments with human tissues/biofluids or Phase I trials, with many early clinical trials now including patients after an additional single ascending dose study in human volunteers.

The use of biomarker technologies and strategies in pharmaceutical development remains the basis for translational medicine, improved patient stratification, and identification of underlying causes of diseases once lumped together based primarily on symptomatology. The approval rates for new drugs have also increased relative to 2010, at least partially due to judicious use of biomarkers to identify the best compounds, as well as answering the regulators' questions more specifically. Patients, regulatory reviewers, and the pharmaceutical industry are seeing safer, more efficacious, and better understood drugs to treat complex diseases. The challenges of escalating drug development costs, increasing duration of clinical development times, high rates of compound failure in Phase II and III clinical trials, blockbuster drugs coming off patent, and novel but unproven targets emerging from discovery all continue to modify the arena. These factors have pressured pharmaceutical research divisions to look for ways to reduce development costs, make better and more informed decisions earlier, reassess traditional testing strategies, and implement new technologies to improve the drug discovery and development processes. Biomarkers remain an important tool for getting new medicines to patients and helping identify molecules with unacceptable liabilities earlier in the process.

Biomarkers have proven to be valuable drug development tools that enhance target validation, thereby helping better understand mechanisms of action and enabling earlier identification of compounds with the highest potential for efficacy in humans. In gene therapy, use of animal models of disease in toxicology studies frequently allows very early monitoring of disease-related biomarkers that are known to be important in disease cause and progression, with the same biomarkers measured in the clinical trials. The biomarker endpoints can be essential for eliminating compounds with unacceptable safety risks or lack of target engagement, enabling the concept of "fail fast, fail early," and providing more accurate or complete information regarding drug performance and disease progression. At the same time that pharmaceutical scientists are focusing on biomarkers in drug discovery and development, and clinical investigators and health care practitioners are using biomarkers increasingly in medical decision-making and diagnosis. Similarly, regulatory agencies have recognized and embraced the value of biomarkers to guide regulatory decision-making about targeting, drug safety, and efficacy. Regulatory agencies in the United States, Europe, Great Britain, Japan, and China have taken leadership roles in encouraging biomarker innovation in the industry and collaboration to identify, evaluate, and qualify novel biomarkers. Moreover, a biomarker strategy facilitates the choice of a critical path to differentiate products in a competitive marketplace.

Biomarkers continue to be a significant focus of specialized scientific meetings and extensive media coverage. The targeted use of biomarkers also is more prominent in scientific society meeting presentations to highlight new therapeutic targets, upstream and downstream applications relevant to a given disease, and as case studies describing how decision-making and compound selection were influenced. We, the coeditors, felt that updating the first edition of *Biomarkers in Drug Development: A Handbook of Practice, Application, and Strategy* was timely, as was the continued emphasis on practical aspects of biomarker identification and use, as well as their strategic implementation, and essential application in improving drug development approaches. We each have experience working with biomarkers in drug development, but we recognized that the specialized knowledge of a diverse group of experts was necessary to create the type of comprehensive book that is needed. Therefore, contributions were invited from authors writing chapters in the first edition, and others who are equally renowned experts in their respective fields. The contributors include scientists from academia, research hospitals, biotechnology and pharmaceutical companies, contract research organizations, and consulting firms and those from the FDA. This second edition also has included more coverage on information technology and computational influences in biomarker development and application. The result is a book that we believe will appeal broadly to pharmaceutical research scientists, clinical and academic investigators, regulatory scientists, managers, students, and all other professionals engaged in drug development who are interested in furthering their knowledge of biomarkers.

As discussed early in the book, biomarkers are not new, yet they also are continuously evolving. They have been used for hundreds of years to help physicians diagnose and treat disease. What is new is an expansion from outcome biomarkers to target and mechanistic biomarkers; the availability of "omics," imaging, and other technologies that allow collection of large amounts of data at the molecular, tissue, and whole-organism levels; and the use of data-rich biomarker information for "translational research," from the laboratory bench to the clinic and back. The potential and value from the clinical observations back to the bench should not be taken lightly. Improvements in data storage, computational tools, and modeling abilities provide us with the insight through the process and the ability to reverse mine even very large data sets. Later chapters are dedicated to highlighting several important technologies that affect drug discovery and development, the conduct of clinical trials, and the treatment of patients.

The book continues with invited leaders from industry and regulatory agencies to discuss the qualification of biomarker assays in the fit-for-purpose process, including perspectives on the development of diagnostics. The importance of statistics cannot be overlooked, and this topic is also profiled with a practical overview of concepts, common mistakes, and helpful tips to ensure credible biomarkers that can address their intended uses. Specific case studies are used to present information on concepts and examples of utilizing biomarkers in discovery, preclinical safety assessment, clinical trials, and translational medicine. Examples are drawn from a wide range of target-organ toxicities, therapeutic areas, and product types. It is hoped that by presenting a wide range of biomarker applications, discussed by knowledgeable and experienced scientists, readers will develop an appreciation of the scope and breadth of biomarker knowledge and find examples that will help them in their own work.

Lessons learned and the practical aspects of implementing biomarkers in drug development programs are perhaps the most critical message to convey. Many pharmaceutical companies have created translational research divisions, and increasingly, external partners, including academic and government institutions, contract research organizations, and specialty laboratories, are providing technologies and services to support biomarker programs. This is changing the traditional organizational models within industry and paving the way toward greater collaboration across sectors and even among companies within a competitive industry. Perspectives from contributing authors representing several of these different sectors are presented.

The book concludes with a perspective on future trends and outlooks on development, including increasing capabilities in data integration, privacy concerns, the reality of personalized medicine, and the addressing of ethical concerns. The field of biomarkers in drug development is evolving rapidly, and this book presents a snapshot of some exciting new approaches. By utilizing the book as a source of new knowledge, or to reinforce or integrate existing knowledge, we hope that readers will gain a greater understanding and appreciation of the strategy and application of biomarkers in drug development and become more effective decision-makers and contributors in their own organizations.

We also note with regret the passing of Dr. Mallé Jurima-Romet, our coeditor for the first edition. Although Mallé was not able to be part of the second edition of the book, her spirit and commitment to the field of biomarkers resides throughout the book. She was a champion of biomarkers and influenced many during her career.

As he has for many years, Dr. Felix de la Iglesia also directed us with advice, commentary, and mentorship. His coaching to always work with sound science, pay attention to the literature, not being afraid to go somewhere just because no one else has ventured into that territory, and to push boundaries all resonate in work. The value of his experience and critical commentary have enhanced this book.

July 2019

Ramin Rahbari
Jon Van Niewaal
Michael R. Bleavins

Part I

Biomarkers and Their Role in Drug Development

1

Biomarkers Are Not New

Ian Dews

Envestia Ltd., Thame, Oxfordshire, UK

Introduction

The word *biomarker* in its medical context is a little over 40 years old. The first ever usage of this term was by Karpetsky, Humphrey, and Levy in the April 1977 edition of the *Journal of the National Cancer Institute*, where they reported that the "serum RNase level … was not a biomarker either for the presence or extent of the plasma cell tumor." Few new words have proved so popular – a recent PubMed search lists more than **810, 676** publications that use it! Part of this success can undoubtedly be attributed to the fact that the word gave a long-overdue name to a phenomenon that has been around at least since the seventh century BC, when Sushustra, the "father of Ayurvedic surgery," recorded that the urine of patients with diabetes attracted ants because of its sweetness. However, although the origins of biomarkers are indeed ancient, it is fair to point out that the pace of progress over the first 2500 years was somewhat less than frenetic.

Uroscopy

Because of its easy availability for inspection, urine was for many centuries the focus of attention. The foundation of the "science" of uroscopy is generally attributed to Hippocrates (460–355 BC) who hypothesized that urine was a filtrate of the "humors," taken from the blood and filtered through the kidneys, a reasonably accurate description. One of his more astute observations was that bubbles on the surface of the urine (now known to be due to proteinuria)

Biomarkers in Drug Discovery and Development: A Handbook of Practice, Application, and Strategy,
Second Edition. Edited by Ramin Rahbari, Jonathan Van Niewaal, and Michael R. Bleavins.

were a sign of long-term kidney disease. Galen (AD 129–200), the most influential of the ancient Greco-Roman physicians, sought to make uroscopy more specific but, in reality, added little to the subject beyond the weight of his reputation, which served to hinder further progress in this as in many other areas of medicine.

Five hundred years later, Theophilus Protospatharius, another Greek writer, took an important step towards the modern world when he investigated the effects of heating urine, thus developing the world's first medical laboratory test. He discovered that heating urine of patients with symptoms of kidney disease caused cloudiness (in fact, the precipitation of proteins). In the sixteenth century, Paracelsus (1493–1541) in Switzerland used vinegar to bring out the same cloudiness (acid, like heat, will precipitate proteins).

Events continued to move both farther north and closer to modernity when in 1695 Frederick Deckers of Leiden in the Netherlands identified this cloudiness as resulting from the presence of albumin. The loop was finally closed when Richard Bright (1789–1858), a physician at Guy's Hospital in London, made the association between proteinuria and autopsy findings of abnormal kidneys.

The progress from Hippocrates's bubbles to Bright's disease represents the successful side of uroscopy, but other aspects of the subject now strike us as a mixture of common sense and bizarre superstition. The technique of collecting urine was thought to be of paramount importance for accurate interpretation. In the eleventh century, Ismail of Jurjani insisted on a full 24-hour collection of urine in a vessel that was large and clean (very sensible) and shaped like a bladder, so that the urine would not lose its "form" (not at all sensible). His advice to keep the sample out of the sun and away from heat continues, however, to be wise counsel even today.

Gilles de Corbeil (1165–1213), physician to King Philip Augustus of France, recorded differences in sediment and color of urine which he related to 20 different bodily conditions. He also invented the matula, or jorden, a glass vessel through which the color, consistency, and clarity of the sample could be assessed. Shaped like a bladder rounded at the bottom and made of thin clear glass, the matula was to be held up in the right (not the left) hand for careful inspection against the light. De Corbeil taught that different areas of the body were represented by the urine in different parts of the matula. These connections, which became ever more complex, were recorded on uroscopy charts that were published only in Latin, thus ensuring that the knowledge and its well-rewarded use in treating wealthy patients were confined only to appropriately educated men. To further this education, de Corbeil, in his role as a professor at the Medical School of Salerno, set out his own ideas and those of the ancient Greek and Persian writers in a work called *Poem on the Judgment of Urines*, which was set to music such that medical students could memorize it more easily. It remained popular for several centuries.

Blood Pressure

One of the first deviations from the usage of urine in the search for markers of function and disease came in 1555 with the publication of a book called *Sphygmicae artis iam mille ducentos annos perditae & desideratae Libri V* by a physician named Józef Struś (better known by his Latinized name, Iosephus Struthius) from Poznán, Poland. In this 366-page work, Struthius described placing increasing weights on the skin over an artery until the pulse was no longer able to lift the load. The weight needed to achieve this gave a crude measure of what he called "the strength of the pulse" or, as we would call it today, blood pressure.

Early attempts at quantitative measurement of blood pressure had to be made on animals rather than on human subjects because of the invasiveness of the technique. The first recorded success with these techniques dates from 1733, when the Reverend Stephen Hales, a British veterinary surgeon, inserted a brass pipe into a horse's artery and connected the pipe to a glass tube. Hales observed the blood rising in the tube and concluded not only that the rise was due to the pressure of the blood in the artery but also that the height of the rise was a measure of that pressure.

By 1847, experimental technique had progressed to the point where it was feasible to measure blood pressure in humans, albeit still invasively. Carl Ludwig inserted brass cannulas directly into an artery and connected them via further brass pipework to a U-shaped manometer. An ivory float on the water in the manometer was arranged to move a quill against a rotating drum, and the instrument was known as a *kymograph* ("wave-writer" in Greek).

Meanwhile, in 1834, Jules Hérisson had described his *sphygmomètre*, which consisted of a steel cup containing mercury, covered by a thin membrane, with a calibrated glass tube projecting from it. The membrane was placed over the skin covering an artery, and the pressure in the artery could be gauged from the movements of the mercury into the glass tube.

Although minor improvements were suggested by a number of authors over the next few years, credit for the invention of the true *sphygmomanometer* goes to Samuel Siegfried Karl Ritter von Basch, whose original 1881 model used water in both the cuff and the manometer tube. Five years later, Scipione Riva-Rocci introduced an improved version in which an inflatable bag in the cuff was connected to a mercury manometer, but neither of these early machines attracted widespread interest. Only in 1901, when the famous American surgeon Harvey Cushing brought back one of Riva-Rocci's machines on his return from a trip to Italy did noninvasive blood pressure measurement really take off.

Sphygmomanometers of the late nineteenth century relied on palpation of the pulse and so could only be used to determine systolic blood pressure. Measurement of diastolic pressure only became possible when Nikolai Korotkoff observed in 1905 that characteristic sounds were made by the constriction of the artery at certain points in the inflation and deflation of the cuff. The greater accuracy allowed by auscultation of these *Korotkoff sounds* opened the way for the massive expansion in research works on blood pressure that characterized the twentieth century.

Imaging

To physicians keen to understand the hidden secrets of the human body, few ideas have been more appealing than the dream of looking through the skin to examine the tissues beneath. The means for achieving this did not appear until a little over a century ago and then very much by accident. On the evening of 8 November 1895, Wilhem Roentgen, a German physicist working at the University of Würzburg, noticed that light was coming from fluorescent material in his laboratory and worked out that this was the result of radiation escaping from a shielded gas discharge tube with which he was working. He was fascinated by the ability of this radiation to pass through apparently opaque materials and promptly set about investigating its properties in more detail. While conducting experiments with different thicknesses of tinfoil, he noticed that if the rays passed through his hand, they cast a shadow of the bones.

Having seen the potential medical uses for his new discovery, Roentgen immediately wrote a paper entitled "On a new kind of ray: a preliminary communication" for the Würzburg Physical Medical Society, reprints of which he sent to a number of eminent scientists with whom he was friendly. One of these, Franz Exner of Vienna, was the son of the editor of the *Vienna Presse*, and hence the news was published quickly, first in that paper and then across Europe. Whereas we are inclined to believe that rapid publication is a feature of the Internet age, the Victorians were no slouches in this matter, and by 24 January 1896, a reprint of the Würzburg paper had appeared in the London *Electrician*, a major journal able to bring details of the invention to a much wider technical audience.

The speed of the response was remarkable. Many physics laboratories already had gas discharge tubes, and, within a month, physicists in a dozen countries were reproducing Roentgen's findings. Edwin Frost produced an X-ray image of a patient's fractured wrist for his physician brother, Gilmon Frost, at Dartmouth College in the United States, while at McGill University in Montreal, John Cox used the new rays to locate a bullet in a gunshot victim's leg. Similar results were obtained in cities as far apart as Copenhagen, Prague, and Rijeka in Croatia. Inevitably, not everyone was initially quite so impressed; *The Lancet* of 1 February 1896, expressed considerable surprise that the Belgians had decided to bring X-rays into practical use in hospitals throughout the country! Nevertheless, it was soon clear that a major new diagnostic tool had been presented to the medical world, and there was little surprise when Roentgen received a Nobel Prize in Physics in 1901.

Meanwhile, in March 1896, Henri Becquerel, Professor of Physics at the Muséum National d'Histoire Naturelle in Paris, while investigating Roentgen's work, wrapped a fluorescent mineral, potassium uranyl sulfate, in photographic plates and black material in preparation for an experiment requiring bright sunlight. However, a period of dull weather intervened, and, prior to performing the experiment, Becquerel found that the photographic plates were fully exposed. This led him to write this: "One must conclude from these experiments that the phosphorescent substance in question emits rays which pass through the opaque paper and reduce silver salts." Becquerel received a Nobel Prize, which he shared with Marie and Pierre Curie, in 1903, but it was to be many years before the use of spontaneous radioactivity reached maturity in medical investigation in such applications as isotope scanning and radioimmunoassay.

The use of a fluoroscopic screen on which X-ray pictures are to be viewed was implicit in Roentgen's original discovery and soon became part of the routine equipment not only of hospitals but even of shoe shops, where large numbers of children's shoe fittings were carried out in the days before the true dangers of radiation were appreciated. However, the greatest value of the real-time viewing approach emerged only following the introduction of electronic image intensifiers by Philips in 1955.

Within months of the introduction of planar X-rays, physicians were asking for a technique that would demonstrate the body in three dimensions. This challenge was taken up by several scientists in different countries, but because of the deeply ingrained habit of reviewing only the national, not the international, literature, they remained ignorant of each other's progress for many years.

Carl Mayer, a Polish physician, first suggested the idea of tomography in 1914. André-Edmund-Marie Bocage in France, Gustav Grossmann in Germany, and Allesandro Vallebona in Italy all developed the idea further and built their own equipment. George Ziedses des Plantes in the Netherlands pulled all these strands together in the 1930s and is generally considered the founder of conventional tomography.

Further progress had to wait for the development of powerful computers, and it was not until 1972 that Godfrey Hounsfield, an engineer at EMI (EMI Records Ltd., a British Transnational conglomerate), designed the first computer-assisted tomographic device, the EMI scanner, installed at Atkinson Morley Hospital, London, an achievement for which he received both a Nobel Prize and a knighthood.

Parallel with these advances in X-ray imaging were ongoing attempts to make similar use of the spontaneous radioactivity discovered by Becquerel. In 1925, Herrman Blumgart and Otto Yens made the first use of radioactivity as a biomarker when they used bismuth-214 to determine the arm-to-arm circulation time in patients. Sodium-24, the first artificially created biomarker radioisotope, was used by Joseph Hamilton to investigate electrolyte metabolism in 1937.

Unlike X-rays, however, radiation from isotopes weak enough to be safe was not powerful enough to create an image merely by letting it fall on a photographic plate. This problem was solved when Hal Anger of the University of California, building on the efficient γ-ray capture system using large flat crystals of sodium iodide doped with thallium developed by Robert Hofstadter in 1948, constructed the first gamma camera in 1957.

The desire for three-dimensional images that led to tomography with X-rays also influenced radioisotope imaging and drove the development of single-photon-emission computed tomography (SPECT) by David Kuhl and Roy Edwards in 1968. Positron-emission tomography (PET) also builds images by detecting energy given off by decaying radioactive isotopes in the form of positrons that collide with electrons and produce γ-rays that shoot off in nearly opposite directions. The collisions can be located in space by interpreting the paths of the γ-rays, and this information is then converted into a three-dimensional image slice. The first PET camera for human studies was built by Edward Hoffman, Michael Ter-Pogossian, and Michael Phelps in 1973 at Washington University. The first whole-body PET scanner appeared in 1977.

Radiation, whether from X-ray tubes or from radioisotopes, came to be recognized as having dangers both for the patient and for personnel operating the equipment, and efforts were made to discover media that would produce images without these dangers. In the late 1940s, George Ludwig, a junior lieutenant at the Naval Medical Research Institute in Bethseda, Maryland, undertook experiments using industrial ultrasonic flaw detection equipment to determine the acoustic impedance of various tissues, including human gallstones surgically implanted into the gallbladders of dogs. His observations were detailed in a 30-page project report to the Naval Medical Research Institute dated 16 June 1949, now considered the first report of its kind on the diagnostic use of ultrasound. However, a substantial portion of Ludwig's work was considered classified information by the Navy and was not published in medical journals.

Civilian research into what became the two biggest areas of early ultrasonic diagnosis – cardiology and obstetrics – began in Sweden and Scotland, respectively, both making use of gadgetry initially designed for shipbuilding. In 1953, Inge Edler, a cardiologist at Lund University, collaborated with Carl Hellmuth Hertz, a graduate student in the department of nuclear physics who was familiar with using ultrasonic reflectoscopes for nondestructive materials testing, and together they developed the idea of using this method in the field of medicine. They made the first successful measurement of heart activity on

29 October 1953, using a device borrowed from Kockums, a Malmö shipyard. On 16 December of the same year, the method was used to generate an echo encephalogram. Edler and Hertz published their findings in 1954.

At around the same time, Ian Donald of the Glasgow Royal Maternity Hospital struck up a relationship with boilermakers Babcock & Wilcox in Renfrew, where he used their industrial ultrasound equipment to conduct experiments assessing the ultrasonic characteristics of various *in vitro* preparations. With fellow obstetrician John MacVicar and medical physicist Tom Brown, Donald refined the equipment to the point where it could be used successfully on live volunteer patients. These findings were reported in *The Lancet* on 7 June 1958, as "Investigation of abdominal masses by pulsed ultrasound."

Nuclear magnetic resonance (NMR) in molecules was first described by Isidor Rabi in 1938. His work was followed up eight years later by Felix Bloch and Edward Mills Purcell, who, working independently, noticed that magnetic nuclei such as hydrogen and phosphorus, when placed in a magnetic field of a specific strength, absorb radio-frequency energy, a situation described as being "in resonance."

For the next 20 years, NMR found purely physical applications in chemistry and physics, and it was not until 1971 that Raymond Damadian showed that the nuclear magnetic relaxation times of different tissues, especially tumors, differed, thus raising the possibility of using the technique to detect disease. Magnetic resonance imaging (MRI) was first demonstrated on small test tube samples in 1973 by Paul Lauterbur, and in 1975 Richard Ernst proposed using phase and frequency encoding and the Fourier transform, the technique that still forms the basis of MRI.

The first commercial nuclear magnetic imaging scanner allowing imaging of the body appeared in 1980 using Ernst's technique, which allowed a single image to be acquired in approximately five minutes. By 1986, the imaging time was reduced to about five seconds without compromising on image quality. In the same year, the NMR microscope was developed, which allowed approximately 10-mm resolution on approximately 1-cm samples. In 1993, functional magnetic resonance imaging (fMRI) was developed, thus permitting the mapping of function in various regions of the brain.

Electrocardiography

Roentgen's discovery of X-rays grew out of the detailed investigation of electricity that was a core scientific concern of the nineteenth century, and it is little surprise that investigators also took a keen interest in the electricity generated by the human body itself. Foremost among these was Willem Einthoven. Before his days, although it was known that the body produced electrical currents, the technology was inadequate to measure or record them with any sort of accuracy. Starting in 1901, Einthoven, a professor at the University of Leiden, conducted a series of experiments using a string galvanometer. In his device, electric currents picked up from electrodes on the patient's skin passed through a thin filament running between very strong electromagnets. The interaction of the electric and magnetic fields caused the filament or "string" to move, and this was detected by using a light to cast a shadow of the moving string onto a moving roll of photographic paper.

It was not, at first, an easy technique. The apparatus weighed 600 lb, including the water circulation system essential for cooling the electromagnets, and was operated by a team of five technicians. Over the next two decades, Einthoven gradually refined his machine and used it to establish the electro-cardiographic (ECG) features of many different heart conditions, work that was eventually recognized with a Nobel Prize in 1924.

As the ECG became a routine part of medical investigations, it was realized that a system that gave only a "snapshot" of a few seconds of the heart's activity could be unhelpful or even misleading in the investigation of intermittent conditions such as arrhythmias. This problem was addressed by Norman Holter, an American biophysicist, who created his first suitcase-sized "ambulatory" monitor as early as 1949, but whose technique is dated in many sources to the major paper that he published on the subject in 1957, and other authors cite an even later, 1961 publication.

Hematology

The scientific examination of blood in order to learn more about the health of the patient can be dated to 1642, when Anthony van Leeuwenhoek first observed blood cells through his newly invented microscope. Progress was at first slow, and it was not until 1770 that leucocytes were discovered by William Hewson, an English surgeon, who also observed that red cells were flat rather than spherical, as had earlier been supposed.

Association of blood cell counts with clinical illness depended on the development of a technical method by which blood cells could be counted. In 1852, Karl Vierordt at the University of Tübingen developed such a technique, which, although too tedious for routine use, was used by one of his students, H. Welcher, to count red blood cells in a patient with "chlorosis" (an old word for what is probably our modern iron-deficiency anemia). He found, in 1854, that an anemic patient had significantly fewer red blood cells than did a normal person. Platelets, the third major cellular constituent of blood, were identified in 1862 by a German anatomist, Max Schultze.

Remarkably, all these discoveries were made without the benefit of cell staining, an aid to microscopic visualization that was not introduced until 1877 in Paul Ehrlich's doctoral dissertation at the University of Leipzig. The movement of blood cell studies from the research laboratory to routine support of patient care needed a fast and automatic technique for separating and counting cells, which was eventually provided by the Coulter brothers, Wallace and Joseph. In 1953 they patented a machine that detected the change in electrical conductance of a small aperture as fluid containing cells was drawn through. Cells, being nonconducting particles, alter the effective cross section of the conductive channel and so signal both their presence and their size.

An alternative technique, flow cytometry, was also developed in stages between the late 1940s and the early 1970s. Frank Gucker at Northwestern University developed a machine for counting bacteria in a laminar stream of air during World War II and used it to test gas masks, the work subsequently being declassified and published in 1947. Louis Kamentsky at IBM Laboratories and Mack Fulwyler at the Los Alamos National Laboratory experimented with fluidic switching and electrostatic cell detectors, respectively, and both described cell sorters in 1965. The modern approach of detecting cells stained with fluorescent antibodies was developed in 1972 by Leonard Herzenberg and his team at Stanford University, who coined the term *fluorescence-activated cell sorter* (FACS).

Blood and Urine Chemistry

As with hematology, real progress in measuring the chemical constituents of plasma depended largely on the development of the necessary technology. Until such techniques became available, however, ingenious use was made of bioassays, developed in living organisms or preparations made from them, to detect and in some cases quantify complex molecules. A good example of this is the detection of human chorionic gonadotrophin (hCG) in urine as a test for pregnancy. Selmar Aschheim and Bernhard Zondek in Berlin, who first isolated this hormone in 1928, went on to devise the Aschheim–Zondek pregnancy test, which involved five days of injecting urine from the patient repeatedly into an infantile female mouse which was subsequently killed and dissected. The finding of ovulation in the mouse indicated that the injected urine contained hCG and meant that the patient was pregnant.

In the early 1940s, the mouse test gave way to the frog test, introduced by Lancelot Hogben in England. This was a considerable improvement, in that injection of urine or serum from a pregnant woman into the dorsal lymph sac of the female African clawed frog (*Xenopus laevis*) resulted in ovulation within 4–12 hours. Although this test was known to give a relatively high proportion of false negatives, it was regarded as an outstanding step forward in diagnosis. One story from the 1950s recounts that with regard to the possible pregnancy of a particular patient, "opinions were sought from an experienced general practitioner, an eminent gynecologist, and a frog; only the frog proved to be correct."

Pregnancy testing, and many other “biomarker” activities, subsequently moved from out-and-out bioassays to the “halfway house” of immunological tests based on antibodies to the test compound generated in a convenient species but then used in an *ex vivo* laboratory setting, and in 1960 a hemagglutination inhibition test for pregnancy was developed by Leif Wide and Carl Gemzell in Uppsala.

Not all immune reactions can be made to modulate hemagglutination, and a problem with the development of immunoassays was finding a simple way to detect whether the relevant antibody or antigen was present. One answer lay in the use of radiolabeled reagents. Radioimmunoassay was first described in a paper by Rosalyn Sussman Yalow and Solomon Berson published in 1960.

Radioactivity is difficult to work with because of its safety concerns, so an alternative was sought. This came with the recognition that certain enzymes (such as ABTS or 3,3′,5,5′-tetramethylbenzidine) which react with appropriate substrates to give a color change could be linked to an appropriate antibody. This linking process was developed independently by Stratis Avrameas and G.B. Pierce. Since it is necessary to remove any unbound antibody or antigen by washing, the antibody or antigen must be fixed to the surface of the container, a technique first published by Wide and Porath in 1966.

In 1971, Peter Perlmann and Eva Engvall at Stockholm University, as well as Anton Schuurs and Bauke van Weemen in the Netherlands, independently published papers that synthesized this knowledge into methods to perform enzyme-linked immunosorbent assay (ELISA).

A further step toward physical methods was the development of chromatography. The word was coined in 1903 by the Russian botanist Mikhail Tswett to describe his use of a liquid–solid form of a technique to isolate various plant pigments. His work was not widely accepted at first, partly because it was published in Russian and partly because Arthur Stoll and Richard Willstätter, a much better-known Swiss–German research team, were unable to repeat the findings.

However, in the late 1930s and early 1940s, Archer Martin and Richard Synge at the Wool Industries Research Association in Leeds devised a form of liquid–liquid chromatography by supporting the stationary phase, in this case water, on silica gel in the form of a packed bed and used it to separate some acetyl amino acids derived from wool. Their 1941 paper included a recommendation that the liquid mobile phase be replaced with a suitable gas that would accelerate the transfer between the two phases and provide more efficient separation: the first mention of the concept of gas chromatography. In fact, their insight went even further, in that they also suggested the use of small particles and high pressures to improve the separation, the starting point for high-performance liquid chromatography (HPLC).

Gas chromatography was the first of these concepts to be taken forward. Erika Cremer working with Fritz Prior in Germany developed gas–solid chromatography, while in the United Kingdom, Martin himself cooperated with Anthony James in the early work on gas–liquid chromatography published in 1952. Real progress in HPLC began in 1966 with the work of Csaba Horváth at Yale. The popularity of the technique grew rapidly through the 1970s, so that by 1980, this had become the standard laboratory approach to a wide range of analytes. The continuing problem with liquid or gas chromatography was the identification of the molecule eluting from the system, a facet of the techniques that was to be revolutionized by mass spectrometry.

The foundations of mass spectrometry were laid in the Cavendish Laboratories of Cambridge University in the early years of the twentieth century. Francis Aston built the first fully functional mass spectrometer in 1919 using electrostatic and magnetic fields to separate isotope ions by their masses and focus them onto a photographic plate. By the end of the 1930s, mass spectrometry had become an established technique for the separation of atomic ions by mass.

The early 1950s saw attempts to apply the technique to small organic molecules, but the mass spectrometers of that era were extremely limited by mass and resolution. Positive theoretical steps were taken, however, with the description of time-of-flight (TOF) analysis by W.C. Wiley and I.H. Maclaren and quadruple analysis by Wolfgang Pauli.

The next major development was the coupling of gas chromatography to mass spectrometry in 1959 by Roland Gohlke and Fred McLafferty at the Dow Chemical Research Laboratory in Midland, Michigan. This allowed, for the first time, an analysis of mixtures of analytes without laborious separation by hand. This, in turn, was the trigger for the development of modern mass spectrometry of biological molecules.

The introduction of liquid chromatography–mass spectrometry (LC–MS) in the early 1970s, together with new ionization techniques developed over the last 25 years (i.e. fast particle desorption, electrospray ionization, and matrix-assisted laser desorption/ionization), has made it possible to analyze almost every class of biological compound class right up into the megadalton range.

Fashionable "Omics"

In Benet Street, Cambridge, stands a rather ordinary pub which on Saturday, 28 February 1953, enjoyed 15 minutes of fame far beyond Andy Warhol's wildest dreams. Two young men arrived for lunch and, as James Watson watched, Francis Crick announced to the regulars in the bar that "we have found the secret of life." The more formal announcement of the structure of DNA appeared in *Nature* on 2 April 1953 in a commendably brief paper of two pages with six references. Watson and Crick shared a Nobel Prize with Maurice Wilkins, whose work with Rosalind Franklin at King's College, London, had laid the groundwork. Sadly, Franklin's early death robbed her of a share of the prize, which is never awarded posthumously.

Over the next two decades, a large number of researchers teased out the details of the genetic control of cells, and by 1972 a team at the Laboratory of Molecular Biology of the University of Ghent, led by Walter Fiers, were the first to determine the sequence of a gene (a coat protein from a bacteriophage). The same team followed up in 1976 by publishing the complete RNA nucleotide sequence of the bacteriophage. The first DNA-based genome to be sequenced in its entirety was the 5368-base-pair sequence of bacteriophage Φ-X174 elucidated by Frederick Sanger in 1977. The science of genomics had been born.

Although the rush to sequence the genomes of ever more complex species (including humans in 2001) initially held out considerable hope of yielding new biomarkers, focus gradually shifted to the protein products of the genes. This process is dated by many to the introduction in 1977 by Patrick O'Farrell at the University of Colorado in Boulder of two-dimensional polyacrylamide gel electrophoresis (2-D PAGE). The subject really took off in the 1990s, however, with technical improvements in mass spectrometers combined with computing hardware and software to support the extremely complex analyses involved.

The next "omics" to become fashionable was metabolomics, based on the realization that the quantitative and qualitative pattern of metabolites in body fluids reflects the functional status of an organism. The concept is by no means new, the first paper addressing the idea (but not using the word) having been "Quantitative Analysis of Urine Vapor and Breath by Gas–Liquid Partition Chromatography" by Robinson and Pauling in 1971. The word *metabolomics*, however, was not coined until the 1990s.

The Future

Two generalizations may perhaps be drawn from the accelerating history of biomarkers over the last 2700 years. The first is that each new step depends on an interaction between increasing understanding of the biology and technical improvement of the tools, leading to a continuous spiral of innovation. The second is the need for an open but cautious mind. Sushustra's recognition of the implications of sweet urine has stood the test of time; de Corbeil's *Poem on the Judgment of Urines* has not. The ultimate fate of more recent biomarkers will only be revealed by time.

2

Biomarkers: Facing the Challenges at the Crossroads of Research and Health Care

Gregory J. Downing

Innovation Horizons, LLC, Washington, DC, USA

Introduction

Across many segments of the biomedical research enterprise and the health care delivery sectors, the impact of biomarkers has been transforming in many ways: from business and economics to policy and planning of disease management. The pace of basic discovery research progress has been profound worldwide, with the intertwining of innovative technologies and knowledge providing extensive and comprehensive lists of biological factors now known to play integral roles in disease pathways. These discoveries have had a vast impact on pharmaceutical and biotechnology industries, with tremendous growth in investment in biomarker research reaching into the laboratory technology and services sector. These investments have spawned new biomedical industry sectors, boosted the roles of contract research organizations, supported vast new biomarker discovery programs in large corporate organizations, and prompted the emergence of information management in research. Similarly, growth in the academic research programs supporting biomarker research has greatly expanded training capacity, bench and clinical research capacity, and infrastructure while also fueling the growth of intellectual property. By many reports, private-sector applications of biomarkers in toxicity and early efficacy trials have been fruitful in developing decision-making priorities that are introducing greater efficiency in early- to mid-stage medical product development. Despite the heavy emphasis on private and publicly funded research, the reach of the impact of biomarkers into clinical practice interventions is challenging to quantify. The costs of development remain high for many drugs, and the numbers of new chemical entities reaching the marketplace have continued to remain relatively low compared to prior years and expectations following robust research expansion of the 1980s and 1990s. Industry concerns about the sustainability of research and development programs have grown in the backdrop of the

Biomarkers in Drug Discovery and Development: A Handbook of Practice, Application, and Strategy, Second Edition. Edited by Ramin Rahbari, Jonathan Van Niewaal, and Michael R. Bleavins.

clinical challenges that are attendant on biomarker applications in clinical trials. Understanding of the clinical implications of disease markers as a consequence of relatively slow evidence development has taken much longer to discern than many had predicted. There have been many challenges in establishing a translational research infrastructure that serves to verify and validate the clinical value of biomarkers as disease endpoints and their value as independent measures of health conditions. The lack of the equivalent of the clinical trial infrastructure for biomarker validation and diagnostics has slowed progress compared to therapeutic and device development. Evidence development processes and evaluations have begun to emerge for biomarkers, and wide adoption of them in clinical practice measures has not yet matured. For some, the enthusiasm and economic balance sheets have not been squared, as the clinical measure indices that had been hoped for have been viewed as moderately successful by some and by others as bottlenecks in the pipelines of therapeutic and diagnostic development.

Brief History of Biomarker Research, 1998–2008: The First Decade

During the last decade of the twentieth century, biomedical research underwent one of the most dramatic periods of change in history. Influenced by a multitude of factors – some scientific, others economic, and still others of policy – new frontiers of science emerged as technology and knowledge converged and diverged – bringing new discoveries and hope to the forefront of medicine and health. These capabilities came about as a generation's worth of science that brought to the mainstream of biomedical research the foundation for a molecular basis of disease: recombinant DNA technology. Innovative applications of lasers, novel medical imaging platforms, and other advanced technologies began to yield a remarkable body of knowledge that provided unheralded opportunities for discovery of new approaches to the management of human health and disease.

Here we briefly revisit a part of the medical research history, which led to the shaping of new directions, that is captured simply by the term *biomarker*, a biological indicator of health or disease. Looking back to the 1980s and 1990s and the larger scheme of health care, we see that many new challenges were being faced during this period. The international challenges and global economic threats posed by human immunodeficiency virus (HIV) and AIDS provided the impetus for one of the first steps in target-designed therapies and the use of viral and immune indicators of disease. Unusually innovative and, strategically directed efforts in discovery and clinical research paradigms were coordinated at the international level using clinical measures of disease at the molecular level. The first impact of biomarkers on discovery and translational research, both privately and publicly funded, as related to biological measures of viral load, CD^{4+} T-lymphocyte counts, and other parameters of immune function and viral resistance came to be a mainstay in research and development. Regulatory authority was put in place to allow "accelerated approval" of medical products using surrogate endpoints for health conditions with grave mortality and morbidity. Simultaneously, clinical cancer therapeutics programs had some initial advances with the use of clinical laboratory tests that aided in the distinction between responders and nonresponders to targeted therapies. The relation of *Her2/neu* tyrosine kinase receptor in aggressive breast cancer and response to (Herceptin) [1], and similarly, the association of imatinib (Gleevac) responsiveness with the association of the presence of Philadelphia chromosome translocation involving *BCR/Abl* genes in chronic myelogenous leukemia [2], represented some of the cases where targeted molecular therapies were based on a biomarker test as a surrogate endpoint for patient clinical response. These represented the entry point of pharmaceutical science moving toward co-development, using diagnostic tests to guide selection of therapy around a biomarker.

Diverse changes were occurring throughout the health care innovation pipeline in the 1990s. The rise of the biotechnology industry became an economic success story underpinned by successful products in recombinant DNA technology, monoclonal antibody production, and vaccines. The device manufacturing and commercial laboratory industries became major forces. In the United States, the health care delivery system underwent changes with the widespread adoption of managed care programs, and an effort at health care reform failed. For US-based academic research institutions, it was a time of particular tumult for clinical research programs, often supported through clinical care finances, downsized in response to financial shortfalls. At a time when scientific opportunity in biomedicine was, arguably, reaching its zenith, there were cracks in the enterprise that was responsible for advancing basic biomedical discovery research to the clinic and marketplace.

In late 1997, the director of the National Institutes of Health (NIH), Harold Varmus, met with biomedical research leaders from academic, industrial, governmental, and clinical research organizations, technology developers, and public advocacy groups to discuss mutual challenges, opportunities, and responsibilities in clinical research. In this setting, some of the first strategic considerations regarding "clinical markers" began to emerge among stakeholders in clinical research. From a science policy perspective, steps were taken to explore and organize information that brought to light the need for new paradigms in clinical development. Some of these efforts led to the framing of definitions of terms to be used in clinical development, such as *biomarkers* (a characteristic that is measured and evaluated objectively as an indicator of normal biological processes, pathogenic processes, or pharmacologic responses to a therapeutic intervention) and *surrogate endpoints* (a biomarker that is intended to substitute for a clinical endpoint and is expected to predict clinical benefit or harm, or lack of benefit or harm, based on epidemiologic, therapeutic, pathophysiologic, or other scientific evidence) and descriptions of the information needs and strategic and tactical approaches needed to apply them in clinical development [3]. A workshop was held to address statistical analysis, methodology, and research design issues in bridging empirical and mechanism-based knowledge in evaluating potential surrogate endpoints [4]. In-depth analyses were held to examine information needs, clinical training skills, database issues, regulatory policies, technology applications, and candidate disease conditions and clinical trials that were suitable for exploring biomarker research programs. As a confluence of these organizational activities, in April 1999, an international conference was hosted by the NIH and US Food and Drug Administration (FDA) [5]. The leadership focused on innovations in technology applications, such as multiplexed gene analysis using polymerase chain reaction technologies, large-scale gel analysis of proteins, and positron-emission tomography (PET) and magnetic resonance imaging (MRI). A summary analysis was crafted for all candidate markers for a wide variety of disease states, and a framework was formed for multiple disease-based public–private partnerships in biomarker development. A series of research initiatives supported by industry, NIH, and FDA were planned and executed in ensuring months. New infrastructure for discovery and validation of cancer biomarkers was put in place. Public–private partnerships for biomarker discovery and characterization were initiated in osteoarthritis, Alzheimer disease, and multiple sclerosis. Research activities in toxicology markers for cardiovascular disease and metabolism by renal and hepatic transformation systems were initiated by FDA. These events did not yield a cross-sector strategic action plan, but did serve as a framework for further engagement across governmental, academic, industrial, and non-governmental organizations. Among the breakthroughs was the recognition that new statistical analysis methods and clinical research designs would be needed to address multiple variables measured simultaneously and to conduct meta-analyses from various clinical studies to comprehend the effects of a biomarker over time and its role as a reliable surrogate endpoint. Further, it was recognized that there would be needs for data management, informatics, clinical registries, and repositories of biological specimens, imaging files, and common reagents.

Over the next several years, swift movement across the research and development enterprise was under way. It is obvious that future biomarker research was driven in the 1990s and early years of the twenty-first century by the rapid pace of genome mapping and the fall in cost of large-scale genomic sequencing technology, driven by the Human Genome Project. It is now apparent that biomarker research in the realm of clinical application has acquired a momentum of its own and is self-sustaining.

Table 2.1 Major scientific contributions and research infrastructure supporting biomarker discovery.

Human genome project
Mouse models of disease (recombinant DNA technology)
Information management (informatics tools, open-source databases, open-source publishing, biomarker reference services)
Population-based studies and gene–environment interaction studies
Computational biology and biophysics
Medical imaging: structural and functional
High-throughput technologies: *in vitro* cell–based screening, nanotechnology platforms, molecular separation techniques, robotics, automated microassays, high-resolution optics
Proteomics, metabolomics, epigenomics
Pharmacogenomics
Molecular toxicology
Genome-wide association studies
Molecular pathways, systems biology, and systems engineering

The major schemes for applications of biomarkers can be described in a generalized fashion in four areas: (i) molecular target discovery, (ii) early-phase drug development, (iii) clinical trials and late-stage therapeutic development, and (iv) clinical applications for health status and disease monitoring. The building blocks for biomarker discovery and early-stage validation over the last decade are reflected in Table 2.1. Notable to completion of the international Human Genome Project was the vast investment in technology, database development, training, and infrastructure that have been applied throughout industry toward clinical research applications.

Science and Technology Advances in Biomarker Research

In the past two decades of biomarker research, far and away the most influential driving force was completion of the Human Genome Project in 2003. The impact of this project on biomarker research has many facets beyond establishment of the reference data for human DNA sequences. This mammoth undertaking that was initiated in 1990 led to the sequence for the nearly 25 000 human genes and to making them accessible for further biological study. Beyond this and the other species genomes that have been characterized, human initiatives to define individual differences in the genome provided some of the earliest large-scale biomarker discovery efforts. The human haplotype map (HapMap) project defined differences in single-nucleotide polymorphisms (SNPs) in various populations around the world to provide insights into the genetic basis of disease and into genes that have relevance for individual differences in health outcomes. A collaboration among 10 pharmaceutical industry companies and the Wellcome Trust Foundation, known as the SNP consortium, was formed in 1999 to produce a public resource of SNPs in the human genome [6]. The SNP consortium used DNA resources from a pool of samples obtained from 24 people representing several racial groups. The initial goal was to discover 300 000 SNPs in two years, but the final results exceeded this, as 1.8 million SNPs had been released into the public domain at the end of 2002 when the discovery phase was completed. The SNP consortium was notable, as it would serve as a foundation for further cross-industry public–private partnerships that would be spawned as a wide variety of community-based efforts to hasten the discovery of genomic biomarkers (see below).

The next phase of establishing the basic infrastructure to support biomarker discovery, particularly for common chronic diseases, came in 2002 through the International HapMap Project, a collaboration among scientists and funding agencies from Japan, the United Kingdom, Canada, China, Nigeria, and the United States [7]. A haplotype is a set of SNPs on a single chromatid that are associated statistically. This rich resource not only mapped over 3.1 million SNPs but also established additional capacity for identifying specific gene markers in chronic diseases and represented a critical reference set for enabling population-based genomic studies to be done that could establish a gene–environmental basis for many diseases [8].

Within a short time of completing the description of the human genome, a substantial information base was in place to enable disease–gene discoveries on a larger scale. This approach to referencing populations to the well-described SNP maps is now a major undertaking for defining gene-based biomarkers. In recent years, research groups around the world have rapidly been establishing genome-wide association studies (GWASs) to identify specific gene sets associated with diseases for a wide range of chronic diseases.

This new era in population-based genetics began with a small-scale study that led to the finding that age-related macular degeneration is associated with a variation in the gene for complement factor H, which produces a protein that regulates inflammation [9]. The first major implication in a common disease was revealed in 2007 through a study of type II diabetes variants [10]. To demonstrate the rapid pace of discovery of disease gene variants: within 18 months following the study, there were 18 disease gene variants associated with defects in insulin secretion [11].

The rapid growth in GWASs is identifying a large number of multigene variants that lead to subclassification of diseases with common phenotype presentations. Among the databases being established for allowing researchers public access to these association studies is dbGaP, the database of genotype and phenotype. The database, which was developed and is operated by the National Library of Medicine's National Center for Biotechnology Information, archives and distributes data from studies that have investigated the relationship between phenotype and genotype, such as GWASs. The dbGAP contains at least 36 population-based studies that include genotype and phenotype information. Worldwide, dozens if not hundreds of GWASs are under way for a plethora of health and disease conditions associated with genetic features. Many of these projects are collaborative, involve many countries, and are supported through public–private partnerships. An example is the Genomics Association Information Network (GAIN), which is making genotype–phenotype information publicly available for a variety of studies in mental health disorders, psoriasis, and diabetic nephropathy [12]. For the foreseeable future, substantial large-scale efforts will continue to characterize disease states and catalog genes associated with clinically manifested diseases.

As technology and information structures advance, other parameters of genetic modifications represent new biomarker discovery opportunities. The use of metabolomics, proteomics, and epigenomics in clinical and translational research is now being actively engaged. A large-scale project to sequence human cancers, the Cancer Genome Atlas, is focused on applying large-scale biology in the hunt for new tumor genes, drug targets, and regulatory pathways. This project is focused not only on polymorphisms but also on DNA methylation patterns and copy numbers as biomarker parameters [13]. Again, technological advances are providing scientists with novel approaches to inferring sites of DNA methylation at nucleotide-level resolution using a technique known as high-throughput bisulfite sequencing (HTBS). Large-scale initiatives are also under way to bring a structured approach to relating protein biomarkers into focus for disease conditions. Advances in mass spectrometry, protein structure resolution, bioinformatics for archiving protein-based information, and worldwide teams devoted to disease proteomes have solidified in recent years. Although at a more nascent stage of progress in disease characterization, each of these emerging new fields is playing a key complementary role in biomarker discovery in genetics and genomics.

Supporting this growth in biomarker discovery is massive investment over the last 10 years worldwide by public and private financers that has spawned hundreds of new commercial entities worldwide. Private-sector financing for biomarker discovery and financing has become a major component of biomedical research and development (R&D) costs in pharmaceutical development. Although detailed budget summaries have not been established for US federal funding of biomarker research, in a survey by McKinsey and Co., biomarker R&D expenditures in 2009 were estimated at US$ 5.3 billion, up from US$ 2.2 billion in 2003 [14].

Table 2.2 Major international policy issues related to biomarker research.

Partnerships and collaborations: industry, team science
Expanded clinical research capacity through increases in public and private financing
Open-source publishing, data-release policies
Standards development and harmonization
FDA Critical Path Initiative
Regulatory guidance for medical product development
Biomarkers Consortium
Evidence-based medicine and quality measures of disease
International regulatory harmonization efforts
Public advocacy in medical research
Genetic Information Non-discrimination Act of 2008 (US)

Policies and Partnerships

Although progress in biomarker R&D has accelerated, the clinical translation of disease biomarkers as endpoints in disease management and as the foundation for diagnostic products has had more extensive challenges. A broad array of international policy matters over the past decades have moved to facilitate biomarker discovery and validation (Table 2.2). In the United States, the FDA has taken a series of actions to facilitate applications of biomarkers in drug development and use in clinical practices as diagnostic and therapeutic monitoring. A voluntary submission process of genomic data from therapeutic development was initiated by the pharmaceutical industry and the FDA in 2002 [15]. This program has yielded many insights into the role of drug-metabolizing enzymes in the clinical pharmacodynamic parameters of biomarkers in drug development. In July 2007, guidelines for the use of multiplexed genetic tests in clinical practice to monitor drug therapy were issued by the FDA [16]. More recently, the FDA has begun providing label requirements indicating those therapeutic agents for which biomarker assessment can be recommended to avoid toxicity and enhance the achievement of therapeutic responses [17]. In 2007, Congress authorized the establishment of a private–public resource to support collaborative research with the FDA.

One of the major obstacles to clinical genomic research expressed over the years has been a concern that research participants may be discriminated against in employment and provision of health insurance benefits as a result of the association of genetic disease markers. After many years of deliberation, the US Congress passed legislation known as the Genetic Information Non-discrimination Act of 2008, preventing the use of genetic information to deny employment and health insurance.

The past decade has seen many new cross-organizational collaborations and organizations developed to support biomarker development. For example, the American Society for Clinical Oncology, the American Association for Cancer Research, and the FDA established collaborations in workshops and research discussions regarding the use of biomarkers for ovarian cancer as surrogate endpoints in clinical trials [18]. The FDA Critical Path Initiative was developed in 2006, with many opportunities described for advancing biomarkers and surrogate endpoints in a broad range of areas for therapeutic development [19, 20]. Progress in these areas has augmented industry knowledge of application of biomarkers in clinical development programs and fostered harmony with international regulatory organizations in the ever-expanding global research environment. This program has been making progress on expanding the toolbox for clinical development – many of the components foster development and application of biomarkers. As an example of international coordination among regulatory bodies, the FDA and the European Medicines Agency (EMA) for the first time worked together to develop a framework allowing submission, in a single application to the two agencies, of the results of seven new biomarker tests that evaluate kidney damage during animal testing of new drugs. The new biomarkers are kidney injury molecule-1 (KIM-1), albumin, total protein, β_2-microglobulin, cystatin C, clusterin, and trefoil factor-3, replacing blood urea nitrogen (BUN) and creatinine in assessing acute toxicity [21]. The development of this framework is discussed in more detail by Goodsaid later in this book.

In 2007, Congress established legislation that formed the Reagan–Udall Foundation, a not-for-profit corporation to advance the FDA's mission to modernize medical, veterinary, food, food ingredient, and cosmetic product development; accelerate innovation; and enhance product safety. Another important policy in biomarker development occurred with the above-mentioned establishment of the Critical Path Institute in 2006 to facilitate precompetitive collaborative research among pharmaceutical developers. In working closely with the FDA, these collaborations have focused on toxicology and therapeutic biomarker validation [22]. In 2007, a new collaboration building on public–private partnerships with industry was formed to develop clinical biomarkers. In 2006, the Biomarkers Consortium was established as a public–private initiative with industry and government to spur biomarker development and validation projects in cancer, central nervous system, and metabolic disorders in its initial phase [23]. These programs all support information exchange and optimize the potential to apply well-characterized biomarkers to facilitate pharmaceutical and diagnostic development programs.

Other policies that are broadening the dissemination of research findings relate to the growing directions toward open-source publishing. In 2003, the Public Library of Science began an open-source publication process that provides instant access to publications [24]. Many scientific journals have moved to make their archives available 6–12 months after publication. In 2008, the National Institutes of Health implemented a policy that requires publications of scientific research with US federal funding to be placed in the public domain within 12 months of publication [25]. All of these policy actions are favoring biomarker research by accelerating the transfer of knowledge from discovery to development. New commercial management tools have been developed to provide extensive descriptions of biomarkers and their state of development. Such resources can help enhance industry application of well-characterized descriptive information and increase efficiency of research by avoiding duplication and establishing centralized credentialing of biomarker information [26].

New business models are emerging among industry and patient advocacy organizations to increase the diversity of financing options for early-stage clinical development [27]. Private philanthropies conducted with key roles of patient groups are supporting research in proof-of-concept research and target validation with the expectation that these targeted approaches will lead to commercial interests in therapeutic development. Patient advocacy foundations are supporting translational science in muscular dystrophy, amyotrophic lateral sclerosis, juvenile diabetes, multiple myeloma, and Pompe disease, often with partnerships from private companies [28, 29].

Challenges and Setbacks

While progress in biomarker R&D has accelerated, the clinical translation of disease biomarkers as endpoints in disease management and as the foundation for diagnostic products has had more extensive challenges [30]. For example, we have not observed a large number of surrogate endpoints emerging as clinical trial decision points. Notable exceptions to this include imaging endpoints that have grown in substantial numbers. In most cases for drug development, biomarkers are being applied in therapeutic development to stratify patients into subgroups of responders, to aid in pharmacodynamic assessment, and to identify early toxicity indicators to avoid late-stage failures in therapeutic development.

There are difficulties in aligning the biomarker science to clinical outcome parameters to establish the clinical value in medical practice decision-making. As applied in clinical practice, biomarkers have their most anticipated applications in pharmacotherapeutic decisions in treatment selection and dosing, risk assessment, and stratification of populations for disease preemption and prevention. In the United States, challenges to the marketplace are presented by the lack of extensive experience with pathways for medical product review and reimbursement systems that establish financial incentives for biomarker development as diagnostic assays. Clinical practice guidelines for biomarker application in many diseases are lacking, leaving clinicians uncertain about what roles biomarker assays play in disease management. In addition, few studies have been conducted to evaluate the cost-effectiveness of inclusion of biomarkers and molecular diagnostics in disease management [31]. The lack of these key pieces in a system of modern health care can cripple plans for integration of valuable technologies into clinical practice.

Scientific setbacks have also occurred across the frontier of discovery and development. Among notable instances was the use of pattern recognition of tandem mass spectrometric measurements of blood specimens in ovarian cancer patients. After enthusiastic support of the application in clinical settings, early successes were erased when technical errors and study design issues led to faulty assumptions about the findings. Across clinical development areas, deficiencies in clinical study design have left initial study findings unconfirmed, often due to overfitting of sample size to populations and improper control for selection and design bias [32].

Commercial development of large-scale biology companies has also struggled in some ways to identify workable commercial models. Initial enthusiasm about private marketing of genomic studies in disease models faltered as public data resources emerged. Corporate models for developing large proteomic databases faltered based on a lack of distinguished market value, little documented clinical benefit, and wide variability in quality of clinical biospecimens.

Evidence development to support clinical utility of many biomarkers to be used as clinical diagnostics is difficult to establish, as clinical trial infrastructure has not yet been established to validate candidate biomarkers for clinical practice. An obstacle to this has been access to well-characterized biospecimens coupled with clinical phenotype information. This has led to calls for centralized approaches to biospecimen collection and archiving to support molecular analysis and biomarker research [33]. Furthermore, a wide variety of tissue collection methods and DNA and protein preparation for molecular analysis has been at the root of many problems of irreproducibility. Standards development and best practices have been represented as cornerstones in facilitating biomarker validation [34, 35]. Similarly, reacting to the lack of reproducibility of findings in some studies, proposals have been made for standards in study design for biomarker validation for risk classification and prediction [36].

Looking Forward

The next decade of biomarker research is promising, with a push toward more clinical applications to be anticipated. Key factors on the horizon that will be integral to clinical adoption are summarized in Table 2.3. The confluence of basic and translational research has set the stage for *personalized medicine*, a term of art used widely now, which indicates that health care practices can be customized to meet specific biological and patient differences. The term was

Table 2.3 Looking ahead: implementing biomarkers in clinical care.

Intellectual property policy
Phenotypic disease characterization
Clinical translation: biomarker validation and verification
Clinical trials stratification based on biological diversity
Surrogate endpoints: managing uncertainty and defining boundaries in medical practice
Data-sharing models
Dynamic forces in industry financing
Co-development of diagnostics and therapeutics
Clinical infrastructure for evidence development and clinical utility of diagnostics
Health information exchange and network services
Consumer genomic information services

not part of the lexicon in 1997 but speaks to the consumer-directed aspects of biomedical research. Genomic services offered to consumers have emerged with the use of GWASs, although clinical value and impact are not known. It is clear that the emergence of biomarkers in an ever-changing health care delivery system will in some fashion incorporate the consumer marketplace.

Prospects for biomarkers that continue to play a major role in the transformation of pharmaceutical industry research remain high as new technology platforms, bioinformatics infrastructure, and credentialed biomarkers evolve. The emergence of a clearer role for federal regulators and increased attention to appraisal of value from genomic-based diagnostics will help provide guideposts for investment and a landscape for clinical application. One can anticipate that the impact of genomics in particular will probably provide a clinical benefit in chronic diseases and disorders, where multiple biomarker analyses reflect models and pathways of diseases. An emerging clinical marketplace is evolving for the development and application of biomarker assays as clinical diagnostics. The pathway for laboratory-developed tests will probably evolve to include FDA oversight of certain tests with added complexity of multiple variables integrated into index scoring approaches to assist in therapeutic selection. Clinical practice guidelines are beginning to emerge for the inclusion of biomarkers to guide stratification of patients and therapeutic decision-making. The early impact of these approaches is now evident in oncology, cardiovascular, and infectious disease and immune disorders.

Advancing clinical biomarker to improve safety and quality of health care as a mainstay remains many years away. The clinical evaluation processes for diagnostics and targeted molecular therapy as a systematic approach have not yet been firmly established. The use of electronic health information and integration of information from health plans and longitudinal data collection and randomized clinical trials will need integration and coordination for effective implementation in medical decision-making.

In 2008, the first impact was felt for over-the-counter or electronically available consumer services based on genetic tests. Utilizing public and private genome-wide association databases and private-sector resources using powerful search engines coupled with family history information and SNP analysis developed a consumer service that identifies possible health risks. Although the medical benefit of such services remains undocumented, the successful entry of several services and the growth of online commercial genomic services indicate interest among health-conscious citizens for understanding inherited disease risk. Another noteworthy factor that will probably play an important role in the next decade of clinical biomarker adoption is the development of standards and interoperability specifications for the health care delivery system and consumers. Interoperable health record environments will probably provide more flexibility and mobility of laboratory information and provide advantages to consumer empowerment in prevention and disease management.

One of the most important and sweeping challenges with biomarkers is the development of intellectual property policies that will bring opportunity and entrepreneurship in balance with meeting unmet market needs and clinical value. Because of the likelihood that single gene mutations or simple protein assays are not by themselves a discovery equivalent to diagnostic tests or new clinical markers, the arraying of the convergence of circles of technologies and knowledge will need new approaches to management for the combined aspects of technology to be brokered as real value in health care. Indeed, the challenges ahead in noting the importance overall for this notion is underscored more globally by Alan Greenspan in noting that "arguably, the single most important economic decision our lawmakers and courts will face in the next twenty-five years is to clarify the rules of intellectual property" [37].

Overall, previous decade's worth of work has charted a robust and vibrant course for biomarkers across the biomedical research and development landscape. Clinical applications of biomarkers in medical practice are coming more into focus through diagnostics and molecularly targeted therapies, but a long period of time may pass before biomarker-based medicine becomes a standard in all areas of health care practice.

References

1 Ross, J.S., Fletcher, J.A., Linette, G.P. et al. (2003). The *Her-2/neu* gene and protein in breast cancer 2003: biomarker and target of therapy. *Oncologist* 8: 307–325.

2 Deininger, M. and Druker, B.J. (2003). Specific targeted therapy of chronic myelogenous leukemia with Imatinib. *Pharmacol. Rev.* 55: 401–423.
3 Biomarkers Definitions Working Group (2001). Biomarkers and surrogate endpoints: preferred definitions and conceptual framework. *Clin. Pharmacol. Ther.* 69: 89–95.
4 De Grottola, V.G., De Gruttola, V.G., Clax, P. et al. (2001). Considerations in the evaluation of surrogate endpoints in clinical trials: summary of a National Institutes of Health Workshop. *Control. Clin. Trials* 22: 485–502.
5 Downing, G.J. (ed.) (2000). *Biomarkers and Surrogate Endpoints: Clinical Research and Applications*. Amsterdam: Elsevier Science.
6 Sachidanandam, R., Weissman, D., Schmidt, S. et al., The International SNP Map Working Group (2001). A map of human genome sequence variation containing 1.42 million single nucleotide polymorphisms. *Nature* 409: 928–933.
7 The International HapMap Consortium (2003). The International HapMap Project. *Nature* 426: 789–796.
8 The International HapMap Consortium (2007). A second generation human haplotype map of over 3.1 million SNPs. *Nature* 449: 851–861.
9 Klein, R.J., Zeiss, C., Chew, E.Y. et al. (2005). Complement factor H polymorphism in age-related macular degeneration. *Science* 308: 385–389.
10 Sladek, R., Rocheleau, G., Rung, J. et al. (2007). A genome-wide association study identifies novel risk loci for type 2 diabetes. *Nature* 445: 881–885.
11 Perry, J.R. and Frayling, T.N. (2008). New gene variants alter type 2 diabetes risk predominantly through reduced beta-cell function. *Curr. Opin. Clin. Nutr. Metab. Care* 11: 371–378.
12 GAIN Collaborative Research Group (2007). New models of collaboration in genome-wide association studies: the Genetic Association Information network. *Nat. Genet.* 39 (9): 1045–1051.
13 Collis, F.S. and Barker, A.D. (2007). Mapping the cancer genome: pinpointing the genes involved in cancer will help chart a new course across the complex landscape of human malignancies. *Sci. Am.* 296: 50–57.
14 Conway M., McKinsey and Co. (2007). Personalized medicine: deep impact on the health care landscape.
15 Orr, M.S., Goodsaid, F., Amur, S. et al. (2007). The experience with voluntary genomic data submissions at the FDA and a vision for the future of the voluntary data submission program. *Clin. Pharmacol. Ther.* 81: 294–297.
16 FDA (2007). Guidance for Industry and FDA Staff: pharmacogenetic tests and genetic tests for heritable markers. https://www.fda.gov/media/71422/download (accessed 3 May 2019).
17 Frueh, F.W., Amur, S., Mummaneni, P. et al. (2008). Pharmacogenomic biomarker information in drug labels approved by the United States Food and Drug Administration: prevalence of related drug use. *Pharmacotherapy* 28: 992–998.

18 Bast, R.C., Thigpen, J.T., Arbuck, S.G. et al. (2007). Clinical trial end-points in ovarian cancer: report of an FDA/ASCO/AACR public workshop. *Gynecol. Oncol.* 107 (2): 173–176.

19 FDA (2007). The critical path to new medical products. https://grants.nih.gov/grants/guide/notice-files/not-od-08-033.html (accessed 5 March 2019).

20 FDA (2004). Challenge and opportunity on the critical path to new medical products. http://www.fda.gov/oc/initiatives/criticalpath/whitepaper.html (accessed 23 August 2008).

21 FDA (2008). European Medicines Agency to consider additional test results when assessing new drug safety. https://wayback.archive-it.org/7993/20170114031739/http://www.fda.gov/NewsEvents/Newsroom/PressAnnouncements/2008/ucm116911.htm (accessed 3 May 2019).

22 Woolsey, R.L. and Cossman, J. (2007). Drug development and the FDA's critical path initiative. *Clin. Pharmacol. Ther.* 81: 129–133.

23 The Biomarkers Consortium (2008). On the critical path of drug discovery. *Clin. Pharmacol. Ther.* 83: 361–364.

24 Plos (2008). Public Library of Science. http://www.plos.org (23 September 2008).

25 National Institutes of Health (2008). Revised policy on enhancing public access to archived publications resulting from NIH-funded research. NOT 08–033. http://grants.nih.gov/grants/guide/notice-files/not-od-08-033.html (5 March 2019).

26 Thomson Reuters (2008). BIOMARKERcenter. http://scientific.thomsonreuters.com/products/biomarkercenter/ (accessed 23 September 2008).

27 Kessel, M. and Frank, F. (2007). A better prescription for drug-development financing. *Nat. Biotechnol.* 25: 859–866.

28 PriceWaterhouseCoopers (2007). Personalized medicine: the emerging pharmacogenomics revolution. Blogal Technology Centre, Health Research Institute, San Jose, CA.

29 Trusheim, M.R., Berndt, E.R., and Douglas, F.L. (2007). Stratified medicine: strategic and economic implications of combining drugs and clinical biomarkers. *Nat. Rev. Drug Discovery* 6 (4): 287–293.

30 Phillips, K.A., Van Bebber, S., and Issa, A. (2006). Priming the pipeline: a review of the clinical research and policy agenda for diagnostics and biomarker development. *Nat. Rev. Drug Discovery* 5 (6): 463–469.

31 Phillips, K.A. and Van Bebber, S.L. (2004). A systematic review of cost-effectiveness analyses of pharmacogenomic interventions. *Pharmacogenomics* 5 (8): 1139–1149.

32 Ransohoff, D.W. (2005). Lessons from controversy: ovarian cancer screening and serum proteomics. *J. Natl. Cancer Inst.* 97: 315–319.

33 Ginsburg, G.S., Burke, T.W., and Febbo, T. (2008). Centralized biospecimen repositories for genetic and genomic research. *JAMA* 299: 1359–1361.

34 National Cancer Institute (2007). National Cancer Institute best practices for biospecimen resources. https://biospecimens.cancer.gov/bestpractices/ (accessed 3 May 2019).
35 Thomson Reuters (2008). Establishing the standards for biomarkers research.
36 Pepe, M.S., Feng, Z., Janes, H. et al. (2008). Pivotal evaluation of the accuracy of a biomarker used for classification or prediction: standards for study design. *J. Natl. Cancer Inst.* 100 (20): 1432–1438.
37 Greenspan, A. (ed.) (2007). *The Age of Turbulence: Adventures in a New World*. New York, NY: Penguin Press.

3

Enabling Go/No Go Decisions

J. Fred Pritchard[1] *and M. Lynn Pritchard*[2]

[1] *Celerion, Lincoln, NE, USA*
[2] *Branta Bioscience, LLC, Littleton, NC, USA*

Understanding Risk

Developing a drug product is a "risky" business. The investment of time and money is high, while the chance of a successful outcome is low compared to other industries that create new products. Yet the rewards can be great, not only in terms of monetary return on investment (ROI) but also in the social value of contributing an important product to the treatment of human disease.

Risk is defined as "the possibility of loss or injury" [1]. Therefore, inherent in the concept is a sense of probability of occurrence of something unwanted. Everyday decisions and actions that people take are guided by conscious and unconscious assessments of risk. We are comfortable with schemes whereby we sense that a situation is of high, medium, or low risk. We often deal with the concept of a relative risk where we compare the likelihood of loss or injury to other options or situations. Some risks can be defined in more absolute terms such as a population measure based on trend analysis of prior incidence statistics (e.g. current risk of postmenopausal Caucasian women in the United States being diagnosed with breast cancer). These types of population-based risk data, while often much debated in the scientific and popular press, do affect decision-making at the individual level.

As opposed to an individual's skill at assessing risks, decisions and actions taken during drug development require conscious risk assessment by groups of people. There are many stakeholders involved in the development of a drug product. These include the specialists that perform the scientific studies required in drug development (a group hereafter called the "scientists"). Also included are the investors and managers that make decisions about how finite resources will be used (the "sponsors"). Clinical investigators who administer the drug (the "principal investigators") and the healthy volunteers and patients

Biomarkers in Drug Discovery and Development: A Handbook of Practice, Application, and Strategy,
Second Edition. Edited by Ramin Rahbari, Jonathan Van Niewaal, and Michael R. Bleavins.

(the "subjects") who agree to participate in clinical trials are stakeholders as are the regulatory authorities (the "regulators") and Institutional Review Boards or Ethics Committees (the "IRBs/ECs") that approve use of the experimental drug in humans. Each stakeholder has his or her unique perspective of risk. The prime focus of some is the business risk involved including how much work and money are invested to progress the drug at each phase of development. On the other hand, IRBs/ECs, regulators, investigators, and subjects are primarily concerned with the safety risk to the patient relative to the potential benefit, and these are encoded into regulations to ensure consistency in oversight. Drug candidates that represent new targets for therapy provide a hope for improved efficacy but do require more interactions with regulators and investigators adding to development expense. In addition, because the target is unproven, there is a greater relative risk for therapeutic failure compared to proven pharmacological targets of disease intervention. Therefore, when attempting to express the risks involved in developing a drug, it is important to understand the varying perspectives of each major group of stakeholders.

Each stakeholder will be asked to assess risk based on current data on one or many occasions during drug development. Their assessment will become part of the decision-making process that drives drug development in a logical and hopefully collaborative manner. Effective decision-making requires integrating these varying risk assessments in a balanced manner.

Decision Gates

A "decision gate" is defined by one or more key questions that must be answered for the drug to proceed further in development. Drug development is a process that proceeds through several decision gates from identification of a potential therapeutic agent through to marketing a new drug product [2]. Answers to these questions must meet a set of criteria that have been previously agreed by decision-makers before they will open the gate. It is "go/no go" because the future product life of the drug hangs in the balance. While these questions can vary depending on the therapeutic agent being developed, common gates and questions are listed in Table 3.1.

Different go/no go decision gates require agreement to proceed by different groups of stakeholders. This is particularly true when the question "Is the drug candidate safe to give to humans?" is addressed. The sponsor must decide whether to submit an investigational new drug (IND) application based on the data and information collected from animal tests and possibly *in vitro* assays involving human tissues and enzymes. In most countries, the same data is also evaluated by the regulator agency(ies) (e.g. United States Food and Drug Administration [FDA], Medicines and Healthcare Products Regulatory Agency [MHRA], European Medicines Agency [EMA]) who must have time to

Table 3.1 Go/no go decision gate in drug development.

Decision gate	Question	Decision-maker(s)	Role of biomarker
Disease target	Does a drugable target exist that impacts disease progression?	Scientist	Defining mechanism of action
Lead candidate	Does a suitable drug candidate exist with properties predicted to impact disease in a positive way?	Scientist	Impact on disease
		Sponsor	Drug delivery to site of action
First-in-human	Can the drug candidate be given safely to humans?	Sponsor	Impact on disease
		Regulators	Safety measures of clinical relevance
		IRB/EC	
		Investigators	
Clinical proof-of-concept	Does the drug work in humans as it was designed?	Sponsor	Confirming mechanism of action in humans
			Impact on disease
			Defining dose-limiting toxicity
Begin Phase III	Can dosage, target patient populations, and pivotal efficacy and safety study designs be justified?	Sponsor	Impact on disease/dose–response
		Regulators	Patient selection
			Patient safety
Marketing application	Has safe and effective use of the drug been proven?	Regulators	Validated markers that may contribute to the confirmation of safety and efficacy
Post-marketing safety	Are there emerging safety issues that need further action?	Sponsors	Predict patients more likely to experience rare events
		Regulators	

object if they feel the safety of the subjects in the first few clinical trials may be unduly compromised. Finally, the data are reviewed again by the IRBs/ECs and the principal investigator who look specifically at how safety will be evaluated and managed, and represent the interests of the volunteers. It is as if the gate has four locks, each with a key owned by a separate entity who will decide independently whether they open their lock. One cannot pass through the gate until all four locks are open.

This discussion focuses on the nonclinical and clinical decisions to be made during drug development because this is where biomarkers have their role. The go/no go decisions involved in developing a reliable process for manufacturing drug substance and drug product are critically important, but out of scope here. Table 3.1 identifies which stakeholders hold the keys to the locks for each of the common go/no go decision gates listed in drug development. For some, regulators hold a key to the lock; for others, the decision is based on business drivers and the sponsor holds a key. The "First-in-Human" decision gate has the most stakeholder locks.

Disciplined planning and decision-making is required to leverage the value of the decision gate approach, and some useful tools have emerged that help focus those who hold a key to a lock on a go/no go decision gate. Critical to disciplined decision-making is creating a clear set of acceptable answers/findings that will unlock the decision gate and are agreed and understood by all involved in making the decision. Many companies start evolving a target product profile or TPP, or something similar, even at the earliest go/no go decision gates of drug development. In this way, a common way of thinking is preserved throughout the life of the product. An example of a template for a TPP affecting the "First-in-Human" decision gate is depicted in Table 3.2. The development program plan is assembled by determining what studies need to be done and their design that would provide information critical to answering the key questions at each go/no go decision gate.

Regulators also use TPPs in helping sponsors to define what criteria are necessary for approval. The FDA has formalized this concept in a guidance [3]. Such a TPP is a living document that embodies the notion of beginning with the goal in mind. It forms a written basis around which regulators and sponsors can have meaningful discussions that will progress drug development. In its final form, the TPP should resemble the proposed labeling for the drug.

The value of a well-reasoned drug development plan based on a decision gate approach can only be leveraged if there is discipline in the decision-making process. What method will be used to make a decision: a democratic vote of a committee or a single decider who is advised by others? Stakeholders need to clearly understand what their role is in making the decision at each gate. Are they a decider, a consultant, or someone who just needs to know what the decision is in order to effectively do their job? Go/no go decisions need to be made when all information required to answer the key questions is available. Go/no

Table 3.2 Example of a target product profile defining criteria required to move through the decision gate: "Safe to give to humans?"

	Efficacy	Animal safety	PK/ADME	CMC	Phase I study design
Best achievable (enhance investment)	Rodent model: $ED_{90} < 0.3$ mg/kg	Rodent: NOAEL > 50 mg/kg	Dog half-life >8 h	API stable for at least 6 mo	Cohorts of healthy normal subjects: staggered (SD to MD) dose escalation
	Human receptor: $IC_{50} < 1$ μM	Dog: NOAEL > 50 mg/kg	Dog BA >90%	Cost of goods for API <$10 000/kg	Normal safety monitoring
		No genotoxicity	No active metabolites	GMP tablet	
		No CV effects	No *CYP2D6* inhibition		
Base case (invest in next phase)	Rodent model: $ED_{90} < 1$ mg/kg	Rodent: NOAEL > 10 mg/kg	Dog half-life >4 h	API stable for at least 3 mo	Cohorts of healthy normal subjects: sequential dose escalation
	Human receptor: $IC_{50} < 10$ μM	Dog: NOAEL > 10 mg/kg	Dog BA >50%; <10% active metabolites	Cost of goods for API <$20 000/kg	Intensive CV monitoring
		No genotoxicity	*CYP2D6* inhibition: $K_i > 10$ μM	GMP tablet	
		No CV; effects <100 mg/kg			
Minimum required (proceed with caution)	Rodent model: $ED_{90} < 2$ mg/kg	Rodent: NOAEL > 5 mg/kg	Dog half-life >2 h;	API stable for at least 3 mo	Patients: multisite dose escalation
	Human receptor: $IC_{50} < 50$ μM	Dog: NOAEL > 5 mg/kg	Dog BA >30%; <30% active metabolites	Cost of goods for API <$30 000/kg	Intensive CV monitoring
		Equivocal genotoxicity at highest exposure	*CYP2D6* inhibition: $K_i > 1$ μM	GMP tablet	FDA review after each dose escalation
		No CV effects <30 mg/kg			

go decisions should not be revisited in the absence of significant relevant new information. If undisciplined reassessment of the same results occurs, there is either a lack of decision discipline or key information was missing when the original decision was made representing a failure in planning. Once a new drug candidate has progressed through a decision gate, the organization should be committed to expend even greater resources (money and scientist's time) to do the studies to address criteria for the next decision gate along the development path.

Role of Biomarkers in Decision-Making

Biomarker data can be critical to a go/no go decision. By definition, biomarkers may reflect the biology or the progression of disease and/or the effect of drug treatment.

Therefore, information provided by properly selected biomarkers can greatly influence the decision on whether or not to progress through a go/no go decision gate. Since there are several stakeholders from multiple disciplines who may participate in such decisions, each brings a slightly different perspective on how to describe or characterize different types of biomarkers.

Drug developers (sponsors and scientists with a strong dose of regulator input) tend to think in terms of how the biomarker is used during the course of bringing a new therapeutic to market: efficacy biomarkers, safety biomarkers, and markers to select or exclude patients for study. Efficacy biomarkers are those that establish proof-of-mechanism or "Clinical Proof-of-Concept" (CPoC). Safety biomarkers provide sentinels of toxicity and generally can be applied broadly across many types of therapy. They are especially useful when they predict early evidence of organ toxicity during dose escalation studies conducted initially in clinical development. Examples of these are listed in Table 3.3. Finally, biomarkers deployed to aid patient selection for clinical studies, such as genomic tests to determine likely responders to targeted treatments, over time have led to the development of more sophisticated and targeted "companion diagnostics."

As a new drug candidate moves into later clinical testing, regulators view biomarkers through the lens of "qualification" for the purposes of supporting regulatory decisions about a drug's pathway to approval as well as the diagnostic or treatment consequences in medical practice after the drug is approved [4]. This perspective characterizes biomarkers as diagnostic, prognostic, predictive, or response biomarkers.

Investigators, practicing physicians, and patients value considering biomarkers by type of disease. In 2015, US FDA conducted a survey announced in the Congressional Federal Register. The results were compiled into a list [5] which is periodically updated and provides a useful reference on what biomarkers exist

Table 3.3 Examples of safety biomarkers used in preclinical and clinical development.

	Preclinical	Clinical	Comment
Cardiac rhythm	*In vitro* Ikr	ECGs in FIH/FIP	Robust ECGs in FIH → waiver of TQT?
	In vivo QTc	Thorough QTc study (ICH E14)	
	(ICH S7B)		
Myocardial injury	Histology	Cardiac tropinin (cTn)	LDH and CK lack specificity, sensitivity
	cTnT, cTnI (dog, rat)		
Acute renal tubular injury	Histology, KIM-1, clusterin, albumin, trefoil-factor 3	Preclinical markers require validation in clinic	BUN and creatinine clearance lack specificity, sensitivity for early nephrotic damage
Acute glomerular and renal tubular injury affecting reabsorption	Histology, total protein, cystatin-C B2-microglobulin, RPA-1 (rats)		
Liver injury	Histology, total bilirubin (specific), ALT (sensitive), AST, ALP, prothrombin time		Impact on disease/dose–response
			Patient selection
			Patient safety
Testicular toxicity	Histology, LH, testosterone	Semen analysis, inhibin-B	FSH inconsistent

that might be leveraged to better influcnce treatment choices and how a patient with a specific disease may be responding to drug therapy. The particular classification by disease referenced here is also valuable to the drug developer as it puts specific biomarkers into context of how they could be deployed in specific disease-focused clinical programs to aid decision-making and approval, as well as eventual use.

Regardless of how one slices the ever-expanding universe of biomarkers, biomarkers continue to grow and influence the critical decisions involved in how new drugs are developed, approved, and used. Since the focus of this book is on biomarkers in drug development and this chapter deals with risk-based decision-making, the remainder of this review reflects this perspective.

Approximately 12% of drug candidates entering clinical testing result in a marketed product [6], and this success rate has not improved over many years.

This means that at some point during clinical development, nearly 90% of drug candidates fail. If failure occurs late in clinical development, hundreds of millions of dollars and a great deal of time and effort will have been invested with little or no return. Proper incorporation of biomarkers in drug development strategy enables the concept of "fail fast, fail early." Early failure actually lowers overall risk because it enables one to move resources and utilize available patients on other promising therapies. The challenge, therefore, is to identify relevant biomarkers early enough to implement them for "go/no go" decisions at the critical early clinical stages of the development process.

Efficacy Biomarkers

Efficacy biomarkers may be the only way to address the decision gate asking the question "Is there evidence that the drug is working in humans using the same mechanism of action defined by animal studies?" This question is answered by the "Clinical Proof-of-Concept" (CPoC) study; often a Phase IIa study in a small number (50–200) of patients. However, occasionally, CPoC can be addressed during Phase I development if mechanism of action can be demonstrated in healthy volunteers (e.g. effects on blood pressure, body weight, lipids). The changes in the data collected during this study do not have to achieve a statistical significance; rather, there needs to be enough indication of potential efficacy to convince the decision-makers to open the gate and spend the resources to progress the drug further. Often the minimal result for progression can be defined ahead of time (e.g. using a TPP approach), making the actual decision-making process much easier and more objective.

Early in clinical development, multiple biomarker assays or technologies may be utilized in order to more fully understand the safety and actions of the drug in humans. One may use a collection of biomarker methods most of which are Fit for Purpose (FFP) but not likely to be clinically validated surrogate markers of effect. For early decision-making, it is not necessary to have results that meet the same standards of accuracy, precision, and reproducibility as those required for drug approval. Indeed, it is during these early trials that promising tests for use in later clinical development or marketing are first identified from among the pool of available experimental biomarkers.

As a new drug product moves into later clinical testing, regulators take a more proactive role in defining the expectations for biomarkers to be used as primary and secondary study endpoints. US FDA's Critical Path Initiative [7–9] has been a major force in defining what criteria should be used to qualify response biomarkers of drug effect and predictive biomarkers of impact of therapy on disease. Properly developed and utilized, such markers can provide the pivotal evidence for drug approval.

New technologies are now rapidly emerging that show similar promise providing efficacy signals earlier in treatment of certain slowly progressing diseases. Imaging-based biomarkers (discussed in more detail elsewhere in this volume) provide objective endpoints that may be confidently evaluated in a reasonable time frame. Imaging techniques tend to be expensive, but can be cost-effective when used in well-defined situations where subjective assessment has been the only approach available. Examples of therapeutic areas where imaging is reshaping clinical trial design include evaluating the presence of β-amyloid plaque in patients with Alzheimer's disease with now regulatory-approved positron emission tomography (PET)-imaging agents [10], quantifying tumor burden before and after cancer therapy using now-accepted methods for radiologically calculating solid tumor volume [11], and characterizing affected joints and tissues in rheumatoid arthritis [12].

Safety Biomarkers

The primary rationale for toxicity testing in animals is to identify potential target organs of toxicity for new drugs, foods, cosmetics, and chemicals in our environment. Dose-dependent, histological evidence of drug damage to an animal organ provides a rationale for setting dose escalation strategies and maximum targeted doses for first-in-human studies. Blood or urine biochemical markers that reflect organ damage or changes in function are particularly useful in defining the safety profile of new drug candidates relative to their efficacy response.

Table 3.3 provides some common examples of safety biomarkers deployed in drug development. Historically, most drug failures due to safety have involved damage to liver or kidney (the key organs for drug metabolism and elimination), the cardiovascular system, and occasionally the central nervous system or reproductive organs. Consequently, biomarkers that can predict such effects early in the etiology of the toxicity are extremely valuable in drug development, and some of these have been encoded into regulatory guidelines published by the International Conference on Harmonisation (ICH) for safety pharmacology [13]. As indicated in Table 3.3, there are several exciting avenues of emerging technology that will provide more accurate and early predictors of organ system toxicity that will hopefully improve the record of drug failures late in development due to safety issues.

Biomarkers and Patient Selection

Patient selection is another area where decisions are made using biomarker information both in clinical trials and during patient care. Identifying potential responders or nonresponders prior to Phase III clinical trials can have a profound effect on the size of a clinical trial. Eliminating nonresponders can dramatically reduce the intersubject variability in response, thereby reducing the number of subjects in each group of a pivotal clinical trial required to demonstrate effectiveness. The result is a saving in time and large savings in the costs of later stage clinical development.

The Regulatory Side of Biomarkers – Validation and Qualification

Traditional clinical trial endpoints, such as morbidity and mortality, often require extended time frames and may be difficult to evaluate. Fully validated surrogate markers can be used as primary endpoints for pivotal Phase IIb and Phase III safety and efficacy trials. Lowering blood pressure and lowering low-density lipoprotein (LDL) cholesterol are both examples of surrogate biomarker endpoints upon which several drugs have been approved, because it has been demonstrated that effects on these markers directly impact morbidity and mortality several years later.

As regulators hold keys to the locks on decision gates later in clinical drug development and approval, regulation has emerged to help guide and standardize how potentially useful biomarker approaches are applied. Under the Critical Path Initiative, originally launched in 2004 [7–9], FDA recognized the importance of biomarkers, with one of the stated goals being "developing new biomarkers and disease models to improve clinical trials and medical therapy." This eventually led to FDA setting up the biomarker qualification program and associated guidance for groups seeking to gain general acceptance for the use of specific biomarkers [14]. FDA and EMA have been coordinating efforts to harmonize the approach to qualifying biomarkers. Prior to these efforts, sponsors worked on validating biomarkers on their own or with the cooperation of academic groups.

Biomarkers undergoing rigorous regulatory scrutiny are ones that are used to make decisions about treatment. Since the original approval of Herceptin® (trastuzumab) for patients with metastatic breast cancer whose tumors overexpress Her2, there now are several highly targeted drugs approved requiring a "companion diagnostic" in order to identify the patients likely to respond to therapy. In fact, FDA keeps track of the approved or cleared companion diagnostics, along with their corresponding medicines, on their website [15].

Future Roles of Biomarkers in Decision-Making

The evolution of biomarkers in guiding smarter drug development will be driven by two elements: (i) the increased focus on customizing therapy to the individual needs of the patient whether they are involved in clinical trials of a new therapy or engaged in the local current standard of medical care and (ii) implementing aspects of emerging technologies to create faster, better, and less expensive ways of capturing useful biomarker data that can enrich the data upon which critical drug development decisions are made.

As we learn more and more about the cellular biology of disease and the genetic, epigenetic, and impact of environmental factors in progression and response to medical intervention, the challenge for the drug developer will be determining how to deploy ways to leverage this knowledge to enrich the information available at each critical go/no go decision gate that they face. Success will involve being conversant in the science leavened with what technologies can be deployed with an appropriate degree of accuracy and in a timely manner that will provide meaningful data and information about how the drug is behaving, particularly in the target patient population. Maintaining and consulting knowledge sources such as regulatory websites will be even more important to drug developers as all stakeholders attempt to consolidate understanding and coordinate best practice on a global scale. There is no doubt in these authors' minds that future editions of this book will dedicate significant space to how emerging technologies have been adapted to the highly regulated environment of drug development to assist all the stakeholders responsible for critical decision-making to make informed choices of what drug, what dose, and what patient should be considered in plans for future development, approval, and use.

References

1 Merriam-Webster (1988). *Webster's Ninth New Collegiate Dictionary*. Springfield, MA: Merriam-Webster, Inc.

2 Pritchard, J.F., Jurima-Romet, M., Reimer, M.L.J. et al. (2003). Making better drugs, decision gates in non-clinical drug development. *Nat. Rev. Drug Discovery* 2: 542–553.

3 FDA (2007). Guidance for Industry and Review Staff: Target Product Profile — A Strategic Development Process Tool. https://www.fda.gov/regulatory-information/search-fda-guidance-documents/target-product-profile-strategic-development-process-tool.

4 Amur, S., LaVange, L., Zinch, I. et al. (2015). Biomarker qualification: toward a multiple stakeholder framework for biomarker development, regulatory acceptance, and utilization. *Clin. Pharmacol. Ther.* 98: 34–46.

5 FDA (2015). FDA survey to identify potential biomarkers for qualification. Compiled survey results. https://www.fda.gov/media/93053/download (accessed 03 May 2019).

6 DiMasi, J.A., Grabowski, H.G., and Hansen, R.W. (2016). Innovation in the pharmaceutical industry: new estimates of R&D costs. *J. Health Econ.* 47: 20–33.
7 FDA (2004). Critical path opportunities report. wayback.archive-it.org/7993/20180125032208/https://www.fda.gov/ScienceResearch/SpecialTopics/CriticalPathInitiative/CriticalPathOpportunitiesReports/ucm077262.htm (accessed 03 May 2019).
8 FDA (2006). Critical path opportunities report. http://wayback.archive-it.org/7993/20180125142845/https://www.fda.gov/downloads/ScienceResearch/SpecialTopics/CriticalPathInitiative/CriticalPathOpportunitiesReports/UCM077254.pdf (accessed 03 May 2019).
9 FDA (2006). Critical path opportunities list. http://wayback.archive-it.org/7993/20180125035449/https://www.fda.gov/downloads/ScienceResearch/SpecialTopics/CriticalPathInitiative/CriticalPathOpportunitiesReports/UCM077258.pdf (accessed 03 May 2019).
10 Kozauer, N. and Katz, R. (2013). Regulatory innovation and drug development for early-stage Alzheimer's disease. *N. Engl. J. Med.* 368: 1169–1171.
11 Eisenhauer, E.A., Therasse, P., Bogaerts, J. et al. (2009). New response evaluation criteria in solid tumours: revised RECIST guideline (version 1.1). *Eur. J. Cancer* 45: 228–247.
12 Boutry, N., Morel, M., Flipo, R.-M. et al. (2007). Early rheumatoid arthritis: a review of MRI and sonographic findings. *Am. J. Roentgenol.* 189 (6): 1502–1509.
13 International Conference on Harmonization (ICH). List of Guidance for Safety Pharmacology. http://www.ich.org/products/guidelines/safety/article/safety-guidelines.html (accessed 2016).
14 FDA. Biomarker qualification program. https://wayback.archive-it.org/7993/20180125052703/https://www.fda.gov/Drugs/DevelopmentApprovalProcess/DrugDevelopmentToolsQualificationProgram/BiomarkerQualificationProgram/default.htm (accessed 03 May 2019).
15 FDA. List of cleared or approved companion diagnostic devices (in vitro and imaging tools). http://www.fda.gov/medicaldevices/productsandmedicalprocedures/invitrodiagnostics/ucm301431.htm (accessed 2016).

4

Developing a Clinical Biomarker Method with External Resources: A Case Study

Ross A. Fredenburg

Amathus Therapeutics, Inc., Cambridge, MA, USA

The purpose of this chapter is to share a personal experience in developing a set of cerebral spinal fluid (CSF) biomarkers by outsourcing with several contract research organizations (CROs) to supplement internal capabilities within a biotechnology company. A portion of this work was published in American Association of Pharmaceutical Scientists Journal in 2012 [1] for those interested in examining the specifics of the methodology more closely. I will mention the experiments completed during the development within the following pages, not as a review of the cited article but rather as an illumination of the lessons learned during the intensive effort to prepare methods for the analysis of clinical tissue samples.

The opinions offered have been informed by subsequent experiences with outsourced biomarker development at different types of organizations: academic, small niche companies, and large all-in-one corporations. An attempt has been made to express general themes that are useful in any situation, but every interaction requires a customized approach. It is hoped that this discussion can serve as a good starting point for others developing biomarkers outside their own organizations.

In the fall of 2010, Link Medicine Corporation, a small biotechnology company in Cambridge, MA, was in need of an accessible marker of Alzheimer's disease (AD) progression. The biomarker was important to monitor during a clinical trial of their autophagy inducer molecule for treatment of the debilitating disease. This biomarker was an essential measure needed to determine if the novel therapeutic mechanism of the development molecule was having the intended effect in patients. A clear indication of efficacy was needed in this drug development program to determine the viability of the therapeutic mechanism, to keep the company in business through future funding rounds, and to provide the support necessary to advance the program in clinical development.

Biomarkers in Drug Discovery and Development: A Handbook of Practice, Application, and Strategy, Second Edition. Edited by Ramin Rahbari, Jonathan Van Niewaal, and Michael R. Bleavins.

Cerebrospinal fluid samples were being collected from Alzheimer's disease subjects and confidence was low in the available methodology for quantitation of the triad of proteins whose levels had proven to move in tandem with the progression of AD ($A\beta_{42}$, Tau, and pTau; [2]). Although widely used, the available technique displayed troubling issues with linearity of the signal in early development studies. The pressure was on the internal Link Medicine research team to efficiently develop an alternate methodology that could maximize the chances of detecting the small change in disease state expected in the planned 28-day clinical trial. To add to the challenge, the company had limited internal laboratory resources and no capability to directly process and analyze human clinical trial samples. A competent partner was needed, as was a robust method. A small team was formed to clear this important hurdle.

This brings us to the first lesson learned from this intense experience: ***Select the method carefully before transferring to an outside group.*** Many scientists have experienced the frustration of realizing that the established kit in which they have invested time and resources requires substantial optimization. If at all possible, focus on existing technology and educate yourself on the performance characteristics of the method. Often, investigators at other drug discovery organizations can be sources of valuable information about which methods have been accepted for use in a sophisticated research environment. This was a key first step in our selection of the enzyme-linked immunosorbent assay (ELISA) kits that we studied in detail for the following several months. If you are in a small organization with a limited professional network, strongly consider engaging an experienced consultant that may have the requisite contacts to point you in the right direction. The key point is to avoid reinventing the wheel if at all possible.

Of course, not all disease indications have established biomarkers and methods in place. Working with external resources does not preclude the development of novel biomarkers, but additional time and resources should be included in the development plan. In the early stages of any bespoke method development, close attention to details and oversight are often required, and a certain amount of serendipity is very helpful. Overseeing a research plan from afar increases the difficulty in detecting these details – luck does not travel far over a phone line! The first stages are best completed with internal resources; once the framework is in place, the method can be transferred to a group skilled in execution and consistency of analysis. During this initial phase of biomarker method development, it may be appropriate to familiarize yourself with a number of different methodological approaches to the same endpoint. Selection of key performance parameters (those particular to your company's application and expectations) can clarify the process of vetting a number of options early on. In our initial survey, we evaluated a bead-based multiplex technology as well as traditional ELISAs and electro-chemiluminescent (ECL) antibody-based methods via a series of external contracts. The technology for detecting Aβ-proteins and Tau in CSF was well developed at that time, so new method development was not necessary, but the field was clearly struggling with the consistency of the results generated by the available methods. Once your group has identified a suitable method and generated a satisfactory initial data set, the time is right to engage with a partner that can carry the process through validation and ultimately into clinical samples analysis. This is not a trivial task, and the selection of partner has a major impact on the success of the venture.

The second major theme: ***The importance of selecting the right CRO for the development and validation of your biomarker methods cannot be overstated.*** Think of the process in the same terms as hiring your own internal group (this will be a recurring theme in this chapter – do not think of the CRO as a fast-food restaurant). Check company references, investigate the experiences of those in your network for candid reviews, and look at the curriculum vitae (CVs) of the scientists who will be performing the experiments. Ensure that the group and individuals have the requisite skill sets for your application. Does the CRO have experience with your chosen technology, do they have knowledge of the biology underlying the biomarker in question, and are they capable of handling the downstream sample analysis demands (tracking, regulatory compliance, safety considerations, processing capacity, reporting format, etc.) of your company's clinical development plan? Even if it is unlikely that your biomarker method will be executed in a good laboratory practice (GLP) environment, it is a good idea to work with a partner that possesses a quality assurance unit (QAU). These types of service providers are accustomed to the demands of GLP and will generally deliver higher quality data sets and more traceable study records than those companies who have never worked in a highly regulated environment. For our biomarker program, we followed a "GLP-like" approach, in which the QAU of the partner companies would review all experimental data as if in a GLP-compliant environment. This provided us with the control of the process that we desired, as well as a level of transparency that was helpful in driving the research from afar. We engaged two separate companies to work on different methods. This was intended to gain first-hand experience both with the methods and with contract organizations before we selected the final site of clinical sample analysis. The two partners were quite different, and our experience with both will be discussed below. Conducting an audit of every potential partner's facilities and procedures can be worth the expense and time, particularly if you are considering entrusting clinical sample analysis that could determine the future of your company.

If your process of gathering and vetting candidate partners is successful, you will have a short list of potential options. Before the first pipet is raised, however, an important step must be completed. The structure of the contract that is signed between the parties will have an impact on the progression of the research that is enabled by the agreement. Certainly, the definition of the scope of work is important to the ongoing interaction between sponsor and vendor. If written too specifically, then every inevitable change to the development process will need to be addressed individually, and this can slow progression. A broad, open-ended scope of work will be difficult to cost for the vendor, and therefore this generally does not work unless the sponsor is willing to support a number of full-time equivalents (FTEs) for a set period of time. The latter strategy is expensive and makes the most sense in a program in which many stages of the work are integrated into a larger contractual relationship. Recently, CROs have started to entertain risk-sharing relationships with success payments or even royalties staged later in the process. This can lower the initial cost for the sponsor, and in the best of cases instill a sense of ownership in the vendor that is often absent in fee-for-service relationships. In our program, we negotiated a contract with one of the partners to allow that company to offer the optimized method to other sponsors (under standard agreements the methods developed within CROs are the property of the sponsor). That decision reduced the cost of the contract but had a tangible impact on the interaction between our team and the external group. This dynamic will be discussed further below. The most common contract contains sufficient detail to allow accurate cost estimation but also clear definition of the process to add to or change the scope of work. Both teams need to commit to the process by maintaining some flexibility and a certain amount of trust to permit smooth and uninterrupted flow of the science that all of this effort is supporting. If possible, the scientists should be isolated from the business end of the relationship. Assigning personnel to handle the negotiations in another "room" while the experiments and scientific discussion proceed is a good idea. An experienced CRO also understands the value of supporting the client despite changes in scope and the contractual wrangling that may accompany those changes, and a sophisticated sponsor has a contingency budget in anticipation of unforeseen deviations from the plan.

Once the ink has dried on the agreement, the real work begins. If a good relationship has been germinated between the sponsor and vendor, then the transfer of reagents and technology will have already begun before final execution of the contract and experiments can commence immediately. A project leader should be assigned to the program by the vendor and sponsor, and a kick-off meeting should be scheduled to make sure everyone who will be involved in the process is familiar with one another and that the expected lines of communication between the teams are firmly established. Even with ideal preparations, it is only after the first week or two of work under the contract that a real sense of the true nature of the interaction is revealed. Once data starts to flow, it will become apparent how comfortable the vendor is with data interpretation and how comfortable are you as the sponsor sharing the responsibility of designing the successive experiments. Each interaction is different, so a key skill for a leader of outsourced projects is adaptability. The interaction between sponsor and vendor is complex and dynamic and is in many ways more challenging than a direct supervisory role. The best work is generated by a partner that is invested in the scientific process, and the differing business models of the sponsor and service companies can often complicate the process. CROs are paid by the experiment: the definition of the term "fee-for-service." Biotechnology and pharmaceutical companies are paid for success over the long term. As discussed above, contractual strategies can be applied to help structure the overall interaction, but significant effort should be applied to the process of team building among the scientific personnel at each site. Beginning with the kick-off meeting, periodic site visits are often effective in establishing trust, confidence, and familiarity. These visits also present a means to check in on issues identified by an audit, but the overall purpose is to demonstrate that you as the sponsor are engaged and interested in establishing a solid and interactive connection with the team at the contract laboratory. In between site visits, periodic update meetings should be scheduled to keep the teams aligned on progress, to clearly deliver guidance on the experimental plan, and to maintain the level of energy and enthusiasm needed to drive the program efficiently. A good approach is to use e-mail communication for daily data sharing, phone calls or video conferences for weekly meetings, and site visits (two to four times per year) for long-term contracts. Video can help maintain the connection established via an initial site visit, but should not be expected to replace in-person contact and requires extra information technology (IT) support to avoid technical glitches. In our work with the two CROs that were developing the improved ELISA methods, several site visits were performed and regular teleconferences were held to keep the process moving. In other projects I have had the ability to hold in-person weekly check-ins – finding a local group to support your program does have many advantages, but it is not always a viable option due to instrumentation or expertise.

As the development process moves along, the choices made while building the sponsor–vendor relationship will have an impact. Some results will confuse the team, mistakes will be made, and alignment may falter. These stresses challenge the project leader to exert control while maintaining the positive tone of the interaction. Keeping on top of the data stream is very important to maintaining the level of engagement required in most circumstances, so don't underestimate the time it will take to oversee the outsourced project. Look for input from the vendor whenever possible, especially in cases where you have identified a group with specific expertise, but remember that the sponsor needs to set the tone of the overall program. In our development projects at two sites (let's call them North and South), we discovered that the North group was quite adept and eager to help guide the project, while the South group was content to wait for direction from our internal team. The independent attitude of the former partner (North) may have arisen from our decision to defray the cost by offering the optimized method to the vendor at the end of the interaction; they were essentially developing their own method for future marketing to other interested companies. Mission accomplished, apparently. The South vendor was working on a traditional fee-for-service basis and therefore functioned as a service group. At first, we were happy to have a partner with some initiative, and the easing of responsibility from our small internal team was helpful. The interaction with the South scientists was a bit frustrating, as we needed to make every decision in order to get the program moving. As each aspect of the program progressed, an interesting change occurred. The North team became more difficult to oversee, and certain internal issues in their team were not rectified as quickly as if we (the client) were fully engaged. Decisions were made without the full input of our team, and we started to feel that the full authority necessary to yield a definitive method that met our specific needs was being lost. The South group performed well and, once the experiments started to roll, we were able to achieve a rhythm of data interpretation, experimental design, and assay execution that propelled the development forward. Our internal team developed a rapport that supported crisp decision-making, and having access to a partner that would respond faithfully to our directions was a great complement to the research initiative. We discovered several method improvements in the South laboratory that were used to accelerate North's development. We also worked on the interaction with the North team (more site visits!) and concluded the development in full alignment with that team. In the end, clinical sample analysis was performed by the "North" CRO, and we published the work that was completed by the "South" CRO [1]. The lesson here is one of flexibility, adaptability, and persistence. We applied many of the strategies outlined above to find the best partners, but we were still challenged during the development to change course and vary our approach to respond to the reality of the dynamic situation.

Now that we have discussed many broad principles of working with external partners to develop biomarker methods, let me provide some examples of the improvements to the methods achieved through a well-functioning interaction between sponsor and vendor. A key first step in the evaluation of a method for detecting a naturally occurring analyte in mammalian tissue is the assessment of the impact of the background composition of that tissue on the quantification of the analyte. How does the "hay" hinder the discovery of the "needle"? Indeed, we found that the CSF matrix we were testing interfered with the $A\beta_{42}$ method. A careful analysis of the effect of matrix dilution led to the establishment of a minimum required dilution (MRD) of sixfold prior to analysis of CSF samples. This modification led to nearly a doubling of measured $A\beta_{42}$ concentration in standard CSF samples. We also noticed a clear positional bias within the assay plates and worked with the CRO to mitigate the impact of that artifact by designing the plate layout intentionally to avoid introducing bias into the analysis of paired condition. This is not something that most CROs would have examined nor noticed on their own; a sponsor needs to be vigilant and as close to the data as possible. After observing poor spike recovery even at the optimal MRD, we profiled the influence of several types of polypropylene tubes. We found that a significant amount of $A\beta_{42}$ protein was adhering to even the polypropylene tubes we had selected to be the lowest in protein binding, and that this adsorption varied between individual CSF donors. Our team worked closely with the CRO to identify the optimal additive to prevent the adsorption of analyte protein to the assay tube and thereby eliminated a key source of variability in the clinical data set. A defined set of human CSF samples from different donors was tested both at the North and South CROs under the optimized set of conditions, employing two similar, but not identical, methods. The precision of the quantitation of $A\beta_{42}$ was markedly improved in comparison to published results from different sites of analysis (unpublished results, [3]). This highlights the utility of rigorous method development to both the advancement of a company's goals and the collective understanding of the field.

Once the method was adequately developed, we proceeded to the stage of method validation. Precision, accuracy, selectivity, specificity, parallelism, linearity of dilution, and tissue sample stability were all assessed for the optimized method in a "GLP-like" environment. Here is where the efforts in establishing a strong working relationship up to that point truly paid off. These experiments flowed well because both the sponsor and vendor teams understood the method fully and communicated well by that point in the interaction. Once validated, the optimized method was rolled out for the analysis of our valuable clinical samples. The data generated were deemed superior to other reported CSF data sets, in the consistency of the protein levels throughout the day for each patient and the overall precision within each treatment group (unpublished data). This was the delivery that we originally were entrusted to provide by our company.

It is hoped that this case study has helped the reader think about how to approach an outsourced biomarker method development project. Externalization can expand the reach of a small and nimble team, and with the right attitudes and aptitudes involved the experience can be a rewarding one in many ways.

References

1 Cullen, V.C., Fredenburg, R.A., Evans, C. et al. (2012). Development and advanced validation of an optimized method for the quantitation of Aβ42 in human cerebrospinal fluid. *AAPS J.* 14: 510–518.

2 Molinuevo, J.L., Blennow, K., Dubois, B. et al. (2014). The clinical use of cerebrospinal fluid biomarker testing for Alzheimer's disease diagnosis: a

consensus paper from the Alzheimer's Biomarkers Standardization Initiative. *Alzheimers Dement.* 10: 808–817.

3 Mattsson, N., Andreasson, U., Persson, S. et al. (2011). The Alzheimer's Association external quality control program for cerebrospinal fluid biomarkers. *Alzheimers Dement.* 7: 386.e6–395.e6.

Part II

Identifying New Biomarkers: Technology Approaches

5

Imaging as a Localized Biomarker: Opportunities and Challenges

Jonathan B. Moody[1], Philip S. Murphy[2], and Edward P. Ficaro[1]

[1] *INVIA Medical Imaging Solutions, Ann Arbor, MI, USA*
[2] *GlaxoSmithKline Research and Development, Stevenage, UK*

Introduction

Medical imaging is used to visualize patient anatomy noninvasively for the diagnosis of disease, but perhaps more significant for drug development is its ability to *quantify* physiological and physicochemical processes in a living subject. It is this capability that gives rise to the concept of imaging biomarkers.

Medical imaging technologies include a range of modalities used in the routine clinical setting: X-ray computed tomography (CT), positron emission tomography (PET), single photon emission computed tomography (SPECT), magnetic resonance imaging (MRI), ultrasound (US), and various types of optical imaging. Beyond their diagnostic utility in clinical practice, all of these technologies are generally applicable to quantitative imaging, which has been defined as *the extraction of quantifiable features from medical images for the assessment of normal findings or the severity, degree of change, or status of a disease, injury, or chronic condition* [1]. Thus, a quantitative imaging biomarker (QIB) that combines the general concepts of biomarkers and quantitative imaging can be defined as:

> … an objective characteristic derived from an *in vivo* image as an indicator of normal biological processes, pathogenic processes, or response to a therapeutic intervention [2].

In this chapter, we will discuss the use of medical imaging to quantify imaging biomarkers. After a brief overview of quantitative imaging in biological research, we present general characteristics of imaging biomarkers. The discussion will focus on the opportunities and challenges of using imaging biomarkers in drug development. Quantification introduces substantial

Biomarkers in Drug Discovery and Development: A Handbook of Practice, Application, and Strategy,
Second Edition. Edited by Ramin Rahbari, Jonathan Van Niewaal, and Michael R. Bleavins.

challenges to imaging biomarker implementation and practice. To address these challenges, the Quantitative Imaging Biomarkers Alliance (QIBA – http://www.rsna.org/qiba) was formed in 2007 by the Radiological Society of North America (RSNA) with the mission to "improve the value and practicality of quantitative imaging biomarkers." We conclude by reviewing QIBA's activities and recent progress toward achieving this goal.

Overview of Imaging in Biological Research

From their inception, the various noninvasive imaging modalities have been widely used in basic biological research. The broad utility of imaging in biological research reflects the breadth of fundamental physical processes that underlie all imaging technologies. The same types of probes and methods of detection that are generally used in biochemistry and molecular and cellular biology, such as optical, nuclear (γ emission), nuclear magnetic resonance (NMR) spectroscopy, X-ray, and others, also form the basis of noninvasive imaging technologies. When the transmission electron microscope was introduced in the 1930s, the morphology of submicroscopic cellular structures that had previously been inferred by indirect methods such as X-ray diffraction or structural chemistry could be directly visualized for the first time in electron micrographs [3]. By analogy, the development of medical imaging technologies (for example, US [4, 5], SPECT [6], PET [7, 8], CT [9], and MRI [10, 11]) has provided the ability to visualize *in vivo* biological processes previously accessible only by indirect or invasive means. As the science of biological imaging has grown, the development of new imaging applications has been driven by the unique ability of noninvasive imaging to reveal information about anatomical structure and physiological function as well as various disease states in living subjects.

It will be helpful to briefly review the use of imaging in biological research to understand the context of imaging biomarker discovery. We briefly describe some common clinical imaging technologies and then highlight some of the image-based methods for measuring biological processes that have been reported in the literature. General introductions to imaging technologies may be found in [12, 13], and a comprehensive treatment of the theoretical and physical basis of imaging is given in [14]. Several detailed examples of biological imaging applications in drug discovery have been described in a recent monograph [15].

Clinical Imaging Technologies

Clinical X-ray CT is commonly used for noninvasive 3-dimensional (3D) imaging of anatomical structure. The CT signal is generated by the transmission of X-rays through the body. The main source of CT image contrast is the attenuation of X-rays by tissues of different atomic densities. Mineralized tissues appear brighter due to higher X-ray attenuation, and soft tissues have intermediate intensity with X-ray attenuation similar to that of water. CT can provide high spatial resolution (>0.5 mm), and temporal resolution can be high enough to image the motion of the human heart with appropriate gating methods. Spatial and temporal resolution as well as sensitivity (signal-to-noise ratio) are proportional to the intensity of the X-rays, which must be balanced between the requirements of image quality and radiation safety. CT data acquisition is straightforward and rapid with minimal operator variability.

MRI is also widely used for anatomical imaging. Based on NMR spectroscopy, the MRI signal is generated by detecting the intrinsic spin of atomic nuclei using specific combinations of applied magnetic fields [16, 17]. The proton nuclei of hydrogen in tissue water and lipids are the most commonly imaged due to their relaxation properties and large abundance in the body. In principle, the NMR basis of signal detection also permits imaging or spatially localized spectroscopy of other biologically relevant nuclei such as ^{13}C, ^{19}F, ^{23}Na, and ^{31}P. Various aspects of the MRI signal depend on the molecular and cellular environment of the nuclei which leads to excellent image contrast between soft tissues with good spatial resolution (~1–5 mm). Temporal resolution is relatively poor although imaging heart function is possible with gating techniques. The chemical sensitivity of MRI is also relatively poor, generally requiring tens of micromolar to millimolar volumes of nuclei for signal detection. MRI resolution and sensitivity depend on the main magnetic field strength, the concentration of target nuclei and the duration of image acquisition which must be balanced between the required image quality and data acquisition efficiency.

Nuclear imaging includes both PET and SPECT modalities and is based on the detection of γ rays (high energy photons) emitted after intravenous infusion of a radiolabeled tracer. Image contrast in nuclear imaging relies on the use of tracers that act as biomolecular probes targeting specific *in vivo* molecular processes. In general, the chemical sensitivity of nuclear imaging methods is the highest of all clinical modalities, with the picomolar sensitivity of PET being about 2–3 orders of magnitude higher than that of SPECT [18]. The relatively poor spatial resolution of PET (~5 mm) and SPECT (~15 mm) has been mitigated somewhat through the development of multimodality image acquisition combining PET/CT or SPECT/CT [19], and more recently PET/MRI [20], which provide high-resolution anatomical images in combination with nuclear medicine images of molecular processes. Like CT, nuclear imaging utilizes ionizing radiation, and the resolution and sensitivity of nuclear imaging are proportional to the radiation dose received by the subject.

US is similar to MRI in that images are formed without the use of ionizing radiation. Ultrasound is primarily used to image anatomy and blood flow [21]. The US signal is generated by transmitting high-frequency acoustic waves into the body and then measuring the resulting reflected and scattered waves. Image contrast arises from bulk tissue properties that influence wave propagation [12].

Good spatial resolution (~0.5–2 mm) and excellent temporal resolution can be obtained. Spatial resolution is directly proportional to the frequency of the applied wave, while sensitivity is inversely proportional to wave frequency.

Table 5.1 compares the strengths and limitations of each imaging modality [13]. In all modalities, the acquired raw data typically requires image reconstruction and post-processing using computational methods of varying complexity [17, 22]. In some cases, a series of images may also be used to calculate parametric images representing image-based measures corresponding to particular biological processes (e.g. MRI images of water diffusion [23], or PET images of cerebral blood flow or glucose metabolism [24]).

Image-Based Measures of Biological Processes

Many different noninvasive imaged-based measures of physiological function and biological processes have been studied in biomedical research [15, 25–28]. The different characteristics of imaging modalities enable measures that span multiple levels of detail in the imaging subject from anatomy and physiological processes to cellular and molecular aspects of biological processes. Apart from differences in imaging modality, such image-based measures may be classified into three general types according to the way image contrast is generated: by using endogenous biophysical mechanisms, exogenous contrast agents, or labeled tracers [29]. Endogenous image contrast is associated with the intrinsic mechanism of signal generation. Examples using MRI include different relaxation mechanisms such as T_1, T_2, T_2*, macromolecular cross-relaxation, chemical exchange, and transport mechanisms such as blood flow, perfusion, and diffusion. In some cases, the intrinsic contrast is insufficient to distinguish the biological structure or process of interest, and an exogenous contrast agent is infused to enhance the image signal. Examples of contrast agents include gadolinium- [30] and iron oxide–based [31] agents for MRI, iodine-based agents for CT [32], and microbubbles for US [12]. Table 5.2 lists a few examples of endogenous and contrast-enhanced biological measures that have been explored using MRI.

Imaging with the use of labeled tracers or molecular probes is known as *molecular imaging* [13, 44] and also involves infusion of an exogenous agent. In this case, the imaging tracer is designed for cellular uptake or molecular binding with high affinity for a specific *in vivo* molecular target or biological process [45]. By definition, tracer-based imaging methods employ concentrations of labeled tracers that are small enough to avoid perturbing the biological system under study [46], as distinguished from contrast agents which are typically designed to remain outside of cells and require much higher physiological concentrations to achieve adequate image contrast. Molecular imaging is able to provide biological information on normal and disease processes not available from other methods. For example, molecular imaging has been used to establish the relationship between inflammation

Table 5.1 Comparison of strengths and limitations of clinical imaging modalities.

Modality	Strengths	Limitations
CT	Limitless depth of penetration High spatial resolution Good temporal resolution High throughput Low operator variability	Relatively poor sensitivity (without CA) Primarily anatomical information Limited soft tissue resolution Limited MI applications Ionizing radiation
MRI	Limitless depth of penetration High spatial resolution Good temporal resolution Quantitative data No ionizing radiation	Relatively poor sensitivity (without CA) Relatively poor temporal resolution Relatively expensive Relatively poor throughput Relatively high operator variability
PET	Limitless depth of penetration Excellent sensitivity Quantitative data Low operator variability	Relatively expensive Cyclotron or radionuclide generator Ionizing radiation Limited spatial resolution Relatively poor throughput
SPECT	Limitless depth of penetration Good sensitivity Quantitative data Low operator variability	Relatively expensive Ionizing radiation Limited spatial resolution Lack of attenuation correction Relatively poor throughput
US	Relatively inexpensive Quantitative data No ionizing radiation Excellent temporal resolution Good sensitivity (with CA) High throughput	Relatively poor sensitivity (without CA) Limited depth of penetration Limited to soft tissues Coupling of instrument to subject Relatively high operator variability

CA, contrast agent; CT, computed tomography; MI, molecular imaging; MRI, magnetic resonance imaging; PET, positron emission tomography; SPECT, single photon emission computed tomography; and US, ultrasound.

"Clinical utility" refers to the capability of applying a modality both clinically in humans in addition to pre-clinically in animal models, enabling translation of measurement methodologies from early discovery to later clinical stages. "Throughput" relates to the overall measurement rate of a modality, including image acquisition and quantification processes. "Operator variability" refers to how much variability is introduced by human operators of imaging equipment and software during the process of imaging biomarker measurement.

Source: Adapted from James et al. 2012 [13].

Table 5.2 Examples of image-based measures of biological processes that have been studied using MRI.

Biological process or parameter	Image-based measure	MRI imaging method
Relative cerebral blood volume, relative cerebral tissue perfusion	T_2 and T_2^* weighted intensity changes	DSC bolus tracking [33]
Cerebral tissue perfusion	Difference between spin inverted and control images	ASL [33]
Task-induced functional activation in the brain	T_2 and T_2^* weighted intensity correlation with stimulus	BOLD [34]
In vivo water distribution, *in vivo* fat distribution	Phase-shifted images from water-fat chemical shift	PDFF MRI [35]
Peripheral edema	Integral increase in T_2 values	T_2 weighted MRI [36]
Liver iron concentration	Weighted sum of bi-exponential T_2 values	T_2 weighted MRI [37]
In vivo temperature distribution	Temperature-dependent proton resonance frequency shift	MRTI [38]
Spatial distribution of pH	Amide proton transfer	CEST/PARACEST [39, 40]
Glycogen concentration	Amide proton transfer	CEST [41]
Intra- and extracellular pH	[1–13C] metabolite concentrations	Hyperpolarized MRSI [42, 43]
Metabolic substrate selection, enzyme flux	[1–13C] metabolite concentrations	Hyperpolarized MRSI [42, 43]

DCS, dynamic susceptibility contrast; ASL, arterial spin-labeling; BOLD, blood-oxygenation-level-dependent; MRI, magnetic resonance imaging; MRSI, magnetic resonance spectroscopic imaging; MRTI, magnetic resonance temperature imaging; CEST, chemical exchange saturation transfer; PDFF, proton density fat fraction.
These techniques use either endogenous contrast mechanisms or exogenous contrast agents.

and microcalcification in early-stage atherosclerotic plaques [47]; it has been used to visualize the *in vivo* spatial distribution and time course of vascular endothelial growth factor receptor expression in a model of peripheral arterial disease [48]; it has been used for real-time imaging of metabolic pathway changes after oncogene-driven progression and regression of liver cancer [49]; and it has been combined with conditional disease models using transgenic mice and genetic profiling to yield a powerful system for *in vivo* preclinical study of disease processes and potential therapeutic targets [50–53]. The most common molecular imaging applications utilize radio-labeled tracers for detection by the nuclear medicine modalities. However, many non-nuclear approaches have also been developed with optical [54], MRI [28, 42–44] and ultrasound modalities [55, 56].

The majority of such image-based measures are distinguished from direct imaging of anatomy by being indirectly connected to the biology. A correlation between the image-based measure and some identifying property of the biological or disease process is typically established invasively or histologically in preclinical animal studies [29]. The invasive or histological correlate often provides an initial rationale for the noninvasive imaging approach, and may guide development and subsequent interpretation of the image-based measurement [13]. Some examples of disease-identifying properties and corresponding image-based correlates are:

- *Lung density*: X-ray attenuation in lung
- *Liver viscoelasticity*: shear wave propagation and attenuation in liver
- *Brain amyloid burden*: brain uptake of targeted amyloid tracer
- *Tumor glycolytic activity*: standardized tumor uptake of fluorodeoxyglucose (FDG) tracer
- *Tumor vascularity*: tumor uptake of gadolinium-based MRI contrast agent
- *Tumor cellularity*: apparent diffusion coefficient of water in tumor
- *Tumor growth*: serially measured tumor volume from an image region of interest

It is helpful to think of such image-based methods as noninvasive assays, in the sense that the measurements extract or infer some biological, physiological, or molecular characteristic from the subject as a whole. While not always clinically diagnostic for disease, such image-based assays can often provide the ability to monitor disease progression and the effects of therapeutic intervention [29]. The associated biological or biophysical correlation provides widely varying degrees of linkage between the measurement and biology. Molecular imaging provides perhaps the most direct linkage since the characteristics of target, signal generation, and the probe's affinity for the target can be determined empirically, leading to relative confidence in the functional or molecular basis of the image-based signal. For example, the use of ^{18}F-FDG, a glucose analog, as a radiotracer in PET imaging relies on the property of glucose uptake and phosphorylation within metabolically active cells. In other cases, a single image-derived parameter corresponds to a complex of multiple functional parameters. For example, kinetic modeling of dynamic contrast-enhanced (DCE) MRI data yields a model parameter, K^{trans}, that reflects general vascularity but can be influenced by several factors such as blood flow, vascular permeability, and vascular surface area, potentially complicating interpretation of this image-based measure.

Even when an explicit biophysical model of the image signal is available, as is the case for magnetization transfer MRI [57], it may be costly or impractical to acquire the imaging measurements necessary to routinely apply that model. For magnetization transfer MRI, a related phenomenological model has proven useful to relate a simpler measurement (the magnetization transfer ratio,

MTR) to the diffuse macromolecular changes and demyelination associated with multiple sclerosis [58]. Another example is the measurement of tracer or contrast agent uptake by organs or tumors. The most rigorous measurement of tracer kinetics involves the dynamic acquisition of a series of images followed by computational modeling to estimate rate parameters of interest. This procedure may be impractical in some settings, and a much simpler method may be used to acquire a single static image after tracer infusion and a specified uptake period, followed by estimation of tracer concentration in the image region-of-interest [59].

The lack of specificity of some image-based measures warrants caution in their use prior to adequate technical validation (Section "Essential Characteristics of Imaging Biomarkers"). Early *ex vivo* NMR measurements of bulk T_1 and T_2 relaxation in tumors [60] were initially believed to distinguish malignancy from nonmalignancy, and even inspired the development of the first whole body MRI system [61]. However, these results have not been subsequently validated *in vivo*, although several other MRI methods for characterizing tumors and evaluating therapy have since been developed [62].

Despite the exploratory nature of many biomedical imaging applications, the diversity and potential of many image-based biological measures often furnish much of the interest and enthusiasm for particular imaging biomarkers. However, numerous additional factors must be addressed to establish the validity, reliability, and suitability of a particular image-based measure as a tool for evaluating novel therapeutics.

Anatomy of an Imaging Biomarker

Like the drug development process itself, the implementation and practice of imaging biomarkers is a highly collaborative multidisciplinary effort. An imaging biomarker may be conceptually defined by two aspects: image acquisition and image analysis. Image acquisition refers to all activities and technologies related and leading to the acquisition of image data, such as imaging hardware operation, maintenance and quality control, subject preparation and positioning, contrast agent or labeled tracer preparation, dosing and injection protocols, and physiological monitoring. Image analysis includes all post-acquisition activities such as image reconstruction, processing, feature extraction, computational modeling, quantification, and interpretation. These two aspects of an imaging biomarker require significant input from a diverse array of scientific and engineering disciplines, including biology, pharmacology, pathology, chemistry, physics, biomedical and nuclear engineering, computer science and mathematics, as well as medicine. Each of these specialties has several critical roles in the implementation and use of imaging biomarkers:

- *Biologist/physiologist/molecular biologist/pathologist*: help establish the biological relationship between the image-based measurement and the physiological function or disease process; help establish the rationale for studying particular image-based measures; and help interpret the quantitative results.

- *Radiologist/physician/pathologist/pharmacologist*: help establish the clinical relationships between physiology, disease processes, imaged-based measurements, and the relevant therapeutic or safety outcomes during treatment; and help understand all possible actions of the treatment (beneficial and deleterious) [63] to aid in the interpretation of imaging results.
- *Biochemist/radiochemist*: provide understanding of physiological processes and molecular targets at the biochemical level; and design, synthesize, characterize, and test optical, magnetic, and molecular probes and labeled tracers.
- *Mathematician/statistician/computer scientist*: provide the fundamental mathematical, statistical, and quantitative basis for image acquisition, reconstruction and analysis; and provide software tools for performing many tasks associated with image acquisition, reconstruction, quantification, analysis, visualization, and interpretation.
- *Imaging physicist/biomedical and nuclear engineer/technologist*: provide the physical basis for imaging technologies, signal generation and detection, hardware design and optimization, and quality control; provide the technological expertise for isotope production to be used for generating radiotracers for labeling imaging agents; and operate the imaging equipment, performing image acquisition, processing, and analysis tasks.

It is perhaps not surprising that the broad interdisciplinary nature of imaging biomarkers has led to a recognition that basic terminology is often used inconsistently among scientists of different specialties. This barrier to effective communication within biomarker development teams has been addressed by the development of standard glossaries of related concepts and definitions associated with general biomarkers [64] and clinical trials [65]. Table 5.3 gives a brief list of key terms with definitions adopted by the imaging biomarkers community [1].

Essential Characteristics of Imaging Biomarkers

In Section "Image-Based Measures of Biological Processes" we highlighted some uses of imaging in biological research. In such studies, when a correlation has been demonstrated between an image-based measurement and a particular physiological characteristic or disease process, it is tempting to identify that measurement as an imaging biomarker. However, this is only a first step. The technical performance of the biomarker must be rigorously established using appropriate performance metrics. Further, the goal of medicine to positively affect a clinically relevant outcome (how a patient feels, functions, or survives

Table 5.3 Key formal definitions of terms adopted by the quantitative imaging biomarkers alliance (QIBA) from the International Vocabulary of Metrology (VIM 2008).

Term	Definition
Quantity	A property of a phenomenon, body, or substance, where the property has a magnitude that can be expressed as a number and a reference (a reference can be a measurement unit, a measurement procedure, a reference material, or a combination of such)
Quantity value	A number and a reference together that express the magnitude of a quantity
Measurand	The quantity intended to be measured (including any property or state of the imaging subject relevant to the quantity)
Measurement	The process of experimentally obtaining one or more quantity values that can reasonably be attributed to a quantity
Metrology	Science of measurement and its applications

See also reference [1] for additional terms.

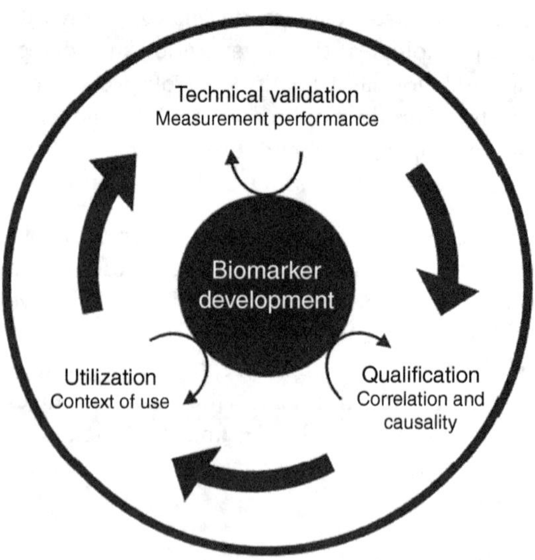

Figure 5.1 Three interrelated stages of biomarker development. For a given biomarker, each stage is accomplished iteratively and the current state of each stage informs following stages. The overall process implies a continual reevaluation of each development stage as further data and results become available in ongoing research. See Section "Essential Characteristics of Imaging Biomarkers" for further details. Source: Adapted from reference Institute of Medicine 2010 [66].

[65]) dictates that ultimately an image-based measure should have some degree of correlation with such a clinical outcome. In addition, the context of use of a proposed biomarker will determine how it will be appropriately used. A general process of three interrelated stages has been proposed to evaluate these characteristics of biomarkers [66] (Figure 5.1), a process which becomes essential as particular image-based measures progress from basic biomedical research to medical product development and clinical care.

Technical Validation

In order for an image-based measure to be an effective tool in drug development, the correlation linking the image signal to a particular identifying property of a disease process or pathway must be characterized, which is termed *technical* or *analytical validation* [2, 66]. Proper validation of a biomarker's technical performance includes:

- Definition of the measurand (quantity intended to be measured) including specification of the relevant properties and state of the imaging subject such as:
  - biological conditions including species
  - subject demographics
  - state and stage of disease
- Specification of the measuring device:
  - imaging modality and acquisition settings
  - image reconstruction and analysis methods
- Adequate optimization of the measuring device for measurement of the defined measurand (parameter *estimation*, Section "Imaging Biomarkers and Diagnostic Imaging")
- Evaluation of statistically appropriate metrics of measurement performance [67]:
  - bias, precision, and linearity over the measurement range of interest
  - test–retest (within-subject) variability
  - intra- and interobserver variability

Determining the bias of an image-based measurement requires knowledge of the true quantity value which is often not available, and comparison with a commonly accepted reference standard may instead be necessary [67]. There are often multiple sources of variability associated with image acquisition, processing, and analysis. Determining the overall measurement variability and its relation to the inherent biological variability of the measurand are essential for appropriate interpretation and utilization of biomarker measurements [68].

Technical validation will generally be an iterative process. Initially published research studies of an image-based biological measure may provide only partial validation by presenting evidence of its correlation with a biological property and by occasionally reporting some measure of reliability in a limited sample size, often only in animal studies. Additional development will often be necessary to fully characterize and optimize the measurement performance in image phantoms, animal models, and patients [68, 69]. Establishing the technical performance of an imaging biomarker for use beyond a single imaging center is difficult due to the diversity of imaging devices, varying image acquisition protocols and analysis algorithms, and the frequent need to integrate biomarkers with existing local practice. Section "Quantitative Imaging Biomarkers Alliance (QIBA)" discusses recent efforts by the QIBA in broad collaboration with relevant stakeholders to develop standardized procedures for characterizing and documenting the technical performance of particular imaging biomarkers. A fit-for-purpose strategy can also be helpful [64], in which less stringent validation criteria are tolerated at earlier phases of drug development [66]. In this way, the evolving levels of technical validation during early biomarker development can still contribute appropriate and useful evidence for early drug development decision-making [70].

Qualification

An identifying disease property that is correlated with a biomarker measurement may have a natural association with a clinical outcome of interest to the patient, such as reduction of pain, restoration of function, or increased survival. Such an association is a necessary but not sufficient condition for the biomarker to mediate a *causal pathway* (i.e. a chain of biological events related to one another by cause and effect) of the disease process [66]. Thus, biomarker *qualification* seeks to assess or establish evidence for both of the following:

- The correlation between the biomarker measurement and a relevant disease endpoint;
- The degree to which the biomarker measurement captures any relationship between a given therapeutic intervention and the disease endpoint [71].

Evidence of the second type identifies the degree to which the identifying property measured by the biomarker participates in a causal pathway of the disease process and is also affected by the intervention [72, 73]. Although evidence on correlations is relatively straightforward to obtain, casual relationships may require more data on disease pathways and mechanisms of action of the intervention than are usually available, and a probabilistic reasoning is necessary that "emphasizes epidemiological and statistical relationships, acknowledging that all contributing factors are generally not fully understood" [66, 74].

Many relevant endpoints other than a formal clinical outcome may be appropriate in drug development (e.g. endpoints related to target validation, safety assessments, clinical trial enrichment, etc.), and in such cases biomarker qualification seeks to establish the biomarker–intervention–endpoint link to an extent appropriate for the drug development purpose [66, 75].

Utilization

While *qualification* objectively evaluates the evidence on biomarker–intervention–endpoint correlation and causality, the *utilization* stage analyzes the context of use of a biomarker [66]:

- The biomarker measurand (Section "Technical Validation")
- The measuring device and measurement methods (Section "Technical Validation")

- Prevalence, morbidities, and mortalities of the disease
- The class of therapeutic intervention and risks and benefits of the intervention
- The biomarker's purpose of use and drug development context
- The interpretation and specific decisions based on the biomarker measurements

Given the available technical validation and qualification evidence, the contextual analysis aims to determine the suitability of the biomarker for the proposed use.

The regulatory concept of biomarker *qualification* [65] combines general *qualification* as described above and *utilization* analysis, and is a process that provides regulatory assurance that a qualified biomarker can be used as a reliable tool in drug development. The US Food and Drug Administration (FDA) has developed formal guidance for Biomarker Qualification under the Drug Development Tools (DDT) Qualification Program [75]. This process allows a biomarker to achieve regulatory acceptance independent of any particular drug development program. An analogous procedure is also provided by the European Medicines Agency [76].

Further elements of logistical and economic context may also need to be considered [66]:

- The necessary image acquisition hardware and methods and image analysis software tools must be readily available.
- The prescribed image acquisition protocol and analysis methods must be appropriately standardized across hardware platforms, analysis software and imaging centers.
- The imaging biomarker must be cost-effective and readily measurable [29].

Development Stages

Much of the existing literature on image-based biological measures (Section "Image-Based Measures of Biological Processes") could be characterized as biomarker discovery: the discovery and elucidation of potentially useful associations between image signals and disease processes. The three stages that follow discovery and further characterize biomarker development, technical validation, qualification, and utilization (Figure 5.1), are highly interrelated: for technical validation, aspects of the context of use must be defined and some evidence for correlation must already exist; technical optimization also requires specification of the clinical context as part of the process of *task-based assessment of image quality* [14]; evidence of qualification depends on the current state of technical performance and will generally depend on a proposed context of use. Each stage will be developed iteratively as well as inform the other stages [66]. The stage of *technical validation* is distinguished by its objective focus on metrology, while the *qualification* stage objectively evaluates the available evidence on correlation and causality. The *utilization* stage is further distinguished by the more subjective task of the contextual assessment [66]. The development cycle is envisioned to be an ongoing process such that periodic reevaluation of each stage will be necessary as new technology and data become available.

The aims of technical validation and qualification of biomarkers, though costly for any one organization, are presently being accomplished by a growing array of public–private consortia [77]. For imaging biomarkers, there is a general lack of adequate technical validation which is being addressed by the "Quantitative Imaging Biomarkers Alliance", while collaborative work on qualification of imaging biomarkers is being actively pursued by QIBA and the Biomarkers Consortium [78], the Critical Path Institute (C-PATH) [79, 80], and the European Innovative Medicines Initiative Safer and Faster Evidence-based Translation Consortium (SAFE-T) [81].

Imaging Biomarkers and Diagnostic Imaging

The foremost application of medical imaging is the routine clinical diagnosis of disease. Historically the development of imaging systems capable of noninvasively scanning humans followed shortly after the initial technological development of the various imaging modalities [6, 8–10, 61, 82–84]. Moreover, imaging applications emphasizing medical diagnosis have driven many of the technological improvements to the present day.

In general, clinical assessments using diagnostic images are made visually, and measurements are recorded on nominal or ordinal scales. Although the measurement scales utilized by non-imaging biomarkers may be any one of nominal, ordinal, interval, or ratio scales [64, 85], imaging biomarkers have been intentionally defined as quantities measured only on interval or ratio scales [1]. In principle, the *qualitative* visual measures that are used for diagnostic purposes could also serve as imaging biomarkers. However, the restriction to *quantitative* image-based measures eliminates a major source of variability (i.e. reader variability) that can significantly degrade biomarker technical performance.

The essential quantitative nature of imaging biomarkers highlights a key issue for assessment and optimization of "image quality" for biomarker applications. It is widely recognized that the most reliable and objective approach to assess general image quality and optimize the imaging process is to first specify the purpose or task for which the image is produced and secondly to determine how well that task is performed [86, 87]. Two general categories of tasks are *classification* tasks, which include the typical diagnostic tasks of deciding whether a patient has or does not have disease, and *estimation* tasks, which involve estimating quantities of interest from medical images [87]. An example of a *classification* task would be a radiologist using images to detect or localize abnormal lesions and attempting to answer the questions: are the lesions malignant or nonmalignant, and if malignant, what is the grade or stage of the cancer? Examples of *estimation* tasks would include a cardiologist assessing the health of a patient's heart by estimating myocardial perfusion or an oncologist assessing the response of a patient's tumor to cytotoxic therapy by estimating metabolic uptake of [^{18}F]-FDG.

Diagnostic imaging platforms and analysis tools historically have been concerned with *classification* tasks and have been optimized for those tasks. Imaging biomarkers, being inherently quantitative, are concerned with *estimation* tasks. Thus, the requirements and standards for diagnostic imaging systems are somewhat different from those of imaging biomarkers. For example, the demands of quantification for imaging biomarkers may place different limits on image resolution and noise characteristics than are required for diagnostic purposes. Since quantitative *estimation* is essential for imaging biomarkers, image acquisition and analysis systems must be optimized for such purposes. Such optimization may involve differences in hardware design, image quality control processes, and image reconstruction and processing methods.

Key Advantages of Imaging Biomarkers

Imaging biomarkers provide the unique ability to visualize anatomy as well as to spatially localize physiological and molecular processes *in vivo*. Additionally, most imaging modalities have the capability to acquire repeated or temporally resolved images which allow the study of various dynamic biological processes. *In vivo* image–based measurements occur in the biological microenvironments of disease processes or drug targets, accounting for both systemic and local modulators. As *in vivo* assays, imaging biomarkers also share with *in vitro* biomarkers the ability to determine drug–target interactions, to help optimize drug dose, and to longitudinally evaluate individual response to therapy *in vivo* [88].

A broad range of sensitivities and spatial resolutions are also available, providing great flexibility in the choice of applications (Figure 5.2). High-resolution imaging by CT, MRI, or US can evaluate gross effects of therapeutic interventions at the anatomical level of organ systems, e.g. changes in tumor volume. Nuclear medicine, CT, MRI, or US perfusion imaging biomarkers can probe functional effects at a regional tissue and vascular level, e.g. changes in tumor vascularity or cellularity. Nuclear medicine, MRI, and optical molecular imaging biomarkers can provide read-outs down to the level of cellular and molecular effects, e.g. evaluating tumor glycolytic activity. This range of spatial resolutions may sometimes be an advantage when considering the process of biomarker qualification. Molecular imaging applications may attain the level of detail necessary to test for proof-of-mechanism, while qualification of imaging biomarkers may be facilitated by a coarser biological "resolution" at the macroscopic level of organ systems which may potentially be more directly relatable to a clinical outcome.

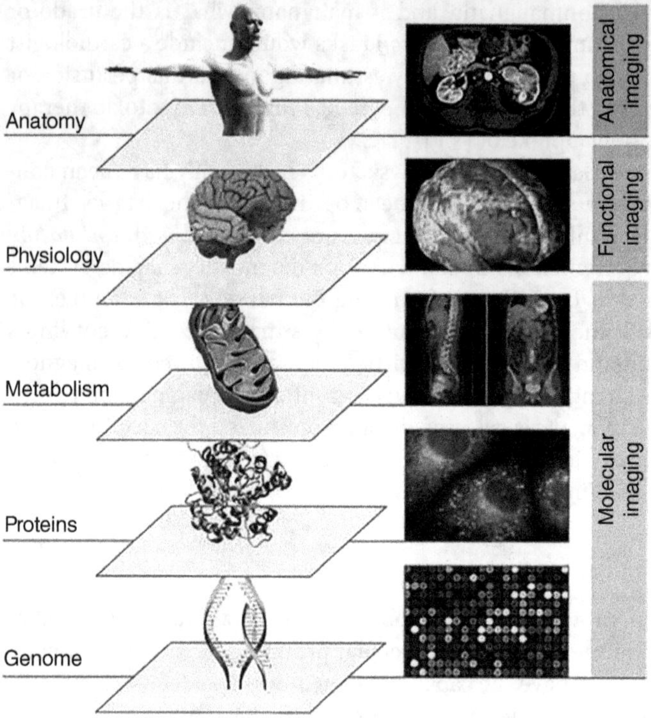

Figure 5.2 Quantitative imaging biomarkers can provide important information at multiple levels of biological "resolution" from anatomical and functional levels of organ systems down to the level of cellular and molecular processes. Source: Lambin et al. 2012 [89]. Reprinted with permission of Elsevier). *(See insert for color reproduction of the figure.)*

The clinical foundation of the various imaging modalities ensures the widespread availability of imaging equipment, particularly CT and US modalities. In addition to the clinical relevance of imaging, there is also the possibility for translation between preclinical and clinical applications, which provides the capability of using similar read-out parameters from early preclinical safety and efficacy studies in animal models to eventual clinical evaluations in humans. Furthermore, longitudinal assessment of individual subjects allows the potential for reduced statistical uncertainty, and consequently a reduction in the number of study subjects. Clearly, this can benefit drug development through faster study completion and reduced costs. However, in practice, there are many additional factors related to the implementation of imaging biomarkers that can influence the net cost savings, and the efficiency achieved in using a validated and qualified imaging biomarker must be carefully evaluated from the context of use in each biomarker application.

Limitations of Imaging Biomarkers

The complexity of image signal generation and the broad generality of medical imaging modalities to obtain images at anatomical, physiological, and molecular levels, all contribute to an acute difficulty in technical validation of imaging biomarkers. In particular, there are two inherent challenges not typically faced by *in vitro* biomarkers [68]:

- The need for extensive human involvement in making the measurements, e.g. equipment operators during image acquisition, and individuals involved in the various processes of image quantification and analysis.
- The lack of a true value of the measurand to be compared with the biomarker measurements for testing of bias during technical validation.

The challenge of measurement variability introduced by human involvement is reflected in the restriction of imaging biomarkers to image-derived quantitative measurements made on interval or ratio scales. This variability must be addressed by appropriate statistical characterization of test–retest, intra-, and interobserver variability [67], standardization and optimization of acquisition and analysis protocols [68, 69], and systems for the assessment of compliance. The challenge of evaluating measurement bias without a true value can be approached by the testing of image phantoms and by applying *in silico* methods for simulation and analysis of the imaging process [1, 68].

Technical validation as described in Section "Technical Validation" incorporates both standardization and optimization of the imaging process. The strengths and limitations of different imaging modalities directly contribute to our ability to standardize and optimize image quality for a specific biomarker task. In general, the limitations of the underlying imaging technologies and methodologies may be addressed by ongoing basic research and technology development already occurring independently of imaging biomarker development. The widespread diagnostic use of medical imaging motivates continual research and vendor improvements in hardware, software, and acquisition methods. Such improvements have recently included faster and more efficient imaging for MRI, innovations that reduce radiation dose in CT, new hardware designs that improve sensitivity and spatial resolution of nuclear imaging modalities, and new capabilities afforded by combining modalities, as in SPECT/CT, PET/CT and most recently PET/MRI. Furthermore, many image analysis algorithms and tools have been borrowed from ongoing computer vision research. All such advances have the potential to benefit imaging biomarker performance. However, careful attention must be paid to the differing criteria for system optimization between diagnostic and biomarker applications (Section "Imaging Biomarkers and Diagnostic Imaging"). Large-scale coordination of the resources necessary to achieve these aims are currently being mobilized by the "Quantitative Imaging Biomarkers Alliance".

There are also inherent limitations that may depend on the particular imaging biomarker measurement method. For example, despite the impressive preclinical accomplishments of molecular imaging to date, expansion of such applications to the clinic has been limited by the relatively few molecular imaging agents that have been approved for clinical use in humans [44, 90]. More generally, within the spectrum of potential imaging biomarkers, those that rely on endogenous contrast mechanisms have the advantage of avoiding the added complexity and safety concerns introduced by the use of exogenous contrast or molecular imaging agents. However, such endogenous contrast mechanisms are often inherently non-specific, relying on correlations between the measurements and biological processes that may be indirect or difficult to interpret. Another inherent limitation is the use of ionizing radiation by nuclear imaging and CT modalities. Recent epidemiological studies have raised concerns in the medical community that exposure to radiation in pediatric CT scans can lead to increased risk of cancer in adulthood [91, 92]. Although the radiation dose of a single scan is relatively low, clinical use of these modalities and cumulative radiation exposure to patients have both increased dramatically in recent decades [93]. Diagnostic imaging has generally been guided by the principle that the risks of ionizing radiation should always be weighed against the benefits of the imaging procedure to the patient [94]. However, there is ongoing scientific controversy on the risks and biological effects of low dose radiation [95, 96]. In spite of the lack of consensus, progress has been made in reducing radiation doses through technological improvements in the sensitivity of imaging devices [97, 98] and acquisition methods [99, 100].

Importance of Context: Disease, Therapeutic Strategy, Development Stage, and Alternative Approaches

It was noted above that the regulatory concept of biomarker *qualification* as defined by current FDA Guidance on the Qualification Process for DDT [75] incorporates both formal qualification (Section "Qualification") and utilization analysis (Section "Utilization"). This process essentially formalizes biomarker utilization analysis through the creation of a Context of Use Statement consisting of two components: a concise biomarker "use statement" and a comprehensive description of "conditions for qualified use," which includes contextual elements such as those listed in Section "Utilization". The process considers a given proposed biomarker use in light of available data on technical validation and qualification, and attempts to judge whether such data support the proposed use [66].

Tumor size as measured by MRI, CT, nuclear medicine, or ultrasound imaging was one of the earliest and most extensively used imaging biomarkers, and provides an example of applying utilization analysis [66]. Tumor size can be measured by different imaging modalities with or without various contrast agents or radiotracers. Different image-based methods can be used such as estimation of one or more linear measures [101], geometric modeling of tumor volume, or direct estimation of tumor volume by delineation of tumor boundaries, which has been a recent focus of validation work [69]. The variability and appropriateness of these modalities and methods for a given proposed biomarker use will be determined by the available technical validation data. Such data may also depend on biological factors such as tumor type and source organ that can influence the appearance and quantification of the tumor in different imaging modalities. Appropriate technical validation will specify a set of measurement conditions and estimate the measurement bias and variability that can be expected under these conditions.

Tumor size provides a measure of the patient's tumor burden, which is plausibly associated with the clinical outcome of overall survival. However, such a correlation does not always guarantee that treatment-related changes in tumor size will also correspond to treatment-related changes in outcome [102]. For example, the treatment may have additional unintended harmful effects on overall survival through causal pathways not related to tumor size and that counter the benefit of any reduction in tumor size [66]. Biological factors in addition to tumor size may influence clinical outcome, such as tumor metabolism or vascularity, tumor cell density in the case of gastrointestinal stromal tumors, or nodal involvement in the case of locally advanced breast cancer [66]. Although knowledge of all causal pathways of a disease process and a given treatment may be rare, the completeness of such knowledge will indicate the appropriateness of using the biomarker as a surrogate endpoint to confirm efficacy in Phase III trials [73]. Imaging in basic biological research, in particular molecular imaging, can play an important role in elucidating such disease processes.

Another important aspect of biomarker context is the purpose of a given biomarker in drug development. Biomarker use may be categorized by the stage of drug development [66]:

- *Discovery*: biomarkers associated with disease that may be screened for as a target for intervention to prevent or treat the disease.
- *Early product development*: biomarkers used for target validation, compound screening, pharmacodynamic assays, safety assessments, and subject selection for clinical trials, and as endpoints in Phases I and II clinical trials.
- *Surrogate endpoints for product approval*: biomarkers used for Phase III clinical trial confirmation of efficacy and to substantiate claims for product marketing.

Alternatively, the FDA DDT Guidance defines four descriptive categories of biomarker use that are generally applicable in drug development [75]:

- *Diagnostic biomarker*: a disease characteristic that categorizes a person by the presence or absence of a specific physiological or pathophysiological state or disease.

- *Prognostic biomarker*: a baseline characteristic that categorizes patients by degree of risk for disease occurrence or progression of a specific aspect of a disease.
- *Predictive biomarker*: a baseline characteristic that categorizes patients by their likelihood of response to a particular treatment relative to no treatment.
- *Pharmacodynamic biomarker*: a measurable characteristic whose change shows that a biological response has occurred in a patient who has received a therapeutic intervention and for which the magnitude of the change is considered pertinent to the response.

These four categories can help define an appropriate context of use for a given biomarker. Note that a given biomarker can, in different drug development contexts, be placed into more than one category. Although a formal biomarker Context of Use Statement is only necessary when a biomarker is being considered for regulatory acceptance, these general considerations are very helpful for informal biomarker evaluation for internal decision-making [66].

Scope of Imaging Biomarkers in Drug Development

Imaging technologies have played an important role across the spectrum of drug discovery and development [15, 28]. Apart from basic biological research, the roles of imaging biomarkers in drug development generally fall into two categories: supporting internal decision-making and regulatory submissions. Within the category of imaging biomarkers used for internal decision, there are three basic applications that imaging biomarkers have in common with *in vitro* biomarker technologies: confirming that a drug candidate hits the intended biological target; testing whether hitting the target alters the disease process (proof-of-mechanism); and testing whether altering the disease process affects the clinical status of the patient (proof-of-therapeutic-concept) [103, 104]. Imaging biomarkers have provided early critical information at each of these development signposts through *in vivo* molecular imaging studies of competitive binding or receptor occupancy [104], imaging of target distribution and radio-labeled drug distribution [105, 106], lead optimization, and preclinical and early clinical efficacy testing [29, 36]. Other tasks in early clinical development that imaging biomarkers have addressed include appropriate dose and regimen selection and *in vivo* pharmacokinetic/pharmacodynamic studies [107, 108], patient stratification, and study population enrichment [103].

Importantly, imaging biomarkers for industry internal decision-making have demonstrated value for drug development without comprehensive technical validation or qualification [109]. Indeed, many such biomarkers may be proprietary, and specific to a particular development program, therapeutic target, or mechanism [104]. However, achieving a degree of technical validation appropriate to a given task and the associated risk is crucial for successful fit-for-purpose biomarker application. The use of novel imaging biomarkers is most effective if development and characterization activities, such as PET radiotracer development or MRI pulse sequence and method development, are begun as early as possible and ideally in tandem with the corresponding drug development program [70, 104].

The FDA DDT Qualification Process has played an important role in promoting regulatory acceptance of biomarkers as general tools independent of any particular drug application. As of 2019, eight clinical biomarkers have received regulatory qualification, of which one is an imaging biomarker: image-derived total kidney volume for use as a prognostic biomarker for patient selection in clinical trials of Autosomal Dominant Polycystic Kidney Disease [75]. Eight additional imaging biomarkers have been submitted to the FDA for consideration and are at various stages of qualification [75]:

- Standardized Uptake Value (SUV) of [^{18}F]-fluorodeoxyglucose by tumors as measured by quantitative FDG-PET/CT (pharmacodynamic/response biomarker of treatment response of lungcancer and lymphoma)
- Tumor volume measured by x-ray computed tomography (pharmacodynamic/response biomarker of treatment response of primary and metastatic solid tumors)
- Baseline hippocampal volume measured by MRI in Alzheimer's Disease (prognostic biomarker for patient selection in studies of Alzheimer's disease)
- Lung and airway structural and functional parameters as assessed by CT (pharmacodynamic/response biomarker of treatment response in COPD and asthma patients)
- Quantitative MRI of articular cartilage thickness (pharmacodynamic/response biomarker of treatment response for knee osteoarthritis)
- MRI assessment of iron-corrected liver inflammation and fibrosis (prognosticbiomarker for selection of patients at risk for nonalcoholicsteatohepatitis)
- MRI assessment of anatomic features of the terminal ileum and large bowel (pharmacodynamic/response biomarker of treatment response for Crohn's disease)

The significance of the FDA's DDT program lies in part with its attempt to create a general biomarker qualification process which, given the breadth of biomarker technologies and applications, is inherently difficult to generalize.

Quantitative Imaging Biomarkers Alliance (QIBA)

The QIBA (http://www.rsna.org/qiba) was organized to address the challenges of developing quantitative imaging guidelines and standards necessary for widespread adoption of imaging biomarkers. QIBA is primarily a volunteer organization consisting of academic researchers, healthcare professionals, and industry stakeholders who contribute expertise toward developing quantitative imaging biomarker solutions and promoting their adoption in drug development and clinical practice. QIBA has grown dramatically in recent years and has become a focal point of ongoing large-scale coordination, standardization, and dissemination of leading edge imaging biomarkers. QIBA is organized in three tiers [110]: the top-level Steering Committee is responsible for managing the strategic direction, processes, and infrastructure, as well as oversight of all QIBA activities and committees. The Steering Committee manages Modality Coordinating Committees for the four major imaging modalities (CT, MRI, nuclear medicine, and ultrasound) as well as a process committee charged with developing QIBA-specific policies and procedures. Finally, multiple Biomarker Committees are organized under each Modality Coordinating Committee with the task of developing, testing, and documenting specific imaging biomarkers. The top-level Steering Committee is appointed by the RSNA, and membership of all other committees is open to any party with relevant interests. The QIBA process encourages active participation of all stakeholders and is open to proposals for the formation of new Biomarker Committees to target promising imaging biomarkers.

QIBA Process

A major achievement of QIBA to date has been the creation and support of a uniform approach to developing imaging biomarker solutions. This approach focuses on proposed imaging biomarkers that have been sufficiently characterized by prior research and hold the highest potential for further broad scale development and standardization. The process includes the following steps [111]:

1. Define the clinical question and propose an existing quantitative imaging method for further development as an imaging biomarker.
   a. The imaging biomarker is formally proposed through the appropriate Modality Coordinating Committee and evaluated by the QIBA Steering Committee according to five selection criteria (Table 5.4).
   b. If approved by QIBA, a biomarker committee is established and publicized, and volunteers are recruited. The new Biomarker Committee will be supported by RSNA staff and resources, and is responsible for the following steps of the process.
2. Produce a QIBA Profile (Section "QIBA Profiles").
   a. A literature search is performed, and sources of measurement error and variation in the selected imaging biomarker are identified; potential solutions and standards are specified.
   b. Any necessary groundwork is planned and performed to fill gaps in knowledge, e.g. measurement uncertainty, bias, linearity, etc.
   c. Imaging protocols and analytical methods are harmonized between different hardware and software platforms.
   d. A QIBA Profile for the imaging biomarker is drafted (Stage 0), opened for public review/comment (Stage 1), then revised according to comments received, and published (Stage 2) (Table 5.5).

Table 5.4 QIBA criteria for evaluating proposed imaging biomarker opportunities.

Is it transformational?
Does it address a significant medical biomarker need with a considerable impact on public health? Does it address a critical gap in the biomarker qualification/validation process?
Is it translational?
Will the result improve the objectivity of metrics used in multicenter studies or adoption into clinical care?
Is it feasible?
Can the result be achieved in 3–5 years with a likelihood of achieving the expected outcome?
Is it practical?
Does it leverage existing resources and/or warrant access to Radiological Society of North America resources and support?
Is it collaborative?
Would it benefit from QIBA's multi-stakeholder approach and is it feasible under QIBA's policies?

Source: Data from Radiological Society of North America [111].

3. Confirm, promulgate, and maintain the QIBA Profile.
   a. The published Profile undergoes technical confirmation by feasibility testing and implementation at multiple imaging centers using multiple imaging systems (Stage 3).
   b. The technically confirmed Profile undergoes further field testing to verify the Profile's claimed performance (Stage 4).
   c. The claim-confirmed Profile is disseminated for vendor implementation, research integration, and clinical user and stakeholder education.
   d. The Biomarker Committee maintains and revises the Profile as needed in response to changes in imaging technology and available data.

Guided by the experience gained by the various Biomarker Committees, the QIBA Process Committee is responsible for ongoing refinement of this general process and associated procedures. Given the interdisciplinary nature of imaging biomarkers, QIBA has also identified a need for consistent and correct usage of general terminology related to statistical methodology and measurement science [1, 2]. Instead of inventing new terms, QIBA has proposed the use of existing internationally standardized terminology for the QIBA Profile development process (a few key terms are shown in Table 5.3; additional terms and concepts may be found in reference [1]).

This process is focused on mobilizing efforts toward *technical validation* of imaging biomarkers (Section "Technical Validation"), which previously had been a notable weakness of imaging biomarkers. By directly addressing this necessary biomarker characteristic, QIBA is providing evidence to better inform further qualification and facilitate improved utilization of imaging biomarkers (Section "Development Stages").

QIBA Profiles

The central role of a Biomarker Committee is to produce a QIBA Profile document targeting a specific imaging biomarker. The QIBA Profile provides a standardized statement of the technical characteristics and context for use of the given biomarker. Following a fit-for-purpose development strategy [64], different stages of Profile development are defined (Table 5.5) to provide an indication of the current stage of technical validation. The Profile also functions as a guide for all individuals involved in the implementation, quality assurance, and eventual use of the biomarker. A core feature is the Profile Claim which follows a standard template stating the measurand, the clinical context, and some metrics of the bias and precision of the biomarker measurement [112]. This indicates the expected uncertainty of the measurement within the specified clinical context, which is essential for characterizing the biomarker's technical performance [67] and for comparing different biomarkers that measure the same measurand [68, 69]. Each QIBA Profile provides a statement of a specific Profile Claim and all specifications necessary for achieving that Claim. In addition, the Profile specifies conformance requirements and assessment procedures to facilitate the publication of formal QIBA Conformance Statements by imaging sites and product vendors [112, 113]. Conformance Statements allow a biomarker user to evaluate whether imaging staff or products can be expected to deliver the claimed performance.

Table 5.6 provides a snapshot of 18 current Biomarker Committees and corresponding QIBA Profiles that are in various stages of development. In addition, new Biomarker Committees are forming for ^{99m}Tc- and ^{123}I-labeled SPECT biomarkers. A majority of the current Profiles were initiated recently and are in the initial stages of development, reflecting the rapidly growing momentum of QIBA activities. The centralization of biomarker development within Biomarker Committees as well as the potential for cross-modality and cross-biomarker learning (e.g. test plans and tools, imaging phantoms, etc.) within the QIBA organization are expected to further accelerate ongoing biomarker Profile development. Note that some biomarkers, such as tumor vascularity and perfusion, and liver viscoelasticity, measure similar disease processes and are being developed for more than one imaging modality. This reflects the complementary strengths and limitations of the modalities (Table 5.1) and the fact that different modalities may be useful in different biomarker contexts.

Table 5.5 Definition of the stages of QIBA profile development.

Stage name	Stage meaning
Stage 0: initial version	Profile is in development for selected imaging biomarker
Stage 1: version for public comment	Profile is ready for public comment and is believed to be practical Profile describes key factors that affect the claim and proposes procedures to address them Profile claim and requirements will likely change based on comments received
Stage 2: publicly reviewed (consensus profile)	The Biomarker Committee (BC) has formally addressed the issues raised during public comment. All profile changes based on received public comments are documented Implementation of profile is encouraged The BC may later modify the claim or requirements if necessary to complete technical confirmation (Stage 3)
Stage 3: technically confirmed	The BC has confirmed that the profile requirements are technically practical The BC asserts that deployments will find it practical to conform to the profile The BC may later modify the claim or requirements if necessary to complete Claim Confirmation (Stage 4)
Stage 4: claim confirmed	The Profile details have been implemented in more than one facility and each participating system and person successfully met the specifications; the overall performance was determined and the claim was achieved. The BC asserts that conformant deployments can achieve the Profile Claim
Stage 5: clinically confirmed	The BC has collected data broadly in the field on patients that confirm the claim

QIBA, quantitative imaging biomarker alliance.
For more details, see the Processes section of the QIBA Wiki (www.qibawiki.rsna.org).

Of the biomarkers in Table 5.6, two have achieved QIBA Stage 3 status (Technically Confirmed): tumor volume change over time by CT volumetry and tumor glycolytic activity by FDG-PET. In addition, five biomarkers have achieved QIBA Stage 2 (Publicly Reviewed Consensus Profile): brain amyloid burden by PET, liver viscoelasticity by MR elastography, small lung nodule volume assessment by low-dose CT, brain dopaminergic degeneration by SPECT, and tumor vascularity by DCE MRI. Of these, tumor volumetry and glycolytic activity have also been submitted to the FDA DDT program for regulatory qualification (Section "Scope of Imaging Biomarkers in Drug Development"). As discussed in Section "Importance of Context: Disease,

Table 5.6 Current QIBA biomarker committees that are developing formal QIBA profiles for 18 quantitative imaging biomarkers (as of 2018).

Biomarker committee	Quantitative imaging biomarker	Primary clinical application
CT volumetry	Tumor volume change over time	Cytotoxic therapy response (3)
CT small lung nodule	Lung nodule volume change over time	Low-dose lung cancer screening and monitoring (2)
CT lung density	Lung tissue density, volume	Assess asthma, emphysema, COPD (0)
CT angiography	Plaque characterization	Assess atherosclerosis (1)
MR dynamic contrast enhanced (DCE)	Tumor vascularity	Anti-angiogenic therapy response (2)
MR diffusion weighted imaging (DWI)	Tumor microstructure, cellularity	Cytotoxic therapy response (1)
MR dynamic susceptibility contrast (DSC)	Cerebral blood volume, flow	Assess stroke (0)
MR Musculoskeletal Imaging	Articular cartilage composition	Assess knee osteoarthritis (0)
MR arterial spin labeling (ASL)	Cerebral perfusion	Assess dementia (0)
MR elastography	Liver viscoelasticity	Assess liver fibrosis (2)
Functional MRI	Functional brain mapping	Presurgical assessment, treatment planning (1)
MR proton density fat fraction	Tissue fat concentration	Assess liver steatosis (0)
PET amyloid	Brain amyloid burden	Alzheimer's disease progression (2)
FDG-PET	Tumor glycolytic activity	Cytotoxic therapy response (3)
SPECT	Brain dopaminergic degeneration	Assess Parkinson's disease (2)
US shear wave speed	Liver viscoelasticity	Assess liver fibrosis (0)
US 3D volume blood flow	Large vessel blood flow rate	Assess renal artery blood flow (0)
Contrast enhanced US	Tumor perfusion	Anti-angiogenic therapy response (0)

3D, three-dimensional; COPD, chronic obstructive pulmonary disease; CT, computed tomography; FDG, fluorodeoxyglucose; MRI, magnetic resonance imaging; PET, positron emission tomography; QIBA, Quantitative Imaging Biomarkers Alliance; SPECT, single photon emission computed tomography; and US, ultrasound.
The QIBA Profile stage of each biomarker is indicated in parentheses under primary clinical application (see Table 5.5 for stage definitions).

Therapeutic Strategy, Development Stage, and Alternative Approaches", in some drug development contexts, tumor metabolism, vascularity, or cellularity may be more appropriate biomarkers to assess treatment response than tumor size.

QIBA publishes an online e-newsletter [114] and holds periodic conferences to encourage face-to-face meetings with stakeholders involved in biomarker Profile development.

Conclusions – Challenges and Future Opportunities

One of the most salient features of imaging science, which is relevant to both diagnostic imaging and imaging biomarkers, is that the state of imaging technology does not stand still. Continual research and development has produced a steady stream of new imaging technologies as well as incremental improvements that address the limitations of existing technologies. Development and implementation strategies for imaging biomarkers must account for this continuous change and attempt to balance the needs of technical validation with the adoption of new innovative technologies. Like software, imaging biomarkers are being "versioned" to account for changes to the underlying technologies. However, this will also require a continuous process of development and maintenance [66], for which open collaborations such as QIBA will be most appropriate. Moreover, with the continued introduction of new imaging innovations, in the future, we might expect the discovery and clinical translation of additional image-based biological measures to promising new imaging biomarkers. This will become increasingly feasible as innovative systems, tools, and collaboration make imaging biomarker development more routine.

References

1 Kessler, L.G., Barnhart, H.X., Buckler, A.J. et al. (2015). The emerging science of quantitative imaging biomarkers: terminology and definitions for scientific studies and regulatory submissions. *Stat. Methods Med. Res.* 24 (1): 9–26.

2 Sullivan, D.C., Obuchowski, N.A., Kessler, L.G. et al. (2015). Metrology standards for quantitative imaging biomarkers. *Radiology* 277 (3): 813–825.

3 Frey-Wyssling, A. (1953). *Submicroscopic morphology of protoplasm*, vol. 2. Amsterdam: Elsevier.

4 Wild, J.J. and Reid, J.M. (1952). Application of echo-ranging techniques to the determination of structure of biological tissues. *Science* 115 (2983): 226–230.

5 Holmes, J.H., Howry, D.H., Posakony, G.J., and Cushman, C.R. (1955). The ultrasonic visualization of soft tissue structures in the human body. *Trans. Am. Clin. Climatol. Assoc.* 66: 208–225.

6 Kuhl, D.E. and Edwards, R.Q. (1963). Image separation radioisotope scanning. *Radiology* 80: 653–662.

7 Wrenn, F.R., Good, M.L., and Handler, P. (1951). The use of positron-emitting radioisotopes for the localization of brain tumors. *Science* 113 (2940): 525–527.

8 Brownell, G.L. and Sweet, W.H. (1953). Localization of brain tumors with positron emitters. *Nucleonics* 11 (11): 40–45.

9 Hounsfield, G.N. (1973). Computerized transverse axial scanning (tomography). 1. Description of system. *Br. J. Radiol.* 46 (552): 1016–1022.

10 Lauterbur, P.C. (1973). Image formation by induced local interactions: examples employing nuclear magnetic resonance. *Nature* 242 (5394): 190–191.

11 Mansfield, P. and Grannell, P.K. (1973). NMR diffraction in solids. *J. Phys. C.* 6 (22): L422–L426.

12 Schaeffter, T. (2005). Imaging modalities: principles and information content. In: *Imaging in Drug Discovery and Early Clinical Trials* (Progress in Drug Research, Vol. 62) (ed. M. Rudin), 15–81. Basel: Birkhauser Verlag.

13 James, M.L. and Gambhir, S.S. (2012). A molecular imaging primer: modalities, imaging agents, and applications. *Physiol. Rev.* 92 (2): 897–965.

14 Barrett, H.H. and Myers, K.J. (2004). *Foundations of Image Science*, 1100. Hoboken, NJ: Wiley-Interscience.

15 Rudin, M. (ed.) (2005). *Imaging in Drug Discovery and Early Clinical Trials* (Progress in Drug Research, Vol. 62), 392. Basel: Birkhauser Verlag.

16 Bernstein, M.A., King, K.F., and Zhou, X.J. (2004). *Handbook of MRI Pulse Sequences*, 1ste, 1040. Academic Press.

17 Haacke, E.M., Brown, R.W., Thompson, M.R., and Venkatesan, R. (1999). *Magnetic Resonance Imaging: Physical Principles and Sequence Design*, 1ste, 914. Wiley-Liss.

18 Rahmim, A. and Zaidi, H. (2008). PET versus SPECT: strengths, limitations and challenges. *Nucl. Med. Commun.* 29 (3): 193–207.

19 Zaidi, H. (ed.) (2005). *Quantitative Analysis in Nuclear Medicine Imaging*, 583. New York, NY: Springer.

20 Judenhofer, M.S., Wehrl, H.F., Newport, D.F. et al. (2008). Simultaneous PET-MRI: a new approach for functional and morphological imaging. *Nat. Med.* 14 (4): 459–465.

21 Jensen, J.A. (2007). Medical ultrasound imaging. *Prog. Biophys. Mol. Biol.* 93 (1–3): 153–165.

22 Defrise, M. (2001). A short reader's guide to 3D tomographic reconstruction. *Comput. Med. Imaging Graph Off. J. Comput. Med. Imaging Soc.* 25 (2): 113–116.

23 Iima, M. and Bihan, D.L. (2015). Clinical intravoxel incoherent motion and diffusion MR imaging: past, present, and future. *Radiology* https://doi.org/10.1148/radiol.2015150244.

24 Huisman, M.C., van Golen, L.W., Hoetjes, N.J. et al. (2012). Cerebral blood flow and glucose metabolism in healthy volunteers measured using a high-resolution PET scanner. *EJNMMI Res.* 2: 63.

25 Frost, J.J. and Wagner, H.N. (1990). *Quantitative Imaging: Neuroreceptors, Neurotransmitters, and Enzymes*, 232. New York, NY: Raven Press.

26 Tofts, P. (ed.) (2003). *Quantitative MRI of the Brain: Measuring Changes Caused by Disease*, 1e, 650. West Sussex: Wiley.

27 Phelps, M.E. (2004). *PET: Molecular Imaging and Its Biological Applications*, 1e. New York, NY: Springer-Verlag.

28 Beckmann, N. (ed.) (2006). *In Vivo MR Techniques in Drug Discovery and Development*, 1e, 592. New York, NY: Taylor & Francis.

29 Rudin, M., Beckmann, N., and Rausch, M. (2005). *Evaluation of drug candidates: Efficacy readouts during lead optimization. In: Imaging in Drug Discovery and Early Clinical Trials* (Progress in Drug Research, Vol. 62), 185–255. Basel: Birkhauser Verlag.

30 Caravan, P., Ellison, J.J., McMurry, T.J., and Lauffer, R.B. (1999). Gadolinium(III) chelates as MRI contrast agents: structure, dynamics, and applications. *Chem. Rev.* 99 (9): 2293–2352.

31 Lee, N., Yoo, D., Ling, D. et al. (2015). Iron oxide based nanoparticles for multimodal imaging and magnetoresponsive therapy. *Chem. Rev.* 115 (19): 10637–10689.

32 Frenzel, T., Lawaczeck, R., Taupitz, M. et al. (2015). Contrast media for X-ray and magnetic resonance imaging: development, current status and future perspectives. *Invest. Radiol.* 50 (9): 671–678.

33 Thomas, D.L., Lythgoe, M.F., Pell, G.S. et al. (2000). The measurement of diffusion and perfusion in biological systems using magnetic resonance imaging. *Phys. Med. Biol.* 45 (8): R97–R138.

34 Jezzard, P. and Ramsey, N.F. (2003). Functional MRI. In: *Quantitative MRI of the Brain: Measuring Changes Caused by Disease*, 1ste (ed. P. Tofts), 415–453. Wiley.

35 Reeder, S.B., Cruite, I., Hamilton, G., and Sirlin, C.B. (2011). Quantitative assessment of liver fat with magnetic resonance imaging and spectroscopy. *J. Magn. Reson. Imaging* 34 (4): 729–749.

36 Major, T.C., Dhamija, S., Black, N. et al. (2008). The t- and l-type calcium channel blocker (ccb) mibefradil attenuates leg edema induced by the l-type ccb nifedipine in the spontaneously hypertensive rat: a novel differentiating assay. *J. Pharmacol. Exp. Ther.* 325 (3): 723–731.

37 St. Pierre, T.G., Clark, P.R., Chua-anusorn, W. et al. (2005). Noninvasive measurement and imaging of liver iron concentrations using proton magnetic resonance. *Blood* 105 (2): 855–861.

38 Stafford, R.J. and Hazle, J.D. (2006). Magnetic resonance temperature imaging for focused ultrasound surgery: a review. *Top. Magn. Reson. Imaging* 17 (3): 153–163.

39 Zhou, J., Payen, J.-F., Wilson, D.A. et al. (2003). Using the amide proton signals of intracellular proteins and peptides to detect pH effects in MRI. *Nat. Med.* 9 (8): 1085–1090.

40 Woods, M., Woessner, D.E., and Sherry, A.D. (2006). Paramagnetic lanthanide complexes as PARACEST agents for medical imaging. *Chem. Soc. Rev.* 35 (6): 500–511.

41 van Zijl, P.C.M., Jones, C.K., Ren, J. et al. (2007). MRI detection of glycogen *in vivo* by using chemical exchange saturation transfer imaging (glycoCEST). *Proc. Natl. Acad. Sci. USA.* 104 (11): 4359–4364.

42 Schroeder, M.A., Clarke, K., Neubauer, S., and Tyler, D.J. (2011). Hyperpolarized magnetic resonance a novel technique for the in vivo assessment of cardiovascular disease. *Circulation* 124 (14): 1580–1594.

43 Kurhanewicz, J., Vigneron, D.B., Brindle, K. et al. (2011). Analysis of cancer metabolism by imaging hyperpolarized nuclei: prospects for translation to clinical research. *Neoplasia* 13 (2): 81–97.

44 Jaffer, F.A. and Weissleder, R. (2005). Molecular imaging in the clinical arena. *JAMA* 293 (7): 855–862.

45 Massoud, T.F. and Gambhir, S.S. (2003). Molecular imaging in living subjects: seeing fundamental biological processes in a new light. *Genes Dev.* 17 (5): 545–580.

46 Cobelli, C., Foster, D., and Toffolo, G. (2002). *Tracer Kinetics in Biomedical Research: From Data to Model*, 700. Springer.

47 Aikawa, E., Nahrendorf, M., Figueiredo, J.-L. et al. (2007). Osteogenesis associates with inflammation in early-stage atherosclerosis evaluated by molecular imaging in vivo. *Circulation* 116 (24): 2841–2850.

48 Willmann, J.K., Chen, K., Wang, H. et al. (2008). Monitoring of the biological response to murine hindlimb ischemia with ^{64}Cu-labeled vascular endothelial growth factor-121 positron emission tomography. *Circulation* 117 (7): 915–922.

49 Hu, S., Balakrishnan, A., Bok, R.A. et al. (2011). ^{13}C-pyruvate imaging reveals alterations in glycolysis that precede c-myc-induced tumor formation and regression. *Cell Metab.* 14 (1): 131–142.

50 Grimm, J., Kirsch, D.G., Windsor, S.D. et al. (2005). Use of gene expression profiling to direct in vivo molecular imaging of lung cancer. *Proc. Natl. Acad. Sci. USA* 102 (40): 14404–14409.

51 Koo, V., Hamilton, P.W., and Williamson, K. (2006). Non-invasive in vivo imaging in small animal research. *Cell. Oncol. Off. J. Int. Soc. Cell. Oncol.* 28 (4): 127–139.

52 Pomper, M.G. and Lee, J.S. (2005). Small animal imaging in drug development. *Curr. Pharm. Des.* 11 (25): 3247–3272.

53 Singh, M. and Johnson, L. (2006). Using genetically engineered mouse models of cancer to aid drug development: an industry perspective. *Clin. Cancer Res.* 12 (18): 5312–5328.

54 Weissleder, R. and Pittet, M.J. (2008). Imaging in the era of molecular oncology. *Nature* 452 (7187): 580–589.

55 Lanza, G.M. and Wickline, S.A. (2003). Targeted ultrasonic contrast agents for molecular imaging and therapy. *Curr. Probl. Cardiol.* 28 (12): 625–653.

56 Kaufmann, B.A. and Lindner, J.R. (2007). Molecular imaging with targeted contrast ultrasound. *Curr. Opin. Biotechnol.* 18 (1): 11–16.

57 Henkelman, R.M., Stanisz, G.J., and Graham, S.J. (2001). Magnetization transfer in MRI: a review. *NMR Biomed.* 14 (2): 57–64.

58 Khaleeli, Z., Sastre-Garriga, J., Ciccarelli, O. et al. (2007). Magnetisation transfer ratio in the normal appearing white matter predicts progression of disability over 1 year in early primary progressive multiple sclerosis. *J. Neurol. Neurosurg. Psychiatry* 78 (10): 1076–1082.

59 Shankar, L.K., Hoffman, J.M., Bacharach, S. et al. (2006). Consensus recommendations for the use of 18F-FDG PET as an indicator of therapeutic response in patients in National Cancer Institute trials. *J. Nucl. Med.* 47 (6): 1059–1066.

60 Damadian, R. (1971). Tumor detection by nuclear magnetic resonance. *Science* 171 (3976): 1151–1153.

61 Damadian, R., Goldsmith, M., and Minkoff, L. (1977). NMR in cancer: XVI. FONAR image of the live human body. *Physiol. Chem. Phys.* 9 (1): 97–100, 108.

62 Sorensen, A.G. (2006). Magnetic resonance as a cancer imaging biomarker. *J. Clin. Oncol.* 24 (20): 3274–3281.

63 Katz, R. (2004). Biomarkers and surrogate markers: an FDA perspective. *NeuroRX* 1 (2): 189–195.

64 Lee, J., Devanarayan, V., Barrett, Y. et al. (2006). Fit-for-purpose method development and validation for successful biomarker measurement. *Pharm. Res.* 23 (2): 312–328.

65 FDA-NIH Biomarker Working Group (2016). *BEST (Biomarkers, EndpointS, and other Tools) Resource [Internet]*. Food and Drug Administration (US) http://www.ncbi.nlm.nih.gov/books/NBK326791/ (accessed 3 May 2019).

66 IOM (Institute of Medicine) (2010). *Evaluation of Biomarkers and Surrogate Endpoints in Chronic Disease*. Washington, DC: The National Academies Press.

67 Raunig, D.L., McShane, L.M., Pennello, G. et al. (2015). Quantitative imaging biomarkers: a review of statistical methods for technical performance assessment. *Stat. Methods Med. Res.* 24 (1): 27–67.

68 Obuchowski, N.A., Reeves, A.P., Huang, E.P. et al. (2015). Quantitative imaging biomarkers: a review of statistical methods for computer algorithm comparisons. *Stat. Methods Med. Res.* 24 (1): 68–106.

69 Obuchowski, N.A., Barnhart, H.X., Buckler, A.J. et al. (2015). Statistical issues in the comparison of quantitative imaging biomarker algorithms using pulmonary nodule volume as an example. *Stat. Methods Med. Res.* 24 (1): 107–140.

70 Murphy, P.S., Bergstrom, M., Moody, J.B., and Ficaro, E.P. (2017). Imaging for early clinical drug development: Integrating imaging science with drug research. In: *Biomarkers in Drug Development: A Handbook of Discovery, Practice, Application, and Strategy*, 2nde (eds. R. Rahbari and M.R. Bleavins). Hoboken, NJ: Wiley.

71 Prentice, R.L. (1989). Surrogate endpoints in clinical trials: definition and operational criteria. *Stat. Med.* 8 (4): 431–440.

72 Fleming, T.R. and DeMets, D.L. (1996). Surrogate end points in clinical trials: are we being misled? *Ann. Internal Med.* 125 (7): 605–613.

73 Fleming, T.R. and Powers, J.H. (2012). Biomarkers and surrogate endpoints in clinical trials. *Stat. Med.* 31 (25): 2973–2984.

74 Hill, A.B. (1965). The environment and disease: association or causation? *Proc. R. Soc. Med.* 58: 295–300.

75 Food and Drug Administration (US) (2016). Drug Development Tools Qualification Programs [Internet]. https://www.fda.gov/drugs/development-approval-process-drugs/drug-development-tool-qualification-programs (accessed 3 May 2019).

76 European Medicines Agency (2017). Qualification of novel methodologies for medicine development [Internet]. http://www.ema.europa.eu/ema/index.jsp?curl=pages/regulation/document_listing/document_listing_000319.jsp&mid=WC0b01ac0580022bb0 (accessed 3 May 2019).

77 Woodcock, J., Brumfield, M., Gill, D., and Zerhouni, E. (2014). The driving role of consortia on the critical path to innovative therapies. *Nat. Rev. Drug Discovery* 13 (11): 781–781.

78 Wholley, D. (2014). The biomarkers consortium. *Nat. Rev. Drug Discovery* 13 (11): 791–792.

79 C-PATH. Critical Path Institute [Internet]. https://c-path.org/ (3 May 2019).

80 Stephenson, D. and Sauer, J.-M. (2014). The predictive safety testing consortium and the coalition against major diseases. *Nat. Rev. Drug Discovery* 13 (11): 793–794.

81 SAFE-T SAFE-T – IMI [Internet]. http://www.imi-safe-t.eu/htdocs/biomarker/drug-induced-injury/imi.html (3 May 2019).

82 Kuhl, D.E. and Edwards, R.Q. (1970). The Mark III scanner: a compact device for multiple-view and section scanning of the brain. *Radiology* 96: 563–570.

83 Phelps, M.E., Hoffman, E.J., Mullani, N.A. et al. (1976). Design considerations for a positron emission transaxial tomograph (PETT III). *Nucl. Sci. IEEE Trans. On.* 23 (1): 516–522.

84 Mansfield, P., Pykett, I.L., and Morris, P.G. (1978). Human whole body line-scan imaging by NMR. *Br. J. Radiol.* 51 (611): 921–922.

85 Stevens, S. (1946). On the theory of scales of measurement. *Science* 103 (2684): 677–680.

86 Wagner, R.F. and Brown, D.G. (1985). Unified SNR analysis of medical imaging systems. *Phys. Med. Biol.* 30 (6): 489–518.

87 Barrett, H.H. (1990). Objective assessment of image quality: effects of quantum noise and object variability. *J. Opt. Soc. Am. A* 7 (7): 1266–1278.

88 Shields, A.F. and Price, P. (2007). *In Vivo Imaging of Cancer Therapy* (Cancer Drug Discovery and Development), 1ste, 326. Totowa, NJ: Humana Press.

89 Lambin, P., Rios-Velazquez, E., Leijenaar, R. et al. (2012). Radiomics: Extracting more information from medical images using advanced feature analysis. *Eur. J. Cancer* 48 (4): 441–446.

90 Weissleder, R. (2006). Molecular imaging in cancer. *Science* 312 (5777): 1168–1171.

91 Pearce, M.S., Salotti, J.A., Little, M.P. et al. (2012). Radiation exposure from CT scans in childhood and subsequent risk of leukaemia and brain tumours: a retrospective cohort study. *Lancet* 380 (9840): 499–505.

92 Mathews, J.D., Forsythe, A.V., Brady, Z. et al. (2013). Cancer risk in 680 000 people exposed to computed tomography scans in childhood or adolescence: data linkage study of 11 million Australians. *BMJ* 346: f2360.

93 Einstein, A.J., Moser, K.W., Thompson, R.C. et al. (2007). Radiation dose to patients from cardiac diagnostic imaging. *Circulation* 116 (11): 1290–1305.

94 Amis, E.S., Butler, P.F., Applegate, K.E. et al. (2007). American College of Radiology white paper on radiation dose in medicine. *J. Am. Coll. Radiol.* 4 (5): 272–284.

95 Journy, N., Rehel, J.-L., Ducou Le Pointe, H. et al. (2015). Are the studies on cancer risk from CT scans biased by indication? Elements of answer from a large-scale cohort study in France. *Br. J. Cancer* 112 (1): 185–193.

96 Siegel, J.A., Pennington, C.W., and Sacks, B. (2017). Subjecting radiologic imaging to the linear no-threshold hypothesis: a non sequitur of non-trivial proportion. *J. Nucl. Med.* 58 (1): 1–6.

97 Kaster, T., Mylonas, I., Renaud, J.M. et al. (2012). Accuracy of low-dose rubidium-82 myocardial perfusion imaging for detection of coronary artery disease using 3D PET and normal database interpretation. *J. Nucl. Cardiol.* 19 (6): 1135–1145.

98 Einstein, A.J., Johnson, L.L., DeLuca, A.J. et al. (2015). Radiation dose and prognosis of ultra-low-dose stress-first myocardial perfusion SPECT in patients with chest pain using a high-efficiency camera. *J. Nucl. Med.* 56 (4): 545–551.

99 Goldman, A.R. and Maldjian, P.D. (2013). Reducing radiation dose in body CT: a practical approach to optimizing CT protocols. *Am. J. Roentgenol.* 200 (4): 748–754.

100 Hill, K.D. and Einstein, A.J. (2016). New approaches to reduce radiation exposure. *Trends Cardiovasc. Med.* 26 (1): 55–65.

101 Eisenhauer, E.A., Therasse, P., Bogaerts, J. et al. (2009). New response evaluation criteria in solid tumours: revised RECIST guideline (version 1.1). *Eur. J. Cancer* 45 (2): 228–247.

102 Fleming, T.R., Rothmann, M.D., and Lu, H.L. (2009). Issues in using progression-free survival when evaluating oncology products. *J. Clin. Oncol.* 27 (17): 2874–2880.

103 Frank, R. and Hargreaves, R. (2003). Clinical biomarkers in drug discovery and development. *Nat. Rev. Drug Discovery* 2 (7): 566–580.

104 Hargreaves, R. (2008). The role of molecular imaging in drug discovery and development. *Clin. Pharmacol. Ther.* 83 (2): 349–353.

105 Bergström, M. and Långström, B. (2005). Pharmacokinetic studies with PET. In: *Imaging in Drug Discovery and Early Clinical Trials* (Progress in Drug Research) (ed. M. Rudin), 279–317. Birkhauser Verlag.

106 Willmann, J.K., van Bruggen, N., Dinkelborg, L.M., and Gambhir, S.S. (2008). Molecular imaging in drug development. *Nat. Rev. Drug Discovery* 7 (7): 591–607.

107 Seddon, B.M. and Workman, P. (2003). The role of functional and molecular imaging in cancer drug discovery and development. *Br. J. Radiol.* 76 (suppl_2): S128–S138.

108 Uppoor, R.S., Mummaneni, P., Cooper, E. et al. (2008). The use of imaging in the early development of neuropharmacological drugs: a survey of approved NDAs. *Clin. Pharmacol. Ther.* 84 (1): 69–74.

109 Hargreaves, R. and Wagner, J.A. (2006). Imaging as biomarker for decision-making in drug development. In: *In Vivo MR Techniques in Drug Discovery and Development*, 1ste (ed. N. Beckmann), 31–44. Informa HealthCare.

110 Quantitative Imaging Biomarkers Alliance Process Committee (2016). Committee Procedures – QIBA Wiki [Internet]. http://qibawiki.rsna.org/index.php/Committee_Procedures (accessed 3 May 2019).

111 QIBA Process Committee (2019). QIBA Process [Internet]. https://www.rsna.org/research/quantitative-imaging-biomarkers-alliance/process (accessed 3 May 2019).

112 Obuchowski, N.A., Buckler, A., Kinahan, P. et al. (2016). Statistical issues in testing conformance with the quantitative imaging biomarker alliance (QIBA) profile claims. *Acad Radiol.* 23 (4): 496–506.

113 Quantitative Imaging Biomarkers Alliance Process Committee (2016). Process – QIBA Wiki [Internet]. http://qibawiki.rsna.org/index.php/Process.

114 Quantitative Imaging Biomarkers Alliance (2016). QIBA Newsletters [Internet].

6

Imaging for Early Clinical Drug Development: Integrating Imaging Science with Drug Research

Philip S. Murphy[1], *Mats Bergstrom*[2], *Jonathan B. Moody*[3], *and Edward P. Ficaro*[3]

[1] *GlaxoSmithKline Research and Development, Stevenage, UK*
[2] *Independent Consultant, Uppsala, Sweden*
[3] *INVIA Medical Imaging Solutions, Ann Arbor, MI, USA*

Introduction

Clinical imaging technologies continue to develop at a rapid pace catalyzed by the desire to better understand and monitor human health and disease. Many new imaging technologies or refinements to existing methods will become broadly applied diagnostic tools. For drug development, these technology advancements offer both excitement and challenge. On the one hand, new techniques are able to probe drug action noninvasively with increasing specificity to the disease process. On the other hand, it has never been more challenging to navigate the array of methods, develop and characterize the chosen imaging based biomarker, and then implement the technology within a clinical trial setting.

Early clinical development offers a unique opportunity to apply information-rich, quantitative imaging-based biomarkers. There is no constraint to apply endpoints to be considered by regulators. Instead, endpoints seek to inform internal go/no-go decision points or to optimize the application of a drug under development. Data acquired through early phase clinical studies, however, will influence important decisions to commit resource to larger, longer, and more costly trials. Therefore, the robustness and understanding of an imaging biomarker needs careful scrutiny. Beyond decision-making, the early clinical phase also provides an important framework to enable exploratory methods to be developed and optimized for future studies and further disease understanding.

Early clinical development conducted with informative imaging-based technologies to enable robust decision-making requires the study to be carefully

Biomarkers in Drug Discovery and Development: A Handbook of Practice, Application, and Strategy, Second Edition. Edited by Ramin Rahbari, Jonathan Van Niewaal, and Michael R. Bleavins.

considered and matched with biological/pharmacological features of the imaging solutions. To ensure successful early clinical development, needs must be shared early with imaging experts to ensure methods are carefully selected, prepared, and implemented. This chapter reviews the challenges faced by early clinical development and considers how imaging can best contribute.

An Overview of Early Clinical Development

The aim of clinical development is to establish the safety and efficacy profile of a drug candidate through successful regulatory approval. Early development is broadly defined as the clinical trials that are conducted prior to a pivotal, Phase III study. There is significant variation in the activities typically undertaken in the early clinical trials although the aims have always been to demonstrate a safe, efficacious drug candidate in the intended patient population.

The productivity decline of the pharmaceutical industry has been broadly acknowledged, as exemplified by the analysis of Paul et al. [1], who stated that for the industry to remain viable a substantial increase in the number of cost-effective new medicines is required with cost control in the development stage. Without this reversal of R&D productivity, the current pharmaceutical industry model cannot sustain sufficient innovation. Reasons for the declining productivity are complex and continue to be interrogated. However, in clinical development, it is becoming clear that shifting attrition to early stages of clinical development would have a positive impact on resource allocation by reducing expenditures on late stage failures. An obvious focus in early clinical studies is to make informed and robust decisions, efficiently failing poor drug candidates early and investing to accelerate those that are promising.

Several companies have outlined their responses to the productivity decline with analyses of their pipeline successes and failures and proposed new frameworks to improve success. For example, Morgan et al. [2] described an analysis of Pfizer's early stage pipeline based on Phase II decisions of 44 programs. It was found that in 43% of cases the drug mechanism had not been sufficiently evaluated. Concluding the analysis, the authors defined the key parameters that, when adequately characterized, increased the chance of Phase II success. These parameters, the so-called *three Pillars of survival*, are based on pharmacokinetic (PK)/pharmacodynamic (PD) principles: drug exposure at the site of action, target binding, and pharmacological activity. An early clinical development program integrating the best tools to characterize the PK/PD early, including imaging methods, is likely to improve subsequent success rates.

Reporting additional analyses, Owens et al. [3] described Eli Lilly's *Chorus* model that focuses a clinical development organization on moving to proof of concept rapidly at low cost. Beyond the organizational efficiencies described, the central philosophy is to maximize assessment of the drug early in a human setting, thereby shifting attrition to the less expensive phase of development: the so-called "fail-fast" model. There is a specific focus to evaluate the pharmacological profile early in Phase I/Ib trials to replicate preclinical findings. The right biomarkers are required to confirm pharmacology. The authors argued that such trials are at present often not conducted because biomarkers are not sufficiently validated, especially within the context of small trials. However, if pharmacology cannot be sufficiently assessed, then demonstrating clinical efficacy in longer trials in humans will be unlikely.

In the report by Cook et al. [4], AstraZeneca reviewed their small molecule portfolio over a five-year period to distill the most important parameters that determine success. The factors relevant for the clinical phases of development included a strong link between target and disease, identifying the most responsive patient population, ensuring adequate tissue exposure, and establishing preclinical/clinical PK/PD underpinned by PD biomarkers.

In response to analyses like these, there is a shift in industry toward studying the most important attributes of a drug early in the clinical setting. This will only be achieved with robust, informative biomarkers selected from an array of different technology platforms. Imaging is one technology platform that has a significant potential to directly contribute to drug development productivity.

How Imaging Can Inform Early Development?

With a renewed focus on better early drug development, what critical questions can imaging answer?

A broad tool set is required to assess multiple aspects of drug action in different tissues and disease settings. Both highly drug-specific information, such as demonstration of target engagement, and downstream markers of pharmacological response are important. Imaging techniques commonly applied in later stages of development to quantify structural change over time (e.g. tumor size contributing to progression free survival endpoints) often lack both sensitivity to measure change in short clinical trials and specificity to the disease process and drug mechanism. Defining rapid, mechanistic insights into drug action requires techniques that have sufficient specificity to the drug and sensitivity to detect change often relying upon functional and molecular techniques. Guiding the integration of these complex technologies into early clinical development is challenging but, if successful, will directly support decision-making (Figure 6.1).

An increased application of advanced imaging methods will be pivotal to creating information-rich early development programs. Some of the specific features of imaging which may be of special advantage include the following:

- measurement in the tissue/organ of interest
- accurate and precise measurements in multiple small sub-regions

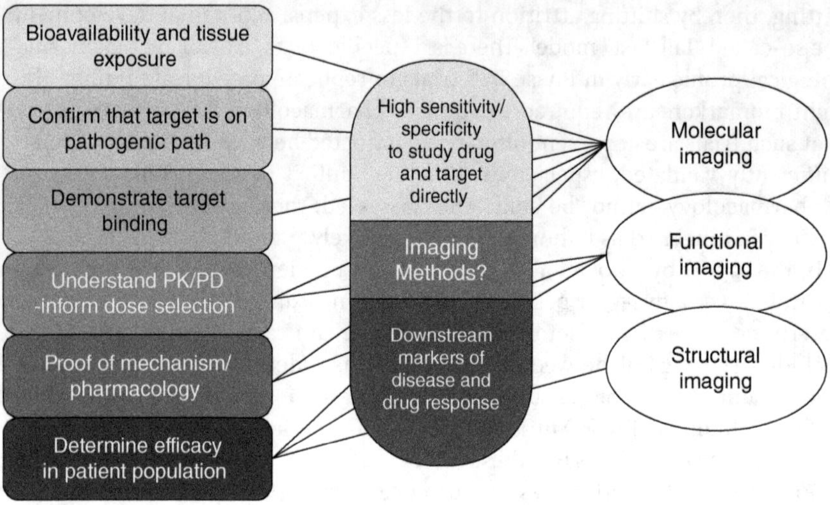

Figure 6.1 Imaging technologies need to be carefully selected in order to integrate the most relevant method for a given drug development question.

- multi-organ coverage or even whole body
- non-invasive
- can be repeated at selected time points
- some tools mechanistically relevant to drug pharmacology

Developing, characterizing, and applying the most informative and tractable methodologies is vital for successful imaging application in drug development. The different techniques that can be applied are considered against some fundamental attributes of the drug that early clinical development aims to investigate. To consider this, the three pillars as defined by Morgan et al. [2] are considered further.

Pillar 1 – Exposure at the Target Site of Action

Direct measurement of drug kinetics and concentration in target tissues is challenging. Blood plasma measurements of drug concentration can be considered a surrogate for tissue concentrations, but assumptions may not hold for all tissues and diseased tissue will inevitably make the relationships more complex.

Tissue Drug Penetration

Penetration of drug into tumor tissue is complex and a potential factor that limits solid tumor response [5, 6]. Although it may be relevant to assume that small organic molecules distribute quickly into the extravascular interstitial space and equilibrate with free drug concentration in plasma, the situation is far more complex for protein drugs such as antibody constructs. Furthermore, the targets for small molecules may reside within the cells and the access to this space, and the ensuing drug concentration at the target may include further confounding mechanisms such as restricted passage across the cell membrane and efflux transporter systems. Despite the importance of achieving adequate exposure at the target, even gross measurements of drug concentration are difficult to acquire clinically. It is accepted that plasma concentrations are not a reliable predictor of tumor concentration; however, there are few reports of plasma vs. tumor concentrations [7]. Limited opportunities for biopsy to evaluate multiple tumor sites and to make assessments at different time points after drug administration continue to limit the understanding of drug distribution.

Brain Drug Penetration

Understanding drug penetration into the brain is important in many settings to confirm adequate brain exposure for centrally acting drugs or to interpret potential safety signals of drugs not intended to act centrally. Due to restricted permeability of the brain capillary endothelium and the transporters at the blood–brain barrier (BBB), plasma concentration measurements do not provide a useful surrogate of drug concentration in the brain parenchyma [8]. Given that brain tissue cannot be readily accessed clinically, other measurements are required. While it is assumed that cerebral spinal fluid (CSF) sampling can provide a surrogate of drug concentration in the brain, it has been demonstrated that the relationship is complex [9, 10]. Furthermore, CSF sampling is an invasive procedure which may restrict its application in clinical trials. Given the limitations of CSF measurements, it is clear that a direct measurement of brain tissue drug concentration would be of significant benefit in health and disease. To this end, the imaging field has developed methods to perform quantitative and noninvasive analyses [11].

Positron Emission Tomography as a Method to Study Drug Exposure

As described by Bergstrom et al. [11], a low dose (termed a microdose) of a new chemical entity can be radiolabeled with a positron-emitting radionuclide in a manner that maintains equivalent chemical structure. Following intravenous (i.v.) injection, the distribution and kinetics can be measured using positron emission tomography (PET) imaging. Such direct, minimally invasive imaging methods can provide information not available by other methods, and have become a core component of many drug development programs.

Although important quantitative information on drug concentrations in different tissues can be acquired, it is important to realize that this is an estimate of tissue distribution and that cellular and subcellular distribution is not discerned. Incorporation into cells may be inferred due to the magnitude of the

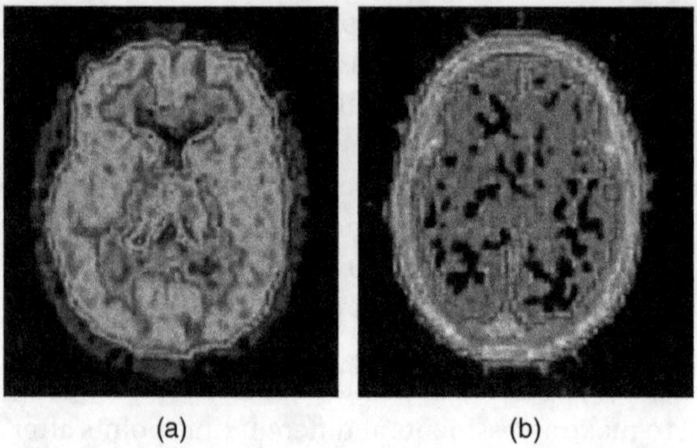

Figure 6.2 PET examples of a drug with high brain uptake (a) and one with limited brain access (b). *(See insert for color representation of this figure.)*

signal in relation to plasma concentration and relative sizes of the interstitial and cellular volumes (Figure 6.2).

It is important to consider the PK properties of a drug and its application like route of administration when considering the option of performing a biodistribution study.

A small organic compound may have a rapid plasma elimination and an expected fast tissue distribution, and the follow-up time allowed by ^{11}C (up to 1.5 hours due to the radioactivity half-life of 20 minutes) may be sufficient to describe the key parameters. ^{18}F with its two hours of radioactivity half-life may allow sequential studies up to eight hours. However, antibodies with clearance and distribution time lines of the order of days would not be meaningful to explore in such short time spans and will require more long-lived radionuclides like ^{111}In, ^{124}I, or ^{89}Zr with multiple day half-lives [12, 13].

For some purposes, like inhalation deposition, it may be less important to have the identical compound as the drug labeled, and drug or particles labeled with ^{99m}Tc may serve as substitutes [14, 15]. However, even deposition might be affected by exact scaffolds and formulations, and proper understanding of disposition may require observation of the actual molecule. Also, PET has a significantly better spatial resolution and quantitation accuracy which might be beneficial in inhalation studies [16, 17].

Radiolabeling for PET Studies

For each new study, consideration needs to be given to the experimental design to ensure the imaging will provide the information required. However, the primary variable from program to program is the molecule itself and the feasibility of radiolabeling. For small molecules to maintain their chemical equivalence, the radioactive, positron-emitting radionuclide ^{11}C is potentially incorporated as a substitute to an existing ^{12}C. A significant proportion of pharmaceutical products contain fluorine, enabling the potential to replace ^{19}F with the positron-emitting ^{18}F. Illardi et al. [18] stated that over 20% of the top 200 marketed drugs (in 2011) contain fluorine.

There is a broad literature on radiolabeling methods and extensive reviews outlining options and examples [19, 20]. For each new chemical entity requiring labeling, it comes down to the skill and creativity of the radiochemist to identify a synthetic strategy to incorporate the label at a suitable position. Radiolabeling must maintain the equivalence of the original molecule, and do so with few steps to ensure a sufficient amount of radioactivity is achieved. Finally, the labeled molecule must be formulated for injection into a patient. The complexity of development of the chemistry and the process required to implement within a clinical study in a good manufacturing practice (GMP) compliant environment should not be underestimated. For some molecules, there is no tractable radiolabeling scheme, and with others the complexity of radiolabelling will add cost and long timelines to prepare. Consideration of such approaches should be incorporated into early planning of clinical activities to ensure that a PET method is available to synchronize with subsequent clinical studies. For example, a PET sub-study can be integrated within a Phase I study. A PET radiochemistry solution delivered too late to impact early clinical studies will add cost to the program and contribute minimally to decision-making.

Creating strong links between a pharmaceutical company and expert PET centers is vital to ensure that programs can efficiently incorporate novel PET strategies. When a decision to pursue a PET radiolabeling program is agreed, the steps required must be carefully considered. This includes evaluation of the tractability of radiolabeling, method testing and optimization, preclinical evaluations, and implementation of GMP manufacturing for a clinical study. In addition to challenges around the radiochemistry synthesis, the regulatory processes and timelines need particular consideration. Saleem et al. [21] outlined the steps required to progress a radiolabeled new molecular entity to the clinic with timelines and identified bottlenecks.

The continually evolving expertise of the radiochemistry community has expanded the options to radiolabel small molecules with carbon-11 or fluorine-18 to produce a detectable, equivalent drug analogue *in vivo*. It seems likely that this field will continue to evolve to allow both new labeling strategies previously unavailable such as the [^{18}F] trifluoromethylation as described by Huiban et al. [22], and more efficient platforms such as microfluidics [23] to reduce time, cost, and complexity associated with these studies.

A suite of methods is available to measure new chemical entities and formulations within different tissue compartments. Direct incorporation of a label into a new chemical entity is a demanding and costly process [20–24]. New radiolabeling methodologies together with refinement of the regulatory framework are likely to lead to a broader use of these methods in drug development. At this point, the high-cost and long timelines limit use but even as it stands now, it is important to note that these costs are minimal relative to the cost of late stage drug failures.

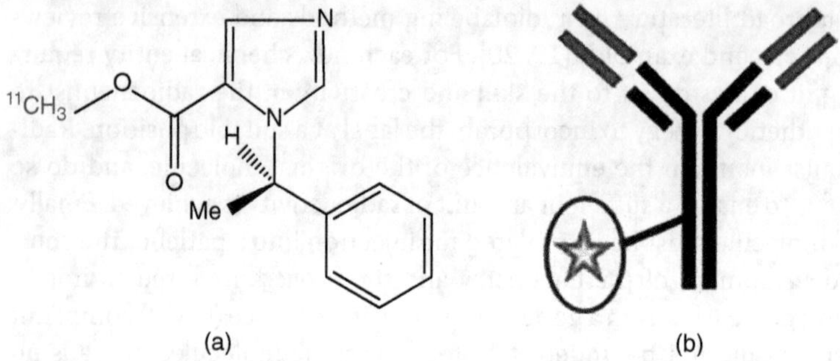

Figure 6.3 Illustration of a radiolabeled small organic molecule (a) and an antibody (b). *(See insert for color representation of this figure.)*

Generic Labeling of Protein Pharmaceuticals

Whereas the labeling method of small organic molecules is typically developed for each individual structure, the labeling of protein pharmaceuticals can often be performed in a standardized way, using ^{111}In, ^{124}I, or ^{89}Zr [24]. Using similar approaches for different protein drugs, a linker is attached to the protein. At a suitable time point prior to application in a study, the radionuclide is added and caught by the conjugation "cage" (Figure 6.3).

In most cases, there is approximately a 1 : 1 relation between label and therapeutic molecule. Larger ratios can be obtained, increasing the specific activity, but at the risk of influencing the properties.

Although labeling, including the first modification by attaching a conjugation scaffold to the protein and then associating the radiolabel, means a modification of the original compound, this modification is usually small enough to ensure intact pharmacological properties. However, for each such labeled entity, it is essential to certify that the binding and distribution properties have not been changed. It is often obvious that the binding to the target should be explored, but for antibodies it may also be important to certify unchanged properties with respect to the Fc receptors which may be important for distribution.

Another very important aspect with regard to labeling of antibodies is to know the features of binding and internalization. Many antibodies are quickly internalized after binding to the target, either directly into the target cell or indirectly to neighboring cells including macrophages through activation of the Fc receptors. With the so-called "residualizing" radionuclides ^{111}In and ^{89}Zr, the radiolabel may stay inside the cell for multiple days even if the protein it was attached to is degraded. ^{124}I on the other hand may be exported rapidly from the cell in the degradation of its bound protein. Hence, a biodistribution study may include not only the passive distribution into tissue but also the cellular processes of binding and internalization, and the results may be different depending on the choice of radionuclide.

Preclinical Qualifications of a Labeled Drug

Although it seems obvious to use imaging with a labeled drug as a means to study tissue distribution, when this question is pertinent, there are additional concerns of pharmacology that need consideration. It is always advisable to perform preclinical studies in order to address potential upcoming questions in a clinical imaging study of biodistribution.

A key aspect is to understand the potential contribution to the imaging signal coming from metabolites of the drug. Radionuclide-based techniques cannot in the *in vivo* setting discriminate between radioactivity coming from the radiolabeled intact compound and radioactive metabolites distributing in the tissue. In some cases, for example, when studying uptake of a radiolabeled small molecule in the brain, a significant portion of the radioactivity in plasma will within the time span of a study be constituted by radioactive metabolites of the drug. An analysis of the plasma composition of radioactivity is commonly performed and feasible in a study. However, depending on the properties of the metabolites, these may or may not distribute into the brain and confound the signal. A preclinical study may reveal whether this is likely to affect the results of the study.

Another aspect that needs consideration is the dose of drug to be applied. The microdosing concept allows a more rapid entry into a human study with limited preclinical toxicity evaluation required [11]. However, there may be dose-dependent factors which govern the tissue and body distribution. This includes binding to specific protein carriers in plasma, metabolic pathways degrading the substance, carrier system moving the drug across barriers, and efflux system removing the drug from cells. Additionally, the application of very low amounts of drug may lead to a preferential binding to its specific target, which then results in a high signal in target rich structures. If this occurs, the images and data will be biased by target expression and not only biodistribution relevant to a therapeutic situation. For these reasons, it is in most cases advisable to administer the radiolabeled drug together with "cold" compound at relevant doses. If a preclinical study has given indications that "microdose" levels can be translated to pharmacological levels, then such a subtherapeutic dose concept may be applied.

The route of application of a radiolabeled entity is commonly i.v. One reason is that the radioactivity is directly and immediately distributed in the blood compartment and hence across the whole body. This safeguards against a high local radioactivity and a risk of local high radiation dose. Other administration routes such as inhaled or oral administration have been used [16, 17, 25], but at the expense of a much reduced amount of administered radioactivity and hence an inaccurate measurement of tissue distribution. However, the local radioactivity concentration at deposition sites is often adequate for accurate measurements.

An oral or inhaled administration of the "cold" compound and an i.v. administration of the radiolabeled compound can sometimes be merged through

PK modeling where the exchange between plasma and tissue obtained in the imaging study is modeled and applied to the plasma PK profile of the "cold" compound [26].

In summary, with due caution, imaging technologies can characterize exposure at the site of action providing data that can't be acquired through plasma and tissue sampling. Multiple methods can be considered depending upon the drug candidate of interest and the relevant clinical questions.

Pillar 2 – Binding to the Pharmacological Target

Direct evidence that target binding is occurring can be obtained using PET combined with radioligands which have adequate properties of binding affinity and specificity (Figure 6.4).

Although the literature has shown an emphasis on the development of radioligands for neuroreceptors, PET is applicable to study multiple entities which bind ligands, such as enzymes and even soluble ligands. Soluble ligands include radiolabeled monoclonal antibodies binding to vascular endothelial growth factor (VEGF) [27] or placental growth factor (P1GF) [28] where imaging studies have been successfully performed in both animals and humans.

Target Expression

With the availability of specific radioligands which bind preferentially to a defined target, in principle it is possible to define target expression, at least in qualitative terms and sometimes quantitatively.

A study with a sub-pharmacological dose of the radioligand may demonstrate that the target is differentially expressed between tissues and between patients. Such studies can be of great value as a means to understand normal physiology and disease. Although imaging studies in populations of patients may aid in the understanding of target expression and target heterogeneity, such studies may be too expensive or logistically complicated to apply routinely for therapy selection. Nonetheless, results of imaging studies may ultimately guide the qualification of other biomarkers.

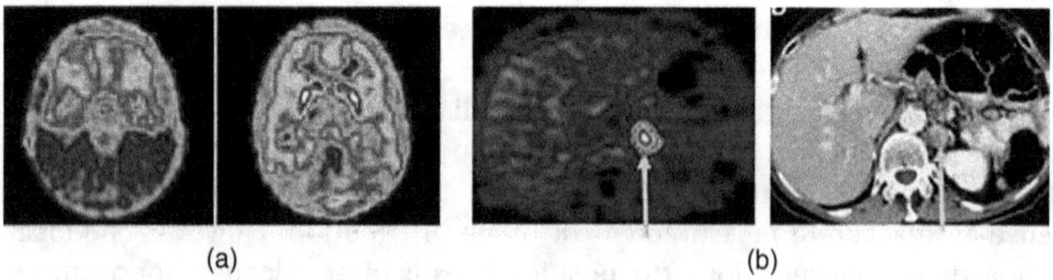

Figure 6.4 Illustration of a neuroreceptor ligand (a) with high binding to striatum and low in cerebellum, and a ligand binding to an enzyme expressed in adrenocortical tumor (b). *(See insert for color representation of this figure.)*

Studying a single radioligand alone, although being a good indicator that the target is expressed, may not give a quantitative value of target density. However, if through modeling the specific binding can be extracted, relative values can then be defined with reasonable accuracy and used in modeling of tissue pharmacology. Although more seldom reported, it is in principle possible to perform dose-escalation studies to establish target density and binding affinity *in vivo* in humans [29].

Target Interaction

With the availability of a tool to define tissue expression of a specific target, it is often possible to use this method to prove that the administration of a drug at a certain dose will act on the target. The typical setup is to perform a baseline study with the radioligand at low dose and measure the specific binding to the target. After decay of the radioactivity, the drug is given at the desired dose and route of administration, and a new imaging study is performed at a suitable time after drug administration. A reduction of the specific signal from the second imaging study is an indication that the therapeutic drug has acted on the target, and the degree of reduction is an indication of the fraction of target molecules which have been bound or inhibited by the drug.

This option may be of great importance as a basis to understand that the drug acts not only on the target but also as the basis to understand appropriate dosing and scheduling. A dose-escalation study can supply vital information for PK/PD modeling. A study defining the time course of target engagement can predict an optimal application of the drug. As a prelude to a human study, this concept can be applied in a preclinical setting and PK/PD modeling with allometric scaling that can aid in the planning of the human study or indicate that a human study with imaging is not needed.

Sometimes a human study for assessment of target engagement related to dose ("dose-occupancy study") can be planned ahead with respect to which doses to probe, but often it is more effective to plan the study with an "adaptive design" where one or a few doses are first tested and evaluated and the results guide the further doses to probe [30]. However, this "adaptive design" puts high demands on a quick turnover and analysis of data.

The proof of target engagement using this type of imaging requires the availability of an imaging probe which is specific for the same target the drug is interacting with. It may be an advantage for the radioligand to be a different molecule than the drug; the drug, for example, may not have optimal performance as an imaging tool or may have multiple molecular targets. It is, however, important that the radioligand binds not only to the desired target but also to the same binding site of the drug target. This binding should always be evaluated in a preclinical setting early in the planning of human or preclinical studies with drug challenge.

The applicability and relevance of a target engagement study with dose escalation are highly dependent upon an understanding of the mechanisms of drug interaction and ligand binding.

It is usually assumed that a high degree of "occupancy" is necessary with receptor antagonists, but it strongly strengthens the concept if through other means we can define which "occupancy" is targeted. With agonist drugs, the degree of "occupancy" may be very low at pharmacologically relevant doses, and the precision of the imaging may not be sufficient. If possible, it is desired that the degree of target engagement be estimated and placed in relation to the precision of the method to predict cohort sizes needed for a clinical study.

Methods have been incorporated into multiple neuroscience programs. Examples include the following:

- antipsychotic dopaminergic drugs acting on dopamine D1, D2, and D3 receptors [31]
- antipsychotic serotonergic drugs acting on 5HT1 and 5HT2 receptors [32]
- NK1 antagonists promoted as antidepressants acting on NK1 receptors [33]
- psychoactive drugs acting on Monoamine oxidases A and B [34]

Studies aiming to interrogate the target remain technically challenging to achieve clinically, particularly within the timelines relevant to early drug development. Early consideration is imperative with parallel imaging method development and qualification required to synchronize with the clinical program.

Pillar 3 – Expression of Pharmacology

Evidence of pharmacology downstream from target engagement can be demonstrated through several different imaging methodologies selected with careful consideration of the disease setting and anticipated drug activity. An example of a downstream measure of pharmacology is the application of dynamic contrast–enhanced magnetic resonance imaging (DCE-MRI) to the study of anti-angiogenic and anti-vascular therapies [35]. This represents a good example of how a relatively straightforward imaging technique can define changes related to therapeutic mechanism within days of treatment. Changes reported with DCE-MRI have been observed across different drugs and different doses (including some reports of PK/PD relationships) and have been monitored over time in multiple lesions. Although the method is not without challenges (e.g. multi-site standardization), the many applications demonstrate the broad utility of a downstream PD marker. More methods with these characteristics are required to ensure pharmacology can be demonstrated early across different diseases and drug mechanisms.

To identify, develop, and apply markers of pharmacology, the following factors should be considered:

- Is the method sufficiently characterized in the disease being investigated?

- Is the method likely to demonstrate a change measurable within a short trial duration, for example, 12 weeks?
- Is there some evidence that the PD endpoint will predict longer term clinical benefit in future trials?
- Will the response likely be sufficiently sensitive and with variability suitably controlled to enable statistically relevant observations in reasonable cohort sizes?
- Can it be assumed that a relationship to PK can be measured?
- Do the participating centers have adequate equipment and technical experience to ensure a high-quality implementation of the imaging?

In most cases, imaging PD endpoints are not sufficiently characterized for a given setting to enable straightforward implementation, leading to clear, objective decision-making. First, the method needs to possess sufficient technical validation and qualification to minimize methodological ambiguity [36]. Second, consideration will need to be given to how the results will be interpreted in the overall study – for example, is there sufficient confidence in the method that failure to measure a PD response with the imaging will enable a clear decision about the drug and/or dose? Much of this will depend upon how the imaging endpoint is considered – a primary decision-making endpoint, a secondary endpoint contributing to decision-making, or an exploratory endpoint that is largely disconnected from decision-making but may support conclusions from the primary/secondary endpoints. Biomarker application will depend on both biomarker validation and qualification and a subjective assessment of the drug development context [36]. Given the range of drug development needs, technology opportunities, and gaps in knowledge, careful preparation of these methods for application in early trials is paramount.

Further Examples of Pharmacodynamics Imaging

One of the most common PD imaging endpoints in oncology is the measure of tumor size, so well established that it will not be further discussed here [37]. Second to this is the application of [^{18}F]-fluorodeoxyglucose (FDG) PET as a response marker. In some specific cases, robust responses to treatment have been observed already within days, such as in gastrointestinal stromal tumors (GISTs) treated with Gleevec [38]. In diverse settings, FDG-PET has and will be included in oncology programs due to the fact that FDG is often included in the characterization of tumor burden and because of observations of treatment responses with different types of treatment.

Good practice guidelines exist with respect to the performance of such studies, and study results can be well described with respect to degree of change in relation to the precision of the method [39–41] (Figure 6.5).

FDG-PET is increasingly used in other indications such as inflammatory diseases, based on the observations of increased FDG uptake [42].

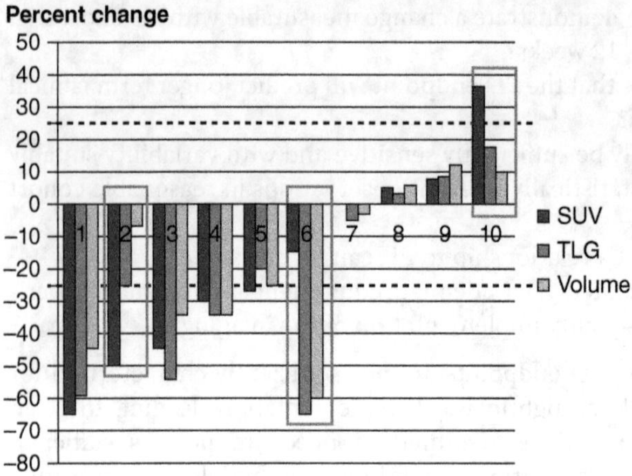

Figure 6.5 FDG-PET may supply results on three parameters: maximum FDG uptake as standardized uptake value (SUV), total lesion glycolysis as TLG, and tumor volume. A "waterfall plot" indicates the proportions of patients (1–10)/lesions with different degrees of response. *(See insert for color representation of this figure.)*

Other pharmacodynamic measures acquired through imaging include cerebral blood flow evaluated with magnetic resonance imaging (MRI) or H_2O-PET. Imaging probes for specific disease processes such as amyloid burden in the brain have been used for the assessment of treatments which aim at clearing the aberrant proteins or inhibiting their release [43]. An increasing range of functional and molecular markers that can be used to define pharmacological response is expected.

Preparing Imaging Methods for Application as Pharmacodynamic Endpoints

Relative to the breadth of clinical imaging science, too few potentially impactful methods have been applied and proven utility in drug studies. In the longer term, precompetitive consortia offer a framework to deliver better techniques [44]. In the shorter term, drug development organizations will often need to embark on method development for their specific program needs. Some methods may be ready for implementation without further characterization of the technique. However, in many cases, gaps in our understanding necessitate a technical validation or qualification program [36]. In other cases, a clinical drug study offers an ideal framework to implement a method as an exploratory endpoint with the primary objective to develop the technique for future studies. Some general scenarios can be considered:

1. *The method is well characterized and ready to be implemented*: the technique has been sufficiently applied in previous studies and the method can be implemented without further development activities. The focus on implementation will be to ensure the imaging is implemented to high quality,

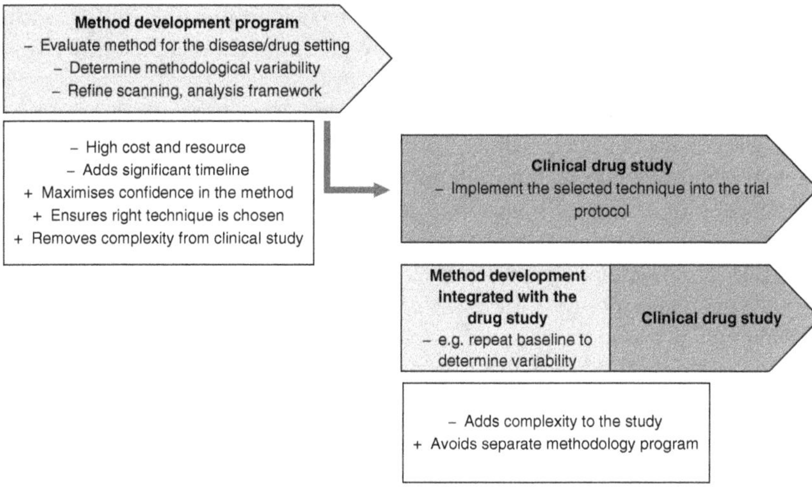

Figure 6.6 Options are available to develop imaging methods before integrating into clinical drug studies.

ensuring standardization of acquisition across centers and across subject visits and ensuring a clear analysis framework with well-defined imaging endpoints.

2. *The method is reasonably well understood but knowledge gaps exist*: interpretation of results will only be possible with additional characterization of the method. In this case, a pre-study validation might be conducted in the intended patient population to measure methodological variability or comparison of an imaging endpoint with histology [36]. Such a validation study will also enable the refinement of the scanning methods and analysis framework to prepare for a subsequent drug study. It also represents an opportunity to establish standardized scanning procedures at the center or centers participating in a subsequent intervention study.
3. *It is not known which technique should be selected*: in this case, either a methodology development plan needs to be established or different methods or techniques could be implemented as exploratory endpoints with the sole purpose of establishing methods for future trials. Preclinical validation and qualification studies may be desirable.

Different approaches can be considered to prepare imaging techniques for early drug development. Many factors will dictate the chosen approach including cost, timelines, and how important the imaging endpoint is to the program decision-making (Figure 6.6).

While clinical drug programs continue to seek new, informative methodologies, there are still relatively few imaging methods being incorporated as PD endpoints. This is at a time when the "suite" of technologies has never been broader. This rapid change in the technology base is in itself a problem with imaging methods evolving rapidly, and many methods are insufficiently characterized and not enabled across centers.

Imaging itself can offer multiparameter insights into drug action. With the advent of multi-modality assessments particularly now through multi-modality scanners, it is likely that imaging will provide an increasingly complex package of evidence – for example, functional and structural evidence with MRI combined with molecular evaluations with PET. Furthermore, the development and application of molecular imaging probes will offer increasingly insightful PD and mechanistic endpoints – for example, probes that can monitor inflammatory cell populations in tissues.

Integrating these multimodality approaches will likely be challenging, yet if done successfully can enable localized, quantitative, and comprehensive evaluations of drug action in the patient.

Guiding an Imaging Solution Toward Application

Integrating imaging methodologies into early clinical development requires a step-wise approach to consider the most appropriate technique and then to assess existing evidence around technical validation and qualification of the technique toward utilization [36].

1. *Method selection*: the first imperative is to establish what will be most beneficial to the drug development program and evaluate which methods would provide the right information. Gathering imaging science expertise broadly together with drug developers will allow all options to be considered. All scientific and practical factors will need to be considered; for example, radiation dose limitations in certain patient populations will limit some methods and may favor techniques that do not involve ionizing radiation.
2. *Time and cost*: as previously outlined, methods vary in their state of readiness for a given clinical trial application. This state has a significant impact on time and cost considerations.
   a. Existing techniques that have been applied to numerous early clinical development programs previously (e.g. FDG-PET in oncology [39, 41, 45]). For such techniques, the focus will be to ensure the method is incorporated appropriately into the protocol – for example, timing of scanning relative to treatment. Integrating such methods will be largely operational, ensuring that the technique can be applied robustly at the trial centers involved in the study. Time will be required to engage with clinical centers to ensure implementation of methods or through an imaging contract research organization (CRO) to support these activities. Such activities are typically performed in parallel to preparation

for the clinical trial and seldom add bottlenecks to the study setup. Imaging-related costs will primarily be defined by the site scanning costs and support for image analysis/review at the site or if required centralized to another location (e.g. an imaging CRO may receive the data in order to coordinate a blinded independent review).

b. Techniques that have some precedence but may not be well understood in the proposed clinical trial setting, or where multiple methods are being considered and need to be compared. In this case, further investigation of the methodology may be required before integration in the drug study. This typically necessitates time and costs dedicated to a technical validation study but will ensure sufficient confidence in the technique for subsequent application.
c. Techniques that require bespoke development and qualification activities. In this case, significant activities are required to prepare the methods for use and will likely necessitate a separate technical validation study before integrating within a clinical trial. For example, new radiochemistry methods can take three years to prepare and implement requiring early planning.

3. *The role of external partnerships in the application of advanced imaging*: strong partnerships between drug development organizations and clinical imaging centers will ensure reactive access to both expertise and scanning infrastructure. An ongoing partnership will ensure that new opportunities can be readily progressed – for example, testing a new MRI method in one center before deploying it within a multicenter trial or evaluating radiochemistry options before embarking on a drug biodistribution study.

   Since imaging is highly subspecialized with different centers having different levels of expertise – e.g. functional neuroimaging, lung imaging – a broad network of experts and centers is typically required to support a broad pharma R&D program.

   In addition, contract research organizations specializing in imaging offer an important function to ensure the imaging components of a study can be operationalized particularly for multicenter studies. Such companies can provide an interface with the site imaging experts to qualify sites for the imaging component, coordinating receipt of imaging data, quality control steps, centralized analysis, and/or radiological review.
4. *How much standardization is required? Multi vs. single center*: even simple imaging methodologies represent cost and complexity to integrate in multicenter trials. When it comes to more complex methods with utility in early phase studies, the challenge is addressed through extensive activities to implement and standardize methods (e.g. DCE-MRI for a rheumatoid arthritis trial) or the methods are simply not feasible to extend beyond a single, specialized center (radiolabeled new chemical entity).

Recruitment is a challenge in most trial settings and may compromise study completion. Solving it by recruiting across many centers can be detrimental to the successful implementation of complex methodologies (e.g. standardization of methods across different scanners, availability of a radioligand). Trial centers suitable for subject recruitment may lack imaging expertise or scanning infrastructure specifications. Current consortia efforts toward imaging biomarker standardization are moving beyond merely specifying standard procedures [39, 40] to defining detailed performance criteria and biomarker-specific conformance testing [45, 46].

A balance always needs to be struck between fewer sites to control standardization complexity and the number of sites required for recruitment. Over time, with more informative endpoints in early development, fewer subjects may be needed to demonstrate pharmacology. The aim of imaging science for drug development should be to provide tools to reliably demonstrate pharmacology in few subjects. However, until improved methods are available, the subject numbers for a trial will often be defined by suboptimal and variable primary endpoints which increase subject numbers.

5. *Integration within the study*: sufficient imaging details will need to be defined within the clinical trial protocol. Important details include timing of imaging relative to other protocol-defined events such as dosing and clear definition of the endpoints to be derived from imaging.

   Successful imaging studies will require many imaging-related parameters to be defined (e.g. image acquisition parameters, patient preparation and positioning, image processing). Definition of these details will promote standardization if imaging is performed across imaging centers and will ensure consistency of baseline assessments to follow-up. These details are typically captured in documents such as *image acquisition guidelines* and referenced from the protocol. Particularly for complex methods, early engagement with site imaging experts is vital to ensure the methods can be successfully implemented in time for conduct of the study.

Conclusions and Future Directions

Imaging offers a noninvasive window into both disease and therapeutic response and has become a broadly embedded platform technology supporting early phase drug development. Areas such as oncology and neuroscience have demonstrated an established need for imaging catalyzed largely by the limited ability to access and study tissue *ex vivo*. Furthermore, where standard methods such as tumor size measurements with computed tomography (CT) are established, the field has evolved to incorporate molecular methods including FDG-PET and new tracers more specific to the pathophysiology and drug response. Such fields are likely to evolve further, particularly with the advent of cell specific imaging. Beyond oncology and neuroscience, it will be important to ensure the role of imaging is accelerated, for example, studies of infectious and inflammatory diseases. To catalyze new opportunities for imaging in drug research, drug developers will need to challenge the imaging scientific community to address current impediments.

Further enhancement of widely used modalities (e.g. increasing magnetic field strength for MRI and new probe molecules for PET) combined with emerging new techniques (e.g. clinical optical imaging with near-infrared probes) will improve sensitivity and specificity. However, investment and time are required to guide these nascent technologies toward quantitative, decision-making tools. A future state is envisaged where small cohorts of subjects in a single study can be measured with multimodal imaging to define drug distribution, target engagement, and measures of downstream pharmacology and efficacy.

Complexity of imaging data will grow, with increasing resolution, multimodality approaches, and use of imaging on different scales (from tissue to whole body). How we handle this data will need to be a focus, ensuring we have quantitative workflows able to extract endpoints from these increasingly complex data sets.

A flow of information-rich impactful imaging techniques will become a reality if drug developers, imaging scientists, clinicians, biologists, and mathematicians coordinate. Many of the imaging needs will only be addressed with broad, noncompetitive consortia with the scale required to move concepts from the laboratory toward clinical application for the benefit of patients.

References

1 Paul, S.M., Mytelka, D.S., Dunwiddie, C.T. et al. (2010). How to improve R&D productivity: the pharmaceutical industry's grand challenge. *Nat. Rev. Drug Discovery* 9 (3): 203–214.

2 Morgan, P., Van Der Graaf, P.H., Arrowsmith, J. et al. (2012). Can the flow of medicines be improved? Fundamental pharmacokinetic and pharmacological principles toward improving Phase II survival. *Drug Discovery Today* 17 (9-10): 419–424.

3 Owens, P.K., Raddad, E., Miller, J.W. et al. (2015). A decade of innovation in pharmaceutical R&D: the Chorus model. *Nat. Rev. Drug Discovery* 14 (1): 17–28.

4 Cook, D., Brown, D., Alexander, R. et al. (2014). Lessons learned from the fate of AstraZeneca's drug pipeline: a five-dimensional framework. *Nat. Rev. Drug Discovery* 13 (6): 419–431.

5 Minchinton, A.I. and Tannock, I.F. (2006). Drug penetration in solid tumours. *Nat. Rev. Cancer* 6 (8): 583–592.

6 Tannock, I.F., Lee, C.M., Tunggal, J.K. et al. (2002). Limited penetration of anticancer drugs through tumor tissue: a potential cause of resistance of solid tumors to chemotherapy. *Clin. Cancer Res.* 8 (3): 878–884.

7 Fuso Nerini, I., Morosi, L., Zucchetti, M. et al. (2014). Intratumor heterogeneity and its impact on drug distribution and sensitivity. *Clin. Pharmacol. Ther.* 96 (2): 224–238.

8 Liu, X., Smith, B.J., Chen, C. et al. (2006). Evaluation of cerebrospinal fluid concentration and plasma free concentration as a surrogate measurement for brain free concentration. *Drug Metab. Dispos.* 34 (9): 1443–1447.

9 Kusuhara, H. and Sugiyama, Y. (2004). Efflux transport systems for organic anions and cations at the blood-CSF barrier. *Adv. Drug Delivery Rev.* 56 (12): 1741–1763.

10 Shen, D.D., Artru, A.A., and Adkison, K.K. (2004). Principles and applicability of CSF sampling for the assessment of CNS drug delivery and pharmacodynamics. *Adv. Drug Delivery Rev.* 56 (12): 1825–1857.

11 Bergstrom, M., Grahnen, A., and Langstrom, B. (2003). Positron emission tomography microdosing: a new concept with application in tracer and early clinical drug development. *Eur. J. Clin. Pharmacol.* 59 (5–6): 357–366.

12 Van Dongen, G.A., Huisman, M.C., Boellaard, R. et al. (2015). 89Zr-immuno-PET for imaging of long circulating drugs and disease targets: why, how and when to be applied? *Q. J. Nucl. Med. Mol. Imaging* 59 (1): 18–38.

13 van Dongen, G.A., Visser, G.W., Lub-de Hooge, M.N. et al. (2007). Immuno-PET: a navigator in monoclonal antibody development and applications. *Oncologist* 12 (12): 1379–1389.

14 Newman, S.P., Pitcairn, G.R., Hirst, P.H., and Rankin, L. (2003). Radionuclide imaging technologies and their use in evaluating asthma drug deposition in the lungs. *Adv. Drug Delivery Rev.* 55 (7): 851–867.

15 Newman, S.P. and Wilding, I.R. (1999). Imaging techniques for assessing drug delivery in man. *Pharm. Sci. Technol. Today* 2 (5): 181–189.

16 Bergstrom, M., Cass, L.M., Valind, S. et al. (1999). Deposition and disposition of [11C]zanamivir following administration as an intranasal spray. Evaluation with positron emission tomography. *Clin. Pharmacokinet.* 36 (Suppl 1): 33–39.

17 Yates, R., Sorensen, J., Bergstrom, M. et al. (2005). Distribution of intranasal C-zolmitriptan assessed by positron emission tomography. *Cephalalgia* 25 (12): 1103–1109.

18 Ilardi, E.A., Vitaku, E., and Njardarson, J.T. (2014). Data-mining for sulfur and fluorine: an evaluation of pharmaceuticals to reveal opportunities for drug design and discovery. *J. Med. Chem.* 57 (7): 2832–2842.

19 Antoni, G. (2015). Development of carbon-11 labelled PET tracers-radiochemical and technological challenges in a historic perspective. *J. Labelled Compd. Radiopharm.* 58 (3): 65–72.

20 Elsinga, P.H. and Dierckx, R.A. (2014). Small molecule PET-radiopharmaceuticals. *Curr. Pharm. Des.* 20 (14): 2268–2274.

21 Saleem, A., Murphy, P., Plisson, C., and Lahn, M. (2014). Why are we failing to implement imaging studies with radiolabelled new molecular entities in early oncology drug development? *Sci. World J.* 2014: 269605.

22 Huiban, M., Tredwell, M., Mizuta, S. et al. (2013). A broadly applicable [18F]trifluoromethylation of aryl and heteroaryl iodides for PET imaging. *Nat. Chem.* 5 (11): 941–944.

23 Pascali, G., Watts, P., and Salvadori, P.A. (2013). Microfluidics in radiopharmaceutical chemistry. *Nucl. Med. Biol.* 40 (6): 776–787.

24 Wangler, C., Schirrmacher, R., Bartenstein, P., and Wangler, B. (2010). Click-chemistry reactions in radiopharmaceutical chemistry: fast and easy introduction of radiolabels into biomolecules for in vivo imaging. *Curr. Med. Chem.* 17 (11): 1092–1116.

25 Fischman, A.J., Bonab, A.A., and Rubin, R.H. (2000). Regional pharmacokinetics of orally administered PET tracers. *Curr. Pharm. Des.* 6 (16): 1625–1629.

26 Bergstrom, M., Yates, R., Wall, A. et al. (2006). Blood-brain barrier penetration of zolmitriptan--modelling of positron emission tomography data. *J. Pharmacokinet. Pharmacodyn.* 33 (1): 75–91.

27 Gaykema, S.B., Brouwers, A.H., Lub-de Hooge, M.N. et al. (2013). 89Zr-bevacizumab PET imaging in primary breast cancer. *J. Nucl. Med.* 54 (7): 1014–1018.

28 Oude Munnink, T.H., Tamas, K.R., Lub-de Hooge, M.N. et al. (2013). Placental growth factor (PlGF)-specific uptake in tumor microenvironment of 89Zr-labeled PlGF antibody RO5323441. *J. Nucl. Med.* 54 (6): 929–935.

29 Hietala, J., Nagren, K., Lehikoinen, P. et al. (1999). Measurement of striatal D2 dopamine receptor density and affinity with [11C]-raclopride in vivo: a test-retest analysis. *J. Cereb. Blood Flow Metab.* 19 (2): 210–217.

30 Zamuner, S., Di Iorio, V.L., Nyberg, J. et al. (2010). Adaptive-optimal design in PET occupancy studies. *Clin. Pharmacol. Ther.* 87 (5): 563–571.

31 Nordstrom, A.L., Farde, L., Nyberg, S. et al. (1995). D1, D2, and 5-HT2 receptor occupancy in relation to clozapine serum concentration: a PET study of schizophrenic patients. *Am. J. Psychiatry* 152 (10): 1444–1449.

32 Gefvert, O., Lundberg, T., Wieselgren, I.M. et al. (2001). D(2) and 5HT (2A) receptor occupancy of different doses of quetiapine in schizophrenia: a PET study. *Eur. Neuropsychopharmacol.* 11 (2): 105–110.

33 Bergstrom, M., Hargreaves, R.J., Burns, H.D. et al. (2004). Human positron emission tomography studies of brain neurokinin 1 receptor occupancy by aprepitant. *Biol. Psychiatry* 55 (10): 1007–1012.

34 Bergstrom, M., Westerberg, G., Nemeth, G. et al. (1997). MAO-A inhibition in brain after dosing with esuprone, moclobemide and placebo in healthy volunteers: in vivo studies with positron emission tomography. *Eur. J. Clin. Pharmacol.* 52 (2): 121–128.

35 O'Connor, J.P., Jackson, A., Parker, G.J., and Jayson, G.C. (2007). DCE-MRI biomarkers in the clinical evaluation of antiangiogenic and vascular disrupting agents. *Br. J. Cancer* 96 (2): 189–195.

36 Moody, J.B., Murphy, P.S., and Ficaro, E.P. (2017). Imaging as a localized biomarker: opportunities and challenges. In: *Biomarkers in Drug Development: A Handbook of Discovery, Practice, Application, and Strategy*, 2e (ed. M.R. Bleavins). Wiley.

37 Eisenhauer, E.A., Therasse, P., Bogaerts, J. et al. (2009). New response evaluation criteria in solid tumours: revised RECIST guideline (version 1.1). *Eur. J. Cancer* 45 (2): 228–247.

38 Van den Abbeele, A.D. and Ertuk, M. (2008). FDG-PET to measure response to targeted therapy: the example of gastrointestinal stromal tumor and imatinib mesylate (Gleevec). *PET Clin.* 3 (1): 77–87.

39 Boellaard, R., O'Doherty, M.J., Weber, W.A. et al. (2010). FDG PET and PET/CT: EANM procedure guidelines for tumour PET imaging: version 1.0. *Eur. J. Nucl. Med. Mol. Imaging* 37 (1): 181–200.

40 Boellaard, R., Oyen, W.J., Hoekstra, C.J. et al. (2008). The Netherlands protocol for standardisation and quantification of FDG whole body PET studies in multi-centre trials. *Eur. J. Nucl. Med. Mol. Imaging* 35 (12): 2320–2333.

41 Westerterp, M., Pruim, J., Oyen, W. et al. (2007). Quantification of FDG PET studies using standardised uptake values in multi-centre trials: effects of image reconstruction, resolution and ROI definition parameters. *Eur. J. Nucl. Med. Mol. Imaging* 34 (3): 392–404.

42 Glaudemans, A.W., de Vries, E.F., Galli, F. et al. (2013). The use of (18)F-FDG-PET/CT for diagnosis and treatment monitoring of inflammatory and infectious diseases. *Clin. Dev. Immunol.* 2013: 623036.

43 Rinne, J.O., Brooks, D.J., Rossor, M.N. et al. (2010). 11C-PiB PET assessment of change in fibrillar amyloid-beta load in patients with Alzheimer's disease treated with bapineuzumab: a phase 2, double-blind, placebo-controlled, ascending-dose study. *Lancet Neurol.* 9 (4): 363–372.

44 Kessler, L.G., Barnhart, H.X., Buckler, A.J. et al. (2015). The emerging science of quantitative imaging biomarkers terminology and definitions for scientific studies and regulatory submissions. *Stat. Methods Med. Res.* 24 (1): 9–26.

45 QIBA. FDG-PET/CT Technical Committee (2013). FDG-PET/CT as an Imaging Biomarker Measuring Response to Cancer Therapy: RSNA. Version 1.05, Publicly Reviewed Version. https://www.rsna.org/uploadedFiles/RSNA/Content/Science_and_Education/QIBA/QIBA_FDG-PET_Profile_v105_Publicly_Reviewed_Version_FINAL_11Dec2013.pdf (3 May 2019).

46 Obuchowski, N.A., Buckler, A., Kinahan, P. et al. (2016). Statistical issues in testing conformance with the quantitative imaging biomarker alliance (QIBA) profile claims. *Acad. Radiol.* 23 (4): 496–506.

7

Circulating MicroRNAs as Biomarkers in Cardiovascular and Pulmonary Vascular Disease: Promises and Challenges

Miranda K. Culley[1,2] *and Stephen Y. Chan*[1,2]

[1] *Center for Pulmonary Vascular Biology and Medicine, Pittsburgh Heart, Lung, Blood, and Vascular Medicine Institute, University of Pittsburgh School of Medicine, Pittsburgh, PA, USA*
[2] *University of Pittsburgh Medical Center, Pittsburgh, PA, USA*

Introduction

In this chapter, we explore the evolving role of circulating microRNAs (c-miRNAs) as biomarkers, specifically in cardiovascular and pulmonary vascular disease. We describe our current understanding of intracellular and extracellular microRNA (miRNA) biogenesis and the detection methods for these miRNA molecules in the circulation. We then discuss the proposed applications and current obstacles to their development as diagnostic and prognostic biomarkers in vascular health and disease.

MicroRNA Biogenesis

miRNAs are endogenous, small noncoding RNA molecules that are encoded in intergenic regions of the genome or within introns of noncoding or coding transcripts. These genes are transcribed by RNA polymerase II into a primary transcript with a stem-loop structure (pri-miRNA). The miRNA maturation process begins in the nucleus with cleavage of the pri-miRNA to pre-miRNA by RNase III Drosha – a double-stranded RNA endonuclease which forms the *Microprocessor complex* when bound to the essential cofactor DGCR8. Importantly, Drosha cleavage defines the terminus of the miRNA and thus determines the specificity of the miRNA for a particular target (Figure 7.1). Next, the pre-miRNA is exported from the nucleus via a transport complex composed of exportin 5 and guanylyl imidodiphosphate, guanosine-5'-triphosphate (GTP)-binding nuclear proteins. In the cytoplasm, pre-miRNA is cleaved at the hairpin loop by Dicer – a cytoplasmic RNase III

Biomarkers in Drug Discovery and Development: A Handbook of Practice, Application, and Strategy, Second Edition. Edited by Ramin Rahbari, Jonathan Van Niewaal, and Michael R. Bleavins.

endonuclease – to yield an RNA duplex which is subsequently loaded on an Ago2 protein to form the RNA-induced silencing complex (RISC). Following release of the "passenger strand," the now mature, single-stranded miRNA can bind to the 3′-untranslated regions of messenger RNA as a part of the RISC complex to downregulate gene expression via translational repression or induction of mRNA degradation [2]. The biogenesis of miRNAs is summarized in Figure 7.1 [1]. In this way, miRNAs are estimated to regulate the expression of more than 60% of protein-coding genes, thus controlling a multitude of physiological and pathological processes [3]. Individual miRNAs have been linked to cardiovascular and pulmonary vascular disease states, including acute myocardial infarction (AMI), acute coronary syndrome (ACS), heart failure (HF), and pulmonary hypertension, among others [4–6].

In 2008, Mitchell et al. [7] reported the substantial expression of extracellular miRNAs in the plasma. Subsequent studies revealed the presence of miRNAs in several other extracellular fluid spaces, including serum, saliva, and urine [8, 9]. While naked or exogenous miRNA are rapidly degraded in the plasma by RNase, these extracellular miRNAs were found to be remarkably stable. These so-called c-miRNAs were subsequently found to be protected from degradation by being packaged in lipid vesicles, with RNA-binding proteins, and in lipoprotein complexes [7].

Specifically, there are three types of lipid vesicles that have been discovered to carry and stabilize miRNAs in the extracellular compartment: exosomes, microvesicles, and apoptotic bodies. Exosomes (40–100 nm) are formed from endosomes and released when multivesicular bodies fuse with the plasma membrane [10, 11]. Microvesicles (50 nm–1 μm) are shed directly from the plasma membrane of cells under both physiologic and pathologic conditions [12]. Apoptotic bodies (1–5 μm) are small membranous particles released from the cell surface during apoptosis [13]. Separately, c-miRNAs are also stabilized in the plasma by RNA-binding proteins. These ribonucleoprotein complexes have been reported to potentially account for 90% of the c-miRNA species. Nucleophosmin 1 [14] and argonaute 2 [15, 16] are two examples of this type of protein that can prevent c-miRNA degradation in the extracellular space. Lastly, some extracellular miRNAs are associated with lipoproteins such as high-density lipoprotein (HDL) [17]. These three mechanisms of extracellular transport are represented in Figure 7.2 [18].

Detection of Extracellular miRNAs

For c-miRNAs to be utilized as clinical biomarkers, the methods for extraction must be reproducible and the methods for detection must be rapid and accurate. Currently, the three major techniques for quantitative expression analysis are reverse transcription followed by quantitative polymerase chain reaction

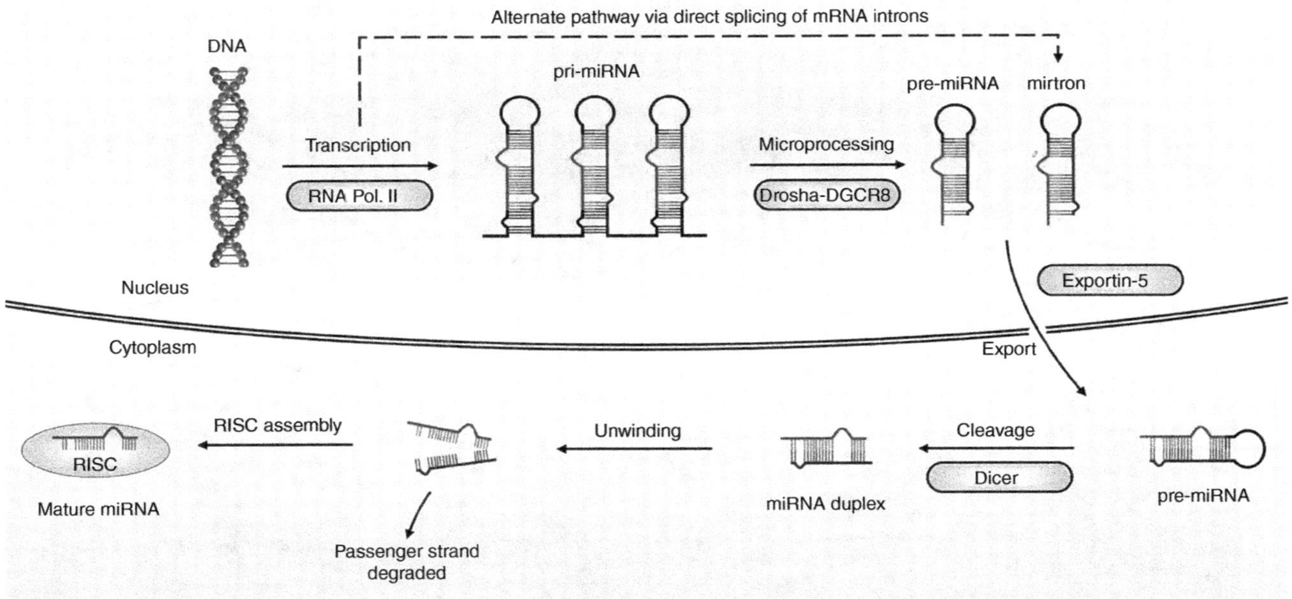

Figure 7.1 Biogenesis of miRNA. miRNAs are transcribed in the nucleus by RNA polymerase II into the primary transcript (pri-miRNA). The pri-miRNA transcript is cleaved to pre-miRNA by the *microprocessor complex*, which is made up of the Drosha endonuclease and DGCR8 cofactor. The pre-miRNA is exported via exportin 5 from the nucleus to the cytoplasm where the transcript is further modified by Dicer endonuclease to yield an RNA duplex. Following the release of the "passenger strand," the now mature, single-stranded miRNA is loaded on to Ago2 protein to form the RNA-induced silencing complex (RISC). This miRNA as a part of the RISC can specifically bind the 3′-untranslated region of messenger RNA to downregulate gene expression via translational repression or induction of mRNA degradation. Source: Adapted with permission of Romaine et al. 2015 [1]. https://heart.bmj.com/content/101/12/921. Licensed under CC-BY-4.0 http://creativecommons.org/licenses/by/4.0/. *(See insert for color representation of this figure.)*

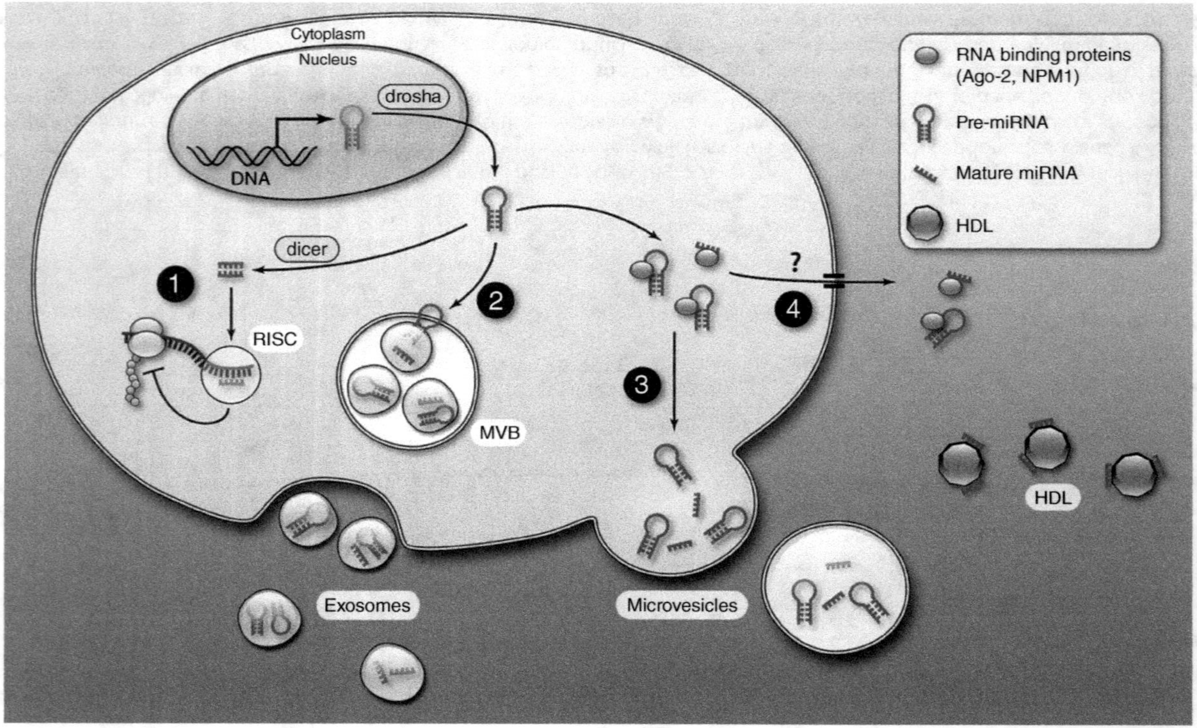

Figure 7.2 Transport of extracellular miRNA. For intracellular miRNA, one strand of this double-stranded mature miRNA duplex is loaded into the RNA-induced silencing complex (RISC) (1). Separately, miRNAs can be incorporated into exosomes, which originate from the endosome and are released from cells when multivesicular bodies (MVBs) fuse with the plasma membrane (2). Cytoplasmic miRNAs can also be released by microvesicles, which are released from the cell through plasma membrane blebbing (3). Circulating miRNAs can be bound to RNA-binding proteins such as Ago2 (4) or associated with high-density lipoproteins (HDL). The mechanisms by which specific miRNAs are packaged and released still remains largely uncharacterized. Source: Creemers et al. 2012 [18]. Adapted with permission of Wolters Kluwer Health, Inc. *(See insert for color representation of this figure.)*

(RT-qPCR), microarray hybridization, and high-throughput sequencing – all adapted from prior techniques of assessing mRNA expression [19]. Below we will briefly discuss the merits and pitfalls of c-miRNA detection with each of these technologies (Figure 7.3).

To adapt mRNA expression profiling techniques for miRNA, there are several intrinsic properties of miRNAs to address. First, traditional qPCR primers and protocols must be altered given both the length and nucleotide composition of miRNA. In the case of the former, the length of the majority of miRNA transcripts is <30 nucleotides, which is often insufficient for annealing with traditional qPCR primers. Additionally, the guanine–cytosine (GC) composition of these small transcripts can cause substantial variability in melting temperatures, which requires alteration of the annealing temperature during qPCR. Second, detection and quantification based on sequence specificity is more difficult due to sequence homology between sequential forms of miRNAs (pri-, pre-, and mature miRNA) during the maturation process and within miRNA families [20]. Furthermore, alternative Drosha and Dicer cleavage as well as post-transcriptional modification can result in miRNA with endogenous variation at the 3′ end called isomiRs [21, 22]. Due to the high degree of sequence homology, certain detection methods cannot discern between different isomiRs of a miRNA transcript. Apart from the intrinsic qualities of miRNA, their relative abundance compared to total RNA mass represents an extremely small fraction, estimated at approximately 0.01%.

This means miRNA must be selectively detected in a background of many more RNA species [23]. Moreover, of the already small miRNA fraction, extracellular miRNAs are at a relatively lower concentration than intracellular miRNA. Thus, when evaluating c-miRNA in a peripheral blood sample, miRNA from erythrocyte lysis or other blood cells could greatly confound measurement profiles of extracellular miRNA species [24]. Understanding these challenges in miRNA profiling is critical for proper interpretation of c-miRNA levels and thus the use of miRNAs as biomarkers. Next, we evaluate the step-wise process of extracellular miRNA expression profiling and strategies to combat the above limitations.

Sampling

The quality of biological sampling and isolation can greatly affect and/or confound downstream miRNA quantitation. In blood, hemolysis and anticoagulants such as heparin are major confounders to the populations of c-miRNAs, perturbing biomarker profiles significantly in plasma or serum [24–26]. Sampling techniques of c-miRNAs quantified from other extracellular fluid compartments, including placenta, saliva, and urine, are prone to similar confounding variables [9]. Thus, to minimize variability and noise in downstream miRNA quantitation, consistency in sampling technique from any compartment is of paramount importance.

RNA Extraction

Most commercial RNA isolation kits are based on chemical extraction using chaotropic salts followed by solid-phase extraction on silica columns [20]. McAlexander et al. [27] compared five separate commercial kits specific for extracellular miRNA isolation (mirVana, miRCURY Cell and Plant, Exiqon miRCURY biofluids kit, TRIzol LS isolation, and Qiagen plasma/serum kit) using a standard source sample of plasma from a single blood draw from the same donor. Their evaluation suggested that Exiqon miRCURY biofluids kit

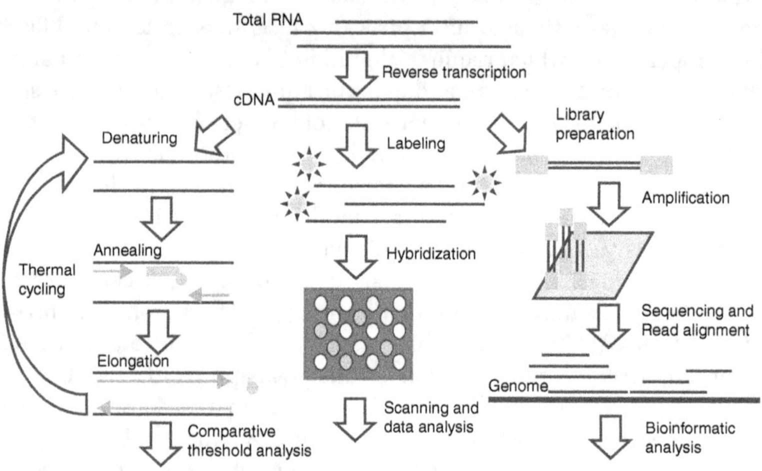

	RT-qPCR	Microarray	RNA-seq
Strengths of the technology	- High yield and high specificity - Commercially available at relatively low cost	- Multiple miRNAs identified in parallel - Useful for healthy versus disease comparison	- High-throughput large-scale data generation - Identifies novel miRNAs and distinguishes between single nucleotide variants and separate isomiRs
Limitations of the technology	- Target gene expression must be normalized to control genes	- Target gene expression must be normalized to healthy controls	- Relatively high cost - Requires tremendous analytical support - High rate of false positives and negatives
miRNA-specific challenges of quantification	- Lack of a standardized endogenous control miRNA - Short length - GC content	- Low abundance relative us cellular RNA - Low specificity - Cannot detect novel miRNAs	- Inefficient library preparation by random hexamers due to the short length of the template transcript
Technological **improvements**	- Combined polyadenylation and reverse transcription - S-poly(T) RT primer - Digital PCR provides absolute quantification	- Increase the number of complementary oligos included on each CGH chip	- Blocking techniques to remove high abundance miRNA - Complementary long padlock proteins - Synthetic and protein-based nanopore sequencing

Figure 7.3 Techniques for miRNA quantification. Technologies for the measurement of c-miRNAs were adapted from methods used to identify and measure mRNAs, including RT-qPCR, microarray hybridization, and RNA sequencing. These three major techniques share similarities in their initial steps: RNA extraction and reverse transcription reaction to convert RNA to cDNA. Therefore, these techniques are similarly limited by the inherent characteristics of miRNAs, short length, and high GC content. RT-qPCR uses thermal cycling during which denaturing, annealing of the DNA-specific primers, and subsequent elongation of these template strands take place. Comparative threshold analysis compares the cycle number at which the fluorescence measurement passes a fixed threshold value for the c-miRNA of interest vs. a positive synthetic miRNA control. RT-qPCR is commercially available, relatively cost-effective, and highly specific but is limited to one transcript quantified per reaction well and by its inability to quantify absolute abundance. Microarray applies labeled sample cDNA to a chip of known oligonucleotide fragments and then employs a scanner to quantify fluorescence to compare specific transcript expression between treated and control samples. This technique allows for the rapid relative quantification of several c-miRNAs in parallel but cannot provide absolute measurements and has relatively low specificity. Lastly, RNA sequencing performs single-nucleotide sequencing in thousands of cDNA transcripts at one time. This high-throughput NGS technique can distinguish between isomiRs and identify novel c-miRNAs in a short amount of time but is limited by the relatively high cost, requirement for advanced biostatistical analysis, and high rate of false positives and negatives, among other challenges. CGH, comparative genomic hybridization. *(See insert for color representation of this figure.)*

outperformed the other methods based on relative recovery of the selected miRNAs, while TRIzol LS exhibited poor recovery in comparison to the other methods [27]. In comparison, however, despite variability in the expression profiles following different RNA extraction protocols, other studies have not identified a "gold standard" approach [28, 29]. Instead, the general consensus has been to consistently use a single, pre-analytical protocol for sampling and extraction to prevent variation between experiments.

RT-qPCR

RNA quantification via RT-qPCR begins with the synthesis of a cDNA strand from a miRNA transcript by reverse transcriptase. The RT reaction can be achieved by two distinct primer methods: poly-A tail and stem-loop. Poly-A tail primer method begins with polyadenylation of all RNA transcripts to lengthen the transcript followed by synthesis of the cDNA strand by reverse transcriptase. During qPCR, cDNA is quantified with a miRNA-specific forward primer and a universal reverse primer. By contrast, the stem-loop primer method uses a miRNA-specific stem-loop reverse transcriptase that stabilizes and lengthens only the specific miRNA transcript first, thus reducing future nonspecific amplification and increasing specificity. A miRNA-specific forward primer that allows for qPCR machine detection via fluorescently labeled probe oligonucleotides is used. The polymerase chain reaction (PCR) itself employs thermal cycling to rapidly amplify a primer-specific DNA molecule. Quantitation of a desired cDNA is based on the comparative threshold cycle (Ct), which is the cycle number at which the fluorescence measurement in the real-time amplification curve passes a fixed fluorescent threshold amount. Therefore, relative changes in miRNA expression can be derived by comparing the Ct value of the target miRNA to the Ct value of the positive control, which is most often synthetic miRNA spikes or a high abundance endogenous miRNA. RT-qPCR is a high-yield, high-sensitivity quantification method that demonstrates relative expression of one transcript in a single reaction well [30]. This is a relatively easy, commercially available assay system that could be used in a clinical setting for specific biomarker detection. However, RT-qPCR is limited because it can only calculate RNA expression levels relative to a control. Relatively small changes in miRNA expression may be biologically significant, making normalization even more critical than that of other RNA functional classes. To this point, c-miRNAs lack a uniformly accepted noncellular endogenous control, making relative miRNA expression profiles more challenging to interpret.

Recently, Niu et al. [31] reported an improved method for RT-qPCR, S-Poly(T) Plus, which relies on a uniquely designed S-Poly(T) reverse transcription primer that contains miRNA-specific bases and an oligo(dT) sequence. This improves binding strength and thermodynamic stability between the RT primer and its miRNA template, ultimately increasing sensitivity. MiRNA quantitation is further optimized for high-throughput screening of clinical samples by using glycogen to increase RNA yield and combining polyadenylation and reverse transcription into one step. These results were confirmed by quantifying miRNA expression profiles in the sera of pulmonary arterial hypertension (PAH) patients with congenital heart disease [31].

Alternatively, digital polymerase chain reaction (dPCR) partitions individual molecules into many replicate reactions, resulting in one molecule per reaction. The primary advantage of dPCR over RT-qPCR is the ability for absolute quantification of a desired transcript without the need for a control. Furthermore, an improved dPCR technology using nanoliter-sized aqueous droplets in oil (ddPCR) is now commercially available. Hindson et al. [32] demonstrated that ddPCR is more precise and yields more easily reproducible results while being equally sensitive.

Microarray

Microarray hybridization uses extracted miRNA which is converted to cDNA using RT, and then tagged with fluorophore-labeled nucleotides at the 3′ end. Hybridization of tagged cDNA to a glass plate with complementary synthetic DNA oligonucleotides allows for detection of fluorescent intensity via a microarray scanner [33]. These microarray chips often have hundreds to thousands of known, pre-arranged DNA sequences that allow for expressional analysis in parallel. Microarray technology works well when comparing two distinct cohorts, providing relative abundance of specific targets in the healthy vs. diseased group. However, microarray is limited to the known complementary oligonucleotides available and does not allow for identification of novel miRNAs. Microarrays also have limited specificity, and thus results often require validation using RT-qPCR. Microarrays, similar to RT-qPCR, lack the capability of performing absolute quantification of miRNA abundance. Lastly, similar to all of these methods that were originally intended and optimized for mRNA detection, the use of microarrays is challenged by the innate properties of miRNAs including the short length, low abundance relative to total cellular RNA, and low specificity [20].

RNA Sequencing

Next generation sequencing (NGS), specifically RNA sequencing, begins with the preparation of a cDNA library using the isolated small RNA sample followed by parallel sequencing of numerous cDNA molecules from the library in a single run. Bioinformatics-based analysis of the reads identifies known and novel miRNAs while providing a relative abundance of each miRNA transcript compared to total number of reads. Unlike the methods previously described, RNA sequencing can distinguish specific single-nucleotide variants and isomiRs [21, 34]. Despite its extensive measurement capabilities, RNA sequencing remains the most expensive option currently available, requires advanced biostatistical analysis for interpretation [35], and can produce false positive and negative results. miRNA-specific limitations include (i) inefficient conversion to cDNA by random hexamers due to the short length of the template transcript, (ii) certain biases introduced by library preparation methods [36], and (iii) inconsistencies introduced by mapping techniques [36]. Additionally, highly abundant miRNAs can prevent the quantitation of lower abundance small RNA species. To address this limitation, NGS can be employed with an additional blocking technique which shows significant reduction of a common blood cell contaminant, hsa-miR-16-5p, and leads to the identification of a greater number of lower abundance miRNA species [37]. Efforts to make RNA sequencing more suited for a clinical setting include the use of complementary long padlock proteins (cLPPs). cLPP is a recent DNA capture technology that uses physically tethered amplification primers to amplify several targets in a single reaction with high specificity and efficiency. In the context of cardiomyopathy and congenital heart disease, this assay was found to be rapid, cost-effective, as well as more specific and more sensitive to mutation detection than traditional NGS assays, making this a promising step toward utilizing specific NGS protocols in a clinical setting [38].

Nanopore Sequencing

Nanopore technology is a novel method that uses single-molecule strategies for quantitative miRNA detection. The nanopore sensor detects changes in electrical ion current across a pore lumen. There are two types of nanopore assays that have been studied in the context of miRNAs: synthetic and protein-based assays. Both are highly sensitive and specific, allowing for discrimination of miRNAs with similar sequences and single-nucleotide polymorphisms (SNPs), as well as directly and label-free [39]. Wanunu et al. [40] described the use of the synthetic assays as single-molecule counters to quantify sequence-specific miRNA from cellular RNA. Wang et al. [41] demonstrated selective detection of sub-picomolar levels of cancer-associated c-miRNAs, miR-155 and let-7, from plasma samples of lung cancer patients using a protein nanopore-based sensor, specifically an α-hemolysin protein pore with a probe designed to identify target miRNAs. This method can be applied to c-miRNAs associated with other disease states and thus is particularly promising in the field of miRNA biomarker discovery. A major limitation of the widespread use of protein nanopores is optimization of the artificial membrane platform that holds the nanopore in place, and this has stalled the use of protein nanopores commercially [42].

In summary, several inherent challenges still exist when quantifying c-miRNAs. However, the techniques for miRNA expression methods are evolving rapidly, becoming more sensitive and specific to (i) detect low abundance extracellular transcripts, (ii) discern separate isomiRs, (iii) measure absolute concentrations without the requirement of control, and (iv) more rapidly test several miRNAs and several samples at one time via high-throughput platforms. All of these improvements may soon result in a cost-effective, gold standard for the use of miRNA as biomarkers in a clinical setting.

Utility of c-miRNAs as Biomarkers in Cardiovascular Disease

The use of circulating biomarkers to improve the accuracy of diagnosis and treatment in cardiovascular diseases has been longstanding. Current biomarkers of myocardial injury and disease are polypeptides, for example, cardiac troponin (cTnI) in ACS and B-type natriuretic peptide (BNP) in congestive heart failure (CHF). However, despite their wide use in clinical practice, these cardiac biomarkers are not without their limitations, particularly in terms of their ability to discriminate among different subtypes of disease, severity of disease, and/or predisposition to disease in the future. The development of new diagnostic biomarkers such as c-miRNAs could aid in differentiating these disease features. Many studies have identified potential c-miRNA biomarkers for a multitude of cardiovascular disease states (Table 7.1), including ACS and CHF which we explore in greater depth below.

For ACS, the most commonly studied c-miRNAs that appear at increased levels include miR-1, miR-133, miR-208, and miR-499. Specifically, levels of miR-1, miR-133a and miR-133b, and miR-499 are upregulated in patients with ST-segment–elevated myocardial infarction (STEMI) [43, 71]. In a separate study, miR-208b and miR-499 were found at increased levels in patients with myocardial infarction (MI). Notably, these two c-miRNAs were further increased in patients with STEMI vs. non-ST-segment–elevated myocardial infarction (NSTEMI), distinguishing between the two types and providing comparable diagnostic utility with troponin levels [72]. In a separate study, Devaux et al. [73] demonstrated that several elevated c-miRNAs associated with AMI – miR-208b, miR-499, and miR-320a – did not outperform troponin as a diagnostic or prognostic biomarker nor did they improve diagnostic power when combined with troponin levels [73]. However, in a large cohort study, miR-1, miR-499, and miR-21 in particular each increased the diagnostic value of high sensitivity (hs)-troponin T in the context of ACS. Further, when these three c-miRNAs were evaluated together, they demonstrated greater diagnostic value than troponin as independent predictors of ACS [44]. Zeller et al. [64] identified novel c-miRNAs specific for unstable angina pectoris and demonstrated that a decrease in miR-132 and miR-150 with a concurrent increase in miR-186 provided higher discriminatory power in combination than hs-cTnI.

Similar to c-miRNA discovery in ACS, c-miRNAs may better discriminate CHF from other causes of dyspnea [78] as well as more precisely differentiate forms of heart failure [57, 79]. For example, Watson et al. [57] compared the expression changes of five c-miRNAs – miR-30c, miR-146a, miR-221, miR-328, and miR-375 – in non-HF patients, patients with heart failure with preserved ejection fraction (HFpEF), and patients with heart failure with reduced ejection fraction (HFrEF). They concluded the discriminatory power of BNP could be improved when considered with combinations of these five c-miRNAs, resulting in better prediction of patients with HFpEF vs. HFrEF [57]. Such a combination of traditional and new c-miRNA biomarkers could greatly enhance current risk prediction algorithms, with the ultimate goal of developing personalized medicine based on "-omics" sciences [26].

Beyond aiding in diagnosis, a number of c-miRNAs have been assessed for the ability to offer cardiovascular prognosis. Several studies have identified potential c-miRNAs to predict prognosis of ACS patients in addition to current prognostic indicators (*N*-terminal of pro b-type natriuretic peptide [NT-proBNP] and left ventricular ejection fraction [LVEF]). For example, miR-208b and miR-34 are markedly increased in patients with more left ventricular remodeling in the six months following an AMI. Further, elevated levels of these two c-miRNAs are associated with increased risk of mortality and heart

Table 7.1 Circulating miRNAs in cardiovascular disease.

c-miRNAs	Expression outcomes	References
miR-1	Upregulation in AMI Upregulation in ACS	[43, 44]
miR-9	Upregulation in T2DM	[45]
miR-16	Upregulation associated with impaired LV contractility six months post-AMI Downregulation in CAD	[46, 47]
miR-17	Downregulation in CAD	[48]
miR-19a	Upregulation in AMI	[49]
miR-21	Upregulation in ACS Downregulation in AF	[44, 50, 51]
miR-22	Upregulation in HF	[52]
miR-23a	Upregulation in metabolic syndrome	[53]
miR-27a	Upregulation associated with impaired LV contractility six months post-AMI	[46]
miR-29	Downregulation in CHF, AF, or both Upregulation in T2DM	[45, 54]
miR-30a	Upregulation in AMI	[55]
miR-30b	Downregulation in HF vs. non-HF dyspnea or control	[56]
miR-30c	Downregulation in HFrEF and HFpEF	[57]
miR-30d	Upregulation in T2DM	[45]
miR-31	Downregulation in CAD	[47]
miR-34a	Upregulation predicts LV remodeling Upregulation associated with mortality six months post-MI Upregulation associated with HF within one year post-MI Upregulation in T2DM	[45, 58, 59]
miR-92a	Downregulation in CAD	[48]
miR-92b	Upregulation in HF	[52]
miR-101	Downregulation associated with impaired LV contractility six months post-AMI	[46]
miR-103	Downregulation in HF vs. non-HF dyspnea or control	[56]
miR-122	Upregulation in hyperlipidemia Upregulation associated with severity of CAD	[60]
miR-124	Upregulation in T2DM	[45]
miR-125b	Downregulation in AMI	[61]

Table 7.1 (Continued)

c-miRNAs	Expression outcomes	References
miR-126	Downregulation in CAD Downregulation associated with susceptibility for T2DM Upregulation associated with incident MI	[48, 62] [63]
miR-130a	Upregulation in hypertension	[53]
miR-132	Downregulation in UA	[64]
miR-133a	Upregulation in AMI Upregulation in CAD Upregulation associated with decreased myocardial salvage, larger infarcts, and reperfusion injury after STEMI Upregulation associated with severity of coronary artery stenosis Upregulation associated with all-cause mortality in ACS patients	[43, 48, 65–68]
miR-133b	Upregulation in AMI	[43]
miR-134	Upregulation in AMI Upregulation associated with increased risk of HF or mortality six months post-MI	[69]
miR-142	Downregulation in HF vs. non-HF dyspnea or control	[56]
miR-144	Upregulation in T2DM	[70]
miR-145	Downregulation in CAD	[47, 48]
miR-146a	Downregulation in HFrEF and HFpEF Upregulation in T2DM	[57] [45]
miR-150	Downregulation in UA Downregulation associated with impaired LV contractility six months post-AMI Downregulation in AF	[46, 50, 64]
miR-155	Downregulation in CAD	[48]
miR-181a	Downregulation in CAD	[47]
miR-186	Upregulation in ACS	[64]
miR-192	Upregulation associated with HF within one year post-MI	[59]
miR-194	Upregulation associated with HF within one year post-MI	[59]
miR-195	Upregulation in AMI Upregulation in hypertension	[53, 55]
miR-197	Downregulation associated with incident MI Upregulation in metabolic syndrome	[63] [53]

Table 7.1 (Continued)

c-miRNAs	Expression outcomes	References
miR-208a	Upregulation in AMI Upregulation in CAD	[48, 71]
miR-208b	Upregulation in AMI Upregulation predicts LV remodeling Upregulation associated with HF or mortality at six months post-MI	[58, 66, 72–74]
miR-210	Upregulation in aortic stenosis	[75]
miR-221	Downregulation in HFrEF and HFpEF Upregulation in AMI Upregulation associated with decreased LVEF	[57] [76]
miR-223	Downregulation associated with incident MI	[63]
miR-296	Upregulation in essential hypertension	[77]
miR-320a	Upregulation in AMI Upregulation in HF	[44–52, 57, 58, 65, 66, 73, 74, 78–92]
miR-320b	Downregulation in AMI	[61]
miR-323	Upregulation in ACS	[80]
miR-328	Downregulation in HFrEF and HFpEF Upregulation in AMI Upregulation associated with increased risk of HF or mortality six months post-MI Downregulation in AF	[57, 69, 93]
miR-337	Upregulation in CAD	[94]
miR-342	Downregulation in HF vs. non-HF dyspnea or control	[56]
miR-370	Upregulation in hyperlipidemia Upregulation associated with severity of CAD	[60]
miR-375	Downregulation in HFrEF and HFpEF Upregulation in T2DM	[57] [45]
miR-423	Upregulation in HF	[52, 78]
miR-433	Upregulation in CAD	[94]
miR-483	Upregulation in response to LVAD placement	[81]
miR-485	Upregulation in CAD	[94]
miR-499	Upregulation in AMI Upregulation in ACS Upregulation associated with reduced systolic function after MI and increased risk of HF or mortality	[43, 44, 67, 72–74]
miR-509	Upregulation in metabolic syndrome	[53]

Table 7.1 (Continued)

c-miRNAs	Expression outcomes	References
miR-571	Upregulation in T2DM	[95]
miR-652	Upregulation associated with readmission for HF	[80]
miR-661	Upregulation in T2DM	[95]
miR-770	Upregulation in T2DM	[95]
miR-892	Upregulation in T2DM	[95]
miR-1202	Upregulation at baseline predicts poor response to LVAD placement	[81]
miR-1303	Upregulation in T2DM	[95]
let-7b	Downregulation in AMI	[55]
let-7e	Upregulation in essential hypertension	[77]
hcmv-miR-UL112	Upregulation in essential hypertension	[77]

ACS, acute coronary syndrome; AF, atrial fibrillation; AMI, acute myocardial infarction; CAD, coronary artery disease; CHF, congestive heart failure; HFrEF, heart failure with reduced ejection fraction; HFpEF, heart failure with preserved ejection fraction; LV, left ventricle; LVAD, left ventricular assist device; LVEF, left ventricular ejection fraction; MACE, major adverse cardiovascular event; T2DM, type II diabetes mellitus; and UA, unstable angina.

failure [58]. Similarly, Devaux et al. [46] demonstrated that a panel of four miRNAs – miR-16, miR-27a, miR-101, and miR-150 – predicted left ventricular contractility following AMI. Pilbrow et al. [80] discovered two novel c-miRNAs that could aid in diagnosis of ACS. In particular, levels of miR-652 in the lowest tertile were significantly associated with readmission for heart failure and, in combination with NT-proBNP and LVEF, miR-652 significantly improved risk stratification. Lastly, for patients with advanced stage heart failure, miR-1202 and miR-483-3p have been proposed for use in monitoring and predicting patient response to the placement of a left ventricular assist device (LVAD) over time [81]. While each of these studies provided new information on c-miRNA levels in ACS and CHF patient populations, there are still discrepancies regarding validation across multiple cohorts, which c-miRNA or c-miRNA panel combinations definitively identify or predict a given disease, and whether these c-miRNAs improve discriminatory power with and without polypeptide biomarkers. For example, major adverse cardiovascular events (MACEs) correlated with elevated miR-133 in patients who had previously had a STEMI; however, miR-133 could not independently predict MACE in this population [65]. Similarly, in two separate studies, miR-133, miR-208, and miR-499-5p were strongly associated with increased risk of mortality or heart failure in patients with ACS but lost their potential for prognostic indication when adjusting for troponin [66, 74].

Utility of c-miRNAs as Biomarkers in Pulmonary Vascular Disease

Pulmonary vascular disease encompasses a group of conditions localized to the vasculature of the lung including pulmonary hypertension, chronic thromboembolic disease, and acute pulmonary embolism. Appropriate diagnosis is hampered by the need for invasive right heart catheterization and is often only possible at later stages of disease. Thus, there is a great need for noninvasive methods for diagnosis, especially in the early stages of disease.

Despite the discovery of multiple c-miRNA targets in the context of pulmonary hypertension, none have been used in a clinical setting. For example, plasma samples from patients with PAH revealed a significant reduction of circulating miR-150, and this decreased expression was associated with poor survival [82]. Other c-miRNA changes in the circulation have been directly correlated with PAH severity; more specifically, miR-451 and miR-1246 were decreased and miR-23b, miR-130/301 family members, miR-210, and miR-191 were increased in the plasma of PAH patients [83–85]. Similarly, the circulating levels of miR-26a in PAH patients correlated with six-minute walk distance, a clinical indicator for disease severity [86]. Chronic thromboembolic pulmonary hypertension (CTEPH) is a form of pulmonary hypertension caused by thromboemboli in the pulmonary arteries. Guo et al. [87] demonstrated decreased levels of let-7b in the circulation of patients with CTEPH and suggested the contribution of this miRNA to disease development by characterizing its target genes, namely endothelin-1 and transforming growth factor-β1. Separate from pulmonary hypertension, pulmonary embolism presents an acute risk for patients, requiring both fast and accurate diagnosis. Xiao et al. [88] identified miR-134 was upregulated in the serum of patients with acute pulmonary embolism. Separately, Kessler et al. [89] demonstrated that miR-1233 levels in the circulation could discriminate between patients with pulmonary embolism, NSTEMI, and healthy controls. Both proposed biomarkers require validation in larger cohort studies and against the discriminatory power of current clinical indicators. These biomarkers are summarized in Table 7.2.

Challenges of Using c-miRNAs as Biomarkers

Many challenges remain in developing c-miRNAs as clinical diagnostic and prognostic indicators in cardiovascular and pulmonary vascular disease (Figure 7.4).

First, as highlighted above, there are several technical challenges to miRNA quantification that require standardization of each step of the process,

Table 7.2 Circulating miRNAs in pulmonary vascular disease.

Pulmonary hypertension	Expression outcomes	References
miR-1	Upregulation in AMI Upregulation in ACS	[43, 44]
miR-17	Downregulation in PAH	[96]
miR-23b	Upregulation in PH	[83]
miR-26a	Downregulation in PAH	[86]
miR-130/301	Upregulation in PH	[84]
miR-150	Downregulation in PAH Predicts survival in PAH	[82, 83]
miR-191	Upregulation in PH	[83]
miR-210	Upregulation in PAH	[85]
miR-424	Upregulation in PAH	[96]
miR-451	Downregulation in PH	[83]
miR-1246	Downregulation in PH	[83]
let-7b	Downregulation in CTEPH	[87]
Pulmonary embolism		
miR-134	Upregulation in acute PE	[88]
miR-1233	Upregulation in acute PE vs. NSTEMI	[89]

PAH, pulmonary arterial hypertension; CTEPH, chronic thromboembolic pulmonary hypertension; PH, pulmonary hypertension; PE, pulmonary embolism; and NSTEMI, non-ST elevation myocardial infarction.

including patient sampling [90], RNA extraction, and choice of detection method. The development of a gold standard technique that is more rapid, sensitive, and selective is critically important for clinical application of miRNA as biomarkers. Additionally, there is currently no recognized control to normalize c-miRNA levels and determine their relative abundance, thus hindering reproducibility and comparison of results among separate sites of assay performance.

In addition, there is wide interindividual variation in the c-miRNA profiles and even variability in the same person depending on physiologic context such as exercise, time of day, and diet. As such, established threshold or normal biologic ranges do not yet exist for c-miRNA expression profiles. Next, disease-specific c-miRNAs are often found at relatively low expression levels, making these circulating species difficult to detect and making their biological functions difficult to measure. Lastly, the majority of c-miRNAs are ubiquitously expressed across tissue compartments, thus making the source of c-miRNAs and their role in disease more challenging to define [4].

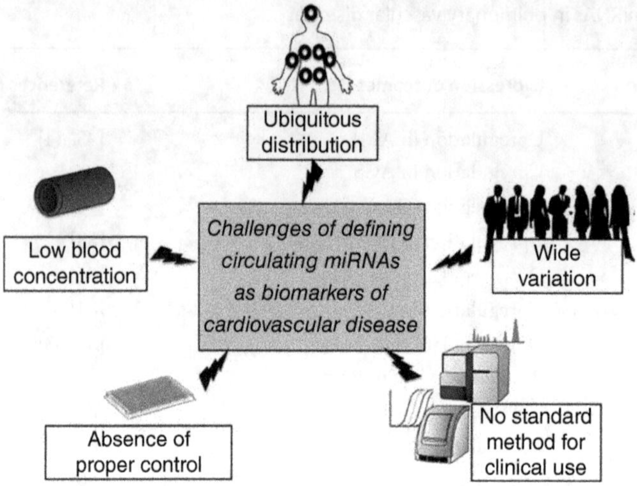

Figure 7.4 Challenges in defining c-miRNA biomarkers for human disease. First, disease-specific extracellular miRNAs are often found at relatively low expression levels. Additionally, the majority of c-miRNAs are ubiquitous, meaning they could be expressed and released by any tissue. This makes the source of pertinent c-miRNA difficult to pinpoint. Third, there is a wide variation in the c-miRNA expression profiling from person to person and even for the same person over time. With no established threshold or biologic ranges for c-miRNA expression in diseased vs. healthy controls, the utility of c-miRNAs rests upon individual patient measurements over time. Lastly, there are technical inconsistencies in sampling, extraction, and detection that have not been standardized for clinical application, and there is no standardized endogenous control for normalizing c-miRNA abundance over time and across patient populations. Taken together, the new field of c-miRNA biomarker discovery still has several limitations, both technical and inherent, to translating this technology to a clinical setting. Source: Min and Chan 2015 [92]. Adapted with permission of John Wiley & Sons.

Conclusions and Future Perspectives

Despite these challenges, the discovery of extracellular miRNAs has changed the field of RNA biology. The relative stability, noninvasive sampling method, and the sensitivity and specificity all serve as compelling arguments for the use of c-miRNAs as clinical biomarkers. However, to realize the use of these molecules as true clinical biomarkers, we believe that the completion of three major endeavors is necessary. First, validation of already proposed c-miRNAs will be essential in much larger cohorts and across a more prolonged longitudinal timeline. Similarly, a cost–benefit analysis should be performed in order to establish whether the clinical quantitation of c-miRNAs is capable of producing a sufficient amount of new information to compensate the costs of utilized methods [91]. Inherent in these validation processes, the technical challenges of miRNA quantitation will need to be addressed via the establishment of standardized protocols that ensure sensitivity, specificity, and reproducibility across patient populations and institutions.

Second, we must refocus our efforts on c-miRNA biomarker discovery to identify miRNAs that could either provide more sensitive detection of disease at its earlier stages or improve the specificity of already existing diagnostic indicators, c-miRNA, or otherwise. Thus far, several groups have set a strong foundation by pinpointing specific c-miRNAs panels for the improved diagnosis of cardiovascular and pulmonary vascular disease. However, c-miRNA recognition could be more efficient. In this way, predictive bioinformatics and the application of network medicine can move the field from fishing for a single, distinct miRNA molecule to a systems-based approach where panels of miRNAs are predicted and then validated in a streamlined manner [92].

Finally, a better framework for understanding the functional role of c-miRNAs is imperative, whether they act as endocrine or paracrine messengers [10] or in another capacity. Until then, our lack of understanding of their biology will limit how the endogenous roles of these miRNAs may be leveraged not only for diagnosis and prognosis but also for therapy.

Acknowledgments

This work was supported by the Angiopathy Training Program (M.K.C.), NIH grants R01 HL124021, HL 122596, HL 138437, and UH2 TR002073 as well as the AHA grant 18EIA33900027 (S.Y.C.).

Disclosures

S.Y.C. has served as a consultant for Zogenix, Vivus, Aerpio, and United Therapeutics; S.Y.C. is a director, officer, and shareholder in Numa Therapeutics; S.Y.C. holds research grants from Actelion and Pfizer. S.Y.C. has filed patent applications regarding the targeting of microRNAs and metabolism in pulmonary hypertension.

References

1 Romaine, S.P.R., Tomaszewski, M., Condorelli, G., et al. (2015). MicroRNAs in cardiovascular disease: an introduction for clinicians. *Heart* .

2 Ha, M. and Kim, V.N. (2014). Regulation of microRNA biogenesis. *Nat. Rev. Mol. Cell Biol.* 15 (8): 509–524.

3 Esteller, M. (2011). Non-coding RNAs in human disease. *Nat. Rev. Genet.* 12 (12): 861–874.

4 Small, E.M. and Olson, E.N. (2011). Pervasive roles of microRNAs in cardiovascular biology. *Nature* 469 (7330): 336–342.

5 Abdellatif, M. (2012). Differential expression of microRNAs in different disease states. *Circ. Res.* 110 (4): 638–650.

6 White, K., Loscalzo, J., and Chan, S.Y. (2012). Holding our breath: the emerging and anticipated roles of microRNA in pulmonary hypertension. *Pulm. Circ.* 2 (3): 278–290.

7 Mitchell, P.S., Parkin, R.K., Kroh, E.M., et al. (2008). Circulating microRNAs as stable blood-based markers for cancer detection. *Proc. Natl. Acad. Sci. USA* 105 (30): 10513–10518.

8 Chen, X., Ba, Y., Ma, L., et al. (2008). Characterization of microRNAs in serum: a novel class of biomarkers for diagnosis of cancer and other diseases. *Cell Res.* 18 (10): 997–1006.

9 Weber, J.A., Baxter, D.H., Zhang, S., et al. (2010). The microRNA spectrum in 12 body fluids. *Clin. Chem.* 56 (11): 1733–1741.

10 Raposo, G. and Stoorvogel, W. (2013). Extracellular vesicles: exosomes, microvesicles, and friends. *J. Cell Biol.* 200 (4): 373–383.

11 Valadi, H., Ekstrom, K., Bossios, A., et al. (2007). Exosome-mediated transfer of mRNAs and microRNAs is a novel mechanism of genetic exchange between cells. *Nat. Cell Biol.* 9 (6): 654–659.

12 Hunter, M.P., Ismail, N., Zhang, X., et al. (2008). Detection of microRNA expression in human peripheral blood microvesicles. *PLoS One* 3 (11): e3694.

13 Zernecke, A., Bidzhekov, K., Noels, H., et al. (2009). Delivery of microRNA-126 by apoptotic bodies induces CXCL12-dependent vascular protection. *Sci. Signal.* 2 (100): ra81.

14 Wang, K., Zhang, S., Weber, J., et al. (2010). Export of microRNAs and microRNA-protective protein by mammalian cells. *Nucleic Acids Res.* 38 (20): 7248–7259.

15 Arroyo, J.D., Chevillet, J.R., Kroh, E.M., et al. (2011). Argonaute2 complexes carry a population of circulating microRNAs independent of vesicles in human plasma. *Proc. Natl. Acad. Sci. USA* 108 (12): 5003–5008.

16 Turchinovich, A., Weiz, L., Langheinz, A., et al. (2011). Characterization of extracellular circulating microRNA. *Nucleic Acids Res.* 39 (16): 7223–7233.

17 Vickers, K.C., Palmisano, B.T., Shoucri, B.M., et al. (2011). MicroRNAs are transported in plasma and delivered to recipient cells by high-density lipoproteins. *Nat. Cell Biol.* 13 (4): 423–433.

18 Creemers, E.E., Tijsen, A.J., and Pinto, Y.M. (2012). Circulating microRNAs: novel biomarkers and extracellular communicators in cardiovascular disease? *Circ. Res.* 110 (3): 483–495.

19 Roden, C., Mastriano, S., Wang, N., et al. (2015). microRNA expression profiling: technologies, insights, and prospects. *Adv. Exp. Med. Biol.* 888: 409–421.

20 Pritchard, C.C., Cheng, H.H., and Tewari, M. (2012). MicroRNA profiling: approaches and considerations. *Nat. Rev. Genet.* 13 (5): 358–369.

21 Ebhardt, H.A., Tsang, H.H., Dai, D.C., et al. (2009). Meta-analysis of small RNA-sequencing errors reveals ubiquitous post-transcriptional RNA modifications. *Nucleic Acids Res.* 37 (8): 2461–2470.

22 Borel, C. and Antonarakis, S.E. (2008). Functional genetic variation of human miRNAs and phenotypic consequences. *Mamm Genome* 19 (7-8): 503–509.

23 Parikh, V.N. and Chan, S.Y. (2013). Analysis of microRNA niches: techniques to measure extracellular microRNA and intracellular microRNA in situ. *Methods Mol. Biol.* 1024: 157–172.

24 Pritchard, C.C., Kroh, E., Wood, B., et al. (2012). Blood cell origin of circulating microRNAs: a cautionary note for cancer biomarker studies. *Cancer Prev. Res. (Phila)* 5 (3): 492–497.

25 Cheng, H.H., Yi, H.S., Kim, Y., et al. (2013). Plasma processing conditions substantially influence circulating microRNA biomarker levels. *PLoS One* 8 (6): e64795.

26 Condorelli, G., Latronico, M.V., and Cavarretta, E. (2014). microRNAs in cardiovascular diseases: current knowledge and the road ahead. *J. Am. Coll. Cardiol.* 63 (21): 2177–2187.

27 McAlexander, M.A., Phillips, M.J., and Witwer, K.W. (2013). Comparison of methods for miRNA extraction from plasma and quantitative recovery of RNA from cerebrospinal fluid. *Front Genet.* 4: 83.

28 Campo Dell'Orto, M., Zangrando, A., Trentin, L., et al. (2007). New data on robustness of gene expression signatures in leukemia: comparison of three distinct total RNA preparation procedures. *BMC Genomics* 8: 188.

29 Debey, S., Schoenbeck, U., Hellmich, M., et al. (2004). Comparison of different isolation techniques prior gene expression profiling of blood derived cells: impact on physiological responses, on overall expression and the role of different cell types. *Pharmacogenomics J.* 4 (3): 193–207.

30 Pimentel, F., Bonilla, P., Ravishankar, Y.G., et al. (2015). Technology in microRNA profiling: circulating microRNAs as noninvasive cancer biomarkers in breast cancer. *J. Lab. Autom.* 20 (5): 574–588.

31 Niu, Y., Zhang, L., Qiu, H., et al. (2015). An improved method for detecting circulating microRNAs with S-poly(T) Plus real-time PCR. *Sci. Rep.* 5: 15100.

32 Hindson, C.M., Chevillet, J.R., Briggs, H.A., et al. (2013). Absolute quantification by droplet digital PCR versus analog real-time PCR. *Nat. Methods* 10 (10): 1003–1005.

33 Shalaby, T., Fiaschetti, G., Baumgartner, M., et al. (2014). MicroRNA signatures as biomarkers and therapeutic target for CNS embryonal tumors: the pros and the cons. *Int. J. Mol. Sci.* 15 (11): 21554–21586.

34 Metzker, M.L. (2010). Sequencing technologies – the next generation. *Nat. Rev. Genet.* 11 (1): 31–46.

35 Qin, L.X., Tuschl, T., and Singer, S. (2016). Empirical insights into the stochasticity of small RNA sequencing. *Sci. Rep.* 6: 24061.

36 Fuchs, R.T., Sun, Z., Zhuang, F., et al. (2015). Bias in ligation-based small RNA sequencing library construction is determined by adaptor and RNA structure. *PLoS One* 10 (5): e0126049.

37 Roberts, B.S., Hardigan, A.A., Kirby, M.K., et al. (2015). Blocking of targeted microRNAs from next-generation sequencing libraries. *Nucleic Acids Res.* 43 (21): e145.

38 Wilson, K.D., Shen, P., Fung, E., et al. (2015). A rapid, high-quality, cost-effective, comprehensive and expandable targeted next-generation sequencing assay for inherited heart diseases. *Circ. Res.* 117 (7): 603–611.

39 Gu, L.Q., Wanunu, M., Wang, M.X., et al. (2012). Detection of miRNAs with a nanopore single-molecule counter. *Expert Rev. Mol. Diagn.* 12 (6): 573–584.

40 Wanunu, M., Dadosh, T., Ray, V., et al. (2010). Rapid electronic detection of probe-specific microRNAs using thin nanopore sensors. *Nat. Nanotechnol.* 5 (11): 807–814.

41 Wang, Y., Zheng, D., Tan, Q., et al. (2011). Nanopore-based detection of circulating microRNAs in lung cancer patients. *Nat. Nanotechnol.* 6 (10): 668–674.

42 Schmidt, J. (2016). Membrane platforms for biological nanopore sensing and sequencing. *Curr. Opin. Biotechnol.* 39: 17–27.

43 D'Alessandra, Y., Devanna, P., Limana, F., et al. (2010). Circulating microRNAs are new and sensitive biomarkers of myocardial infarction. *Eur. Heart J.* 31 (22): 2765–2773.

44 Oerlemans, M.I., Mosterd, A., Dekker, M.S., et al. (2012). Early assessment of acute coronary syndromes in the emergency department: the potential diagnostic value of circulating microRNAs. *EMBO Mol. Med.* 4 (11): 1176–1185.

45 Kong, L., Zhu, J., Han, W., et al. (2011). Significance of serum microRNAs in pre-diabetes and newly diagnosed type 2 diabetes: a clinical study. *Acta Diabetol.* 48 (1): 61–69.

46 Devaux, Y., Vausort, M., McCann, G.P., et al. (2013). A panel of 4 microRNAs facilitates the prediction of left ventricular contractility after acute myocardial infarction. *PLoS One* 8 (8): e70644.

47 Satoh, M., Takahashi, Y., Tabuchi, T., et al. (2015). Circulating Toll-like receptor 4-responsive microRNA panel in patients with coronary artery disease: results from prospective and randomized study of treatment with renin-angiotensin system blockade. *Clin. Sci. (London)* 128 (8): 483–491.

48 Fichtlscherer, S., De Rosa, S., Fox, H., et al. (2010). Circulating microRNAs in patients with coronary artery disease. *Circ. Res.* 107 (5): 677–684.

49 Zhong, J., He, Y., Chen, W., et al. (2014). Circulating microRNA-19a as a potential novel biomarker for diagnosis of acute myocardial infarction. *Int. J. Mol. Sci.* 15 (11): 20355–20364.
50 McManus, D.D., Tanriverdi, K., Lin, H., et al. (2015). Plasma microRNAs are associated with atrial fibrillation and change after catheter ablation (the miRhythm study). *Heart Rhythm* 12 (1): 3–10.
51 Luo, X., Yang, B., and Nattel, S. (2015). MicroRNAs and atrial fibrillation: mechanisms and translational potential. *Nat. Rev. Cardiol.* 12 (2): 80–90.
52 Goren, Y., Kushnir, M., Zafrir, B., et al. (2012). Serum levels of microRNAs in patients with heart failure. *Eur. J. Heart Fail* 14 (2): 147–154.
53 Karolina, D.S., Tavintharan, S., Armugam, A., et al. (2012). Circulating miRNA profiles in patients with metabolic syndrome. *J. Clin. Endocrinol. Metab.* 97 (12): E2271–E2276.
54 Dawson, K., Wakili, R., Ördög, B., et al. (2013). MicroRNA29: A mechanistic contributor and potential biomarker in atrial fibrillation. *Circulation* 127 (14): 1466–1475, 1475e1–1475e28.
55 Long, G., Wang, F., Duan, Q., et al. (2012). Circulating miR-30a, miR-195 and let-7b associated with acute myocardial infarction. *PLoS One* 7 (12): e50926.
56 Ellis, K.L., Cameron, V.A., Troughton, R.W., et al. (2013). Circulating microRNAs as candidate markers to distinguish heart failure in breathless patients. *Eur. J. Heart Fail* 15 (10): 1138–1147.
57 Watson, C.J., Gupta, S.K., O'Connell, E., et al. (2015). MicroRNA signatures differentiate preserved from reduced ejection fraction heart failure. *Eur. J. Heart Fail* 17 (4): 405–415.
58 Lv, P., Zhou, M., He, J., et al. (2014). Circulating miR-208b and miR-34a are associated with left ventricular remodeling after acute myocardial infarction. *Int. J. Mol. Sci.* 15 (4): 5774–5788.
59 Matsumoto, S., Sakata, Y., Suna, S., et al. (2013). Circulating p53-responsive microRNAs are predictive indicators of heart failure after acute myocardial infarction. *Circ. Res.* 113 (3): 322–326.
60 Gao, W., He, H.W., Wang, Z.M., et al. (2012). Plasma levels of lipometabolism-related miR-122 and miR-370 are increased in patients with hyperlipidemia and associated with coronary artery disease. *Lipids Health Dis.* 11: 55.
61 Huang, S., Chen, M., Li, L., et al. (2014). Circulating MicroRNAs and the occurrence of acute myocardial infarction in Chinese populations. *Circ. Cardiovasc. Genet.* 7 (2): 189–198.
62 Zhang, T., Lv, C., Li, L., et al. (2013). Plasma miR-126 is a potential biomarker for early prediction of type 2 diabetes mellitus in susceptible individuals. *Biomed. Res. Int.* 2013: 761617.

63 Zampetaki, A., Willeit, P., Tilling, L., et al. (2012). Prospective study on circulating microRNAs and risk of myocardial infarction. *J. Am. Coll. Cardiol.* 60 (4): 290–299.

64 Zeller, T., Keller, T., Ojeda, F., et al. (2014). Assessment of microRNAs in patients with unstable angina pectoris. *Eur. Heart J.* 35 (31): 2106–2114.

65 Eitel, I., Adams, V., Dieterich, P., et al. (2012). Relation of circulating MicroRNA-133a concentrations with myocardial damage and clinical prognosis in ST-elevation myocardial infarction. *Am. Heart J.* 164 (5): 706–714.

66 Widera, C., Gupta, S.K., Lorenzen, J.M., et al. (2011). Diagnostic and prognostic impact of six circulating microRNAs in acute coronary syndrome. *J. Mol. Cell. Cardiol.* 51 (5): 872–875.

67 De Rosa, S., Fichtlscherer, S., Lehmann, R., et al. (2011). Transcoronary concentration gradients of circulating microRNAs. *Circulation* 124 (18): 1936–1944.

68 Wang, F., Long, G., Zhao, C., et al. (2013). Plasma microRNA-133a is a new marker for both acute myocardial infarction and underlying coronary artery stenosis. *J. Transl. Med.* 11: 222.

69 He, F., Lv, P., Zhao, X., et al. (2014). Predictive value of circulating miR-328 and miR-134 for acute myocardial infarction. *Mol. Cell. Biochem.* 394 (1-2): 137–144.

70 Karolina, D.S., Armugam, A., Tavintharan, S., et al. (2011). MicroRNA 144 impairs insulin signaling by inhibiting the expression of insulin receptor substrate 1 in type 2 diabetes mellitus. *PLoS One* 6 (8): e22839.

71 Wang, G.K., Zhu, J.Q., Zhang, J.T., et al. (2010). Circulating microRNA: a novel potential biomarker for early diagnosis of acute myocardial infarction in humans. *Eur. Heart J.* 31 (6): 659–666.

72 Devaux, Y., Vausort, M., Goretti, E., et al. (2012). Use of circulating microRNAs to diagnose acute myocardial infarction. *Clin. Chem.* 58 (3): 559–567.

73 Devaux, Y., Mueller, M., Haaf, P., et al. (2015). Diagnostic and prognostic value of circulating microRNAs in patients with acute chest pain. *J. Intern. Med.* 277 (2): 260–271.

74 Gidlof, O., Smith, J.G., Miyazu, K., et al. (2013). Circulating cardio-enriched microRNAs are associated with long-term prognosis following myocardial infarction. *BMC Cardiovasc. Disord.* 13: 12.

75 Røsjø, H., Dahl, M.B., Bye, A., et al. (2014). Prognostic value of circulating microRNA-210 levels in patients with moderate to severe aortic stenosis. *PLoS One* 9 (3): e91812.

76 Coskunpinar, E., Cakmak, H.A., Kalkan A.K., et al. (2016). Circulating miR-221-3p as a novel marker for early prediction of acute myocardial infarction. *Gene* 591 (1): 90–96.

77 Li, S., Zhu, J., Zhang, W., et al. (2011). Signature microRNA expression profile of essential hypertension and its novel link to human cytomegalovirus infection. *Circulation* 124 (2): 175–184.

78 Tijsen, A.J., Creemers, E.E., Moerland, P.D., et al. (2010). MiR423-5p as a circulating biomarker for heart failure. *Circ. Res.* 106 (6): 1035–1039.

79 Wong, L.L., Armugam, A., Sepramaniam, S., et al. (2015). Circulating microRNAs in heart failure with reduced and preserved left ventricular ejection fraction. *Eur. J. Heart Fail* 17 (4): 393–404.

80 Pilbrow, A.P., Cordeddu, L., Cameron, V.A., et al. (2014). Circulating miR-323-3p and miR-652: candidate markers for the presence and progression of acute coronary syndromes. *Int. J. Cardiol.* 176 (2): 375–385.

81 Morley-Smith, A.C., Mills, A., Jacobs, S., et al. (2014). Circulating microRNAs for predicting and monitoring response to mechanical circulatory support from a left ventricular assist device. *Eur. J. Heart Fail* 16 (8): 871–879.

82 Rhodes, C.J., Wharton, J., Boon, R.A., et al. (2013). Reduced microRNA-150 is associated with poor survival in pulmonary arterial hypertension. *Am. J. Respir. Crit. Care Med.* 187 (3): 294–302.

83 Wei, C., Henderson, H., Spradley, C., et al. (2013). Circulating miRNAs as potential marker for pulmonary hypertension. *PLoS One* 8 (5): e64396.

84 Bertero, T., Lu, Y., Annis, S., et al. (2014). Systems-level regulation of microRNA networks by miR-130/301 promotes pulmonary hypertension. *J. Clin. Invest.* 124 (8): 3514–3528.

85 White, K., Lu, Y., Annis, S., et al. (2015). Genetic and hypoxic alterations of the microRNA-210-ISCU1/2 axis promote iron-sulfur deficiency and pulmonary hypertension. *EMBO Mol. Med.* 7 (6): 695–713.

86 Schlosser, K., White, R.J., and Stewart, D.J. (2013). miR-26a linked to pulmonary hypertension by global assessment of circulating extracellular microRNAs. *Am. J. Respir. Crit. Care Med.* 188 (12): 1472–1475.

87 Guo, L., Yang, Y., Liu, J., et al. (2014). Differentially expressed plasma microRNAs and the potential regulatory function of Let-7b in chronic thromboembolic pulmonary hypertension. *PLoS One* 9 (6): e101055.

88 Xiao, J., Jing, Z.C., Ellinor, P.T., et al. (2011). MicroRNA-134 as a potential plasma biomarker for the diagnosis of acute pulmonary embolism. *J. Transl. Med.* 9: 159.

89 Kessler, T., Erdmann, J., Vilne, B., et al. (2016). Serum microRNA-1233 is a specific biomarker for diagnosing acute pulmonary embolism. *J. Transl. Med.* 14 (1): 120.

90 Farina, N.H., Wood, M.E., Perrapato, S.D., et al. (2014). Standardizing analysis of circulating microRNA: clinical and biological relevance. *J. Cell. Biochem.* 115 (5): 805–811.

91 Bienertova-Vasku, J., Novak, J., and Vasku, A. (2015). MicroRNAs in pulmonary arterial hypertension: pathogenesis, diagnosis and treatment. *J. Am. Soc. Hypertens.* 9 (3): 221–234.

92 Min, P.K. and Chan, S.Y. (2015). The biology of circulating microRNAs in cardiovascular disease. *Eur. J. Clin. Invest.* 45 (8): 860–874.

93 McManus, D.D., Lin, H., Tanriverdi, K., et al. (2014). Relations between circulating microRNAs and atrial fibrillation: data from the Framingham Offspring Study. *Heart Rhythm* 11 (4): 663–669.

94 D'Alessandra, Y., Carena, M.C., Spazzafumo, L., et al. (2013). Diagnostic potential of plasmatic microRNA signatures in stable and unstable angina. *PLoS One* 8 (11): e80345.

95 Wang, C., Wan, S., Yang, T., et al. (2016). Increased serum microRNAs are closely associated with the presence of microvascular complications in type 2 diabetes mellitus. *Sci. Rep.* 6: p. 20032.

96 Schlosser, K., White, R.J., and Stewart, D.J. (2015). Discordant regulation of microRNA between multiple experimental models and human pulmonary hypertension. *Chest* 148 (2): 481–490.

Part III

Characterization, Validation, and Utilization

8

Characterization and Validation of Biomarkers in Drug Development: Regulatory Perspective

Federico Goodsaid

Regulatory Pathfinders, San Juan, PR, USA

Introduction

Biomarkers can be considered tools to be qualified in close association with individual submissions for drug approvals. The US Food and Drug Administration (FDA) Drug-Diagnostic Companion Diagnostic Guidance [1] described a specific example of this association, where the approval of a drug and a test are closely linked with each other in both their product concepts and the time lines of their regulatory approvals. Several examples of biomarkers approved through this co-development process are shown in the FDA's Table of Pharmacogenomic Biomarkers in Drug Labeling [2]. Biomarkers in this table include both genetic and translational entries. Genetic biomarkers are often integrated in drug development as clinical or nonclinical markers of drug efficacy or safety for the purpose of patient selection in clinical trials, response prediction through stratification or enrichment, or dose optimization. Translational biomarkers have applications similar to those of genetic biomarkers but may also be useful for response monitoring and as early indicators of toxicity or adverse reactions.

Novel approaches are being tested continuously for the successful integration of biomarkers in drug development. Their applications range from early compound selection through post-marketing applications. However, integration of these novel biomarkers into routine nonclinical and clinical practice and regulatory submissions have often been slow. Hesitation in the application of these tests is often associated with fear not only about how comprehensive data supporting these applications are but also about the regulatory interpretation of the context of use for these applications [3]. Biomarker tests can be integrated into drug development when we have a consensus about the context in which we are measuring the biomarker and the evidence supporting this

Biomarkers in Drug Discovery and Development: A Handbook of Practice, Application, and Strategy, Second Edition. Edited by Ramin Rahbari, Jonathan Van Niewaal, and Michael R. Bleavins.

measurement. These levels of consensus need to be reflected in the regulatory review of biomarker data.

Regulatory Paths in Biomarker Evaluation and Qualification

The path from an exploratory biomarker to a biomarker qualified for a specific application context can be long and unpredictable [4]. Application of these biomarkers requires an objective record for their nonclinical or clinical context and supporting qualification evidence. Information from the development of exploratory biomarkers has been shared between the pharmaceutical industry and the FDA through voluntary exploratory data submissions (VXDSs) [5]. Submissions of exploratory biomarker data have allowed reviewers at the FDA to share with scientists in the pharmaceutical industry study designs, sample isolation and storage protocols, technology platforms, analysis algorithms, biological pathway interpretation, and electronic data submission formats. This experience has been valuable in training FDA reviewers for the analysis and interpretation of biomarker data.

VXDS stressed the need for a regulatory path from exploratory biomarkers to biomarkers qualified for a specific context. Such a path was tested at the FDA through a pilot process for biomarker qualification [6]. This process was focused on the specific needs of the regulatory environment to ensure scientifically accurate and clinically (or preclinically) useful decision-making. In the first use of this new joint-agency review process put in place by US and European drug regulators, the FDA and the European Medicines Association (EMA) [7] allow drug companies to submit the results of seven new tests as evidence of nephrotoxicity by new drugs. The qualification of these biomarkers covers voluntary submission of these data for rat studies. On a case-by-case basis, the FDA will also consider possible application of these biomarkers in phase I human trials. The tests measure levels of seven key proteins or biomarkers that scientists from the FDA and EMA believe provide important new safety information about the effect of drugs on the kidney. When reviewing investigational new drug (IND) applications, new drug applications (NDAs), or Biologics License Applications (BLAs), both regulatory agencies consider the test results in addition to blood urea nitrogen (BUN) and creatinine.

The development of the new renal toxicity biomarkers was led by the Predictive Safety Testing Consortium (PSTC) [8], whose members include scientists from over a dozen pharmaceutical companies. The PSTC was organized and led by the Critical Path Institute [9]. Researchers from Merck and Novartis identified the nonclinical nephrotoxicity biomarkers, tested them to prove their sensitivity, specificity, and positive and negative predictive value, and then shared their findings with the other consortium members for further

study. The consortium then submitted applications for their qualification to FDA and EMA.

This was a unique example of how a group of drug companies can work together to propose and generate qualification data for new safety tests and then present them jointly to the FDA and EMA for qualification. The FDA and EMA laid the groundwork for such joint-agency reviews in 2004 with the development of the VXDS framework. The VXDS review served as the baseline model around which to design the pilot process for biomarker qualification in 2006. A similar biomarker qualification data submission (BQDS) meeting was held in this pilot process to allow an exchange of questions with the sponsor about scientific and clinical information submitted for qualification.

The pilot process for biomarker qualification allowed the PSTC to submit a single application for biomarker qualification to both regulatory agencies, and then to meet jointly with scientists from both agencies to discuss it in detail and address additional scientific questions posed by the regulators. Each regulatory agency reviewed the application separately and made independent decisions on whether each would allow the new biomarkers to be used. The new biomarkers qualified by FDA and EMA were kidney injury molecule 1 (KIM-1), albumin, total protein, β_2-microglobulin, cystatin C, clusterin, and trefoil factor-3. Testing for these proteins will help scientists assess whether a drug is likely to cause damage to the kidneys, a toxic side effect of some drugs. Both FDA and EMA require drug companies to submit the results of two other tests, BUN and serum creatinine, to show whether such kidney damage has occurred.

The seven new tests have provided important advantages over these two tests. For example, in the rat model, once kidney damage has begun to occur, it takes a week before the two current tests (BUN and creatinine) can detect it [10]. The new tests are more sensitive and can reveal cellular damage within hours [11]. BUN and serum creatinine show that damage has occurred somewhere in the kidneys, but the new tests can also pinpoint which part(s) of the kidney have been affected [12].

Additional studies have been ongoing over the past seven years to provide clinical data for a qualification that may allow promising drugs to advance into clinical trials that otherwise would have been abandoned, because currently there are no tests available to detect early-onset renal injury. The new tests were developed and will be carried out initially in rats, but they were selected because other studies have shown that similar biomarkers are produced in human kidney cells [13]. If these studies are successful, the PSTC will present a new application seeking acceptance of the human biomarkers over the next two years.

The need for an accurate, comprehensive, and efficient process for biomarker qualification is closely linked with our ability to quickly integrate new biomarkers in drug development and regulatory review. The biomarker qualification pilot process at the FDA tested the scientific, clinical, and regulatory components for a biomarker qualification process. Experience gained with the pilot

process was useful in the development of a formal regulatory process [14] for biomarker qualification.

Evidentiary Recommendations

Over the first decade of biomarker qualification at the FDA, three nonclinical and three clinical biomarker qualifications have been completed [15]. Of these, all the nonclinical qualification requests were submitted during the pilot phase of this program in 2007–2008. All the clinical qualifications have been for enrichment biomarkers, which the Enrichment Guidance [16] states that do not require qualification. Over 30 qualification submissions have been received at Center for Drug Evaluation and Research (CDER) throughout the past decade. Some have been withdrawn, while 25 [17] remain in a prolonged cycle of confirmation and advice meetings with the FDA, leading to incrementally more complex and expensive qualification studies.

The global futility of biomarker qualification at the FDA is highlighted by the absence over a period of 10 years of a single qualification for a surrogate biomarker. The absence of qualified surrogates has had a major impact on the development of new therapies in the treatment of rare diseases [18], where measurement of conventional endpoints such as the six-minute walk [19] is particularly difficult in patient populations with severe physiological damage.

The most challenging part of this process has been the definition of incremental contexts of use and the corresponding evidence with which biomarkers may be qualified. We should not misrepresent the industry goal: qualified biomarkers capable of development and approval for new drugs in the clinic. We also cannot misrepresent the goal as far as public health is concerned, which is to obtain better biomarkers for routine clinical use as quickly as the data will allow. Intermediate qualification contexts and data need to be defined so that investment in biomarker qualification studies will be productive both for clinics and for the pharmaceutical industry. Initial studies proposed by consortia are unlikely to match a clear context for qualification for a full clinical application of biomarkers. What intermediate contexts for qualification can we define, and what study characteristics can we propose for qualification in these intermediate contexts of use? Several authors [20, 21] have proposed evidentiary recommendations for biomarker qualification.

Unlike the incremental process for biomarker qualification embodied in the pilot process for biomarker qualification at the FDA [7], papers on evidentiary recommendations often propose all-or-nothing qualification contexts, where if the ultimate goal is a clinical qualification, no intermediate qualification contexts are expected to be defined or qualified. This approach is not only time-consuming but also not likely to encourage the investment needed to generate data for biomarker qualification. At each stage, whether the context

of use for a biomarker is to be *in vitro*, in a nonclinical animal model, or in the clinic, a company or consortium proposing a qualification will probably seek a quick return on the qualification of a biomarker after data are available to qualify a biomarker in a specific context in drug development. An effective process for biomarker qualification should include incremental application context steps, so that these incremental steps can quickly benefit the drug development process.

Harmonization

The application of biomarkers to drug discovery and development has the potential of improving the efficacy and speed of bringing more effective and safer new drugs to market. This requires that biomarkers which may be applicable to such uses be qualified for a specific application context. To achieve this, both the process of qualification and the evidentiary criteria and standards for qualification will need to be described and defined. The International Committee on Harmonization (ICH) E16 [22] is a harmonization effort to define the context, structure, and format of the biomarker qualification submission. It is based on the previous experience by the FDA and EMA regarding biomarker qualification. This harmonization effort does not address the evidentiary requirements for biomarker qualification.

The structure, format, and content of a submission of biomarker data for qualification depend on the context in which the biomarker is intended to be used. The first step in drafting a submission for qualification of a biomarker is to determine its context of use, preceding specific decisions on applicable structure and format. The context of use for a biomarker is (i) the general area of biomarker application, (ii) the specific applications/implementations, and (iii) the critical factors that define where a biomarker is to be used and how the information from measurement of this biomarker is to be integrated in drug development and regulatory review. To demonstrate the alignment between proposed context and data, the initial context proposal must be supported by data available at the initial application step or expected to be available throughout the data evaluation process in biomarker qualification. There is a convergent relationship between an initial qualification context and the data supporting it. The initial gap between proposed context and data may need to be filled throughout the qualification process. Initial context proposals, however, should project a significant improvement over currently available biomarkers and/or endpoints.

The context of a biomarker drives data requirements to demonstrate its qualification for the intended application. The structure of a submission document ensures that the context and data can be submitted in a package consistent for consortia submitting qualifications as well as for reviewers

in regulatory agencies evaluating a qualification package. The structure of a qualification submission is independent of the context of this submission, but must also be flexible enough to deal with the specific requirements of each context. On the other hand, the format of data required to qualify a biomarker may vary significantly with the context in which it is to be used. It is therefore only possible to harmonize general regulatory guidelines on data format for biomarker qualification submissions.

Summary

The qualification of novel biomarkers for drug development and regulatory review is made possible by the development and testing of regulatory mechanisms for biomarker qualification. The urgency for getting more and better biomarkers to improve new drug development is clear to the pharmaceutical industry and regulatory agencies. Harmonized processes for biomarker qualification are being actively developed through ICH's harmonization procedures.

References

1 FDA (2014). In vitro companion diagnostic devices. https://www.fda.gov/media/81309/download (accessed 03 May 2019).
2 FDA (2016). Table of pharmacogenomic biomarkers in drug labeling. http://www.fda.gov/drugs/scienceresearch/researchareas/pharmacogenetics/ucm083378.htm (accessed 27 November 2016).
3 Wagner, J.A. (2008). Strategic approach to fit-for-purpose biomarkers in drug development. *Annu. Rev. Pharmacol. Toxicol.* 48: 631–651.
4 Rifai, N., Gillette, M.A., and Carr, S.A. (2006). Protein biomarker discovery and validation: the long and uncertain path to clinical utility. *Nat. Biotechnol.* 24 (8): 971–983.
5 Goodsaid, F. and Frueh, F.W. (2007). Implementing the U.S. FDA guidance on pharmacogenomic data submissions. *Environ. Mol. Mutagen.* 48 (5): 354–358.
6 Goodsaid, F. and Frueh, F. (2006). Process map proposal for the validation of genomic biomarkers. *Pharmacogenomics* 7 (5): 773–782.
7 Goodsaid, F. and Frueh, F. (2007). Biomarker qualification pilot process at the US Food and Drug Administration. *AAPS J.* 9 (1): E105–E108.
8 Goodsaid, F.M., Frueh, F.W., and Mattes, W. (2008). Strategic paths for biomarker qualification. *Toxicology* 245 (3): 219–223.
9 Anonymous (2008). Public consortium efforts in toxicogenomics. *Methods Mol. Biol.* 460: 221–238.

10 Duarte, C.G. and Preuss, H.G. (1993). Assessment of renal function-glomerular and tubular. *Clin. Lab. Med.* 13: 33–52.

11 Vaidya, V.S., Ramirez, V., Ichimura, T. et al. (2006). Urinary kidney injury molecule-1: a sensitive quantitative biomarker for early detection of kidney tubular injury. *Am. J. Physiol. Renal Physiol.* 290 (2): F517–F529.

12 Zhang, J., Brown, R.P., Shaw, M. et al. (2008). Immunolocalization of Kim-1, RPA-1, and RPA-2 in kidney of gentamicin-, mercury-, or chromium-treated rats: relationship to renal distributions of iNOS and nitrotyrosine. *Toxicol. Pathol.* 36 (3): 397–409.

13 Dieterle, F., Maurer, E., Suzuki, E. et al. (2008). Monitoring kidney safety in drug development: emerging technologies and their implications. *Curr. Opin. Drug Discovery Dev.* 11 (1): 60–71.

14 FDA (2014). Qualification process for drug development tools. https://www.fda.gov/media/79473/download (accessed 03 May 2019).

15 FDA (2019). Biomarker qualification program. https://wayback.archive-it.org/7993/20180125052703/https://www.fda.gov/Drugs/DevelopmentApprovalProcess/DrugDevelopmentToolsQualificationProgram/BiomarkerQualificationProgram/default.htm (accessed 03 May 2019).

16 FDA (2012). Enrichment strategies for clinical trials to support approval of human drugs and biological products. https://wayback.archive-it.org/7993/20170403220022/https://www.fda.gov/ucm/groups/fdagov-public/@fdagov-drugs-gen/documents/document/ucm332181.pdf (accessed 03 May 2019).

17 FDA (2016). Biomarker qualification at CDER (presentation).

18 FDA (2016). Scientific dispute regarding accelerated approval of Sarepta Therapeutics' Eteplirsen (NDA 206488) – Commissioner's decision. http://www.accessdata.fda.gov/drugsatfda_docs/nda/2016/206488_summary%20review_Redacted.pdf (accessed 27 November 2016)

19 FDA (2016). Summary minutes of the Peripheral and Central Nervous System Drugs Advisory Committee meeting (25 April 2016). https://www.fda.gov/media/99235/download (accessed 27 November 2016).

20 Altar, C.A., Amakye, D., Bounos, D. et al. (2008). A prototypical process for creating evidentiary standards for biomarkers and diagnostics. *Clin. Pharmacol. Ther.* 83 (2): 368–371.

21 Wagner, J.A., Williams, S.A., and Webster, C.J. (2007). Biomarkers and surrogate end points for fit-for-purpose development and regulatory evaluation of new drugs. *Clin. Pharmacol. Ther.* 81 (1): 104–107.

22 International Committee on Harmonization (ICH) E16 (2010). Genomic biomarkers related to drug response: context, structure and format of qualification submissions. https://www.ich.org/products/guidelines/efficacy/efficacy-single/article/biomarkers-related-to-drug-or-biotechnology-product-development-context-structure-and-format-of-qu.html (accessed 19 October 2008).

9

Fit-for-Purpose Method Validation and Assays for Biomarker Characterization to Support Drug Development

Jean W. Lee[1], Yuling Wu[2], and Jin Wang[3]

[1] *BioQualQuan, Camarillo, CA, USA*
[2] *MedImmune, Gaithersburg, MD, USA*
[3] *Amgen, Inc., Thousand Oaks, CA, USA*

Introduction

Novel biomarkers have been discovered from recent research in genomic and proteomic studies. Differences in gene expression and protein abundance or modification provide the basis for novel biomarkers to be tested further with *in vitro* and preclinical disease models. Interactions of putative biomarkers at interwoven pathways and the intervening drug actions are relevant to polygenic disorders associated with the complex interplay between genetics, epigenetics, and the environment. Activities of novel biomarker development have increased, especially in diabetes, cancer, rheumatoid arthritis, and cardiovascular disease. In addition to clinical biomarkers proven previously, putative biomarkers have been included in clinical trials of new, mechanistic-based drug candidates for exploratory, demonstrative, or characterization applications in the process of developing mechanism-specific biomarkers toward the ultimate goal of surrogacy [1–4]. For example, to the commonly employed measurement of blood glucose, hemoglobin A1c, and circulating insulin, novel biomarkers could be included in diabetes trials, depending on the mechanism of action. For peroxisome proliferator–activated receptor gamma (PPARγ) agonists in type 2 diabetes, free fatty acid was linked to insulin resistance as mechanism-specific, and adiponectin was identified as a proximal biomarker correlated with insulin sensitivity [5, 6], while glycosylated hemoglobin (Hb_{a1c}) was qualified as a surrogate marker [7]. Other mechanisms and the relevant biomarkers (in parentheses), such as lipid and bone metabolism (free fatty acids, leptin, osteocalcin), incretins (glucagon, GLP-1), and inflammation (cytokines, hsCRP, PAI-1, fibrinogen, adhesion molecules) are also involved in the complex disease of diabetes [8–12]. In addition to circulating prostate-specific antigen

Biomarkers in Drug Discovery and Development: A Handbook of Practice, Application, and Strategy, Second Edition. Edited by Ramin Rahbari, Jonathan Van Niewaal, and Michael R. Bleavins.

(PSA), carcinoembryonic antigen (CEA), and other tumor markers, the mRNA of specific genes, various growth factors, and imaging are also used in clinical trials to monitor drug responses for cancer treatments [13–17].

The use of novel biomarkers has become a prominent component of decision-making processes in drug development. They are used in the *in vitro* and preclinical models and early clinical phase for quick-hit and early-attrition decisions [3, 18–20]. Characterization of a novel biomarker in the translational phase requires data collection to show preclinical sensitivity and specificity and linkage to clinical outcomes in multiple clinical studies in humans [4]. The purposes at this phase are different from the exploratory or demonstrative phase. In exploratory and demonstrative studies, pharmacodynamic (PD) correlations are typically unknown, data are used mainly for internal decision-making, and the output will generally not be subject to regulatory review. The extent of method validation can thus be limited to a few basic components to expedite the process and preserve resources without unduly affecting commercialization [21]. By contrast, the purposes of characterization application are to provide pivotal data to establish linkage to clinical outcome and to monitor patient progress upon treatment (Figure 9.1). The data are often used for critical decisions (such as supporting dose selection and patient stratification, and demonstration of drug safety or efficacy), submitted to review by regulatory agencies, or used for post-marketing patient monitoring [22, 23]. In addition, the same method used for characterization is likely to

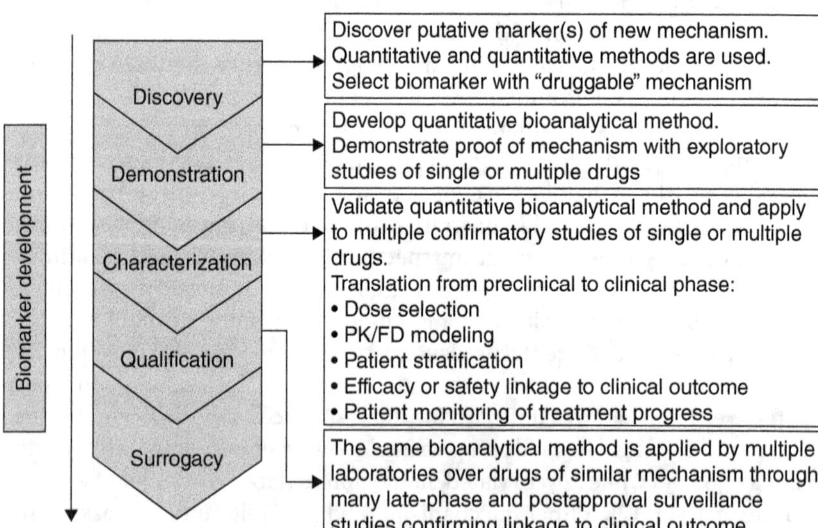

Figure 9.1 Path from putative to confirmed mechanism of a novel biomarker through the processes from discovery to surrogacy. The side boxes illustrate the relationship of the processes to drug development applications and the methods required.

be used in the qualification phase toward surrogacy, confirmed over multiple drugs of similar mechanism, and used during surveillance studies. Therefore, biomarker characterization studies would require more intense rigor in assay method validation and in traceable and more detailed documentation than that of previous phases to meet the study objectives in a defined context of use.

A biomarker is useful if there are distinctions in concentrations in the disease state from that of a healthy status (or disease stabilization, in the case of some oncology trials). The assay performance of the biomarker analytical method must be able to meet this goal of disease status distinction and estimation of the drug effect. Therefore, contrary to the general perception of many analytical laboratories, the required assay acceptance criteria do not depend solely on the deliverable method's accuracy and precision. Instead, consideration should be on the assay suitability for the intended applications based on three major factors: (i) intended use of the data during various stages of drug development, (ii) nature of the assay methodology and the different types of data that it provides, and (iii) biological variability of the biomarker that exists within and between populations. The first factor helps shape the assay tolerance or acceptance criteria for biomarkers.

A fit-for-purpose method validation approach at various phases of biomarker application has been described by [21]. In this chapter, we focus on how the fit-for-purpose method validation and assay application during drug development clinical trials can contribute to biomarker characterization.

General Processes

Quantification of protein expression is important when considering the translation of a candidate protein biomarker. If there are changes in posttranslational modification of the biomarker, it would be preferable to be able to quantify levels of both protein expression and posttranslational modification, utilizing technologies such as ligand-binding assays (LBA), enzymatic assays, liquid chromatography–mass spectrometry (LC-MS) in multiple-reaction mode (MRM-LC-MS), protein assay, and real-time polymerase chain reaction (RT-PCR) genomics. LBAs can be performed in high throughput at a relatively low cost per samples, which is most advantageous for application through phases of characterization and qualification development of a novel biomarker. Therefore, LBAs have been widely used for protein biomarker quantification and will be the typical method for discussion in this chapter.

The developmental processes of a novel biomarker and a new chemical drug entity are intertwined [24, 25]. Biomarker development may happen concurrently with the development of a single or multiple drug candidates, a refined new chemical or biological entity, or for extended indications and/or additional mechanisms. The progress of a potential biomarker does not always coincide with or parallel that of a new drug candidate development. To utilize and integrate information from development of drugs in therapeutic programs and the development of relevant putative biomarkers, a detailed work plan can be prepared for the novel biomarker and the study objectives identified in the corresponding study protocols of each drug candidate. Innovative companies often organize biomarker work groups to facilitate timely input and communication among therapeutic areas and supporting teams.

As depicted in Figure 9.2, novel target (BMKa) and/or proximal biomarkers (BMKb) of a specific mechanism can be used for multiple drug candidates to prove the mechanism of action by drug intervention during the exploratory and demonstration phases. If the circulatory biomarkers are reflecting their actions in the target cells/organs, measurements of blood levels of the target and/or proximal biomarkers can be included during clinical trials for further characterization purposes. In addition, distal biomarkers (BMKi,j) of a specific disease indication that have been used for other drug compounds can be applied for a new drug of a different mechanism toward the same indication. In general, distal biomarkers are closer to the disease symptom magnification for efficacy and off-target indirect effects, while the target and proximal biomarkers are closer to the direct exposure effect. Thus, the establishment of PD relationships to pharmacokinetics (PK) is usually simpler for the target or proximal biomarkers, while linkage to a clinical outcome may be shown more readily by distal biomarkers. PK–PD modeling is used to correlate the dose-exposure effects.

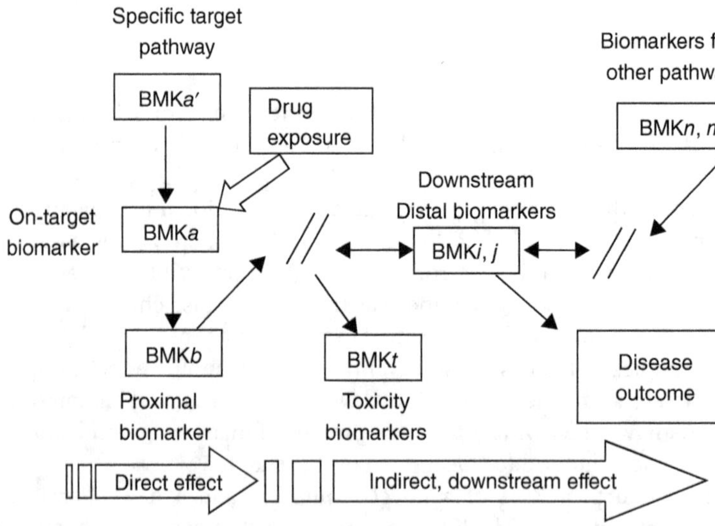

Figure 9.2 On- and off-target biomarkers of a specific mechanism of drug intervention. Possible relationship of biomarkers from a specific pathway to disease outcome. Other pathways may have common distal biomarkers that may be used as predictive index of the same disease outcome.

PK–PD modeling based on various mechanisms of action, direct and indirect drug effects, and reactions of biological cascades has been developed [26]. For proof of biology, the biochemical coverage of biomarkers from target hit to distal to clinical outcome would provide a thorough understanding of drug effect and characterization of a novel biomarker, whether it is a target, proximal, or distal. Additionally, off-target negative effects should be monitored carefully by toxicity biomarkers (BMKt). Such information is vital in the selection of lead candidates and appropriate dose ranges for clinical trials [27].

The application of a novel biomarker in clinical trials for the purpose of biomarker characterization is often included as a secondary objective in a clinical protocol, while toxicity and commonly recognized efficacy biomarkers may be included in the primary objectives. Characterization of a specific novel biomarker may be carried out on different new chemical entity development programs over several therapeutic areas. Therefore, it is important for the assay laboratory to have both a "big picture" perspective on the development of a certain novel biomarker and the specific purpose of the application of that biomarker in a clinical trial. The time required for biomarker assay development and method validation, the operational and logistical issues, including pre-analytical factors, and the limitations in data interpretation are key elements for careful planning and execution to support biomarker development and its applications.

Method Validation and Assay Application in Drug Development

Population Range and Assay Range

In the process of biomarker assay development and validation, samples from the target populations of healthy persons and patients are used to establish expected ranges and biological variations for a given biomarker. Knowledge of these ranges and assay performance in these matrices are necessary for a number of reasons. The initial concentration ranges of healthy populations are usually available from literature or vendor brochures for previously identified biomarkers. During the characterization phase, the range-finding experiments should include sufficient samples from individuals of the target populations (and extended populations if applicable) to enrich the existing information. Biological variation exists in some biomarkers. It is important to note the variable levels over time (i.e. diurnal, seasonal, and food effects) to plan the correct conditions and time of sample collection appropriately. In addition, biological variation can come from the changes in the binding-protein components in the biological matrices, affecting the biomarker measurement in some patient samples. The effect of matrix from patient samples is discussed further later in the chapter.

The results of the range-finding studies and the expected drug effect on the biomarker are used in the design of method validation and assay application in clinical studies. Knowledge of variation and modulation will power the sample size necessary to confidently detect relevant changes in a biomarker after therapeutic intervention. It is desirable to have the biomarker assay range cover the range of the target populations and the expected drug effect. However, this is often not possible. For example, most protein biomarkers are analyzed by LBA, with the assay working range dependent on the antibody reagents used, which is often different from the biological range. Additionally, the biomarker levels of one disease are often different from those of another. For example, inflammatory biomarkers such as IL-6 and TNFα in serum samples for healthy and rheumatoid arthritis patients are in the pg/ml range, whereas those of sepsis patients are in the ng/ml range. Use of the inflammatory biomarkers for sepsis will require dilutions into the working range. Thus, application of these biomarkers to sepsis studies would require dilutional linearity during method validation, and the variability introduced by dilution should be monitored during in-study assays.

Sample Integrity: Sample Collection at the Clinic

Sample integrity means that the measurement of a biomarker after collection, transport, and storage should yield a result as close to that which could be measured in as fresh a sample as possible. Sample characteristics can vary over time, in the fed state, or by age- and gender-specific factors, and by the methods used to prepare the sample. Sample handling can alter a sample irrevocably, so care must be used in specimen collection. Delays in specimen processing, addition of anticoagulants and other additives, type of collection vessel and storage tube, and a myriad of other pre-analytical circumstances can affect a biomarker. A careful review of potential confounding variables is essential prior to method validation. Pre-analytical variables have been shown to compromise data utility in proteomic biomarker discovery and validation [28, 29].

Serum, plasma, or urine samples are the most common matrices during the characterization phase. For diseases involving the brain or central nervous system, cerebral spinal fluids are commonly used. Accessible cells from sputum, bronchial lavage, and blood cells are used for receptor functional measurements. The preservation of samples must be explicitly defined in the study protocol. Control and standardization of sample collection are especially necessary to minimize variability from multiple clinical sites. For both tissue collection and *ex vivo* stimulation assays, it is best to validate sample collection procedures using conditions that closely mimic the planned sampling protocol and to examine sample stability as soon as possible. If sample handling is complicated, it would be prudent to establish tolerance boundaries surrounding sample collection and processing for the clinical sites to abide and document. In addition, staff training and oversight procedures are essential for the monitoring and maintenance of sample integrity.

In some biomarker studies, normalization is used to correct concentration variability. For example, urine creatinine concentration is used to normalize hydration variability in urine samples. In that case, the assay of the normalizing factor should also be validated on stability and reliability prior to the clinical study.

During biomarker characterization, multiple clinical studies from multiple drug candidate development programs may be involved. It is important for the biomarker team to coordinate the study protocols to include standardized procedures for sample collection, shipping, and analysis and for sample storage and disposal. Sample collection and storage stability should be tested by the bioanalytical laboratory to cover the conceivable conditions during in-life study sample analysis. This includes process stability, assessing the impact of short- and long-term storage, and multiple freeze–thaw cycles.

Method Development

What Will Be Measured?

Many biomarkers of interest are proteins whose function is affected by a change in disease progression and drug intervention. The ultimate goal of a biomarker assay is typically to evaluate the effect on the biomarker activity *in vivo*, and a method that measures this activity accurately is most desirable. Direct measurement of changes to target cells, tissues, or organs in response to a drug is rarely feasible, and given current drug potencies, activity assays rarely have adequate sensitivity and precision. Moreover, cell-based and activity assays are usually laborious, low in throughput, and often lack definitive reference standards. LBAs are often chosen for biomarker analysis over activity measurements because of their sensitivity, versatility in application, relatively high throughput, and low cost.

Many biomarkers are endogenous protein that can be physicochemically heterogeneous, with multiple forms present simultaneously. The concentration and biological activity of each component are often unknown and may vary with health status, over time, and between individuals. However, it is desirable to know, even in the case of a well-defined regulatory peptide, the related species of the analyte existing in the biological matrix of measurement. For example, measurements of proinsulin, insulin, and C-peptide provide different information for clinical interpretation and assessment. For large molecules, the heterogeneity of the molecules poses challenges to identification of the exact species being measured. Therefore, various approaches in designing the analytical method, including choices of the reference standard and critical reagents, are worth considering.

LBA methods measure the binding reaction of the mixed components in the matrix. The majority of the data generated by LBA is relative quantification [21]. Limitations of LBA methods are usually a lack of chemical information or correlation of the binding activity to the biological activity. In addition, the reference standard used for data regression may be a recombinant protein or one of several components which may or may not be representative of the entire endogenous mixture. However, as the purpose during biomarker characterization is to monitor population and longitudinal effects of drug treatment and disease progression, this type of relative quantification is adequately "fit for use."

Reference Standard and Alternative Matrix for Standard Preparation
The reference material may not be fully representative of the endogenous analyte because the endogenous analyte may exist in various forms in the biological matrix. A well-defined reference standard is used as a scalar for the relative measurement of the endogenous species. For example, the recombinant form of a biomarker can be produced and characterized with defined molecular size and purity to enable concentration and molar equivalence calculations. Documentation of the reference material is necessary for its use.

Most biomarkers are endogenous molecules with variable measurable amounts in the matrix. It is difficult to find "blank" control matrix for preparation of reference standards. An initial standard curve can be constructed by spiking the reference standard into a protein buffer solution. An initial screen on multiple matrix lots against the buffer standard curve may identify some blank matrix lots for standard preparation. When that is not possible, an alternative matrix can be prepared by depleting the endogenous analyte using methods such as charcoal stripping, high-temperature incubation, acid or alkaline hydrolysis, or affinity chromatography. Commonly, a protein-containing buffer or the matrix from another species with a non-cross-reactive homolog is used. The use of alternative matrices requires studies of matrix effects and parallelism during method development and validation to understand the impact that differences from patient samples will have on assay results.

Parallelism experiments are performed through serial dilutions of a high-concentration sample with the standard matrix. Multiple individual matrix lots (three or more) should be tested to demonstrate lot-to-lot consistency. There are several ways to assess data of parallelism [21]. The concentrations of diluted and undiluted samples observed should all be within the standard curve range. An example is depicted in Figure 9.3: the mean of the concentrations calculated (concentration observed times the dilution factor) of each lot was determined. The ratios of the individual concentrations calculated over the mean of the lot were plot against 1/dilution factor. Parallelism was demonstrated for the test lots because the ratio was not affected by the variable amounts of standard matrix introduced by dilution. When parallelism is not possible due to the

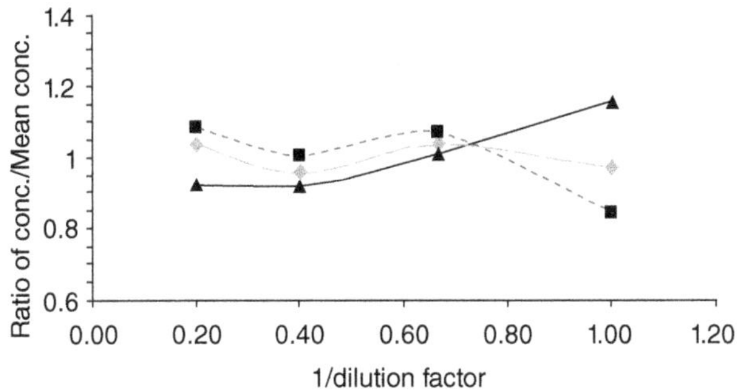

Figure 9.3 Parallelism of three individual matrix lots. Each lot is represented by a separate symbol. The sample was diluted with the standard diluent at dilution factors of 1.5, 2.5, and 5. The concentrations of the diluted and undiluted samples observed were all within the standard curve assay range. The mean of the calculated concentrations (observed concentration times the dilution factor) of each lot was determined. The ratios of the individual calculated concentration over the mean of the lot were plotted against 1/dilution factor. The data showed that the ratios were all within the range of 0.8–1.2, which was the defined acceptance criterion of the assay. Therefore, the parallelism test results were acceptable for the test lots.

unavailability of samples with sufficiently high concentrations of analyte, dilutional linearity should be tested in a manner similar to parallelism, except that high-concentration spike samples are used in place of the endogenous samples.

The failure to demonstrate parallelism should be taken into consideration in data assessment and interpretation. For example, it may mean that the values obtained should be treated as quasi-quantitative instead of as relative quantitative measurements [21]. In addition, within-subject comparison on a longitudinal time sequence would be more useful than between subjects or populations, which affects the clinical study design. All of these considerations would be dependent on the a priori goals of the study.

Selection of Critical Reagents in Ligand-Binding Assays

For the development of a first-in-class drug compound that targets a novel biomarker, assay reagents will need to be developed for both the drug candidate and the target protein biomarker. In the case of a macromolecular drug, often an analog of the drug candidate may be used as the binding reagent(s). On the other hand, reagents for most off-target biomarkers may be available from commercial sources. These reagents may be established assay kits (US Food and Drug Administration [FDA] approved for diagnostic use) or kits for research use only.

For the purpose of characterization of the biomarker of interest, fit-for-purpose method validation should be conducted with sufficient rigor to provide reliable data from multiple clinical studies. This should include assay range finding, accuracy and precision, selectivity, specificity, stability, and robustness (reproducibility) during prestudy method validation. Additional data will be collected from in-study assay performance, change control method validation, and long-term storage stability. Except in the case of FDA-approved diagnostic kits, many of the assay performance parameters will have to be established by the bioanalytical laboratory. For FDA-approved kits, because the application will be for drug development instead of diagnosis, some of the assay performance parameters will need to be established. These often include selectivity against the target patient sample matrices and specificity against the drug compound(s). If the drug is expected to decrease the biomarker to a level lower than the lowest standard of the kit, the method may need to be modified to increase assay sensitivity.

Antibody pairs are typically chosen as capture and detection reagents. In general, the more selective antibody would be chosen as the capturing agent, especially if it is more readily available than the other member of the pair. A tertiary detector antibody can be used that is conjugated to a reporter enzyme such as horseradish peroxidase. Alternatively, a biotinylated detector antibody can be used together with a biotin-binding protein (e.g. antibiotin antibody or an avidin-type protein) conjugated to a reporter enzyme. The sensitivity of an assay can be increased by varying the number of reporter enzyme molecules on the detection reagents, or by using multivalent strategies to increase the effective signal from each analyte captured.

Some assays use receptors or their fragments as binding partners, most often in concert with a specific second antibody. This arrangement can improve selectivity for specific ligands (e.g. a cytokine activated from a latent precursor, or a particular subtype of ligand with distinct binding characteristics from its homologs). The binding selectivity of such reagents can offer added biological relevance to quantification, and can suggest results that might otherwise be obtained only via functional assays.

Critical Reagents Characterization, Consistency, Stability, and Documentation

Characterization of a novel biomarker often involves many clinical studies that last for several years. If the studies are to be conducted within the same company or from a joint program from different institutions, plans should be made to provide consistent supplies of the same reference material and assay reagents throughout the studies. If the materials are from in-house sources, sufficient production of the reference standard and primary reagents should be assured. In addition, multiple preparations of labeled detector reagents should be assessed during method validation. If the materials are from commercial sources, negotiation with the vendor should take place to assure a consistent and sufficient supply of the same batch material, if possible. Often, multiple batches of materials (reference standards and critical reagents) should be tested during method validation. Documentation should be obtained for the stability

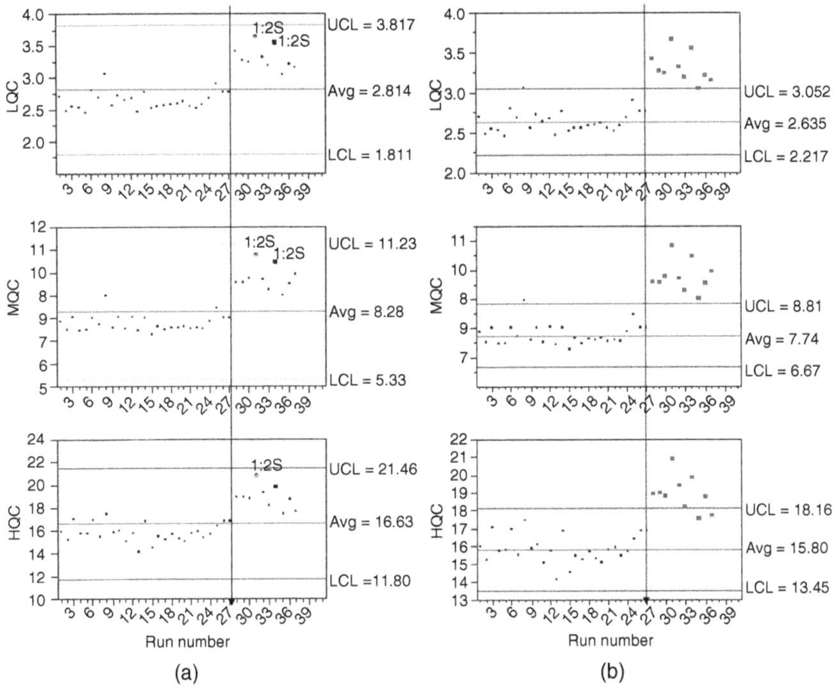

Figure 9.4 Use of sample controls for trend analysis on variability. Low, middle, and high sample controls of a biomarker were monitored in Levey Jennings control charts. There was a noticeable shift in trend in all the sample control levels after run 27. The average, upper, and lower control limits of analytical runs up to run 27 (b) were compared to the overall parameters (a). (*See insert for color reproduction of the figure.*)

of the respective lots over the time span of a clinical study. Documentation of chain of custody should be similar to that of a good laboratory practices (GLP) study. For late-phase studies using multiple clinical sites, the use of a central sample repository offers numerous advantages in controlling the process of specimen collection and labeling of multiple biomarker assays.

In-study sample control (SC) charts such as those shown in Figure 9.4 can be used for trend analysis of assay performance and stability. A big pool of endogenous samples at low, middle, and high concentrations of a biomarker were prepared during method validation and monitored throughout the in-study. The Levey Jennings control charts showed a noticeable shift in trend at all the SC levels after run 27, indicating a systemic bias. The average, upper, and lower control limits of analytical runs up to 27 (right panels) were different from the overall values (left panels). This shift was traced to a change in commercial kit lot.

Method Validation on Performance Parameters

The foremost assay performance parameters to be established for quantitative methods are precision and accuracy. In addition to these two parameters, sensitivity, selectivity, specificity, stability, and reproducibility should also be demonstrated. These parameters are established during prestudy method validation to demonstrate reliably that the method is fit for the purpose of biomarker characterization over multiple clinical studies. If more than one bioanalytical laboratory is used, each additional laboratory should also demonstrate its ability to perform by conducting a full or partial method validation.

Accuracy and Precision

Method performance is consistently better understood and validated through the appropriate use of control samples. To ensure data quality, assay performance is evaluated during method validation with validation samples (VSs) and monitored during sample analysis with quality control samples (QCs). VSs are used in method qualification or validation to define intra- and interrun accuracy and precision, providing data to demonstrate the robustness and suitability of the assay for its intended application. On the other hand, QCs are essential in determining run acceptability during specimen analysis. However, many laboratories have not made the distinction, using the term QC for both VS (for prestudy method validation) and QC (for in-study run acceptance).

For a well-defined biomarker assay, at least six accuracy-and-precision validation runs should be performed to provide statistical data to calculate these assay parameters [30, 31]. Each run should consist of standards prepared by spiking the reference standard into blank biological matrix or an alternative blank matrix. QC/VS at low, middle, and high concentrations should be prepared in the biological matrix or alternative matrix by spiking the reference standard. In addition, SCs should be prepared by pooling authentic samples at the low and high levels of the biomarker to reflect the performance of the endogenous biomarker in assay precision, stability, and reproducibility. At least two sets of QC/VS and SC should be run with the standards in each accuracy and precision experiment.

Accuracy and precision can be evaluated from the total error of VS data from the validation runs in a way similar to that of the macromolecular protein drug as analyte [30]. However, given biological variability and other factors in biomarker research, more lenient acceptance criteria may be used for biomarker PD than that for PK studies. Still, it should be recognized that accuracy and precision data of VS in buffer provide only a relative quantification, which may be quite different from measurements in the authentic matrix.

Concentrations of the endogenous SCs will be determined by multiple validation runs. The true values are then defined after sufficient data collection from pre- and in-study validation. For example, the mean of 30 runs and 2 standard deviations can be used to define the target concentration and acceptable range of the SCs. Because the reference material may not fully represent the endogenous biomarkers, the SCs should be used for stability tests (such as minimum tests of exposure to ambient temperature and freeze–thaw cycle). In addition, the SC can be used for in-study, long-term storage stability analysis and for assessment of lot-to-lot variability of key assay reagents.

Regression models are essential for data calculation from sigmoid curves like those in LBAs. The most commonly used four- or five-parameter logistic regression should be evaluated in conjunction with weighting factors during method development. Final decisions on which curve-fitting model to use should rest on that which offers the best fit for all the standards in the precision profile. In some cases, a less than optimal fit will suffice to allow for greater assay sensitivity.

Sensitivity

To provide data for PK/PD studies, assay sensitivity is usually defined by the assay lower limit of quantification (LLOQ), which is the lowest concentration that has been demonstrated to be measurable with acceptable levels of bias and precision and total error (the sum of bias and precision). However, low-concentration clinical samples may fall below the LLOQ (i.e. the method lacks the required sensitivity). In some instances, the investigator may want to use values below the LLOQ but above the limit of detection (LOD), to obtain a numerical estimate of the changes while recognizing the high variability below the LLOQ region.

LOD is often used as the analytical "sensitivity" of the assay in a commercial diagnostic kit. A common practice of diagnostic kits to determine the LOD is the use of extrapolated concentrations from a response signal of +3 SD (or −3 SD for a competitive assay) of the mean background signal from 30 or more blank samples. The National Committee for Clinical Laboratory Standards (CLSI) has recommended an approach for determining LOD [32, 33]. This statistically sound approach evaluates the limit of blank (LOB) as type I error and LOD as type II error from sufficient numbers of blank and low concentration samples with normally distributed analyte concentrations. One approach of calculation is described briefly as follows: LOB is estimated from the blank result at the 5th percentile position. LOD is calculated as

$$\text{LOD} = \text{LOB} + c\beta\ \text{SDs}$$

SDs is the estimated standard deviation of the low-concentration sample measurement, and $c\beta$ is a correction factor for the population bias:

$$c\beta = \frac{1.645}{1 - 1/[4(N - K)]}$$

Here N is the number of determinations and K is the number of sample sets.

The LOD determined by this approach would avoid the inclusion of analytical false-negative and false-positive values. Due to assay conditions being variable, each bioanalytical laboratory should determine the LOD using the CLSI approach instead of the LOD stated in a kit brochure. If the LOD is used as the lower limit for data inclusion instead of the LLOQ, the user should justify such a choice and be aware of the risk being taken with the higher variability at the LOD to LLOQ range, interpreting the data with caution.

Selectivity and Matrix Effect

Selectivity is the ability of the assay to discriminate the analyte unequivocally in the presence of components that may be expected to be present in the sample and alter assay results. Lack of selectivity can result in signal inhibition or enhancement. The results can appear as false or negative positives (an over or underestimation of the analyte concentration). In general, signal suppression from binding proteins occurs more often for LBAs than that of enhancement, resulting in a negative bias.

Interference in an assay is the product of concentration of the interfering substance times the cross-reactivity (or binding inhibition). However, the concentration–response relationship of LBA is nonlinear, and often the magnitude of cross-reactivity (or inhibition) is not evenly dispersed over the entire assay range. Thus, a standard curve prepared in a protein buffer solution has higher optical density (OD) responses than that in biological matrices, while the differences in responses are variable over the standard curve. Moreover, a standard curve prepared in biological matrices from one person would have different responses than that of other people. Therefore, the estimate of interference should not simply extrapolate across the entire assay range or be represented by an IC_{50} value.

Matrix interference should be surveyed in samples from normal and diseased donors, especially in samples from the anticipated patient population. Spike recovery using samples from persons in the target population should be performed. A later selectivity evaluation of LBA involves a parallelism test of high-concentration endogenous samples from many people diluted at ≥ 4 dilution factors with the standard diluent.

Specificity

Specificity reflects the ability of an assay to distinguish between the analyte of interest and other structurally similar components of the sample. For LBAs, a specificity test should include test of interference from components that may specifically affect the binding reaction. For example, interference from the drug compound may alter ligand interactions with the capture reagent. If the ligand reagent of an LBA is the target drug or its analog, the presence of the drug compound at certain concentrations can interfere with the assay. In addition, biomarker homologs or endogenous molecules of the same family may

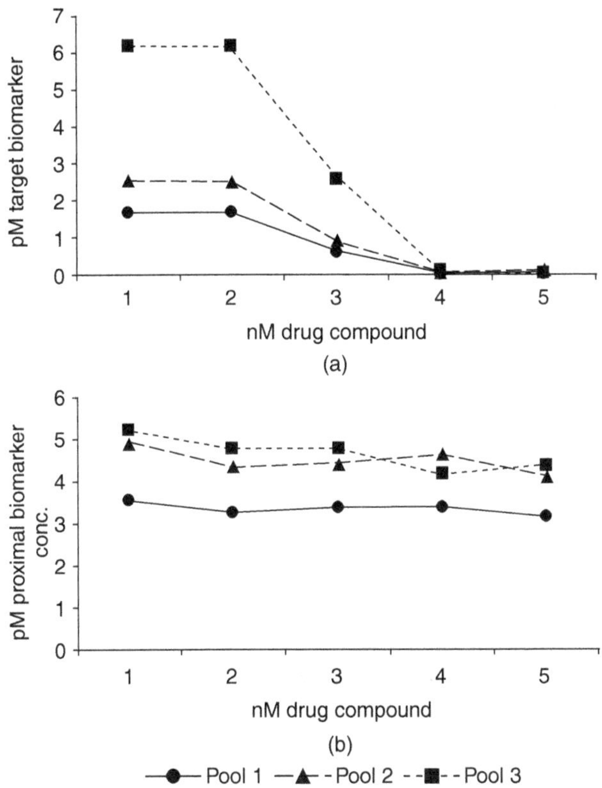

Figure 9.5 Method validation specificity tests of a target biomarker (a) and a proximal biomarker (b) against a drug compound.

contribute to specificity problems. If reference material of the potential interferent is available, specificity can be tested using various amounts of the test material spiking into various levels of VSs [34].

Specificity for the biomarker in the presence of the drug under study should also be tested, using samples from endogenous pools (such as the SCs), with and without the addition of the drug compound at various concentrations spanning the expected therapeutic range. Figure 9.5 illustrates a specificity experiment of a target biomarker (a) and its proximal biomarker (b), each tested with three sample pools at pM levels against various nanomolar drug concentrations. As expected, the test drug inhibited the quantification of the target biomarker at concentrations greater than 2 nM. For the proximal biomarker, the test drug did not affect the assay at all test concentrations, which exceeds the highest expected concentration of the drug in the serum. The results illustrated that the drug affected only its target protein.

Stability

In addition to sample integrity during collection, stability of protein biomarkers and assay reagents is required to show that the assay is not compromised by this pre-analytical factor. Stability information is gathered during method validation on short-term storage and process stability, and is extended through in-study validation for long-term storage stability. Reagent storage stability is tested on reference material stock solutions and ligand-binding reagent stock solution if information from the supplier is lacking. Working solutions will be prepared from the stored stock solution and assay performance compared to those from a freshly reconstituted solution. Process stability is evaluated at different time durations and various conditions to mimic possible sample handling in the laboratory from the results of SCs subjected to stress from multiple freeze–thaw cycles and temperature exposures (e.g. four cycles and up to 16–72 hours to at 2– 8°C and ambient temperature).

Long-term storage stability of the SCs should be started during validation and continued to support the drug development program. The acceptance criteria of storage stability are according to those established during method validation, prior to conducting the experiment. For long-term stability evaluation, in addition to comparison against the initial concentration, trend analysis can be more useful than the acceptance criteria strictly applied. This is due to the challenges of a kit lot change, potentially shifting the mean value of the SCs, and a new preparation of the standard curve, introducing systemic bias to add an error component that is not caused by the storage conditions (similar to that illustrated in Figure 9.4).

Assay Application

Before sample analysis begins, a written method standard operating procedure (SOP) must be in place. Often, a validation report should also be issued. Documentation of sample assay should follow the SOP in a GLP compliance manner (see the discussion later on regulatory issues), with traceable records. The data should be stored in a secure place for auditing.

Standard Operating Procedure

A validated method should have defined procedures, assay performance parameters collected from experiments using the same method during prestudy validation, and acceptance criteria established reflecting the acceptability of the method performance. These elements constitute a method SOP and are discussed in the following section.

Method Procedures: A written method procedure should clearly describe the following:

- the intended purpose of the method
- the principle of the method
- a validation performance summary

- materials and reagents, including lot numbers and expiration dates
- solutions preparation
- standards and QC, SC preparations
- sample pretreatment (processing)
- analytical steps
- data regression
- acceptance criteria
- references

During the development of a novel biomarker, a series of method modifications can take place. It is important that specific names (version number) and effective dates be given to the same method as that used in a clinical study. A historical log should be kept on the modifications made for each version, and its application in specific studies identified. Sometimes, for various reasons, if a method change has to happen within a clinical study, a crossover comparison of the old and new methods applied to a subset of clinical samples must be designed with a statistical approach to show method equivalence.

Regression Model

The response–concentration relationship of the LBA method is nonlinear. Curve fitting would require regression model selection based on multiple runs of standard curves from the method validation accuracy and precision experiments. Typically, a four- or five-parameter logistic model will fit most methods with a choice of weighting factors (such as 1/response or 1/variance). The initial selection is based on the residual of the standard mean. It can be chosen to obtain the best fit over the entire range and with regard to the low end if sensitivity is an issue. The minimal total error of the VS data generated in the validation batches should be used to determine the regression algorithm and weighting. Once the regression model is chosen, it is a basic element of the method that should not be changed during assay implementation. Justification must be given and documented for a regression model change.

Acceptance Criteria

Without sufficient data from the healthy and patient samples, the acceptance criteria for a novel biomarker can initially be set as determined by the assay performance from prestudy method validation. The biological data obtained from in-study validation from subject samples can then be used to refine the initial acceptance criteria set by the prestudy validation. The process of setting acceptance criteria for a protein biomarker measurement follows an evolving path. During the exploratory phase, the acceptance criteria were initially set according to the initial assay performance of prestudy method validation. After use in pilot studies, assessment should be made on the suitability of the assay stringency vs. the effect observed. The biological data obtained from the in-study validation from subject samples during in-study phase can then be used to refine the initial acceptance criteria set by the prestudy validation. For example, an assay with 50% total error may still be acceptable for detecting a twofold treatment effect observed in the in-study phase clinical trial. Setting the same acceptance criteria of a given method may be most convenient for an analytical laboratory. However, for a novel biomarker, it may not be the most appropriate for all of its applications. One could take into account the intended purpose of the application and the possible outcomes in the analytical phase. For example, the effect of one population or indication may be different from another (e.g. change from a twofold treatment effect into only 30%), which may require a more stringent method and/or more subjects to increase the predictive power in such an application for the other population.

Controls in Clinical Studies

For a well-planned assay application, both analytical quality controls and biological SCs should be available to assess the assay variability. Data of analytical variability are gathered from standards, QC and SC of each analytical run of sample analysis. A common set of SCs (or pools of incurred samples from a previous study) analyzed by multiple bioanalytical laboratories can provide an assessment of assay performance among the laboratories. Variability of supplies in reagents and reference standard materials can be detected by tracking the assay performance of a common set of SCs used within and between studies. An example is shown by an SC chart for serum C-terminal telopeptides of type I collagen (CTx) assay in Figure 9.6, using the Westgard rule for monitoring [35] as well as a priori acceptance criteria of 25%. The low and high SCs were prepared by pooling a large volume of authentic samples, and the mean values were established during method validation. Figure 9.6 shows the Levey Jennings charts of more than 117 analytical runs from five clinical studies (A–F) over more than 2.5 years. During this time span, two reference stock materials and three kit batches were used. The plots show a slight positive bias in the SC values after the initial sequence of studies B and C. In addition, a slight negative bias was observed prior to study A. However, since the SC values were still acceptable within ±25%, no action was taken. Other statistical analysis can be used to assess SC data on other assay variability sources, such as operator differences and sample storage stability trend.

At the same time, data from predosed samples can furnish information on biological variability among the subjects and between studies. Figure 9.7 shows the baseline concentration distribution of TRACP 5b, a bone resorption biomarker, in various patient populations from four clinical studies. Statistical tools can be used to construct distribution charts of the predosed samples of each population, and cohorts can be compared. For pilot clinical trials with small number of cohorts, one should note that the data may not be normally

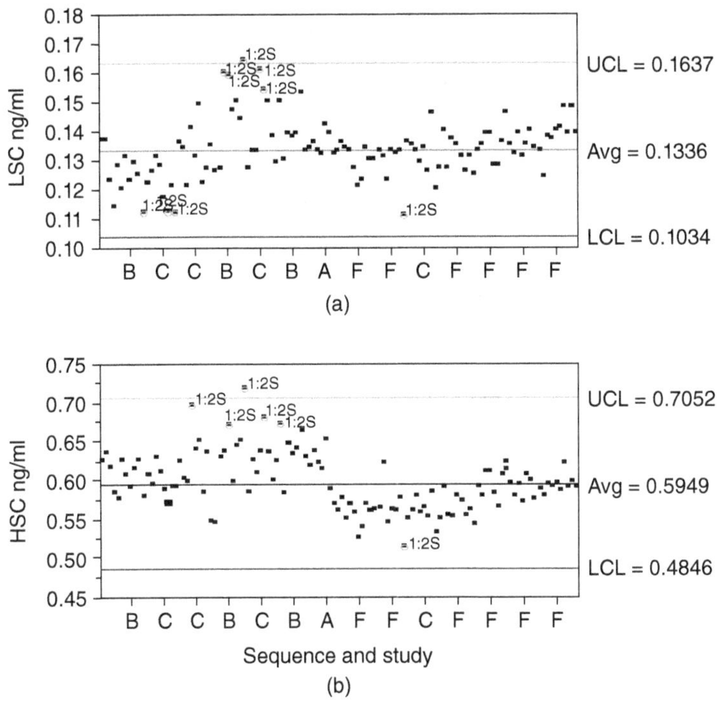

Figure 9.6 Levey Jennings plots of (a) low and (b) high sample controls of serum CTx. Concentrations were plotted against run sequences labeled with the specified studies from A to F. UCL, upper confidence limit; LCL, lower confidence limit. Sample results exceeding the Westgard rule of 1 : 2S (beyond 2 standard deviations) were marked.

distributed (example in Figure 9.7a,c) as compared to those with the late-phase clinical trials (example in Figure 9.7b,d). In addition, multiple collections of predosed samples from the same subject can provide sample collection variables of multiple clinical sites.

Confirming Selectivity in Patient Populations

Biomarkers can reflect the dynamic changes in disease progression and response to therapy, but homeo-static and compensatory biological responses can alter the expected profile in response to drug treatment and concomitant therapies [36, 37]. In the latter case, biomarker assay selectivity is a critical component for interpreting biomarker responses and clinical data, particularly when a concomitant therapy is a potential interferent in the assay. For example, the bone turnover biomarker, serum CTx, is being used in clinical studies involving many different populations. During method validation, serum from healthy males and females and from patients with breast cancer, multiple

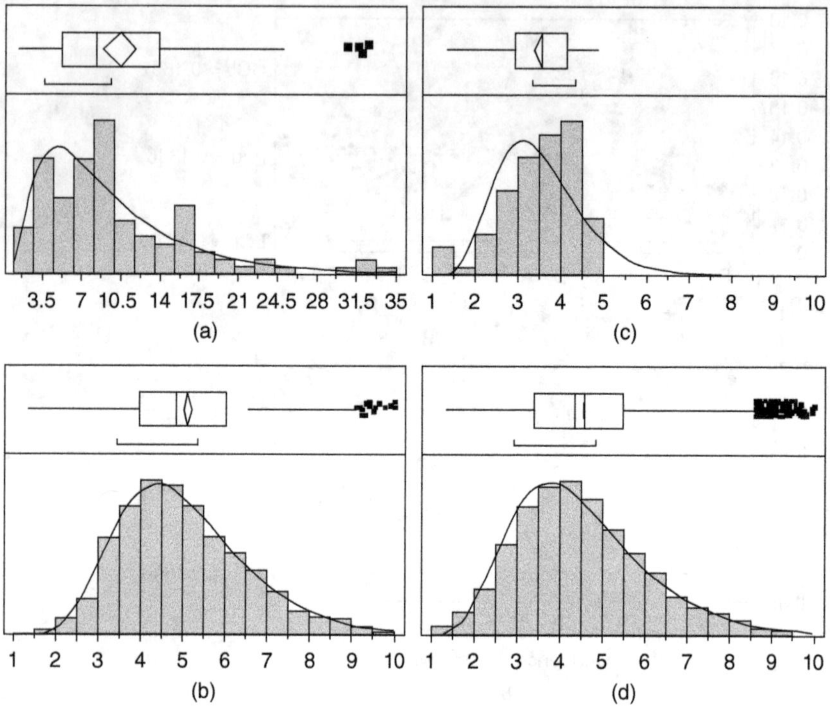

Figure 9.7 Baseline concentration distribution of TRACP5b in various patient populations from four clinical studies: (a) patients with advanced cancer ($n = 105$); (b) patients with prostate cancer ($n = 1427$); (c) postmenopausal patients ($n = 222$); and (d) postmenopausal patients ($n = 7828$). The curved line represents a fit by lognormal distribution. Outlier box plots are shown in the panels above the histograms. *x*-Axis, concentrations of TRACP5b in ng/ml; *y*-axis, relative frequency.

myeloma, osteoporosis, and rheumatoid arthritis were analyzed with and without the addition of CTx at 0.145 ng/ml. The spike recovery was calculated by subtracting the endogenous concentration of the unspiked sample from the spiked sample. The recovery data of the percent difference from the nominal value of the various populations are presented in Figure 9.8. The results show that all normal male and female serum recovered with the acceptance criteria (dashed line of ±25%). Recoveries were more variable with patient samples, although most of them were acceptable. The recovery data set of normal females had the best precision and was used as the control in Dunnett's method to compare with the other groups. The side bar in Figure 9.8 shows that the circles of all groups were overlapping with the control group (bold circle). All groups had p-values less than an α of 0.05, while that of the rheumatoid arthritis group was 0.0587, possibly due to a slight interference from the rheumatoid factor in the samples.

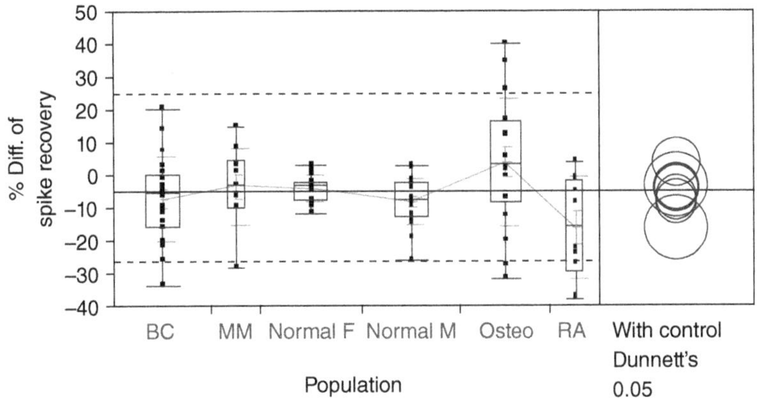

Figure 9.8 Selectivity test of CTx in sera from healthy and patient populations. The populations included healthy male (normal M, $n = 20$) and female (normal F, $n = 20$) and patients with breast cancer (BC, $n = 20$), multiple myeloma (MM, $n = 10$), osteoporosis (Osteo, $n = 18$), and rheumatoid arthritis (RA, $n = 10$). The normal male age range was 45–74, while that of female was 38–67. Serum samples were analyzed without and with the addition of CTx at 0.145 ng/ml. The spike recovery was calculated by subtracting the unspiked sample concentration from that of spiked sample for each person. Percent (%) difference from the nominal value of 0.145 was calculated and the data presented in box plot for each group of population. Each box presents the mean, standard deviations, and range. Side panel: comparisons with a control using Dunnett's method.

Specificity and Sensitivity of Novel Biomarker on Drug Effect

In-study data should be compiled and assessed to determine if (i) assay performances of standards and SCs are within acceptance criteria, (ii) patient sample concentrations are in the ranges expected for predosed samples, and (iii) postdosed patient sample concentrations are consistent with the values expected for previous pilot study (such as PD profile, dose effect, and recovery to normal after treatment change, if applicable). Data should be evaluated for false positive (specificity) and false negative (sensitivity). For example, the variability of the SCs could reflect the assay noise of incurred samples. This variance should be smaller than that of the difference in disease subject vs. normal (specificity) and that of concentration change due to drug vs. placebo effect (sensitivity).

The data mining, interpretation, and linkage of biomarker data to clinical outcome are beyond the scope of this chapter. It should be borne in mind that a useful predictive index of a biomarker is dependent on whether it is used as one of a panel of biomarkers. The statistical power is affected by the sampling number (number of subject and time point) and assay stringency. Therefore, the bioanalytical laboratory should be included in the biomarker team to interpret data use, the model being tested, and the analytical component required for the predictive index desired.

Regulatory Issues

Regulatory guidance for bioanalytical validation indicates that the paramount objective of a bioanalytical method validation is to ensure that the assay is reliable for its intended use [38, 39]. This philosophical viewpoint is central to many guidance documents, conference reports, and "white papers" for bioanalytical validation of assays for conventional small-molecule drugs and macromolecular therapeutics [30, 40–44]. Currently, two basic bioanalytical approaches are being applied to biomarker validation. One is based on the FDA bioanalytical drug assay guidance and is generally referred to as "GLP-like" [41, 42]. The second is based on the CLSI and Clinical Laboratory Improvement Amendments (CLIA) [32, 45]. Presently, no regulatory guidance has been given concerning what experiments should be performed and what data are necessary and appropriate for biomarker assay validation [25]. This has led to inconsistent application of validation procedures and acceptance criteria for biomarkers in clinical trials.

For biomarker assays to be characterized and used for drug development support, they should be GLP-like [19, 21]. However, the guidelines must be flexible, to allow assay- and technology-specific applications to meet specific clinical study objectives on a case-by-case basis. Thus, general guidelines for bioanalytical laboratories should be developed to include both good scientific and clinical sense with flexibility to develop specific applications [25]. At the same time, documentation of change and version control of biomarker methods is important, especially to those who produce elements of the assay or kit.

The FDA has set up an interdisciplinary pharmacogenomic review group (IPRG) to carry out specific review function for the assessment of biomarker qualification data. The team will evaluate study protocols and review study results of novel biomarkers of drug safety, using appropriate preclinical, clinical, and statistical considerations. The team will then develop recommendations and guidance for the submission of biomarker data, assess the original biomarker context proposal through voluntary data submission (VXDS, where the X indicates a wide range of data sources), and then evaluate the qualification study protocol together with the sponsor to reach a consensus protocol [46]. Characterization data of novel biomarkers to support efficacy can be categorically in the voluntary data submission, and discussion with the IPRG will clarify the data requirement for drug labeling.

Documentation

Documents and data on biomarker characterization to support drug development must be kept in a secure place and made accessible for internal or external audits. All processes and data generation should be traceable. It is critically important to define and document the reference standard. Because novel biomarkers do not usually have established reference standards, characterization of the standard material is dependent on the supplier laboratories from commercial or in-house sources. Documentation such as a certificate or record of analysis and a certificate of stability should be obtained. If these are not available, at least the source, identity, potency (or concentration), and lot number of the reference material should be documented. A consistent and sufficient supply of the same reference material lot should be reserved and used for the duration of a study and program if interstudy comparisons are intended.

Similar to the reference material, documentation of the source, identity, potency or concentration, and lot number of critical ligand reagents and their stability should be kept. If possible, the same lot of the capturing ligand should be used throughout a study. It is prudent to test for method robustness using multiple lots of reagents during advanced method validation.

Partial Validation for Change Control

The time span of biomarker characterization can be a long one. Changes in operators, assay equipment, reagent lots, and a variety of other unforeseen events usually happen and will require additional validation. The significance of the alteration in a method will determine the extent of validation necessary, leading to cross-validation, partial validation, or complete revalidation of an assay to control for changes during or between studies. A validation plan should be prepared for partial validations, describing the intended purpose of change, scope, experiments to be carried out, data comparison, and a prior acceptance criterion for method equivalence. A common set of clinical samples analyzed by both methods can be used to show method equivalence. The sampling size should be powered by statistics based on the desired criterion of acceptance and the method variability. If the method change is expected to be extensive, a new validation would be conducted according to the new analytical procedure, followed by a new written validation report.

Conclusions and Perspectives

In this chapter, we have focused on a discussion of quantitative method validation and assay for biomarker characterization to support drug development using the fit-for-purpose approach. The main processes include defining the purpose of application, assay development, validation and use meeting the purpose and statistical treatment, and interpretation of analytical and clinical data.

Due to the heterogeneous nature of many protein biomarkers, discrepancies may exist between the endogenous forms and the reference standard. It is important to understand what is being measured and the clinical meaning of the moieties measured as related to disease progression and drug effect. For example, during the course of development of a LBA method, biophysical methods (such as mass spectrometry coupled to ligand binding solid-phase extraction) can be used to select reagents and identify the chemical species being captured in the ligand solid phase.

In addition, there are differences in the standard curve blank control matrix from the intended biological matrix. Parallelism is a characteristic that must be tested using authentic samples from the target population. Another important consideration is that binding of the biomarker to endogenous binding proteins, receptors, and the xenobiotic drug candidate may cause assay anomaly. Different methods may measure different forms of the biomarker, which often leads to the confounding data in the literature. Specific binding reagents and processing steps can be designed and manipulated to measure "bound" or "free" species of the biomarkers.

In contrast to the exploratory phase, biomarker characterization requires more intense rigor in a quantitative method. For example, method robustness and long-term storage stability must be established; sufficient and consistent supplies of reagents should be assured to sustain the applications through multiple studies; and the same methods should be validated and applied in the participating clinical sites on sample collection and in bioanalytical laboratories on sample assays. Only when a robust method has been established and the assay performance defined, is it possible for the biological variability (clinical variance) to be distinguished from analytical variability and be assigned its clinical significance.

Statistical treatment of biomarker assay performance enables assessing if the method will meet the study purpose before its application. Statistical data from in-study performance continue to add information to the method appropriateness, especially with data from the SCs, which provide trends in authentic sample stability and reagent lot variability. Data from multiple studies and drug candidates (from the first-in-class to later backups, or to another candidate of related pathways) contribute to the accumulated knowledge of the biomarker.

Whereas the exercise of utilizing biomarkers to support drug development of a single or several candidates can be controlled within a pharmaceutical company, the movement of characterization and qualifications of a novel biomarker will depend on the concerted efforts of multiple institutes/organizations. The collaborative effort of sharing and using the same method and reagents may be a challenge. If different methods are to be used, the data interpretation and comparison will add complexity to this challenge. For example, meta-analysis of clinical biomarker data from publications has been problematic because method details and validation parameters were missing to enable proper comparisons. It will be helpful if a consortium organization will handle data repository and communication (such as standardization of publication) of novel biomarkers. This is the goal of the Biomarker Consortium (launched in October 2006) by the FDA, Foundation for the National Institutes of Health, and Pharmaceutical Research and Manufacturers of America through their various working groups.

Acknowledgments

The authors are grateful to the Biomarker Subcommittee in the Ligand Binding Assay Bioanalytical Focus Group of the American Association of Pharmaceutical Scientists for ideas and collaborations in the fit-for-purpose method validation concept. They thank Binodh DeSilva, Daniel Burns, Chad Ray, Han Gunn, and Lennie Uy for their contributions to the content and critical review of this manuscript.

References

1 Pepe, M.S., Etzioni, R., Feng, Z. et al. (2001). Phases of biomarker development for early detection of cancer. *J. Natl. Cancer Inst.* 93: 1054–1061.
2 Wagner, J.A. (2002). Overview of biomarkers and surrogate endpoints in drug development. *Dis. Markers* 18: 41–46.
3 Bjornsson, T.D. (2005). Biomarkers: applications in drug development. *Eur. Pharm. Rev.* 1: 17–21.
4 Wagner, J.A., Williams, S.A., and Webster, C.J. (2007). Biomarkers and surrogate end points for fit-for-purpose development and regulatory evaluation of new drugs. *Clin. Pharmacol. Ther.* 81: 104–107.
5 Berger, J. and Wagner, J.A. (2002). Physiological and therapeutic roles of peroxisome proliferator-activated receptors. *Diabetes Technol. Ther.* 4: 163–174.
6 Wagner, J.A. (2002). Early clinical development of pharmaceuticals for type 2 diabetes mellitus: from pre-clinical models to human investigation. *J. Clin. Endocrinol. Metab.* 87: 5362–5366.
7 Krishnamurti, U. and Steffes, M.W. (2001). Glycohemoglobin: a primary predictor of the development or reversal of complications of diabetes mellitus. *Clin. Chem.* 47: 1157–1165.
8 Lee, Y.H. and Pratley, R.E. (2005). The evolving role of inflammation in obesity and the metabolic syndrome. *Curr. Diabetes Rep.* 5: 70–75.
9 Chu, C.S., Lee, K.T., Lee, M.Y. et al. (2006). Effects of rosiglitazone alone and in combination with atorvastatin on nontraditional markers of cardiovascular disease in patients with type 2 diabetes mellitus. *Am. J. Cardiol.* 97: 646–650.
10 van Doorn, M., Kemme, M., Ouwens, M. et al. (2006). Evaluation of proinflammatory cytokines and inflammation markers as biomarkers for the action of thiazolidinediones in type 2 diabetes mellitus patients and healthy volunteers. *Br. J. Clin. Pharmacol.* 62: 391–402.
11 Haffner, S.M. (2007). Abdominal adiposity and cardiometabolic risk: do we have all the answers? *Am. J. Med.* 120 (9 Suppl 1): S10–S106.

12 Liu, S., Tinker, L., Song, Y. et al. (2007). A prospective study of inflammatory cytokines and diabetes mellitus in a multiethnic cohort of postmenopausal women. *Arch. Intern. Med.* 167: 1676–1685.

13 Sweep, F.C., Fritsche, H.A., Gion, M. et al. (2003). Considerations on development, validation, application, and quality control of immuno(metric) biomarker assays in clinical cancer research: an EORTC-NCI working group report. *Int. J. Oncol.* 23: 1715–1726.

14 Rubin, M.A. (2004). Using molecular markers to predict outcome. *J. Urol.* 172 (5 Pt 2): S18–S21.

15 Quraishi, I., Rishi, M., Feldman, M. et al. (2007). Clinical validation of breast cancer biomarkers using tissue microarray technology. *Appl. Immunohistochem. Mol. Morphol.* 15: 45–49.

16 Yu, L.R. and Veenstra, T.D. (2007). AACR-FDA-NCI cancer biomarkers collaborative. *Expert Rev. Mol. Diagn.* 7: 507–509.

17 Khatami, M. (2007). Standardizing cancer biomarkers criteria: data elements as a foundation for a database. Inflammatory mediator/M-CSF as model marker. *Cell Biochem. Biophys.* 47: 187–198.

18 Bloom, J.C. and Dean, R.A. (eds.) (2003). *Biomarkers in Clinical Drug Development.* New York, NY: Marcel Dekker.

19 Lee, J.W., Weiner, R.S., Sailstad, J.M. et al. (2005). Method validation and measurement of biomarkers in nonclinical and clinical samples in drug development: a conference report. *Pharm. Res.* 22 (4): 499–511.

20 FDA CDER (2006). Using disease, placebo, and drug prior knowledge to improve decisions in drug development and at FDA. Case Studies Across Companies Disease Models at FDA: Overview and Case Studies (Diabetes and Obesity), p. 65.

21 Lee, J.W., Devanarayan, V., Barrett, Y.C. et al. (2006). Fit-for-purpose method development and validation for successful biomarker measurement. *Pharm. Res.* 23 (2): 312–328.

22 Kummar, S., Kinders, R., Rubinstein, L. et al. (2007). Opinion: compressing drug development timelines in oncology using phase '0' trials. *Nat. Rev. Cancer* 7: 131–139.

23 Stoch, S.A. and Wagner, J.A. (2007). Biomarker analysis as a decision-making tool in drug discovery and development: implications for peroxisome proliferator–activator receptors. *Int. J. Pharm. Med.* 21: 271–277.

24 Lee, J.W., Smith, W.C., Nordblom, G.D., and Bowsher, R.R. (2003). Validation of assays for the bioanalysis of novel biomarkers. In: *Biomarkers in Clinical Drug Development* (eds. J.C. Bloom and R.A. Dean), 119–149. New York, NY: Marcel Dekker.

25 Lee, J.W., Figeys, D., and Vasilescu, J. (2007). Biomarker assay translation from discovery to clinical studies in cancer drug development: quantification of emerging protein biomarkers. In: *Genomics in Cancer Drug*

Discovery and Development, Advances in Cancer Research (eds. G.M. Hampton and K. Sikora), 269–298. Amsterdam: Elsevier.

26 Mager, D.E., Wyska, E., and Jusko, W.J. (2003). Diversity of mechanism-based pharmacodynamic models. *Drug Metab. Dispos.* 31: 510–518.

27 FDA (2004). Innovation or stagnation? Challenge and opportunity on the critical path to new medical products. https://web.archive.org/web/20090512144844/www.fda.gov/oc/initiatives/criticalpath/whitepaper.pdf (accessed 28 April 2018).

28 Ferguson, R.E., Hoschstrasser, D.F., and Banks, R.E. (2007). Impact of pre-analytical variable on the analysis of biological fluids in proteomic studies. *Proteom. Clin. Appl.* 1: 739–746.

29 Banks, R.E. (2008). Preanalytical influences in clinical proteomic studies: Raising awareness of fundamental issue in sample banking. *Clin. Chem.* 54: 16–17.

30 DeSilva, B., Smith, W., Weiner, R. et al. (2003). Recommendations for the bioanalytical method validation of ligand-binding assays to support pharmacokinetic assessments of macromolecules. *Pharm. Res.* 20: 1885–1900.

31 Lee, J.W., O'Brien, P., Pan, Y. et al. (2008). Development and validation of ligand binding assays for biomarkers. In: *Ligand-Binding Assays: Development, Validation and Implementation in the Drug Development Arena* (eds. M. Khan and F. JWA). Hoboken, NJ: Wiley.

32 National Committee for Clinical Laboratory Standards (1999). Evaluation of Precision Performance of Clinical Chemistry Devices: Approved Guideline. CLSI Document EP5-A.

33 Tholen DW, Linnet K, Kondratovich M, et al. (2004). Protocols for Determination of Limits of Detection and Limits of Quantitation; Proposed Guideline. NCCLS EP17-P.

34 Lee, J.W. and Ma, H. (2007). Specificity and selectivity evaluations of ligand binding assay of protein therapeutics against concomitant drugs and related endogenous proteins. *AAPS J.* 9: E164–E170.

35 Westgard, J.O. (2003). Internal quality control: planning and implementation strategies. *Ann. Clin. Biochem.* 40: 593–611.

36 Heeschen, C., Dimmeler, S., Hamm, C.W. et al. (2003). Prognostic significance of angiogenic growth factor serum levels in patients with acute coronary syndromes. *Circulation* 107: 524–530.

37 Vural, P., Akgul, C., and Canbaz, M. (2006). Effects of hormone replacement therapy on plasma pro-inflammatory and anti-inflammatory cytokines and some bone turnover markers in postmenopausal women. *Pharm. Res.* 54: 298–302.

38 ICH Guidelines. (1994). Text on Validation of Analytical Procedures, Q2A. International Conference on Harmonization, Geneva, Switzerland.

39 Shah, V.P., Midha, K.K., Dighe, S. et al. (1992). Analytical methods validation: bioavailability, bioequivalence, and pharmacokinetic studies. *Pharm. Res.* 9: 588–592.

40 Shah, V.P., Midha, K.K., Findlay, J.W.A. et al. (2000). Bioanalytical method validation: a revisit with a decade of progress. *Pharm. Res.* 17: 1551–1557.

41 Code of Federal Regulations (2001). CFR Title 21, vol. 1, Good Laboratory Practice for Nonclinical Laboratory Studies, rev. Apr. 1.

42 FDA (2001). Guidance for industry on bioanalytical method validation: availability. *Fed. Regist.* 66: 28526–28527.

43 Findlay, J.W.A., Smith, W.C., Lee, J.W. et al. (2000). Validation of immunoassays for bioanalysis: a pharmaceutical industry perspective. *J. Pharm. Biomed. Anal.* 21: 1249–1273.

44 Miller, K.J., Bowsher, R.R., Celniker, A. et al. (2001). Workshop on bioanalytical methods validation for macromolecules: summary report. *Pharm. Res.* 18: 1373–1383.

45 Code of Federal Regulations (2001) CFR Title 42, vol. 3, Clinical Laboratory Improvement Amendment, rev. Oct. 1

46 Goodsaid, F. and Frueh, F. (2007). Biomarker qualification pilot process at the US Food and Drug Administration. *AAPS J.* 9: E105–E108.

10

Applying Statistics Appropriately for Your Biomarker Application

Mary Zacour

BioZac Consulting, Montreal, Quebec, Canada

Introduction

> Learning how to apply statistical analyses … may take a lifetime of trial (and sometimes error), as it has done in the author's case. There is no evidence that biomedical investigators of the present generation are on a steeper learning curve. [1]

Sitting in the Drill Hall at the University of Toronto with row upon row of shifting, sighing, and coughing co-sufferers, cold fingers clutching one of my two well-sharpened HB pencils while I carefully write out each step of my standard deviation calculation. Statistics 101 Christmas exam. One finger rasping down rows and across columns of tables, p-values, degrees of freedom, these are the memories I share with countless scientists who have studied statistics in preparation for their research careers. Although I left that course and others with respectable marks and a seemingly clear understanding of the null hypothesis and the normal curve, I found myself some years later, scratching my head as I considered a sheet whose bland columns of numbers belied the painstaking hours spent injecting, isolating, pipetting, incubating, and centrifuging, from which they had been distilled. Looked at one way, it seemed to suggest one thing, but was that the correct approach? Looked at another way, the picture was different. The conclusion to be drawn would influence the direction that the entire project would take.

How to apply the right statistical test correctly to a particular research question, given the many choices, assumptions, and caveats, is not always obvious; indeed "gross misunderstandings of the purpose and functions of statistical analysis are apparent in applications to research grant-giving bodies and ethics committees, in manuscripts submitted to journals and sometimes in published

Biomarkers in Drug Discovery and Development: A Handbook of Practice, Application, and Strategy, Second Edition. Edited by Ramin Rahbari, Jonathan Van Niewaal, and Michael R. Bleavins.

papers" [1]. In fact, there seems to be a growing field of forensic biostatistics, pointing out data analysis errors that have crept into respected publications [2–4]. This is not surprising, considering the advent of genomics, proteomics, and other-omic-style research, in which the ability to produce huge quantities of complex data is outstripping the common biologist's understanding of the mathematical issues concerning data interpretation.

Scientists focused on biomarker research in the drug development industry face an additional set of challenges, where the pressures of the business world to use less time and fewer resources, to focus only on the immediate goal, and to quickly release positive findings to investors or other onlookers are often at odds with the expert, thorough, and careful consideration that the more challenging research problems may require. Understandably, biomedical scientists who are often highly expert in their scientific discipline are not necessarily so in the mathematical issues underlying analysis of their data, and many are also without the luxury of a biostatistician with whom to consult at times of indecision. Although some lucky others may feel that there is no need to master the field when they can simply rely on in-house expertise, it is important to recognize that a working knowledge of the factors affecting data analysis facilitates better communication between the statistician and researcher, which, in turn, generally leads to better research design. Furthermore, the question of how to interpret data wisely is not limited to applying the correct statistical test in the correct manner; it also includes understanding the limits and pitfalls of the technical procedures used to generate the data. In this regard, biostatisticians, too, face challenges, such as being unfamiliar with or intimidated by the rapidly evolving and complex technical methods used in the lab. As so well voiced by Ransohoff [5], "how well do biologists … appreciate the nature or seriousness of problems that can be introduced by methods of handling data … ? How well do epidemiologists or biostatisticians understand technical details of specimen collection, handling, and analysis in a way they can use to anticipate and manage specific sources of bias?"

In addition to an understanding of both technical and mathematical issues, there is a third element important to teasing out the elusive meaning of the data, which I like to think of as the HP factor (Hercule Poirot, that is). Born of intellect and experience, modeled by a higher-order view, a wider vision of extraneous factors that may play roles in the outcome, infused with a measure of skepticism and questioning of assumptions, this factor may more commonly be known as "judgment."

Now, I sincerely hope that I am not marked, Rushdie-like, by the statisticians guild for this comment, since this softer science is undoubtedly in the realm of the subjective, and as such, in danger of being classified in the same bin as bias, data selection, and the bubonic plague. To my mind, however, judgment is critical to the success of most endeavors, including data evaluation and statistical analysis. Even though the rules of application of such analyses may be straightforward (in our day, in fact, completed for us by a simple tap of an index finger on a keyboard or mouse), understanding potential errors in our data and/or determining which of many analytical possibilities best fit the situation presented by our own experimental paradigm and data set are not, and this is where judgment often comes in.

There is certainly a large mass of statistical reference material already published and available to answer questions that most biologists might have; however, most texts and articles focused on statistical analysis issues are written by, well ...statisticians. This is a good thing, of course, but sometimes brings with it the negative corollary of a specialized dialect and viewpoint. Joe or Josephine Biologist may find these, at best, difficult to adapt to their own needs, or, at worst, frankly unintelligible.

This chapter is addressed to fellow Joes and Josephines. It is written in plain language by a non-statistician who has often grappled at the interface where experimental results are transduced through an intermediate state as digits, before manifesting themselves, newly metamorphosed, into conclusions that flutter their wings right into our collective scientific knowledge database. Whereas some basic statistics explanations may embellish its text, this is by no means intended to serve as a statistics manual, nor to explore the many intricacies of the field; in fact, statisticians may well be appalled by its gaps and oversimplifications. The publisher, on the other hand, is likely to be pleased at the paucity of equations, since it is common knowledge in the industry that book sales drop with each equation in a text. This chapter addresses data analysis issues pertinent to biomarker research in drug development. More specifically, it is intended to clarify some common misconceptions or confusion concerning which situations are appropriate for alternative choices of statistical analysis and to integrate statistical issues with nonstatistical issues that also affect our data interpretation.

Fundamentals

Is That Significance Significant? (and Conversely, the Significant Insignificance)

Statistics texts all deal with the issue of type I and type II errors, or, respectively, the false finding of a significant difference where none exists and the false finding of no significant difference where one indeed exists. They seem invariably to omit mention, however, of how a statistically significant difference that truly exists and is truly shown can yet remain truly irrelevant. This particular possibility gets no press time in statistics manuals because it remains in the realm of that softer science mentioned above, judgment.

As trained scientists sensitized to the dangers of making inferences based on anecdotal or potentially chance findings, we seem as a group to have acquired the tendency to over exalt the statistically significant finding when it makes its appearance. A *p*-value of 0.05 or less has become our holy grail, the golden chalice that shines from the paragraphs of our grant and investigational new drug (IND) applications, presaging, we hope, the sweet wine to spill forth from them. This, of course, is not unnatural when one considers how often we are faced with budgetary or time constraints on the number of subjects that can be tested and therefore on the ease of statistically confirming such differences when they truly exist. The converse situation of which we need to be mindful, however, is that with enough samples being evaluated, even tiny differences can be statistically significant, and furthermore, even differences that are not tiny might still not be relevant. What constitutes a "significant" increment of an analyte, in the physiological or clinical sense, is likely to vary with the situation; it is not necessarily on the same scale as a statistically significant increment. An additional point to consider, addressed below in the section on measurement error, is whether the magnitude of a statistically significant difference surpasses the magnitude of inaccuracy of measurement.

A qualitative judgment on the significance of significance, rather than simply adjusting our blinders to keep out the sun while logging *p*-values, is certainly worth a moment of silence on our parts. Furthermore, the converse case is also relevant; for example, what if one's inner Hercule is suspicious that a clinical significance may be lurking behind the harmless façade of $p > 0.05$? How could we reconcile the lack of statistical significance in that case? One possibility is that too few samples were evaluated to achieve the statistical power to reveal significance (discussed further later in the chapter). Another is that part of the group of samples tested reacts differently than (an) other part(s), resulting in a total effect across the disparate subgroups that does not reach statistical significance, even though it reflects a clinical situation of great significance.

The latter concept is central to the development of personalized medicine, an emerging field that seeks to improve disease treatment by targeting molecularly distinct subpopulations within a given group, such that response rates to treatments are increased and the administration of ineffective and potentially toxic treatments to inappropriate individuals is diminished. As an example, selection of breast cancer patients according to *ErbB2* tyrosine kinase status has had a significant impact on management of the disease; moderate, but significant, response rates to *ErbB2*-targeted antibody therapeutics in some studies might have been overlooked had study groups not first been selected for its overexpression [6]. Some basic concepts concerning statistical methods of relevance to pharmacogenomics are discussed later, in addition to some of the common errors in their application.

Air the Error

Because statistical descriptions and inferences are based on probability, a certain amount of chance (random) error is always inherent in any statistically based conclusion. This error is routinely quantified by upper and lower confidence limits and mentioned in results sections. Unfortunately, such routine error expression seems to induce a false sense of security in readers that *all* the error inherent in the results has been taken into account.

In fact, there are numerous sources of nonrandom error, generally covered by the blanket term *bias* (i.e. when subjects, specimens, or data in the groups being compared are inherently different or are handled differently in a way that systematically introduces a signal into data for one of the groups compared). Differences such as types of tubes used or length of time until spinning or time samples are maintained frozen prior to analysis would fall into this category. Bias has been deemed such a serious problem in nonexperimental research that some experts consider such studies guilty of bias and erroneous results until proven innocent [7]. The term *nonexperimental research* refers to research where the effects of some perturbation to the system are not being tested, but rather, information is simply gathered (i.e. in epidemiological studies) or quantified (i.e. in laboratory analyses, such as most proteomic analysis). Bias is also a serious and often overlooked problem in experimental research (i.e. where effects of deliberate manipulation of variables are tested). A recent perspective article on how potential flaws in preclinical research may play a major role in late-phase drug failures discusses this in detail, particularly as it relates to animal models [8].

Measurement imprecision, included by some under the term *bias*, but by others considered separately, constitutes an additional significant source of error. In the case of continuous data quantifications, although measurement error almost always occurs, it is almost never quantified, or at least such quantification is in no way accounted for in the result reported. In the case of noncontinuous data, such as in diagnostic tests that may generate a yes/no dichotomy, measurement error can have a different but equally profound effect, potentially contributing to false positives or negatives.

Continuous Data Quantifications

Every assay result is only the best approximation to the true value that a particular assay is able to produce. This is readily acknowledged by all, and, in fact, it is standard practice in drug development to quantify the inaccuracy of assays using known amounts of a reference standard. How the quantification should best be accomplished is beyond the scope of this chapter, but the total error approach wisely incorporates both the components of random error and bias/measurement error in a single measure and is discussed further elsewhere [9–11].

Despite the depth of consideration given to measurement error, there appears to be an element of lip service to its acknowledgment, since error quantifications are generally used in the industry simply to determine acceptance or rejection of an assay's results, rather than to modify or qualify in any way the numerical result produced by the assay. Measurement error is thus consistently brushed under the carpet, with results often reported in ways that dramatically overstate their precision. As an end result, differences from one measure to another in some cases become infused with spurious meaning, when they actually represent nothing more than noise in the signals. One can only speculate on how many drugs that failed during late-phase trials might never have gotten that far had the measurement error in their "promising" preclinical results been accounted for in the data analysis.

Sometimes measurement error is compounded dramatically by introducing more than one erroneous measurement into a final calculation of the result, and again the norm is to ignore this inflated error rather than account for it in the results reported. Let us consider the example of a clinical trial examining the effect of a treatment on the activity of a disease biomarker in white blood cells. In this example, two assays are required: one to quantify the disease biomarker in a lysate prepared from the cells and another to quantify the amount of protein that was culled in each cell preparation. The activity result from the first assay is then normalized against the protein result from the second assay, to give an activity per mass of protein as the final assay result. Looking more closely at the potential error in this example, we should note that standard acceptance criteria for bioassays generally include specification that a certain level of demonstrated measurement error of the method will be considered acceptable, with data still included "as is" in the analysis. Assuming a typical acceptance criterion of up to ±20% inaccuracy, two measurements of a single identical sample of true value X could therefore quantify as differing from each other by 40% of X, and no one would raise an eyebrow. Further, Example 10.1 shows how compounding the same measurement error margin of ±20% over the two assays in this example elevates that potential acceptable difference between identical samples to 220%.

Example 10.1 One assay for enzyme activity and one for protein content of sample are performed, with the final activity reported per mass of protein. Assuming ±20% measurement inaccuracy in each of the two assays, we see that if patient A has a true activity of 10 activity units, acceptably accurate activity assays would reveal values of 8–12 units, and each milligram of protein would be quantified as 0.8–1.2 mg. Therefore, within each assay, an identical sample could have acceptable measurement error, resulting in a 150% increase of one value relative to the other (12 units/8 units, or 1.2 mg/0.8 mg). Normalizing enzyme activity units to milligrams of protein inflates this possible discrepancy between two identical sample results from 150% to a 220% apparent increase in one sample vs. another; that is, the most extreme acceptable values from each assay, normalized against each other, would be 8 units/1.2 mg and 12 units/0.8 mg, for a final resulting range of 6.7–15 units/mg as acceptably accurate results from replicate aliquots of the same sample.

This example demonstrates the compounding of only two measurement errors. Imagine the case when more measurements go into a calculation, each having its own error: for example, when using reporter gene induction in a cell-based system as a model of induction of an *in vivo* response. In this case, there are four measurements that need to be done to control properly, each with its own measurement error range: the reporter gene expression in transfected cells challenged with the experimental treatment, its expression in transfected cells challenged with only the vehicle in which the treatment compound is dissolved, and those same two conditions in cells that were transfected with empty vectors (as a control for the effects of the vector backbone on the system). All too often researchers using models of this sort tend to "simplify" by expressing induction of reporters as simply a fold-increase over the vehicle effect, essentially pretending that the complex and compounded error does not exist.

Noncontinuous Data

Measurement error can also have an important effect on noncontinuous data, an effect which, as noted above, is discrete from random error and is rarely taken into account. In this case, misclassifications can occur, such as false negatives or positives. We can all imagine the personal impact of such misclassifications, such as being falsely diagnosed with a deadly disease, or conversely, not receiving appropriate treatment because of a false-negative test result. Misclassification errors also decrease the power of studies to detect significant differences and have the potential to obscure or even falsify study results, leading to equally devastating post-study impact on a much wider scale.

For example, successful efforts to develop an effective vaccine for malaria, considered to be the deadliest pediatric disease, have been hampered by measurement errors leading to misclassifications. In this case, the fact that vaccine efficacy (VE) is a ratio of the number of infected, vaccinated persons to the number of infected, nonvaccinated persons gives rise to the misconception that false-positive or false-negative diagnosis in the numerator and denominator will cancel out or balance. Actually, since measurement error increases as the lower limit of detection (LOD) is approached, a partial effect of vaccination causing lower-level signals at the time of measurement (such as a slowing of parasite growth, the biomarker for VE) can lead to systematic error in the measurement of the test group that is not matched in the control group, and an overestimation of VE [12].

The end effects of misclassification error of this sort can vary substantially, depending not only on the goal of the testing and the decisions that are dependent on it but also on the overall frequency of the disease being studied. For instance, elevated false-positive rates for a test may have a huge impact in the case of rare diseases, and, conversely, elevated false-negative rates may have a bigger impact with very common diseases. Example 10.2 shows that if a test for a rare disease X (occurring in 1 in 1000 people) gives 99% accurate positive results in people who truly have the disease and 95% accurate negative results in people who truly do not have it, any positive test result it gives will be false in 98% of the cases.

Example 10.2 A diagnostic test has the following known accuracy: It returns a positive result in 99% of patients who truly have the disease and a negative result in 95% of patients who truly do not have the disease. The disease is rare, occurring in the general population at a rate of only 1/1000 people. Although you have no known predisposition or reason to think you have the disease, you decide to get tested and are shocked to learn that the result has come back positive. What are your chances that the diagnosis is a false positive?

$$\text{False positive probability} = 1 - \text{true positive probability}$$

$$\begin{aligned}\text{True positive probability} &= \frac{\text{probability of true positives}}{\text{total of true and false positive probabilities}}\\ &= \frac{0.99 \times 0.001}{(0.99 \times 0.001) + (0.05 \times 0.999)}\\ &= 0.019\end{aligned}$$

$$\text{False positive probability} = 1 - 0.019 = 0.981$$

Your chances are about 98%

that the diagnosis of positive is false.

The test would have to be 99.9% accurate for negative results before it would have a zero-rated false-positive detection rate. Whereas, intuitively, one might think of 95% and 99.9% accuracy as both sounding pretty good, you can see that the seemingly small gap between them could have a surprisingly important impact.

What to Do?

Samuel Taylor Coleridge once said that literary reviewers are usually people who would have been poets, historians, and biographers if they could, but having tried their talents at one or the other, and failed, they turn to critics. As in other areas of life, with the issue of measurement error, too, it is much easier to be a critic than to do something well oneself. What should one do? This is a very good question indeed, and, feeling a little uncomfortable under Samuel's gaze from the sentence above, I shall attempt to come up with some useful suggestions.

First, the appropriate approach is likely to vary according to the situation, and the most crucial issue to consider when choosing an approach to dealing with measurement error is the potential risk of the erroneous measurement to the end user of the drug, device, or process under study. Experiments and/or data interpretation should then be designed to err on the side of caution, if risk is an issue.

Other key determinants may include practicalities such as resources and time available to approach the problem, and questions of clinical or physiological relevance of the potential error margin, thresholds, or measurement ranges of interest. If, for example, a 10- to 100-fold increase in biomarker expression would be required for clinical relevance, and the maximum estimated measurement error is in the range of less than threefold, perhaps there is no issue. The discussion above, concerning whether significant differences are clinically relevant, also applies to the relevance of measurement errors.

If measurement error is of a magnitude that could be clinically relevant, its quantification becomes more important. Perhaps because of the status quo of ignoring measurement error quantification in biomedical methods (for which, after all, error quantification is much more challenging than for analysis of less complex models), this is not a widespread practice in the drug development industry. One way of acknowledging this error while avoiding complex quantifications might be to use the definition of maximum allowable measurement error that is already made routinely for the purposes of acceptance criteria as a threshold of level of difference, below which results would either not be considered to reflect anything but error, or would be more carefully scrutinized in some predefined way. In other words, if you have determined that up to a doubling in endpoint values in a pretreatment vs. posttreatment sample could be attributed solely to compounded measurement error rather than to treatment *per se*, one might decide not to consider effects to be a result of treatment unless they surpass this measurement error threshold. Of course, the fact that a method has been validated according to across-the-board standards such as having better than 15% or 20% total error does not mean that it actually produces results that are this far from the nominal values of the reference standards, but merely that it may do so at times, and that results of assays where these limits are exceeded will be rejected from the study. Error quantifications may well indicate much less error than the maximum allowed for assay acceptance, so given the circumstances it may be worth putting the time in to determine what the actual error is rather than simply assuming that the worst possible error has always occurred. Actual error might be revealed by data mining from previously amassed quality control data and/or on a per-batch basis using internal standards, for example. In these cases, it is as simple as quantifying the proportional values obtained for accuracy rather than simply noting a pass or fail, as seems to be the more usual quality control assessment in industry practice.

In the case where biomarker analysis involves more complex sources of error than a single or a few analytical assays (such as for epidemiological studies or meta-analyses to qualify particular biomarkers for their intended use), more complex uncertainty quantifications would be required. Various such methods are available, including Bayesian [13] and Monte Carlo simulations [14]. Bayesian approaches, which stem from an alternative view of probability compared to the more common frequentist inferential statistics, are touched upon further later. Monte Carlo simulations represent the brute-force type of computation, basically executing a huge number of possible permutations using random draws from all the potential inputs, so as to spit out a final probability function (made up of all the output values from the various random inputs) that comprises the compound error from all the unknown inputs. In a simple example, consider the film *Groundhog Day*, where a man woke every morning to find that it was the same day as the one before. The film consisted of this man going through his interminable repeated days, making different choices when the fixed events of the day occurred, as they did every day; you can think of this man as being stuck inside a Monte Carlo simulation (well, a simple one, since his own behavior choices were the only independent variable with randomly chosen modifications; the range of outcomes were consequences of his varied choices from that one variable). In questions of compound error, of course, there would be many more variables, also all being sampled randomly over and over again. One advantage of choosing a Monte Carlo approach to resolving complex error sources should your experimental paradigm present them is that user-friendly software is readily available, due to its common use for risk analysis in financial and engineering applications.

Once measurement error has been quantified, by whatever means, error distributions could be used to express uncertainty limits around results, analogous to the random error-related confidence intervals common in statistical analyses. These would help to clarify how much of the apparent differences between sets of results might be an artifact of the measuring tool(s) and not necessarily an effect of the treatment that one wishes to measure. In many cases, such as the reporter gene model example above, literature search will reveal readily applicable statistical models for specific types of assay that incorporate compound sources of error into the calculation of statistical significance of differences [15].

Phillips and Lapole suggest a much simpler approach to partially quantify method error uncertainty without any complicated analysis [16]. Like Pablo Picasso, who once said it took him his entire life to learn how to paint like a child, these authors advocate a return to earlier ways: in this case, the grade-school lesson of rounding figures. Rounding can adjust the measurement scale to reflect the limitations of the measurement error. In their words: "Rounding does not create imprecision – the imprecision exists even if we do not accurately report it." Thus even though the measurement instrument may spit out values with several decimal places in them, samples with results in the range of 100 for a test demonstrating 2% inaccuracy (i.e. 98–102) might better be reported rounding to the nearest 5 rather than even to 1's, let alone the decimal points reported by the instrumentation.

Apart from attempting to quantify the error, contain it within a wider measurement scale, or set thresholds of interest that exceed it, as discussed above, we should also not forget to implement strategies to reduce it, when it is possible to do so. For instance, internal controls such as reference standards or paired samples can reduce the impact of the between-assay variance proportion of measurement error.

For noncontinuous data, misclassification (rather than numerical differences) is the end effect of measurement error. In this case, a suggested approach to avoid the negative impact of measurement error might be to assume that error will happen, and design studies to have higher power from the beginning, in order to minimize the impact of error on the power of the study. Statistical tests for noncontinuous data are already inherently less powerful than those for continuous data, and a lower initial power level in the experimental design can further magnify the impact of measurement error. For example, Gordon and Finch illustrate how the same misclassification error rate causes much more power loss in a study that has lower power to start with than in a study that has higher power to start with. In their example, given exactly the same error rate, if the study is designed to have 99% power, it ends up with 95.5% after the error is taken into account, whereas if it is designed to have 80% power, it ends up with only 65.4% [17]. Therefore, specifying a higher power in the experimental design gives an advantage of robustness to the obscuring effects of misclassification error. In real terms, designing an experiment with higher power most often translates into recruiting more subjects; these authors advocate that "paying the cost up front" strategy.

Guilty by Association (and Other Misuses of Correlation Analysis)

"Statistics show that of those who contract the habit of eating, very few ever survive." This William Wallace Irwin quote [18] may bring a small smile to your lips. Should we conclude that eating is a bad habit that causes death? Ridiculous, right? Especially for well-educated biomedical scientists. How about this one: "Greater adiposity is associated with lower everyday physical functioning, such as climbing stairs or other moderate activities, as well as lower feelings of well-being and greater burden of pain" [19]. This quote from a respected medical journal does not seem quite so ridiculous, does it? Whereas the first one is clearly a joke, we might be tempted to nod our heads at the second and experience discomfort in direct proportion to the increases we noted over the past few years in our own level of adiposity.

If we take a closer look at both quotes, we will notice that neither author actually claimed a cause-and-effect relationship. As we all know, however (especially those employed in the marketing department), just putting them together in a small room can't help but to make everyone jump to conclusions. The strength of an association is indeed appropriately quantified by correlation analysis, but this analysis should not be misinterpreted to infer causation. Zou et al. hit the nail on the head when they note that rather than being misused to test hypotheses, correlation analysis should serve to *generate* them [20].

The fact remains, however, that correlation is very commonly used to suggest that A causes B, or, as in the second example above, being fat causes lower physical ability to function and lower feelings of well-being. Such guilt by association certainly is one possibility, but another possibility is that B causes A, or having lower ability to function and poor self-esteem induces weight gain. A third equally plausible hypothesis is that another cause or set of causes is (are) responsible for both the poor feelings/abilities and the weight gain. A fourth is that the observer has just observed and documented the effects of coincidence; if an insufficient sample number were examined, for instance, a fluke might be documented. Even with larger sample numbers, statistical conclusions are always subject to chance, with the type II error (or p-value) describing the percent chances that the "significant difference" is actually just due to chance.

Some extra consideration needs to be given to this when choosing an appropriate biomarker for one's particular purposes. Biomarkers are, after all, by definition, a marker for something else, or an associative entity. If a particular biomarker has been qualified properly, its association with the disease of interest will not be due to chance, but that still does not imply causation. Whether a marker is simply associated, is a cause or an effect of disease, or is a cause or effect of something else causing a disease needs to be considered carefully in the experimental design. The issue of what considerations are important to determine causation is discussed in a landmark paper by Hill [21]. The reader is further referred to an interesting perspective commentary on this subject by Phillips and Goodman [22].

A further frequent misuse of correlation is to test agreement between methods. As pointed out by Bland and Altman [23], even poorly agreeing methods can produce quite high correlations, and for methods to be in perfect agreement their results would plot against each other along the line of equality, whereas results plotting on any straight line would give perfect correlation. Furthermore, the test of significance may show that the two methods are related, but this is irrelevant to the question of agreement, and indeed, it would be amazing if two methods designed to measure the same quantity were not related. One can almost hear their teeth gnashing as they bemoan, "Why has a totally inappropriate method, the correlation coefficient, become almost universally used for this purpose?" It seems clear that this is a case of copycat crime, perhaps a case for the remarkable Hercule Poirot?

Checking the Dessert Menu First

In much the same way that Hercule Poirot might consider the temptations of the dessert menu before settling on his choice of main courses, we all need to consider potential data analysis issues before settling on our experimental design. Some questions need to be asked long before the data are in hand. For example:

1. Is the assay to measure the biomarker as good as it can be, or the best type of assay given the circumstances? Can you make it better? All assays have error, but the more you fine-tune the accuracy, precision, specificity/sensitivity, and robustness via appropriate optimization, the more your result will reflect the marker you wish to measure rather than the shortcomings of your technique. Assay optimization is beyond the scope of this chapter, but should definitely be completed before starting the study. Assay-specific parameters such as accuracy, precision, and specificity of the measurements all have a direct impact on the uncertainty of the statistical analysis outcomes.
2. Have you chosen the best biomarker, or should you rethink from the bottom up? To evaluate this, you must be well aware not only of the qualification that a biomarker is relevant, linked appropriately, and specific to the pathology of interest (important issues, beyond the scope of this chapter), but also of quantitative issues of what to expect in the sample and control groups you have chosen.

   If one wishes to use a biomarker to differentiate between normal healthy persons and those with a disease, for example, one of the many questions to be considered prior to deciding which indeed is the most informative biomarker for the job concerns the potential degrees of measurement between them. In many cases, the biomarker of interest is present in normal healthy persons as well as in those with disease, but with different typical levels in the two populations. In this case, it is very important to establish a priori the descriptive characteristics of the two groups in order to define what a clinically relevant treatment effect might be and how one might judge that such an effect occurred. Published literature, unpublished previous data, and pilot studies all provide potential resource material for defining these descriptive characteristics and should be evaluated carefully in advance. Since both populations will present with a range of values, relevant descriptive characteristics include not only mean values but also the variances, ranges, and the difference between the population distributions (or extent of overlap, if that is the case). One should consider that the biomarker that is more consistent in its levels (i.e. less variance) may be a better choice than one that shows bigger mean differences between populations (but higher variability from person to person within a group). In the latter case, the variability can translate into overlap of two populations, which can be quite a complication when one wishes to discriminate between them.

Another point for reflection in advance concerns integrating the quantitative ability of the measurement technique you intend to use with the quantitative characteristics of the populations you hope to discriminate between, or the effect size that you have determined you will need to achieve clinical relevance. For example, the smaller the differences are between expression levels of a biomarker that you wish to use to discriminate between two populations, the more sensitive the quantitation method you choose needs to be. Consider the example of wishing to measure the treatment effect of drug X on a disease by assessing whether a biomarker is increased from disease-typical levels to the level seen in healthy persons. Would one be able to accomplish this with statistical confidence if the two populations showed close levels or overlap between their two distributions? If the difference between the populations is not big but the measurement error margin is, one may be doomed to an incapacity of discerning what one hopes to, regardless of the potential effects of the treatment. In this case, it is wiser to rethink the fundamental premises of your study design and methodology before starting. Although this may sound like an obvious point, when one considers the common practice of ignoring measurement error and ignoring the compound error introduced by common study practices such as normalizing one assay result to another, it should perhaps give pause for thought.

3. Is your design sufficiently powered to detect what you wish to detect? Statistical power is the ability of a study to enable detection of a statistically significant difference when there truly is one; power in the statistical sense is analogous to the sensitivity of a diagnostic test [24]. Eng says: "One could argue that it is as wasteful and inappropriate to conduct a study with inadequate power as it is to obtain a diagnostic test of insufficient sensitivity to rule out a disease" [25]. In clinical trials, power often receives attention, perhaps thanks to ethics committees that have pointed out that it is frankly unethical to expose people to the risks and discomforts of research if the study has no potential for scientific gain, such as if a study was not designed to include enough people to adequately test the research hypothesis [26]. In laboratory experiments, on the other hand, statistical power is rarely estimated [1] and seems in general over-looked and poorly understood.

   To ensure sufficient power, one needs to think ahead and choose an appropriate sample size. Factors playing a role in how one would define *appropriate* include the following:

   a. the smallest meaningful difference between the two means being compared that the investigator would like the study to detect (this is usually based on subjective criteria such as judgment of clinical importance; additionally, sensitivity and error margins of the measurement method may influence its choice).
   b. the estimated standard deviation within each comparison group (this can be based on a pilot study, prior literature that has been carried out with the same outcome measure, or subjective criteria).

c. the desired power level (this is usually 80% or higher; however, although high power is desirable, it is always in a trade-off with a fixed amount of time and resources for the study).
d. the significance criterion (usually set at 0.05).
e. whether one- or two-tailed analysis is planned.

There are different formulas for computing power or sample size, depending on the experimental design and whether data are continuous or categorical; Example 10.3 shows an illustrative calculation using a formula suited to a standard design involving comparing means of two groups.

Example 10.3 You have qualified your biomarker *X* as an efficacy marker in your disease model. In your studies qualifying *X* as a biomarker and developing a method to assay it, you found that its expression was low in rats with your disease of interest (10 ± 3 units; mean ± SD) but higher in healthy rats (100 ± 30 units). You also noted that interventions that increased *X* were associated with improvements in disease morbidity and that previous groups of animals studied tended to show biological variability of about 30% coefficient of variation between rats in a treatment group. You wish to design an appropriately powered study comparing untreated diseased rats to those treated with a single dose of your new treatment *Y*, to determine if *Y* demonstrates efficacy, and have decided that a mean 5-unit change in *X* is the lowest meaningful change that you wish to be able to detect. You will settle for standard levels of a significance level of 0.05 and a power of 80%. How many rats do you need in each group?

$$n \text{ of each group} = \frac{(\text{SD1}^2 + \text{SD2}^2)(z_{\text{crit}} + z_{\text{pwr}})^2}{D^2}$$

where SD1 is the standard deviation of group 1 and SD2 that of group 2 to be compared to each other (for a minimum difference from 10 to 15 units, this would be 3 and 4.5, respectively); the *z*-values are standard normal deviates corresponding to selected significance criteria and statistical powers (z_{crit} for two-tailed $p < 0.05$ is 1.960, z_{pwr} for 0.80 power is 0.842; these and other values are found in standard statistical tables and software packages); and *D* is the minimum difference that one wishes to detect (5 units).

$$\begin{aligned} n &= \frac{(3^2 + 4.5^2) \times (1.960 + 0.842)^2}{5^2} \\ &= \frac{29.25 \times 7.851}{25} \\ &= 9.2 \end{aligned}$$

Round up to 10 rats in each group.

Numerous web page calculators and statistical programs are available to perform this type of calculation for you; however, it is difficult to use them properly without a working knowledge of the factors affecting power. This subject is well explained by Eng [25, 26].

If your study is already complete and you found a significant difference, you do not have to be concerned as to whether the study was sufficiently powered. This is akin to wondering if you put enough dynamite under the bridge after you have blown it to smithereens. If, however, you found no significant difference and you are feeling uncomfortable about the fact that you did not consider calculating the power before starting the study, you may be tempted to do so after the fact, by plugging the values actually observed in your study for standard deviation and the difference between the populations into the power formula. This practice, dubbed *retrospective power analysis*, is problematic, because it does not tell you what the power to find the smallest meaningful difference was, only what the power was to find statistically significant the difference that actually existed between the groups in your analysis; furthermore, this "observed" power will be inversely related to the p-value observed, so it does not add meaningfully to the information already obtained with the p value [26]. Instead, approaches involving confidence intervals or chi-square tests are recommended to guide the interpretation of negative results [27]. Some practical considerations about power and sample size are discussed further in the section: Taking Care of Business.

4. Have you planned in advance which groups you are going to compare for significant differences? Planned comparisons allow the researcher a greater rejection level α or p-value than do unplanned comparisons. This is because if you first look at your data, then choose your comparisons, you are introducing bias into your analysis, and the rejection level will have to be tightened to compensate for that [28]. Planning ahead also allows you clear-headed time to consider which comparisons you really need when your design includes many different groups. The more comparisons you make, the more the p-value will have to be adjusted against false positives, in effect making it harder for you to see a true positive. Therefore, it is important not to compare groups that are irrelevant to your experimental goals. This is discussed further in the section: t-Test vs. ANOVA.

Taking Care of Business

As touched upon previously, the challenges faced by industry-based scientists often include business-related considerations that are not considered in most statistics manuals. In particular, time and resource constraints are often significant, as well as pressures to proceed at a pace that is intended to satisfy management, investor, or profit-related concerns, but is often incompatible with sufficient depth of study or optimal experimental design. In short, between the business and science aspects of the industry, a strong pressure front constantly flows from the business side to the science side.

At times when shortsightedness and lack of understanding of the impact of financial decisions on scientific practice threaten to nullify the potential uses of our scientific output, we should certainly consider exerting a pressure front in the opposite direction, from science to business. Since we cannot realistically hope to change the pressures of the business world, however, we need to at least consider modifying our experimental design and/or data interpretation approaches to take their effects into account.

For example, let us consider the case of a study design for analysis of a biomarker as a pharmacodynamic (PD) endpoint in a clinical trial, where each patient provides a pretreatment control sample to be compared with a time series of their posttreatment samples. Given an endpoint assay with a high between-experiment variance component, the wisest experimental design might be to study all of a given patient's posttreatment samples within the same assay, together with the matched pretreatment sample.

Instead, scientists are often pressured to run assays and provide PD results as soon as the first or first few of several posttreatment time point samples become available in clinical trials. This business-driven decision is often made despite freezer stability of samples, simply because drug development companies are anxious to get the news about their product's effects in clinical trial announced as quickly as possible; good news has an economic payoff, and similarly, the earlier that bad news is known, the better the chances for diverting precious resources before they are wasted. In the laboratory, however, the corollary of this pressure may be that precious volumes or tissue samples from a pretreatment control are used up in the first analysis, and later analyses on samples not yet collected will have to be calculated in assays that lack this internal per-patient control. As well as introducing bias between the time points that were assayed within a single run and those that incorporate between-assay variability, the added component of variability between assays may add unnecessary noise, obscuring potential treatment effects.

Certainly, as drug development scientists we cannot ignore concerns such as continued supply of cash by investors; indeed, such issues are supremely important to whether a scientific endeavor can even be continued. The point is that the effects of making less than ideal experimental design choices, too, should not be ignored or glossed over (which is, in fact, what tends to happen). Often, no consideration is given in the statistical analysis model to sample processing issues such as that described above, with raw values instead being logged into tables as if they had all been generated in a single assay of balanced design, regardless of the circumstances under which the data were generated.

Obviously, there is no one correct approach to remedy such a situation, and the researcher must decide what best fits the situation, hopefully with the help of a statistician who understands the sample processing issues. For instance, in some cases the best solution might be to convince decision-makers to wait for a controlled comparison of pre- and posttreatment samples, while in others it might be including an internal standard for normalization and changing to a ratio-data design, yet others might be amenable to another analysis strategy. The point is that analysis and/or interpretation of the data should be rethought in the light of particular business decisions that have affected their generation.

In another example of rethinking assay design strategies to respond to the challenges of the business world, one might consider various strategies to increase statistical power in the face of time and resource constraints that typically limit the number of patients one can study. Browner et al. [29] and Eng [25] list a number of strategies for minimizing the sample size while maintaining power.

1. Use continuous measurements instead of categories; this is because, for a given sample size, statistical tests for continuous values are mathematically more powerful than those used for proportions.
2. Seek to decrease the variability of the measurement process (i.e. improve the precision of your processes, optimize your assays; the less standard deviation, the more power).
3. If possible, use paired measurements, matching each patient to his or her own control. Paired measurements are more powerful than unpaired measurements.
4. Add control subjects, even if case subjects are too difficult to obtain. Although it takes more complex equations to calculate power in unbalanced designs than if both groups have the same size, this will still add power.
5. Expand the minimum expected difference – perhaps that which you have specified is unnecessarily small and a larger expected difference could be justified, especially if the study is preliminary, or intended as a screening test. The smaller the expected difference you want to pick up, the greater the number of samples needed to give an equivalent power.
6. Reflect on whether your hypothesis could have merit as a one-tailed, rather than a two-tailed one (although in most cases this is not an option). You can obtain equivalent power using less samples for a one-tailed test. The sample size necessary for a one-tailed design with a 0.05 significance criterion is the same as that for a two-tailed design with 0.10 significance, if all other factors are equal.

Many researchers think of power calculations as a futile exercise, likely to specify an unrealistically high sample size, given the resources on hand for the study. In fact, proper thought to power and the factors that contribute to it could, in many instances, lead to cost and resource savings. Rather than applying an across-the-board animal toxicity experimental design including, for example, 10 animals per group, Example 10.4 shows a case where five animals per group would be sufficient; as this example is built on a data set similar to the one used in Example 10.3, which showed 10 animals per group as being appropriate for the experimental paradigm, it also illustrates that the number of samples that need to be in a test group for appropriate power depends strongly on issues specific to each particular research question.

Thus, industry-based scientists grappling with business cases, profit-and-loss issues, how best to spread their research budget over the necessary items, and other financial-based decisions with an impact on scientific quality would do well to manage their resources appropriately by incorporating power calculations into the design of their laboratory research.

Example 10.4 Imagine an example similar to that described in Example 10.3, except that the biomarker is now a toxicity marker, which you have qualified and shown low under normal conditions in rats (10 ± 3 units; mean $\pm$ SD) but markedly increased in response to toxic reactions (10- to 100-fold). You wish to test a compound for toxicity using the same experimental design as in Example 10.3, but with a few different assumptions:

1. Because of being cautious about the risks of toxicity, you want to ensure higher than 80% power this time: You want to ensure 95% power, so you have more certainty that any finding of no significant difference in the toxicity marker would not be attributable to too few animals being studied.
2. Because you are not concerned if the toxicity marker decreases in association with the treatment, but only if it increases, you adopt a one-tailed design.
3. Because toxicity so markedly increases the biomarker, you decide that the smallest meaningful increase you wish to detect can be greater than in Example 10.3. You decide that 20 units of biomarker would be greater than 3SD over the normal condition and is thus an acceptable threshold, so you set D at $20 - 10 = 10$.

How many rats do you need in each group?

Using the same formula as in Example 10.3, but substituting a z_{pwr} value of 1.645 (the value for 95% power), a z_{crit} value of 1.645 (the one-tailed value for a $p < 0.05$ significance level), a D value of 10, and SD1 and SD2 values of 3 and 6, respectively:

$$n = \frac{(3^2 + 6^2) \times (1.645 + 1.645)^2}{10^2}$$

$$= \frac{45 \times 10.824}{100}$$

$$= 4.87$$

Round up to five rats in each group.

The Odd Risk

Biomarkers are often used in studies that involve nominal data, such as a positive/negative dichotomy for a test result. There are various ways of quantifying the strength of the relationship between groups when it comes to this type of data, which is sometimes also referred to as *proportional data*, in reference to the fact that one measurement is often expressed as relative (proportional) to another (i.e. a percentage of patients responding, or the accuracy or sensitivity of a diagnostic test). Proportional data is often presented in a 2×2 contingency table. The chi-squared statistic is appropriate to test the null hypothesis of independence between row and column variables. Data are often described, however, in terms of difference of proportions, relative risk, or odds ratio. These are not statistical terms, but rather, quantification terms. Often, these are misunderstood, sometimes leading to serious confusion and misinterpretation of published research. When considering the use of diagnostic tests classifying disease states, based on biomarkers, the terminology of odds and risks is actually less useful than true and false positive and negative rates, since it does not capture the aspect of error in disease classification. It is nonetheless important to be able to interpret research correctly where results are described in odds/risks terminology. This subject is well explained in a review by Sistrom and Garvan [30], from which most of the following is paraphrased.

Assuming a data table comparing some outcome measure in two groups, one of which was exposed to some variable and the other of which was not, the difference in proportions is the proportion with the outcome in the exposed group minus the proportion with the outcome in the unexposed group. It is always between −1.0 and 1.0, and equals zero when the response Y is completely independent of the explanatory variable X.

Relative risk is a measure of association between exposure to a factor and risk of a certain outcome, and is defined as risk in the exposed groups divided by risk in the unexposed groups. Public health impact depends not just on relative risk but also on absolute risk, so just looking at relative risk can be misleading. For example, the public health impact of a vaccine that halves the risk of an infection is very different for a rare (say, two cases per million people) or a common disease (say two cases per 10 people); in both cases, the relative risk for vaccinated subjects would be 0.5 compared to unvaccinated subjects, but in the former case, for every million people vaccinated, only one would be saved, whereas in the latter case it would be 100 000 people saved [31].

Odds are the probability that the outcome does occur divided by the probability that it does not occur, with the *odds ratio* being the ratio of the odds of an outcome in one group ("exposed") vs. another ("unexposed"). A key concept about using odds to estimate risk of an event is that the relationship between the odds ratio and the risk ratio depends on the outcome frequency of the event. If the odds of an event are greater in one group than another, it does *not* mean that the risk ratio is increased by the same amount. This is a very common misunderstanding, but in fact, odds ratios and risk ratios are similar only when the outcome being studied is rare (i.e. about 10% or less probability in the unexposed group) [32]. Odds ratios are not a good estimate of risk ratio when the outcome is common in the population being studied. In that case, to obtain correct risks, one needs to apply the equation relating odds to risks, as shown in Example 10.5.

Example 10.5 The equation relating risk ratio (RR) to odds ratio (OR) is as follows:

$$RR = \frac{OR}{(1 - Pr_0) + (Pr_0)(OR)}$$

where Pr_0 is the probability of the outcome in the unexposed group. Schulman et al. [33] compared the frequency of referral for cardiac catheterization for white men (exposed) to that for black men (unexposed). Black men were referred 90% of the time, and the odds ratio for black to white men being referred was 0.6 : 1. The risk ratio for black men to white being referred was

$$\begin{aligned} RR &= \frac{0.6}{(1 - 0.9) + (0.9 \times 0.6)} \\ &= \frac{0.6}{0.64} \\ &= 0.94 \end{aligned}$$

This example shows how an odds ratio of 0.6 for referral to cardiac catheterization for black men vs. white men (as reported by Schulman et al. [33]) does not mean that blacks were referred 40% less often than whites, as was reported in extensive media coverage following publication of that study. Once the correction for a 90% referral frequency for black men was applied, the risk ratio for black men to be referred was 0.94, or 6% less relative to white men rather than the 40% less that the media coverage of the original study had interpreted [34].

Practi(sti)cal Magic: Which Witch is Which?

During my time as a contract research organization scientist, I was astonished at the frequency of requests from pharmaceutical and biotechnology clients to perform inappropriate statistical treatment of the results; it seems that in the drug development industry at least, there is an appreciable level of confusion over when particular tests are appropriate and what underlying assumptions must be met for them to be valid. Accordingly, some refresher basic statistics background is summarized briefly below.

Parametric vs. Nonparametric

Parametric tests are so named because they use parameters, such as mean and standard deviation, in their computations. Student's t-test and the analysis of variance (ANOVA) are examples of this type of test. Parametric tests assume that the data sets are normally distributed and that the various groups have roughly equivalent variance between members of the group. They are prone to falsely indicating a significant difference between the groups if these assumptions are not met. In that case, nonparametric tests provide an alternative. Nonparametric tests rely on ranking of samples relative to each other, and make fewer numerical assumptions. They are less likely to commit type I error, but usually also have less power to reveal true significant differences.

Researchers need to evaluate data with descriptive statistics in order to decide whether the assumptions of the more powerful parametric tests are met or whether nonparametric tests would be the more appropriate choice. Some assumptions apply to both types of test and must also be considered. In that regard, both parametric and nonparametric tests assume that samples included in the analysis are independent of each other. How does one judge whether these assumptions are met, and what should one do about it if they are not?

Independence

To judge whether samples are independent, one needs to ask if within any one group of data included in the analysis, the value of one data point is somehow related to another. If the answer is yes, these samples are not independent. For example, if biomarker expression is measured in two separate tissue samples three separate times each and you enter $n = 6$ data items into the analysis, you have violated the independence criterion. The three replicates of each sample are related to one another (from the same tissue sample); there are actually only two independent measures, not six. Since $n = 2$ is not enough samples to analyze statistically, potential plans to redress this analysis error could include either omitting statistical analysis in favor of descriptive results only or redesigning the experiment to include additional tissue samples.

Often, experiments are designed to measure biomarker expression repeatedly in a single subject, such as in the case of pharmacodynamic measures at different time points following administration of a drug in clinical trial patients. In this case, the independence assumption is not violated, since the related measures will be analyzed as different groups rather than being included within one group. Repeated-measure statistical tests are then applicable, such as the paired t-test (for two measures) or repeated-measure ANOVA (for more than two measures).

To reiterate, then, if samples are not independent of one another and they do not constitute repeated measures, the correct statistical approach is to redesign your experiment. Although more complex analysis strategies do exist for clustering correlated samples within groups, there is no manipulation when using simple tests of significance that will correct for samples in one data set being selected such that the choice of one sample depends on another sample being picked.

Normally Distributed Data

How do you judge if your data are normally distributed? The first step is to perform descriptive statistics and plot the data out as a histogram. In a normal distribution, about 68% of the data falls within 1 standard deviation (1 SD) of the mean, and 95% within 2 SD of the mean. Many statistics programs will compute normality for you using statistical tests. The Kolmogorov–Smirnoff test, for instance, has nothing to do with martinis, being instead one such test for the normal distribution of data. If you have around $n = 25$ or more in each group, you can safely apply the simple eyeball test to your data, accepting data as normal if points fall into a roughly bell-shaped curve when plotted as a histogram [35].

What if you are not sure whether your data are normally distributed? If you have small data sets, for instance, normality can be difficult to determine visually. Similarly, statistical tests do not have much power to make the determination in this case. At this point, one's own personal choice comes into play: If you are conservative, you might choose to play safe with a nonparametric test; if not, you might choose a parametric test. As with other life choices, there are risks associated with either: with parametric tests you risk error of type I, whereas with nonparametric ones you risk type II error. The best solution might be to collect reasonably large data sets so as to avoid this uncertainty.

Assuming you do have a large enough data set to be certain, and what you are certain of is that the data are not distributed normally, well, what then? One solution might be to transform data mathematically such that the transformed data sets become normally distributed, in which case the more powerful parametric tests can then be used. Some commonly used transformations and appropriate types of data for their use [35–38] are listed in Table 10.1. If data are still not normally distributed with equal variances following mathematical transformation, a nonparametric test should be chosen for analysis. Some appropriate nonparametric equivalents of parametric tests are shown in Table 10.2.

Table 10.1 Common data transformations.

Type of data	Example	Transformation
Proportional data	0–100%	Arcsin of the square root transformation
Poisson distributed (counts of random events)	Number of cells of a particular type in a given volume of blood	Square root transformation
Variance proportional to mean squared (i.e. standard deviation proportional to mean)	Serum cholesterol levels in patients	Log transformation
Ratio data (normalized to internal standard on a per-assay basis)	Blots normalized to a housekeeping gene	Log transformation
Highly variable quantities, where variance is proportional to the mean to the fourth power (i.e. standard deviation proportional to mean squared)	Serum creatinine levels in patients	Reciprocal transformation

Table 10.2 Nonparametric equivalents of parametric tests.

Parametric	Nonparametric
Student's *t*-test (unpaired)	Mann–Whitney test (Wilcoxon rank sum)
t-test (paired)	Wilcoxon signed-rank test
t-test (paired), in cases where data are not symmetrical around the median	Sign test
ANOVA (without repeated measures)	Kruskal–Wallis test
ANOVA (with repeated measures)	Friedman's test

Source: Adapted from Ryder and Robakiewicz, 1998 [35].

Equivalent Variances Between Groups

As with the question of whether data are normally distributed, equivalence of variance between groups needs to be evaluated prior to choosing appropriate tests of significance. Sometimes, homogeneity of variance from one group of data to another can be amply judged by plotting a frequency histogram of each group and seeing how spread out the data are, or calculating the variance for each group and qualitatively assessing if the numbers are very different or appear close. It is simple enough, however, to apply the appropriate statistical test, the F-test for two groups, or the F_{max}-test for more than two.

If variances differ between groups and are not rendered equivalent by mathematical transformation (see Table 10.1), you may still be able to use parametric statistics. If, for example, the t-test is appropriate to your experimental design, and variances are unequal between groups but not extremely disparate, you may choose to use the *t*-test for unequal variances (in the case of extreme differences in variance, however, a nonparametric test is the recommended alternative).

If your intended parametric test is ANOVA, on the other hand, the assumption of equivalent variance between test groups is essential [35]. If variances are not equivalent and transformation also does not render them equivalent, you must test statistically using a nonparametric equivalent (see Table 10.2).

Simple Regression and Correlation Analysis

At times, there seems to be some confusion about simple linear regression vs. correlation analysis. This issue is well summed up in a review by Zou et al. [20], from which the following information is, for the most part, culled. The two are similar mathematically, but their purposes are different. Both relate to a function that describes the relationship between a given X and Y; however, regression analysis generally focuses on the form of that relationship and correlation generally on the strength. Furthermore, regression focuses on evaluating the relative impact of a predictor variable on a particular (dependent) outcome, whereas correlation's purpose is to examine the strength and direction of the relationship between two random variables.

Correlation analysis commonly involves either the Pearson coefficient (r) or the Spearman coefficient (r^2), with the former reflecting proportional changes in one variable when the other is changed and the latter using ranks and reflecting instead a monotonic relationship between two variables (i.e. whether one tends to take either a larger or a smaller value than the other, but not necessarily with a proportional change in one variable when the other one is changed). If data sets are skewed or contain outliers, the Spearman coefficient rather than the Pearson is the appropriate choice.

Interpretation of correlation coefficients is often rather qualitative, with the sign indicating the direction of the relationship (positive or negative). Values range from 0.0 (no correlation) to 1.0 (perfect correlation), with 0.5 generally being thought of as moderate, 0.8 as strong, and 0.2 as weak. Statistical significance can also be computed, by formulating a null hypothesis of no correlation and a one-sided alternative hypothesis that the underlying value exceeds or is less than that value, then computing the *z*-test statistic and rejecting the null hypothesis based on the *p*-value.

Simple linear regression analysis results in an r^2 value that is calculated on the basis of Pearson r coefficient and reflects the fraction of the variability in y that can be explained by the variability in x through their linear relationship (or vice versa). As with correlation analysis, a finding of strong linear relationship in a regression analysis does not mean that the variable causes the outcome (as discussed in the section on misuses of correlation), and should not be interpreted that way. Student's *t*-test can be used, for example, to test if there is a linear relationship (i.e. null hypothesis of slope = 0) or whether the y-intercept is a particular value. As a salient point, no extrapolation outside the range of values of the independent variable in the regression analysis should be used to make any predictions.

t-Test vs. ANOVA

Using inappropriate tests of significance for the circumstances, such as testing differences between more than two groups by means of repeated t-tests rather than by ANOVA, is exceedingly common [35]. Don't do it! Repeating pairwise comparisons ignores the experimental effects in all but the two groups being compared and increases the number of false-positive "significant results" expected; multiple means comparisons following ANOVA correct for the increasing chances of false positives by adjusting the acceptable hypothesis rejection level.

If you have only two groups to compare to each other and they meet the assumptions of parametric tests, use Student's t-test. If you are testing more than two groups, however, use ANOVA. Since a standard ANOVA will tell you if there is a statistically significant difference somewhere in your data sets but will not tell you where, you need to apply a multiple means comparison test after the ANOVA, to pinpoint the culprits. There are many tests available, each suited to different types of data, so it is advised that you consult the help menu of your statistical program to decipher which one best fits your own experimental design. As a rule of thumb, do not choose to make any more comparisons than are appropriate for the goals of the experiment, or you risk losing the ability to detect significant differences that may truly exist. This is because multiple means comparison tests apply corrections that tighten the rejection level in proportion to an increased group comparison number (i.e. effectively lowering the p-value required to find a difference significant). In general, you can make one less comparison between groups than the total number of groups, without affecting this rejection level; however, if you exceed this, the rejection level is adjusted. Dunnett's test, for instance, is suited to designs such as a dose–response curve, which include a matched control sample and several different doses; this test compares each group only to the control (not to each other), to answer the question "Which dose(s) give(s) a response" but not "Was the response to one dose different than to another?" Example 10.6 shows how a significant difference between the control and treatment groups could be missed if more comparisons were made than necessary.

Example 10.6 You wish to test for whether any of several doses of your treatment has a significant effect on your outcome. You have collected samples a, b, c, d, (i.e. four groups), where a is the control and b, c, d represent responses to three different concentrations of your treatment. Assuming that your highest dose, d, shows a p-value of 0.02 with respect to control, this difference would be reported as significant by Dunnett's test; that is, three comparisons would be made (*b* vs. *a*, *c* vs. *a*, *d* vs. *a*), one less than the total number of groups, so no adjustment to the rejection level would be necessary to compensate for multiple comparisons, and the *d* vs. *a* calculated p-value of 0.02 ($\leq$0.05) would be interpreted as statistically significant. If instead of Dunnett's test, the Bonferroni test were chosen, six comparisons would be made (the same three as above, plus *b* vs. *c*, *b* vs. *d*, and *c* vs. *d*). Since this would be two more comparisons than the number of groups, the significance threshold of 0.05 would be adjusted to compensate for multiple comparisons (0.05/6 = 0.008), such that a computed p-value of $\leq$0.008 would be required for significance at the so-called 0.05 rejection level (i.e. $p = 0.02$ would not be considered significant).

One-Tailed vs. Two-Tailed

The tails referred to in your statistical package are the non-wagging variety; the term represents the tail edges of the bell-shaped normal distribution. Tests are one- or two-tailed depending on whether they compute the probability of values exceeding the range expected from members of a population in only one direction or potentially in both directions (i.e. greater than or less than).

Using a one-tailed test is justified only if there was an a priori determination that only differences in one specified direction would be assessed, not because one decides after seeing that the one-tailed p is significant but the two-tailed is not, that one only really cares about one direction anyway. Furthermore, having an expectation that the difference to be found will go in a particular direction is not considered adequate justification for performing one-tailed analysis; rather, one-tailed analysis could be considered appropriate if having a large difference in one direction vs. having no difference at all would have equivalent consequences [39]. Under most circumstances, a two-tailed test is the appropriate choice, and if the researcher is in any doubt about which is appropriate, the two-tailed should be chosen.

Paired vs. Unpaired

Paired analysis is appropriate in the following circumstances:

- When you do a before and after measurement in each subject.
- When you match subjects in pairs (i.e. age, etc.), then treat one of the subjects and not the other. Pairs must be made before data are collected.
- When you compare relatives (i.e. sibling studies).
- When you perform an experiment many times, each time with the experimental and control sample treated in parallel (including log-transformed ratio data, as described in Example 10.7).

Example 10.7 You have done Northern blots of gene expression in mutant animals vs. wild-type animals. Each of your blots shows wildly different signal magnitudes from other blots, due to technical reasons such as exposure time; having mutant and wild-type samples on every blot allows you to control for this variability by expressing data as a ratio of mutant to wild-type values.

Ratios are not normally distributed, but logs of ratios are; therefore, you log-transform your data. Ratios become differences when log-transformed (i.e. mutant/wild type = log(mutant) – log(wild type)), so to analyze this type of data you can take the log of each data point and perform a paired t-test on the transformed data.

In judging whether distributions are normal in paired testing, it is the differences between the two members of each pair that must be approximately normally distributed rather than the two distributions themselves. If data from two sets to be compared do not meet one of the criteria above, they should be analyzed by unpaired testing [35].

Miscellaneous Data

Outliers

Within any set of observations, it is not unusual to obtain occasional data points that vary substantially from others found in the same group of subjects. These *outlier data* may be the result of unknown errors or may represent genuine information, and indeed, whether such data should be included or rejected is the subject of much debate. Methods for detecting outlier data and the use of robust methods of statistical analysis to accommodate such data are complex issues, beyond the scope of this chapter; a thorough treatment of the subject is presented by Barnett and Lewis [40], to which the reader is referred.

BLQ Data

Statisticians sometimes warn against *left-censored distributions*, a term evocative of political intrigue and vaguely racy data sets that actually refers to the more bland reality of omitting results that are below the levels of quantitation of the assay (BLQ). BLQ results are fairly common in clinical and preclinical biomarker analyses and can cause serious problems with interpretation of the data if they are not taken into account, since unreported values are unusable in any statistical analysis of the data. Data imputation, a process to replace unreported BLQ values with valid estimates, can also affect the data interpretation, since all values in a data set affect the measures of central tendency, error, and statistical power. It seems clear that estimating a theoretical value for BLQ data points would allow for more accurate conclusions to be drawn from the data than simply removing data through left-censoring, but what value should indeed be assigned to BLQ data points?

Some researchers report BLQ values as zero, perhaps due to the influence of the sometimes black-and-white perspective of assay validation-style cutoffs. Assigning a zero value regardless of the sensitivity of a given test could be misleading, however, since zero is an absolute, whereas BLQ represents a scale from zero up to an upper limit that differs from one test to another in an assay-specific manner. Furthermore, assigning zero values creates other mathematical difficulties, such as not being able to compute fold-differences of another value with respect to the zero or not being able to work with log transforms of data.

One common approach to address these issues in the data imputation process has been to assign arbitrary nonzero values to BLQ data, most commonly either the LOD divided by 2 or divided by the square root of 2, depending on the skewness of the data set in question [41]. Both appreciable bias and loss of power have been demonstrated with this type of approach, however [42, 43]. Succop et al. compared the bias of alternative methods for imputing BLQ values and concluded that by imputing the values based on either median percentiles below the detection limit or based on predicting the BLQ value from an equation model fit to the noncensored data set, both produced a good correlation between predicted and reported low values. In their example data set, bias between the predicted and true values was only 2.9% this way, compared to 348% overestimation bias when LOD/2 was used to impute the BLQ values [44].

Standard curves of analytical assays typically provide an equation model fit that can be used to calculate values down to the zero level, but assay validation typically limits the data reported to levels that have met a cutoff level that is deemed to have accuracy and precision within proscribed acceptance limits (i.e. the lower limit of quantitation). The above study suggests that although equation-predicted values below the levels of quantitation of an assay may not meet acceptance criteria per se, they still represent more accurate estimations for BLQ data than either arbitrary imputation methods or left-censoring of data sets. Succop et al. recommended that analytical labs should provide a numerical result for all samples analyzed, with a flag of those values that are below the detection limit [44].

Issues Specific to Specialized Fields

Genetic Association Studies

Genetic association studies evaluate the association between specific genetic polymorphisms and disease. They often assess relatively small effects against a noisy background of biological and social complexity and, consequently, tend to lack statistical power [45]. Power calculations are critical for genetics researchers who wish to map susceptibility genes, with chi-square being the most commonly used test statistic for association studies. The noncentrality parameter for the chi-square asymptotic distribution developed by Mitra [46] is the key to computing power and sample size for this type of study. Gordon and Finch present detailed instructions for this (including step-by-step use of a web-based calculator), as well as reviewing other concepts of statistical genetics and the best study/analysis design to optimize power for this type of study [17].

Synthesizing evidence from multiple studies has been used as a means of increasing power; however, it seems that meta-analyses in this field have been fraught with serious flaws. As well as more general concerns, there are genetic issues particular to molecular association studies, including checking Hardy–Weinberg equilibrium (HWE, discussed below), handling data from more than two groups while avoiding multiple comparisons, and pooling data in a way that is sensitive to genetic models [4]. As regards the latter two points, there are always at least three possible genotypes to compare, but in practice most studies reduce the number of comparisons by assuming a specific genetic model, such as dominant or recessive, even though there are often no biological justifications for assumption of the model. Using an inappropriate genetic model or inappropriately pooled data produces misleading estimates of odds ratios. Appropriate meta-analysis methods and avoiding assumptions about genetic models are discussed further by Minelli et al. [45] and Thakkinstian et al. [47].

Bias seems to be another serious problem in this type of study. Some methods for detecting and correcting it are discussed by Sterne et al. [48]; publication bias, for example, may be detected by a simple funnel plot of the data. Not just the meta-analyses but also single studies in this field seem excessively prone to various sources of bias, with some critics going so far as to assert that most documented associations, even those that are replicated, represent nothing other than false positives based on bias [49]. The list of prevalent bias sources is impressive and includes biological plausibility bias, ascertainment bias, publication bias, selective reporting bias, spectrum of disease bias, population stratification, biased selection of controls, lack of blinding in the genotyping process, and genotyping error.

Some of these error sources can be addressed simply through researchers being cognizant of them and striving to avoid them. For example, one can choose to blind genotyping rather than failing to do so. In another example, ascertainment bias can be avoided by scrutinizing controls with the same intensity as cases; this type of bias relates to the situation where affected individuals ("cases") have their DNA resequenced to identify rare variants. Identification of such rare variants in an affected group does not necessarily signify a role in disease (as is often assumed when only cases are ascertained by resequencing) since sequencing DNA from any group tends to turn up a few rare mutations. Instead, a strong and statistically convincing *preferential* presence of variants in the cases compared to controls (that had been scrutinized in the same manner) would support the involvement of the variants in disease [50].

In cases where being cognizant of the potential for bias does not allow us to avoid it, designing experiments and/or data analysis with an expectation of inevitable bias can dampen its influence. For example, designing studies to have higher than desired power can combat the power loss expected from genotyping misclassification errors, as discussed earlier. As another example, various statistical techniques that are robust to bias can be used, such as genomic control techniques or family-based methods to address problems of population stratification [17]. *Population stratification* refers to the situation in case–control studies where control groups are not well matched to case groups and in fact have different typical levels of what is being measured simply because they are from different populations. In this situation, association tests like chi-square may falsely indicate associations, even with as high as 100% probability.

In fact, having appropriate control groups is a key determinant of the validity of genetics association studies, and determining whether controls deviate from HWE is a standard way of evaluating this. Theoretically, disease-free control groups from outbred populations should follow HWE, as should combined cases and controls if they both have a particular disease (i.e. such as in studies where different treatments are evaluated). If they do not, it is a signal of some peculiarity, error, or problem with the data sets that could invalidate the key inferences from a genetic association study [51]. For example, if there is a recessive model and the control group has an excess or deficit of one group of homozygotes, this will directly affect calculation of the odds ratio (the control homozygotes divided by the other genotypes is the denominator of the odds ratio). Overall, if there is a significant deviation from HWE, it should induce some thinking about the study; deviations could result from genotyping errors but may also arise from other sources. For example, HWE deviation may suggest that allele-based estimates of genetic effects are biased, or may give further insight into the population from which the data are derived.

A recent review of studies published in high-quality specialized genetics journals demonstrated that HWE is very commonly tested improperly or inadequately [3]. Of 776 associations tested, only 29% reported on HWE, introducing uncertainty as to if it was tested in the others. Furthermore, where HWE testing was described, the test was applied to the correct control group in only 50% of the associations tested. It is a common error to include the disease cases with the control cases in the HWE test when controls are disease-free. Combined cases and controls should be tested only when the controls have the same disease as the cases; otherwise, only controls should be tested. These authors recalculated HWE for the studies reviewed and noted that in most of the samples where HWE was actually violated, this was either not mentioned in the original article or HWE conformity was actually claimed.

Another common problem in the studies reviewed was the unjustified use of an inappropriate test. A number of different statistical tests could be used to test HWE (including chi-square, exact tests, and Bayesian methods), all based on the conditional probability that there would be the number of homozygotes that actually turned out to be in the sample. The chi-square test was the only test applied to HWE calculation in the studies reviewed, despite the fact that the chi-square asymptotic distribution is inadequate to deal with low genotype frequencies and is therefore not justified in studies involving them; in this case, an exact test provides a simple and superior alternative [3, 52].

A final serious concern in the studies reviewed was that only 7% of the studies had an acceptable power to detect HWE deviation, with most studies being much too underpowered to make any claim of lack of deviations. Whereas power to detect HWE is of secondary importance compared to the prime consideration of power to detect a genetic association, undetected modest HWE deviations could affect considerably the inferences of many genetic association studies [3].

Microarray or Other "Omics"-Type Data

Relating "omics"-style biomarker data (gene/protein arrays, multiple m/z peaks from mass spectrometry, etc.) to clinical outcomes is the focus of a large body of research. Making statistically based decisions about what is expressed or produced differentially and how this relates to outcomes is not simple for these types of studies. For one, the data are highly dimensional, with the number of variables often exceeding the number of samples by magnitudes (i.e. in contrast to the desirable format for statistical comparisons of a group of sufficient numbers of samples for each variable tested). Furthermore, data tend to be very "noisy," with widely different magnitudes of signals and the potential for large irrelevant signals to obscure lesser (but relevant) ones. This translates into an inherent tendency for algorithms fit to the data to be more influenced by noise than by salient interest points (a phenomenon known as *overfitting the noise*).

A recent review of analysis of genomics/proteomics data outlines the steps of quality control, clustering, classification, feature ranking, and validation [53]. In Joe/Josephine terms, these could be described as checking and normalizing the data, grouping like things together, statistically evaluating their likely involvement with the outcome, getting rid of the dead wood, and checking the conclusions on an independent data set.

In the first stage, quality control and normalization of signals reduce the technical variability and ensure that biomarkers discovered are statistically significant. Cui and Churchill reviewed statistical tests that have been specifically adapted to cDNA microarray data analysis [54]; many of the issues discussed are similar for highly dimensional data from other study types. They conclude that fold-change is the simplest method for detecting differential expression of genes, but the arbitrary nature of assigned cutoff values, the lack of statistical confidence measures, and the potential for biased conclusions all detract from its appeal. They also stress the need to normalize data properly, such as by using intensity-specific thresholds, since otherwise an excess of low-intensity genes may be identified as being expressed differentially simply because their fold-changes have larger variance than the fold-change values of high-intensity genes.

Expression analysis steps following this generally fall into two categories, supervised or unsupervised, based on whether or not they are led by prior knowledge. Clustering is an *unsupervised method* that groups genes of similar expression profile together; clustered genes may have diverse biological functions, but this is nonetheless useful for exploring new territory and seeing who is dancing together. *Supervised classification methods*, on the other hand, train a mathematical model using prior biological knowledge of certain genes' functional involvement with each other, then use the model for predictions on other genes. This is a useful approach for building on what is already known, but in cases of limited available information it is difficult for a supervised method to achieve accurate predictions.

Flawed statistical analysis is very common in this emerging field, even in studies published in high-impact factor journals. Dupuy and Simon recently found that half of a survey of 90 microarray studies had at least one of three major analysis flaws [2]. These authors also present a list of analysis do's and don'ts specific to this study type, to which the reader is referred for more detail. Analysis flaws like those discussed in this review and elsewhere [55, 56] can lead to completely erroneous conclusions about the utility of given biomarkers or "signatures" of multiple biomarkers as prognostic or diagnostic tools, and may underlie some of the problems reproducing data in this field.

One of the most common flaws present in the published studies was to perform multiple comparisons without making adequate adjustment for the expected increased false-positive rate. Although the standard of setting p at less than or equal to 0.05 is accepted as a reasonable compromise between type I and II errors when data sets are relatively small, in the relatively large, -omic data sets a p-value this high is not sufficiently stringent to identify disease screening markers with an acceptably low false-positive rate. For example, a microarray containing 5000 genes examined for meaningful relationships between the genes and some outcome using this p-value could be expected to yield 250 false-positive relationships due simply to chance.

Dupuy and Simon suggest that the simplest method to combat this type of error is to use a p-value of less than or equal to 0.001, which would give one false positive for every 1000 genes tested. Several other accepted methods are grouped under the term *family-wise error-rate control* (FWER), but they have been criticized for decreasing power substantially [57]; the Bonferroni correction (dividing the nominal significance level by the number of tests) is the simplest of these. The false detection rate (FDR) test, or Benjamini–Hochberg test, is a favored control statistic for microarray analysis and is included in statistical analysis packages such as statistical analysis for microarray (SAM). The FDR achieves the same control of false significant results while greatly improving the statistical power compared to FWER methods [58, 59]. An FDR of 5% is not the same as a p-value of 0.05: $p = 0.05$ means that 5% of tested genes will be false positive, whereas 5% FDR means that 5% of positive genes detected will be false. For example, given a scenario of 5000 genes tested, with 20 of them being found positive, a p-value of 0.05 means that there could have been 250 false positives (due to chance alone), whereas an FDR of 0.05 means that only one false positive was attributable to chance alone.

A second common flaw was to make spurious claims of a meaningful correlation between clusters of genes and clinical outcome, when, in fact, correlation was detected after clustering using a selection of outcome-related differentially expressed genes. Therefore, genes being related to the outcome was a consequence of the selection bias, but was instead reported as independent proof of a correlation between the clinical outcomes and the gene clusters. A number of classification methods decrease noise in the data by requiring selection of the relevant and informative predictor variables before modeling is performed [60], so users need to be aware that if they have used one of these methods there is no point in performing correlation analysis afterward.

The third major flaw concerned the mechanics of classification model validation. Once a model is constructed, it is generally validated using either an external test set or cross-validation, with the end goal being to use the validated model to predict outcomes for unknown samples. In the studies reporting supervised prediction, Dupuy and Simon noted the common flaw of biased, overly optimistic estimation of the prediction accuracy through an incorrect cross-validation procedure. Different errors in cross-validation procedure were noted, but generally they were all related to failing to have test sets (data used to test the model's predictive capacity) completely uninvolved in generating the algorithm in the first place.

Bayesian Information Integration for Personalized Medicine

A key challenge for the success of pharmacogenomics lays in defining analytical methods to integrate and interpret the increasingly complex information that it provides, so that decisions can be made about how people can best benefit from certain therapeutic interventions. Bayesian statistics approaches, often favored by decision theorists, have proven useful in classification models that integrate traditional clinical risk factors with complex genomics data. The point of doing this is to try to predict the outcome of future patients who share certain key clinicogenomic attributes with those from whom the model was derived and to use these predictions for patient-customized diagnostic, prognostic, and treatment purposes.

Most of this chapter is based on the classical frequentist approach to inferential statistics with which most of us are familiar. In this approach, a single hypothesis (the null hypothesis) is formed about the effect of (or, rather, lack thereof) a variable or several variables that are tested using a fairly rigid and well-defined format. The evidence against the hypothesis is all that is considered, and it is in the form of probabilities computed from the distribution of the outcome measure(s). A baseline assumption of randomness is assumed in defining what the probabilities are, and rejection of a null hypothesis is founded on the experimental result being too different from the random probability function to actually be random (i.e. therefore it isn't).

Bayesian statistics, on the other hand, represent a fundamentally different way of thinking about probability, with statistical inferences being made using accumulated evidence of all kinds, both for and against a hypothesis, or several alternative hypotheses. These hypotheses do not even have to be testable, let alone be tested, and certainly not by any particular format. (The name of the first publication of Bayes himself shows an example of such a hypothesis: Divine Benevolence, or an Attempt to Prove that the Principal End of Divine Providence and Government Is the Happiness of His Creatures [61].) Importantly, there is no baseline assumption of randomness: Instead, a "prior probability" is included in the paradigm. Although assigning it a value based on random chances is one option, another is that prior knowledge is incorporated into this value; in this way, Bayesian methods build on past experience to make statistical inferences.

To compute the probability of each particular hypothesis being true, a calculation using Bayes' theorem would multiply all the conditional probabilities that each is true by their prior probabilities and weight the final values relative to each other. *Conditional probabilities* are those from each piece of evidence for or against the hypothesis, and *prior probability* is defined in a very open-ended way; it could be pretty much anything, including subjective feelings, provided that it is decided upon in a consistent manner between all the hypotheses, it sums to 1 for all the hypotheses, and the information used to produce it is not used again later as evidence. As more evidence accumulates, it can be incorporated into updated models [62]. A simple Bayes calculation has already been shown (see Example 10.2); in that case, known disease frequency in the population constituted the prior probability, and the predictive values of the diagnostic test for true and false positives were the conditional probabilities.

Biomarker researchers aiming to define predictive models for personalized medicine have showed how Bayesian methods can strengthen these models by integrating different types of information into them. For instance, Qi et al. [63] noted that both supervised and unsupervised methods have different drawbacks for genomics data analysis, with the latter capturing relevant expression profiles but ignoring important a priori knowledge and the former being limited where little prior knowledge exists. Accordingly, they used a Bayesian approach to perform accurate predictions even when scant prior knowledge is available; their integrated model achieved higher sensitivity and specificity compared to both unsupervised and supervised methods [63].

Similarly, Pittman et al. [64] used Bayesian tree modeling to integrate both clinical information and genomic biomarker status for a sample set of breast cancer patients. They showed that the integrated clinicogenomic model gave substantially greater log-model likelihood (>7) than the model based on genomic biomarker signatures alone, and blew the clinical predictors alone right out of the picture, with a weight of evidence for the clinicogenomic vs. the clinical predictors only of more than 26 log-likelihood units.

Tree models in general involve successive splitting of a given patient sample group with certain known characteristics (i.e. gene signature, clinical risk factors) and outcomes (e.g. cancer status, survival, relapse) into more and more homogeneous subgroups. At each split, the collection of evidence (e.g. clinical or gene factors) is sampled to determine which of them optimally divide the patients according to their outcome, and a split is made if significance exceeds a certain level. Multiple possible splits generate "forests" of possible trees.

Some caveats noted for those using these methods include how to choose among alternative potential models that are identified as being of similar or significant probability, and the issue of uncertainty. In that regard, whereas in some other applications Bayesian approaches are used to choose a single best hypothesis from among several, such an approach is warned against for pharmacogenomic modeling applications [60, 64, 65]. In this case, it is typical to see multiple plausible tree models representing the data adequately, and this is consistent with the physical reality of multiple plausible combinations of genetic and clinical factors that could lead to the same outcome measures.

Rather than choosing one of them, it is critical to define overall predictions by averaging across the multiple candidate models using appropriate weights that reflect the relative fits of the trees to the data observed. The impact of averaging is seen in greater accuracy of the model predictive capacity and, importantly, in accurate estimation of the uncertainty about the resulting prediction (i.e. the prediction uncertainty of such a model is conceptually akin to the measurement uncertainty associated with a laboratory method, as discussed earlier). Nevins et al. underlined well the importance of prediction uncertainty when they pointed out: "A further critical aspect of prognosis is the need to provide honest assessments of the uncertainty associated with any prediction. A predicted 70% recurrence probability, for example, should be treated quite differently by clinical decision-makers if its associated uncertainty is ±30% than if it were ±2%" [65].

Summary: Quick Do's and Don'ts

The field of biomarkers is a wide one, with a huge diversity of potential applications, all of which may have different and complex statistical analysis issues. This chapter has undoubtedly missed many of these and glossed over others, but summarizing concisely what it has covered is still a challenge. Instead, I present below a distillation in the form of one-liners addressing some of the more critical points ("do") and more common misconceptions ("don't").

Do

1. Put on Hercule Poirot's hat and use your judgment when considering analytical issues and your experimental situation.
2. Think ahead. Consider data analysis issues before settling on your experimental design (and long before having the data in hand).
3. Take the limitations of your techniques and experimental error into account in your interpretations.
4. Check that your design is powered appropriately to detect what you wish to detect.
5. Use appropriate tests (i.e. ANOVA when comparing means from more than two groups).
6. Use appropriate controls and check HWE in genetic association studies.
7. Make adequate adjustment for the elevated false-positive rates when dealing with omics-style data.
8. Average over multiple plausible candidate models with appropriate weights (rather than choosing a single one), for best predictive accuracy and uncertainty estimations in pharmacogenomic applications of Bayesian modeling strategies.

Don't

1. Use parametric tests if data do not meet the assumptions of these tests (such as being normally distributed).
2. Use repeated t-tests when comparing means from more than two groups in one experimental design.
3. Compare more groups than are relevant to the goals of your experiment when applying multiple means testing.
4. Use correlation analysis to infer cause and effect or to test agreement between different methods.
5. Simply equate the odds ratio and risk ratio without considering the outcome frequency.

6. Use data to test correlation to outcomes or as a test set for validation of algorithms if they were already involved with those earlier in the process of mathematical modeling of genomics data (i.e. selected based on outcomes in the former case, or involved in algorithm generation in the latter).

References

1 Ludbrook, J. (2001). Statistics in physiology and pharmacology: a slow and erratic learning curve. *Clin. Exp. Pharmacol. Physiol.* 28 (5–6): 488–492.

2 Dupuy, A. and Simon, R.M. (2007). Critical review of published microarray studies for cancer outcome and guidelines on statistical analysis and reporting. *J. Nat. Cancer Inst.* 99 (2): 147–157.

3 Salanti, G., Amountza, G., Ntzani, E.E., and Ioannidis, J.P. (2005). Hardy–Weinberg equilibrium in genetic association studies: an empirical evaluation of reporting, deviations, and power. *Eur. J. Hum. Genet.* 13: 840–848.

4 Attia, J., Thakkinstian, A., and D'Este, C. (2003). Meta-analyses of molecular association studies: methodological lessons for genetic epidemiology. *J. Clin. Epidemiol.* 56: 297–303.

5 Ransohoff, D. (2005). Lessons from controversy: ovarian cancer screening and serum proteomics. *J. Nat. Cancer Inst.* 97 (4): 315–319.

6 Goncalves, A., Borg, J.P., and Pouyssegur, J. (2004). Biomarkers in cancer management: a crucial bridge towards personalized medicine. *Drug Discovery Today* 1 (3): 305–311.

7 Ransohoff, D.F. (2005). Bias as a threat to validity of cancer molecular-marker research. *Nat. Rev. Cancer* 5: 142–149.

8 Gawrylewski, A. (2007). The trouble with animal models. *Scientist* 21 (7): 45–51.

9 Hubert, P., Nguyen-Huu, J.J., Boulanger, B. et al. (2004). Harmonization of strategies for the validation of quantitative analytical procedures: a SFSTP proposal – part 1. *J. Pharm. Biomed. Anal.* 36 (3): 579–586.

10 Findlay, J.W.A., Smith, W.C., Lee, J.W. et al. (2000). Validation of immunoassays for bioanalysis: a pharmaceutical industry perspective. *J. Pharm. Biomed. Anal.* 21: 1249–1273.

11 Hubert, P., Nguyen-Huu, J.J., Boulanger, B. et al. (2006). Validation des procédures analytiques quantitatives: harmonisation des démarches. Partie II – Statistiques. *STP Pharma Prat.* 16: 30–60.

12 Prudhomme O'Meara, W., Fenlon Hall, B., and Ellis McKenzie, F. (2007). Malaria vaccine efficacy: the difficulty of detecting and diagnosing malaria. *Malaria J.* 6: 136.

13 Carlin, B.P. and Louis, T.A. (1996). *Bayes and Empirical Bayes Methods for Data Analysis.* London: Chapman & Hall.

14 Phillips, C.V. and Maldonado, G. (1999). Using Monte Carlo methods to quantify the multiple sources of error in studies. *Am. J. Epidemiol.* 149: S17.

15 Plant, N., Ogg, M., Crowder, M., and Gibson, G. (2000). Control and statistical analysis of in vitro reporter gene assays. *Anal. Biochem.* 278: 170–174.

16 Phillips, C.V. and LaPole, L.M. (2003). Quantifying errors without random sampling. *BMC Med. Res. Methodol.* 3: 9.

17 Gordon, G. and Finch, S.J. (2005). Factors affecting statistical power in the detection of genetic association. *J. Clin. Invest.* 115 (6): 1408–1418.

18 Cohen, M.J. (1998). *The Penguin Thesaurus of Quotations*. Harmondsworth: Penguin Books.

19 Coakley, E.H., Kawachi, I., Manson, J.E. et al. (1998). Lower levels of physical functioning are associated with higher body weight among middle-aged and older women. *Int. J. Obes. Relat. Metab. Disord.* 22 (10): 958–965.

20 Zou, K.H., Tuncali, K., and Silverman, S.G. (2003). Correlation and simple linear regression. *Radiology* 227: 617–628.

21 Hill, A.B. (1965). The environment and disease: association or causation? *Proc. R Soc. Med.* 58: 295–300.

22 Phillips, C.V. and Goodman, K.J. (2006). Causal criteria and counterfactuals; nothing more (or less) than scientific common sense. *BMC Med. Res. Methodol.* 3: 5.

23 Bland, J.M. and Altman, D.G. (1986). Statistical methods for assessing agreement between two methods of clinical measurement. *Lancet* 1 (8476): 307–310.

24 Browner, W.S. and Newman, T.B. (1987). Are all significant *p* values created equal? The analogy between diagnostic tests and clinical research. *J. Am. Med. Assoc.* 257: 2459–2463.

25 Eng, J. (2003). Sample size estimation: how many individuals should be studied? *Radiology* 227: 309–313.

26 Eng, J. (2004). Sample size estimation: a glimpse beyond simple formulas. *Radiology* 230: 606–612.

27 Detsky, A.S. and Sackett, D.L. (1985). When was a "negative" clinical trial big enough? How many patients you need depends on what you found. *Arch. Int. Med.* 145: 709–712.

28 Sokal, R.R. and Rohlf, F.J. (1981). *Biometry*, 2e. New York: W.H. Freeman.

29 Browner, W.S., Newman, T.B., Cummings, S.R., and Hulley, S.B. (2001). Estimating sample size and power. In: *Designing Clinical Research: An Epidemiological Approach*, 2e (eds. S.B. Hulley, S.R. Cummings, W.S. Browner, et al.), 65–84. Philadelphia: Lippincott Williams & Wilkins.

30 Sistrom, C.L. and Garvan, C.W. (2004). Proportions, odds, and risk. *Radiology* 230: 12–19.

31 Motulsky, H. (1995). *Intuitive Biostatistics*. Oxford, UK: Oxford University Press.

32 Agresti, A. (2002). *Categorical Data Analysis*. Hoboken, NJ: Wiley.

33 Schulman, K.A., Berlin, J.A., Harless, W. et al. (1999). The effect of race and sex on physicians' recommendations for cardiac catheterization. *N. Engl. J. Med.* 340: 618–626.

34 Schwartz, L.M., Woloshin, S., and Welch, H.G. (1999). Misunderstandings about the effects of race and sex on physicians' referrals for cardiac catheterization. *N. Engl. J. Med.* 341: 279–283.

35 Ryder, E.F. and Robakiewicz, P. (1998). Statistics for the molecular biologist: group comparisons. In: *Current Protocols in Molecular Biology* (eds. F.M. Ausubel, R. Brent, R.E. Kingston, et al.), A.31.1–A.31.22. New York: Wiley.

36 Bland, J.M. and Altman, D.G. (1996). The use of transformation when comparing two means. *BMJ* 312 (7039): 1153.

37 Bland, J.M. and Altman, D.G. (1996). Transforming data. *BMJ* 312 (7033): 770.

38 Ludbrook, J. (1995). Issues in biomedical statistics: comparing means under normal distribution theory. *Aust. N. Z. J. Surg.* 65 (4): 267–272.

39 Bland, J.M. and Altman, D.G. (1994). One and two sided tests of significance. *BMJ* 309: 248.

40 Barnett, V. and Lewis, T. (1994). *Outliers in Statistical Data*, 3e. New York: Wiley.

41 Hornung, R.W. and Reed, D.L. (1990). Estimation of average concentration in the presence of nondetectable values. *Appl. Occup. Environ. Hyg.* 5: 46–51.

42 Hughes, M.D. (2000). Analysis and design issues for studies using censored bio-marker measurements with an example of viral load measurements in HIV clinical trials. *Stat. Med.* 19: 3171–3191.

43 Thiebaut, R., Guedj, J., Jacqmin-Gadda, H. et al. (2006). Estimation of dynamic model parameters taking into account undetectable marker values. *BMC Med. Res. Methodol.* 6: 38.

44 Succop, P.A., Clark, S., Chen, M., and Galke, W. (2004). Imputation of data values that are less than a detection limit. *J. Occup. Environ. Hyg.* 1 (7): 436–441.

45 Minelli, C., Thompson, J.R., Abrams, K.R. et al. (2005). The choice of a genetic model in the meta-analysis of molecular association studies. *Int. J. Epidemiol.* 34: 1319–1328.

46 Mitra, S.K. (1958). On the limiting power function of the frequency chi-square test. *Ann. Math. Stat.* 29: 1221–1233.

47 Thakkinstian, A., McElduff, P., D'Este, C. et al. (2005). A method for meta-analysis of molecular association studies. *Stat. Med.* 24 (9): 1291–1306.

48 Sterne, J.A.C., Egger, M., and Smith, G.D. (2001). Investigating and dealing with publication and other biases in meta-analysis. *BMJ* 323: 101–105.

49 Ntzani, E.E., Rizos, E.C., and Ioannidis, J.P.A. (2007). Genetic effect versus bias for candidate polymorphisms in myocardial infarction: case study and overview of large-scale evidence. *Am. J. Epidemiol.* 165 (9): 973–984.

50 Hirschhorn, J.N. and Altshuler, D. (2002). Once and again: issues surrounding replication in genetic association studies. *J. Clin. Endocrinol. Metab.* 87 (10): 4438–4441.

51 Khoury, M.J., Beaty, T.H., and Cohen, B.H. (1993). *Fundamentals of Genetic Epidemiology*. New York: Oxford University Press.

52 Emigh, T. (1980). A comparison of tests for Hardy–Weinberg equilibrium. *Biometrics* 36: 627–642.

53 Phan, J.H., Quo, C.F., and Wang, M.D. (2006). Functional genomics and proteomics in the clinical neurosciences: data mining and bioinformatics. *Prog. Brain Res.* 158: 83–108.

54 Cui, X. and Churchill, G.A. (2003). Statistical tests for differential expression in cDNA microarray experiments. *Genome Biol.* 4: 210.

55 Baggerly, K.A., Morris, J.S., Edmonson, S.R., and Coombes, K.R. (2005). Signal in noise: evaluating reported reproducibility of serum proteomic tests for ovarian cancer. *J. Nat. Cancer Inst.* 97 (4): 307–309.

56 Liotta, L.A., Lowenthal, M., Mehta, A. et al. (2005). Importance of communication between producers and consumers of publicly available experimental data. *J. Nat. Cancer Inst.* 97 (4): 310–314.

57 Dudoit, S., Schaffer, J.P., and Boldrick, J.C. (2003). Multiple hypothesis testing in microarray experiments. *Stat. Sci.* 18: 71–103.

58 Benjamini, Y. and Hochberg, Y. (1995). Controlling the false discovery rate: a practical and powerful approach to multiple testing. *J R Stat. Soc. B* 57: 289–300.

59 Benjamini, Y., Drai, D., Elmer, G.L. et al. (2001). Controlling the false discovery rate in behavior genetics research. *Behav. Brain Res.* 125: 279–284.

60 Tong, W., Xie, Q., Hong, H. et al. (2004). Using Decision Forest to classify prostate cancer samples on the basis of SELDI-TOF MS data: assessing chance correlation and prediction confidence. *Toxicogenomics* 112: 1622–1627.

61 Dale, A.I. (2003). *Most Honourable Remembrance: The Life and Work of Thomas Bayes*. New York: Springer-Verlag.

62 Strachan, T. and Read, A.P. (2003). *Human Molecular Genetics*. London: Garland Science.

63 Qi, Y., Missiuro, P.E., Kapoor, A. et al. (2006). Semi-supervised analysis of gene expression profiles for lineage-specific development in the *Caenorhabditis elegans* embryo. *Bioinformatics* 22 (14): e417–e423.

64 Pittman, J., Huang, E., Dressman, H. et al. (2004). Integrated modeling of clinical and gene expression information for personalized prediction of disease outcomes. *Proc. Natl. Acad. Sci. U.S.A.* 101 (22): 8431–8436.

65 Nevins, J.R., Huang, E.S., Dressman, H. et al. (2003). Towards integrated clinico-genomic models for personalized medicine: combining gene expression signatures and clinical factors in breast cancer outcomes prediction. *Hum. Mol. Genet.* 12: R153–R157.

Part IV

Biomarkers in Discovery and Preclinical Safety

11

Qualification of Safety Biomarkers for Application to Early Drug Development

William B. Mattes[1] *and Frank D. Sistare*[2]

[1] *National Center for Toxicological Research, US FDA, Jefferson, AR, USA*
[2] *Merck Research Laboratories, West Point, PA, USA*

Historical Background to Preclinical Safety Assessment

It is often forgotten that the first "blockbuster" drug, sulfanilamide, was discovered and developed in an era devoid of regulatory oversight and guided only by free-market forces. Domagk discovered the antibacterial properties of Prontosil in 1932, and with the subsequent discovery in 1935 that the active moiety was the off-patent and widely available substance sulfanilamide, a number of companies rushed to make preparations for sale to the public. This explosion of therapy options was unfettered by requirements for medicines to be tested for efficacy or safety, although preparations could receive the endorsement of the American Medical Association [1]. Thus, in 1937, when the S.E. Massengill Company of Bristol, Tennessee, sought to prepare a flavored syrup, it simply identified an appropriate excipient to dissolve the drug, prepared 240 gallons of the raspberry-tasting Elixir Sulfanilamide, and marketed it across the nation. Unfortunately, the excipient chosen was diethylene glycol, also used as an antifreeze. We now know this agent to be lethal to a large number of species, causing acute kidney injury at relatively modest doses. Before Elixir Sulfanilamide was identified as the causative agent and pulled from pharmacies, 34 children and 71 adults died. A year later, Congress passed the 1938 Federal Food, Drug and Cosmetic Act, requiring pharmaceutical manufacturers to show product safety before distribution [1]. Thus, society transitioned from its former free-market approach to pharmaceutical development to one of safety testing (initially in animals), careful clinical trials, and government oversight.

Safety testing as it is practiced today owes most of its form to procedures developed by the US Food and Drug Administration (FDA) for testing food [2]. As early as 1949, the publication of "Procedures for the Appraisal of the

Biomarkers in Drug Discovery and Development: A Handbook of Practice, Application, and Strategy, Second Edition. Edited by Ramin Rahbari, Jonathan Van Niewaal, and Michael R. Bleavins.

Toxicity of Chemicals in Foods" began to formalize the practices the agency expected industry to follow in safety testing [3]. These practices included standard study designs and expectations as to what experimental observations would be recorded. They have evolved into the descriptive toxicity tests well known to modern toxicology [4]. Key to the value of these tests is not only their experimental design in terms of dose, route of administration, and duration but also the endpoints evaluated, going beyond the observations of overall animal behavior and health. Thus, a battery of clinical pathology tests examining urine, hematological parameters, and serum chemistry is commonly evaluated [5]. Importantly, a number of tissues from the animal are examined both macroscopically and microscopically after sacrifice, and this histopathological examination allows for the identification of unusual and subtle lesions and changes following compound treatment [6].

In the arena of pharmaceutical product development, these studies, carried out in at least two animal species, one rodent and one non-rodent, are used to assure the safety of human subjects exposed to experimental doses of novel compounds [7]. The types and durations of studies required to support safety in various types of clinical studies have been codified by the International Conference on Harmonisation (ICH) and described in their guidelines on nonclinical safety studies [8]. Even so, human subjects need to be monitored for "adverse events and/or laboratory abnormalities identified in the protocol as critical to safety evaluations" [9].

Limitations Faced in Preclinical Safety Assessment

A critical problem faced by nonclinical safety assessment groups in pharmaceutical drug development is the disparity of responses sometimes seen between the two nonclinical test species in the tools used to assess these responses. Historically, microscopic histopathology is used as a primary tool for identifying compound-induced damage. When damage is identified only at exposures far exceeding those expected in clinical studies, clinical safety is expected. However, microscopic histopathology is not a tool generally applicable to human studies, where clinical pathology measurements play the critical role in assessing adverse responses to drugs. Thus, if damage is observed in one nonclinical species at exposures close to those anticipated for human studies, the crucial question is whether the onset and reversibility of such damage could be monitored with clinical pathology or some other relatively noninvasive technology. Unfortunately, as described here, there are several types of drug-induced organ injury where current clinical pathology assays do not detect damage with sufficient certainty at early stages, and where assurances are needed that discontinuation of drug treatment would be followed by a complete and swift return to normal structure and function.

Kidney Injury

Kidney injury may be produced by a variety of insults, including those induced by drugs or toxicants [10]. Given the known morbidity and mortality associated with acute kidney injury [11, 12], evidence of drug-induced kidney injury in preclinical studies is a matter of serious concern. While on the one hand the kidney is capable of recovery from mild damage [13] if the injurious agent is removed, the very real clinical problem is that traditional noninvasive measures of kidney function are insensitive and confounded by many factors [13–16]. Thus, even modest increases in serum creatinine are associated with significant mortality [12]. There is a real need for noninvasive markers that would detect kidney damage or loss of function at a stage before significant and irreversible damage has occurred. Several markers with just such a potential have been described in numerous reviews [13–20]. However, many of these are described in relatively few clinical studies, most are only recently being examined carefully for their performance in animal models of drug-induced kidney injury, and no consensus understanding between drug development sponsors and regulatory review authorities had been reached as to their utility for regulatory decision-making purposes for several recently proposed renal injury biomarkers. Ultimately, if a microscopic histopathological examination shows evidence of even mild drug-induced kidney injury in a preclinical study at exposures close to those anticipated for clinical use, development of that compound may be stopped, even if human relevance may be questioned, yet unproven.

Liver Injury

The fact that medicines can cause liver injury and failure has been appreciated for some time [21], and drug-induced liver injury remains a serious public health and drug development concern [22–25]. As with other drug-induced organ damage, it may be detected in preclinical studies through microscopic histopathology and clinical chemistry measurements. Since the late 1950s, serum transaminase measurements, in particular that of alanine aminotransferase (ALT), have served as a sensitive and less-invasive measure of liver damage in both animal and human settings [26]. In conjunction with serum cholesterol, bilirubin, alkaline phosphatase, and other factors, ALT has served as a translational biomarker for drug-induced liver injury [27–30]. However, ALT elevations are not always associated with clear evidence of liver injury [31–33], and ALT elevations cannot clearly indicate the etiology of damage [26, 27, 30, 34]. Furthermore, ALT measurements either alone or with bilirubin cannot distinguish patients on a trajectory to severe liver disease and inability to heal or recover from injury, from patients with a full capacity to compensate and return ALT levels to normal despite continuation of drug dosing [35].

Combinations of clinical pathology changes have been used, including ALT and bilirubin, to assure safety in clinical trials [36], but there remains a need for diagnostic assays that reliably link and/or predict the histological observation of liver injury in both a preclinical and clinical setting and discriminate the types and trajectory of apparent injury [30].

Vascular Injury

Although injury to the vascular system is known to be caused by a variety of agents [37], many classes of therapeutic agents produce vascular lesions, in preclinical species with or without clinical signs, and with normal routine clinical pathology data [38]. Often, different preclinical species show a different level and type of response, and in many cases (e.g. minoxidil) the vascular injury reported in preclinical species is not observed in a clinical setting [39]. Drug-induced vascular injury in animals may result from altered hemodynamic forces, from a direct chemical-mediated injury to cells of the vasculature, and/or to an indirect immune-mediated injury of the endothelium and/or medial smooth muscle. The conundrum faced in drug development is that there are no specific and sensitive biomarkers of endothelial and/or vascular smooth muscle injury that are clearly linked to the histological observations in animals and could be used to monitor for injury in clinical settings. Although it is assumed that an inflammatory component may be active at some stage in this process, and biomarkers are proposed for such processes [40], biomarkers that are sufficiently sensitive at early and fully reversible stages of vascular injury have not been fully evaluated [38, 41]. Furthermore, specific markers of vascular injury/inflammation are sought that can discriminate from the multitude of other more benign causes for elevations of inflammatory biomarkers.

Drug-Induced Skeletal Myopathy

With the introduction of hydroxymethylglutaryl-coenzyme A (HMG-CoA) reductase inhibitors (statins), not only was there a successful treatment of hypercholesterolemia and dyslipidemia, but soon also a heightened awareness of the issue of drug-induced myopathy [42]. Statin-induced myotoxicity ranges from mild myopathy to serious and sometimes fatal rhabdomyolysis. Not surprisingly, a variety of drugs have been reported to induce myotoxicities [43]. While skeletal muscle toxicity may be monitored with elevations in serum creatinine kinase (CK), urinary myoglobin, and other markers [42], these biomarkers lack the sensitivity to definitively diagnose early damage or to distinguish the various etiologies of the muscle injury [43]. Thus, there is a need for markers of drug-induced muscle injury with improved sensitivity, specificity, and general utility.

Why Qualify Biomarkers?

More often than not, new biomarkers are judged on the basis of whether they have been subjected to a process of validation. Strictly speaking, *validation* is a process to "establish, or illustrate the worthiness or legitimacy of something" [44]. For judgments of biomarker worthiness or legitimacy, an assessment is needed of both (i) the assay or analytical method to measure the biomarker and (ii) the performance against expectations of the biomarker response under a variety of biological or clinical testing conditions. The term *validation* reasonably applies to the first of these, the process by which the technical characteristics of an assay of a biomarker are defined and determined to be appropriate for the desired measurements [45]. Thus, Wagner has defined validation as the "fit-for-purpose process of assessing the assay and its measurement performance characteristics, determining the range of conditions under which the assay will give reproducible and accurate data" [46]. Even for assay validation, the concept of fit-for-purpose is introduced, which connotes that the process depends on context, and its level of rigor depends on the application of and purpose for the assay. Thus, a biomarker used for an exploratory purpose may not require the more rigorous analytical validation required of a biomarker used for critical decision-making. The elements of biomarker assay validation that would be addressed for different categories of biomarker data and for different purposes have been discussed in this book and elsewhere, and they essentially constitute a continuum of bioanalytical method validation [45, 47]. Such technical bioanalytical assay method validation is a familiar process and does not generally pose a problem for an organization embarked on assay development [45].

The term *validation* has also been applied to a process by which a new test method is confirmed to be broadly applicable to interpretation of biological meaning in a wide variety of contexts and uses, such as in the validation of alternatives to animal tests as overseen by the Interagency Coordinating Committee on the Validation of Alternative Methods (ICCVAM) [48]. This process involving assessment of biological performance expectations is in contrast to that of *qualification*, which Wagner defines as "the fit-for-purpose evidentiary process of linking a biomarker with biological processes and clinical endpoints" [46]. As for assay validation, this fit-for-purpose biological qualification concept marries the nature and extent of testing rigor to the intended application. In the case of biomarkers applied to predicting human outcomes, four general phases have been proposed [49], and as the level of qualification progresses, the utility of a biomarker in clinical use increases [46].

In the case of biomarkers of safety (i.e. those that are used to predict or diagnose adverse responses to drug treatment), this qualification process will necessarily involve certain steps. As with qualification of a clinical disease or outcome marker, these steps link biomarker results with biological processes and endpoints. For example, in qualifying a biomarker for nonclinical use, the levels of a protein biomarker in urine may be correlated with certain chemically induced microscopic histopathology lesions in the kidneys of treated animals and thus serve as a noninvasive diagnostic of the appearance of that lesion. As with a clinical disease or outcome marker, qualification of such a marker could proceed in stages. For the example given, an initial stage may be the correlation described, using a variety of treatments that produce only that lesion measured. Such a correlation could establish the sensitivity of the biomarker. However, establishing the specificity of that biomarker for that particular lesion would require a number of treatments that did not produce the lesion being monitored, but instead produce no lesions in the kidney or anywhere else, produced different lesions in the kidney, and/or produced lesions in different organs. Furthermore, the diversity of chemical treatments (i.e. structural and mechanistic diversity) would also need to be considered in the qualification such that a variety of mechanisms underlying the genesis and progression of the lesion can be evaluated.

Clearly, more data could support a higher level of qualification and thus a higher level of utility. For biomarker qualification, the highest phase or level of qualification is the status *surrogate endpoint*, in which case the biomarker can, in fact, substitute for and serve as the new standard clinical endpoint of how a patient feels, functions, or will survive: for example, in efficacy determinations to support marketing approval decisions. Qualified biomarkers that fall short as surrogate endpoints are nevertheless extremely valuable for both drug development and the general practice of medicine. The key to appreciating the value that will come from opportunities to deploy such biomarkers appropriately is in revealing a thorough understanding of their inherent strengths and limitations. In the early steps of designing studies to test sensitivity and specificity aspects of a biomarker's performance to reveal that thorough understanding, the strategy may be relatively clear. For safety biomarker qualification, the first and most important attributes to benchmark are knowledge of the biomarker link to biology and outcome, sufficient test sensitivity, and minimizing false test negatives. Evaluating the response of a new proposed biomarker in animals against biomarkers in conventional use using an agreed-upon set of well-recognized toxicants known to induce the desired organ injury is a fairly straightforward strategy. Pivotal, however, to the successful execution of such studies is to provoke a sufficient number of cases where the timing of the samples taken and the choice of dose levels will yield subtle and mild treatment-related effects at the boundary between normal and abnormal. A full spectrum of histologic treatment-related lesions from very slight to slight, mild, moderate, marked, and severe will be important in this regard for evaluating biomarker sensitivity.

The approaches taken for qualification of a safety biomarker for clinical uses will necessarily be different from those taken for nonclinical uses. Clearly, one cannot expect to have studies with healthy subjects intentionally exposed to a variety of toxicants, nor can one regularly use microscopic histopathology as a benchmark for clinical toxicity. Nonetheless, the goal is to reproducibly link the biomarker to a clinical outcome currently recognized and widely accepted as adverse. For certain types of drug-induced injury, there are standard-of-care treatments that unfortunately are associated with a known incidence of drug-induced injury. As an example, aminoglycoside antibiotic treatment is associated with a significant incidence of nephrotoxicity [50], mirroring the effects seen in animal models. Similarly, isoniazid treatment has a known risk of hepatotoxicity [51]. Thus, one approach to safety biomarker qualification in the clinic is to monitor novel biomarker levels longitudinally over the course of such a treatment and compare these with the current gold standard commonly used clinical biomarkers and outcomes [52]. Of course, one can complement these studies with those examining biomarker levels in patients with organ injury of a disease etiology [53].

The number of known agents appropriate for testing the sensitivity of safety biomarkers for certain target organ toxicities may be limited. It is generally expected that the number of studies conducted to evaluate sensitivity should reasonably represent a high percentage of the known but limited diverse mechanisms available for testing in animal and human studies. If the mechanisms are varied for each test agent and sensitivity performance remains high, the biomarker will probably find strong use potential. Specificity tests then become very important considerations in a qualification strategy. The number of test compounds that could be deployed to assess the false-positive test rate is far more expansive than the number of known compounds for testing sensitivity. Specificity testing therefore becomes a highly individualized dimension to a biomarker biological qualification strategy for biomarkers that pass tests of sensitivity. The two critical questions to address are whether (i) there are alternative tissue sources to account for alterations in test safety biomarker levels and (ii) whether there are benign, non-toxicologic mechanisms to account for alterations in test biomarker levels. To evaluate specificity, therefore, a prioritized experimental approach should be taken using logical reasoning to avoid the testing burdens of an endless number of possible studies.

The ultimate goal of these studies is the qualification to the level of what Wagner et al. define as a *characterization biomarker* [49], a biomarker "associated with adequate preclinical sensitivity and specificity and reproducibly linked clinical outcomes in more than one prospective clinical study in humans." Such a level of qualification then supports the regulatory use of these biomarkers for safety monitoring in early clinical studies with a new pharmaceutical candidate. A strong case has been made that for safety biomarkers for regulatory decision-making purposes, where the rigor of supporting evidence would be expected to be high, only fully qualified or "characterization" biomarkers are appropriate, and that there is really no regulatory decision-making role for exploratory and "emerging" or "probable valid" biomarkers. In this regard, measurements of such unqualified biomarkers in animal studies used to support the safe conduct of clinical trials would not be expected to contribute unambiguously and sufficiently to study interpretation, should not require submission to regulatory authorities; therefore, the exploration of their utility in such highly regulated studies should be encouraged [54] in order to accelerate the pace of biomarker evaluations and understanding.

Collaboration in Biomarker Qualification

Clearly, the number of preclinical and clinical studies and resources required to qualify a biomarker as a characterization biomarker appropriate for regulatory decision-making are significant. Thus, it is no surprise that in their Critical Path Opportunities List the FDA called for "collaborations among sponsors to share what is known about existing safety assays" [55]. Collaborations of this type have indeed played key roles in addressing technological problems common to a competitive industry. Thus, Sematech, a consortium formed in 1987 and made up of 14 leading US semiconductor producers, addressed common issues in semiconductor manufacture and increased research and development (R&D) efficiency by avoiding duplicative research [56]. Sematech demonstrates that consortia can provide an opportunity for industry scientists to pool their expertise and experience to confront mutual questions collectively. The International Life Sciences Institute has for several years served as a forum for collaborative efforts between industry and academia [57] and for the Biomarkers Technical Committee pursued assay development and evaluation of biomarkers of nephrotoxicity and cardiotoxicity [58]. The Critical Path Institute was incorporated as a "neutral, third party" to serve as a consortium organizer [59] and interface between industry members and the FDA. One of its first efforts was the Predictive Safety Testing Consortium (PSTC) [60, 61], with a specific focus on qualification of biomarkers for regulatory use. The PSTC legal agreement addressed issues such as intellectual property, antitrust concerns, and confidentiality, and thus assured open collaboration in a manner consistent with applicable legal requirements. The PSTC solicited representatives from the FDA and the European Medicines Agency (EMEA) to serve as advisors. As experts in various areas of toxicity, these advisors bring not only their expertise but also the experience of how problems of a given target-organ toxicity are confronted and could be addressed in a regulatory setting. Thus, the development of qualification data was targeted with a keen eye toward what will ultimately support safety decisions in the regulated drug development and regulatory review process.

References

1 Wax, P.M. (1995). Elixirs, diluents, and the passage of the 1938 Federal Food, Drug and Cosmetic Act. *Ann. Intern. Med.* 122: 456–461.

2 Miller, S.A. (1993). Science, law and society: the pursuit of food safety. *J. Nutr.* 123: 279–284.

3 Stirling, D. and Junod, S. (2002). Arnold J. Lehman. *Toxicol. Sci.* 70: 159–160.

4 Eaton, D. and Klaassen, C.D. (2001). Principles of toxicology. In: *Casarett and Doull's Toxicology*, 6e (ed. C.D. Klaassen), 11–34. New York, NY: McGraw-Hill.

5 Weingand, K., Brown, G., Hall, R. et al. (1996). Harmonization of animal clinical pathology testing in toxicity and safety studies. The Joint Scientific Committee for International Harmonization of Clinical Pathology Testing. *Fundam. Appl. Toxicol.* 29: 198–201.

6 Bregman, C.L., Adler, R.R., Morton, D.G. et al. (2003). Recommended tissue list for histopathologic examination in repeat-dose toxicity and carcinogenicity studies: a proposal of the Society of Toxicologic Pathology (STP). *Toxicol. Pathol.* 31: 252–253.

7 FDA (1997). International conference on harmonisation; guidance on general considerations for clinical trials. *Fed. Regist.* 62: 66113.

8 FDA (2008). International conference on harmonisation; draft guidance on M3(R2) nonclinical safety studies for the conduct of human clinical trials and marketing authorization for pharmaceuticals. *Fed. Regist.* 73: 51491–51492.

9 FDA (1997). International conference on harmonisation; good clinical practice: consolidated guideline. *Fed. Regist.* 62: 25691–25709.

10 Schnellmann, R.G. (2001). Toxic responses of the kidney. In: *Casarett and Doull's Toxicology*, 6e (ed. C.D. Klaassen), 491–514. New York, NY: McGraw-Hill.

11 Hoste, E.A., Clermont, G., Kersten, A. et al. (2006). RIFLE criteria for acute kidney injury are associated with hospital mortality in critically ill patients: a cohort analysis. *Crit. Care* 10: R73.

12 Chertow, G.M., Burdick, E., Honour, M. et al. (2005). Acute kidney injury, mortality, length of stay, and costs in hospitalized patients. *J. Am. Soc. Nephrol.* 16: 3365–3370.

13 Vaidya, V.S., Ferguson, M.A., and Bonventre, J.V. (2008). Biomarkers of acute kidney injury. *Annu. Rev. Pharmacol. Toxicol.* 48: 463–493.

14 Trof, R.J., Di Maggio, F., Leemreis, J., and Groeneveld, A.B. (2006). Biomarkers of acute renal injury and renal failure. *Shock* 26: 245–253.

15 Molitoris, B.A., Melnikov, V.Y., Okusa, M.D., and Himmelfarb, J. (2008). Technology insight: biomarker development in acute kidney injury—what can we anticipate? *Nat. Clin. Pract. Nephrol.* 4: 154–165.

16 Ferguson, M.A., Vaidya, V.S., and Bonventre, J.V. (2008). Biomarkers of nephrotoxic acute kidney injury. *Toxicology* 245: 182–193.

17 Devarajan, P. (2007). Emerging biomarkers of acute kidney injury. *Contrib. Nephrol.* 156: 203–212.

18 Bagshaw, S.M., Langenberg, C., Haase, M. et al. (2007). Urinary biomarkers in septic acute kidney injury. *Intensive Care Med.* 33: 1285–1296.

19 Nguyen, M.T. and Devarajan, P. (2007). Biomarkers for the early detection of acute kidney injury. *Pediatr. Nephrol.* 23: 2151–2157.

20 Dieterle, F., Marrer, E., Suzuki, E. et al. (2008). Monitoring kidney safety in drug development: emerging technologies and their implications. *Curr. Opin. Drug Discovery Dev.* 11: 60–71.

21 Zimmerman, H.J. (1999). *Hepatotoxicity: The Adverse Effects of Drugs and Other Chemicals on the Liver*, 2e. Philadelphia, PA: Lippincott Williams & Wilkins.

22 Maddrey, W.C. (2005). Drug-induced hepatotoxicity: 2005. *J. Clin. Gastroenterol.* 39: S83–S89.

23 Arundel, C. and Lewis, J.H. (2007). Drug-induced liver disease in 2006. *Curr. Opin. Gastroenterol.* 23: 244–254.

24 Watkins, P.B. and Seeff, L.B. (2006). Drug-induced liver injury: summary of a single topic clinical research conference. *Hepatology* 43: 618–631.

25 Bleibel, W., Kim, S., D'Silva, K., and Lemmer, E.R. (2007). Drug-induced liver injury: review article. *Dig. Dis. Sci.* 52: 2463–2471.

26 Kim, W.R., Flamm, S.L., Di Bisceglie, A.M., and Bodenheimer, H.C. (2008). Serum activity of alanine aminotransferase (ALT) as an indicator of health and disease. *Hepatology* 47: 1363–1370.

27 Reichling, J.J. and Kaplan, M.M. (1988). Clinical use of serum enzymes in liver disease. *Dig. Dis. Sci.* 33: 1601–1614.

28 Ozer, J., Ratner, M., Shaw, M. et al. (2008). The current state of serum biomarkers of hepatotoxicity. *Toxicology* 245: 194–205.

29 Lock, E.A. and Bonventre, J.V. (2008). Biomarkers in translation; past, present and future. *Toxicology* 245: 163–166.

30 Amacher, D.E. (2002). A toxicologist's guide to biomarkers of hepatic response. *Hum. Exp. Toxicol.* 21: 253–262.

31 Pettersson, J., Hindorf, U., Persson, P. et al. (2008). Muscular exercise can cause highly pathological liver function tests in healthy men. *Br. J. Clin. Pharmacol.* 65: 253–259.

32 Giboney, P.T. (2005). Mildly elevated liver transaminase levels in the asymptomatic patient. *Am. Fam. Physician* 71: 1105–1110.

33 Gaskill, C.L., Miller, L.M., Mattoon, J.S. et al. (2005). Liver histopathology and liver and serum alanine aminotransferase and alkaline phosphatase activities in epileptic dogs receiving Phenobarbital. *Vet. Pathol.* 42: 147–160.

34 Shapiro, M.A. and Lewis, J.H. (2007). Causality assessment of drug-induced hepatotoxicity: promises and pitfalls. *Clin. Liver Dis.* 11: 477–505.

35 Andrade, R.J., Lucena, M.I., Fernandez, M.C. et al. (2005). Drug-induced liver injury: an analysis of 461 incidences submitted to the Spanish registry over a 10-year period. *Gastroenterology* 129: 512–521.

36 Hunt, C.M., Papay, J.I., Edwards, R.I. et al. (2007). Monitoring liver safety in drug development: the GSK experience. *Regul. Toxicol. Pharm.* 49: 90–100.

37 Ramos, K.S., Melchert, R.B., Chacon, E., and Acosta, D. Jr., (2001). Toxic responses of the heart and vascular systems. In: *Casarett and Doull's Toxicology*, 6e (ed. C.D. Klaassen), 597–651. New York, NY: McGraw-Hill.

38 Kerns, W., Schwartz, L., Blanchard, K. et al. (2005). Drug-induced vascular injury: a quest for biomarkers. *Toxicol. Appl. Pharmacol.* 203: 62–87.

39 Mesfin, G.M., Higgins, M.J., Robinson, F.G., and Zhong, W.Z. (1996). Relationship between serum concentrations, hemodynamic effects, and cardiovascular lesions in dogs treated with minoxidil. *Toxicol. Appl. Pharmacol.* 140: 337–344.

40 Blake, G.J. and Ridker, P.M. (2001). Novel clinical markers of vascular wall inflammation. *Circ. Res.* 89: 763–771.

41 Louden, C., Brott, D., Katein, A. et al. (2006). Biomarkers and mechanisms of drug-induced vascular injury in non-rodents. *Toxicol. Pathol.* 34: 19–26.

42 Tiwari, A., Bansal, V., Chugh, A., and Mookhtiar, K. (2006). Statins and myotoxicity: a therapeutic limitation. *Expert Opin. Drug Saf.* 5: 651–666.

43 Owczarek, J., Jasinska, M., and Orszulak-Michalak, D. (2005). Drug-induced myopathies: an overview of the possible mechanisms. *Pharmacol. Rep.* 57: 23–34.

44 Merriam-Webster Online Dictionary (2008).

45 FDA (2001). Guidance for industry on bioanalytical method validation. *Fed. Regist.* 66: 28526–28527.

46 Wagner, J.A. (2008). Strategic approach to fit-for-purpose biomarkers in drug development. *Annu. Rev. Pharmacol. Toxicol.* 48: 631–651.

47 Lee, J.W., Devanarayan, V., Barrett, Y.C. et al. (2006). Fit-for-purpose method development and validation for successful biomarker measurement. *Pharm. Res.* 23: 312–328.

48 Stokes, W.S., Schechtman, L.M., Rispin, A. et al. (2006). The use of test method performance standards to streamline the validation process. *Altex* 23 (Suppl): 342–345.

49 Wagner, J.A., Williams, S.A., and Webster, C.J. (2007). Biomarkers and surrogate end points for fit-for-purpose development and regulatory evaluation of new drugs. *Clin. Pharmacol. Ther.* 81: 104–107.

50 Wiland, P. and Szechcinski, J. (2003). Proximal tubule damage in patients treated with gentamicin or amikacin. *Pol. J. Pharmacol.* 55: 631–637.

51 Tostmann, A., Boeree, M.J., Aarnoutse, R.E. et al. (2008). Antituberculosis drug-induced hepatotoxicity: concise up-to-date review. *J. Gastroenterol. Hepatol.* 23: 192–202.

52 Mishra, J., Dent, C., Tarabishi, R. et al. (2005). Neutrophil gelatinase-associated lipocalin (NGAL) as a biomarker for acute renal injury after cardiac surgery. *Lancet* 365: 1231–1238.

53 Han, W.K., Bailly, V., Abichandani, R. et al. (2002). Kidney injury molecule-1 (KIM-1): a novel biomarker for human renal proximal tubule injury. *Kidney Int.* 62: 237–244.

54 Sistare, F.D. and DeGeorge, J.J. (2008). Applications of toxicogenomics to nonclinical drug development: regulatory science considerations. *Methods Mol. Biol.* 460: 239–261.

55 FDA (2006). Critical Path Opportunities Report and List, p. 28.

56 Irwin, D.A. and Klenow, P.J. (1996). Sematech: purpose and performance. *Proc. Natl. Acad. Sci. U.S.A.* 93: 12739–12742.

57 ILSI (2007). *ILSI About ILSI*.

58 ILSI and HESI (2007). *ILSI: Development and Application of Biomarkers of Toxicity*.

59 Woosley, R.L. and Cossman, J. (2007). Drug development and the FDA's critical path initiative. *Clin. Pharmacol. Ther.* 81: 129–133.

60 Marrer, E. and Dieterle, F. (2007). Promises of biomarkers in drug development: a reality check. *Chem. Biol. Drug Des.* 69: 381–394.

61 Mattes, W.B. (2008). Public consortium efforts in toxicogenomics. *Methods Mol. Biol.* 460: 221–238.

12

A Pathologist's View of Drug and Biomarker Development

Robert W. Dunstan

Abbvie, Worcester, MA, USA

The search for biomarkers is not new. It is as old as medicine itself. Four thousand years ago, a biomarker was limited to sensory observations of the external body – body temperature (touch), skin color (vision), the sound of respiration or gut rumblings (hearing), the odor of one's breath or feces (smell), and even placing a drop of sweat or urine on one's tongue (taste). Sensory observation of internal body fluids followed. Hippocrates (350 BC) used urine color, sediment, and foam to diagnose disease [1].

For medical assessment of disease to advance further, there was a need to amplify or extend beyond physical senses. This occurred with the development of stethoscopes, otoscopes, ophthalmoscopes, and especially, microscopes, all invented and improved over the past 500 years. Next came the advent of instruments to quantify biological changes. In the nineteenth and twentieth centuries, methods were established to culture bacteria and to assess body fluids and tissues by the use of enzymatic, chromatographic, and spectrophotometric techniques. Although these methods changed our ability to define disease, they paled by the development of techniques for the prosecution of DNA, RNA, the proteins they make, and the downstream molecules that proteins can produce (lipids and carbohydrates).

Such technical improvements fired the imagination that molecular biology could develop patient-specific therapies and the concept of personalized medicine was born. Critical to personalized medicine, however, was the need to have methods to consistently diagnose, prognose, and advise on optimal therapeutic interventions. To this end, the word "biomarker" became part of the scientific lexicon in the late 1960s and was codified in 1998 when the National Institutes of Health Biomarker Definitions Working Group defined a biomarker as "a characteristic that is objectively measured and evaluated as an

Biomarkers in Drug Discovery and Development: A Handbook of Practice, Application, and Strategy, Second Edition. Edited by Ramin Rahbari, Jonathan Van Niewaal, and Michael R. Bleavins.

indicator of normal biological processes, pathogenic process or pharmacologic responses to a therapeutic intervention" [2].

Once defined, the search for biomarkers was on. In the pharmaceutical industry, this was at first viewed with the naïve optimism that every drug could and should have its own biomarker to diagnose, prognose, and predict the treatment of choice. We are entering the third decade in this search, and the great transformation by molecular methods with big data analytics has yet to live up to expectations of transforming either the development of new therapeutics or the development of biomarkers to inform their use.

For biomarkers, the search has been intense. As of this writing, a Medline search of citations under the heading of "biomarker and cancer" (so chosen because this is the area of medicine in which the greatest amount of resources had been devoted) from 2000 through 2016 lists 251 496 published manuscripts. Since that time, US Food and Drug Administration (FDA) has approved only 51 biomarkers related to cancer (or almost 5000 articles per approved biomarker, Figure 12.1). Even more troubling is that Goossens et al. [3] reported as of March 2015 there were only 17 predictive tumor biomarkers and 10 prognostic and diagnostic nucleic acid–based tumor biomarkers in clinical use. In sum, with the best of scientific efforts, "very few, if any, new cancer biomarkers have been introduced into clinical practice in the last 20 years. The reason is that most of the newly discovered cancer biomarkers are

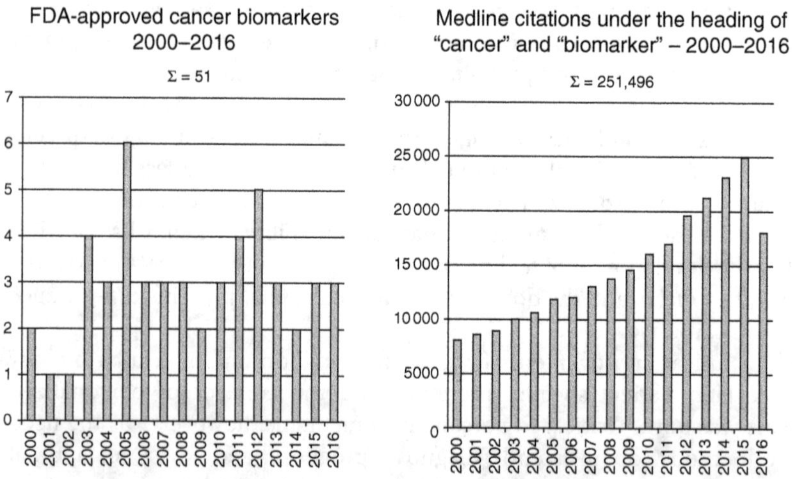

Figure 12.1 A comparison of the number of FDA-approved biomarkers between 2000 and 2016 with the number of papers citing "cancer" and "biomarker" based on Medline citations. Note the larger number of manuscripts compared to the number of actual FDA-approved biomarkers.

inferior in terms of sensitivity and specificity to the classical cancer biomarkers that we currently use. The revolutionary technologies of proteomics, genomics, and other "omics did not deliver on the promise to discover new and improved cancer biomarkers." [4]

In addition, the search for biomarkers has often been fueled by hype. An analysis of 35 highly cited biomarker studies demonstrated many of the claims are exaggerated when compared to subsequent meta-analyses [5]. For the pharmaceutical industry, the acceptance of these early studies can often lead to a large investment in the development of biomarkers that cannot be validated. At the same time, the expense of drug development is nearing unsustainability with the cost of getting a drug approved doubling every nine years [6]. A recent report estimated that it now costs US$ 2.8 billion to get a drug approved and marketed [7].

If biomarkers were to alleviate the cost and/or speed to approval of drug development, then the search for biomarkers has not been very successful. This should not come as a surprise. Drug development is a high-risk business. Good drugs do not "wear out." Synthetic aspirin was invented in 1899 although natural salicylic acid had been used for centuries. Today, this first "wonder drug" is used not only to relieve pain but also to prevent stroke and possibly, cancer. A "new and improved" aspirin would have to be substantively better than the white pill kept in most medicine cabinets, and it would have to generate a profit after the US$ 2.8 billion developmental price tag, to get it to market. This cost would increase further if a biomarker was needed to inform optimal use. In addition, it is also unlikely that there will be a new technology that will transform the process of drug and/or biomarker development. Over the past 50 years, hundreds of revolutionary methods and methodologies have been developed from next generation sequencing to the CRISPR-Cas9 technology to advances in *in vivo* imaging. During this period, the cost of drug development has increased relentlessly (Figure 12.2).

In a recent review, the major reasons why biomarkers fail were placed in four broad buckets: wrong target, wrong molecule, wrong patients, and wrong outcomes [8]. Although these reasons are inherently logical, I would argue the most common reason for biomarker failure is the development of analytical methods with minimal thought applied to the context of disease. In other words, the fault is in study design. There is a broadly accepted belief that large data sets can interrogate disease independently of context, and this is being shown time and time again not to be true.

For genomics, in spite of improvements in technology, data are with limited exceptions, still unable to predict an individual's genetic risk to develop cancer [9]. The use of genomics to parse out conditions with unarguable familial bases such as asthma or schizophrenia has identified only a small proportion

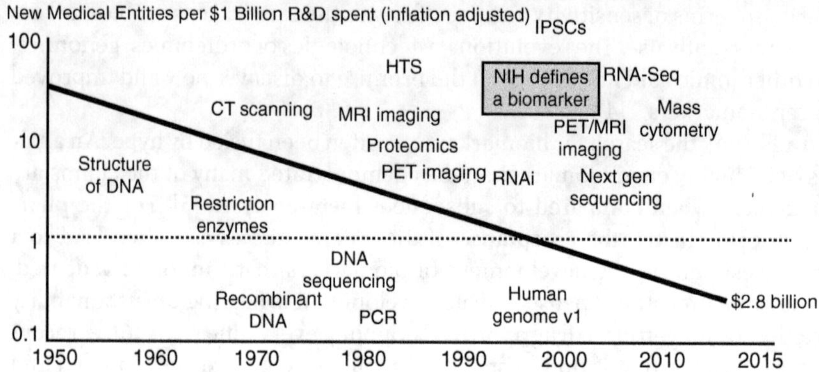

Figure 12.2 Graphing the relationship between the cost of drug development and the impact of emerging molecular and imaging technologies. The graph demonstrates that new technologies and emphasis on biomarker development have done little to decrease the cost of developing a drug in over 65 years. IPSCs, Induced Pluripotential Stem Cells; HTS, High Throughput Screening. Source: Adapted from Scannell et al. 2012 [6] and DiMasi et al. 2016 [7].

of the genetic component of these diseases [10]. For psychological diseases in general:

> No effective treatments have so far been devised on the basis of genetic information and, given what we now know, it seems very unlikely that further research into the genetics of psychosis will lead to important therapeutic advances in the future. Indeed, from the point of view of patients, there can be few other areas of medical research that have yielded such a dismal return for effort expended. The trend is to use enormous samples to find genes of miniscule effects. There is concern that GWAS will ultimately implicate the entire genome. [11]

With regard to cancer and genomics, comprehensive sequence analysis of nearly 1 million tumor samples over the past decade has identified >2 million coding point mutations, >6 million noncoding mutations, >10 000 gene fusions, ~61 000 genome rearrangements, ~700 000 abnormal copy number segments, and >60 million abnormal expression variants [12]. In short, cancer is a genetic disaster, and the thought that technology can design treatments for millions of genotypes is decades away, if at all possible. That is one reason for the new emphasis on immuno-oncology.

For transcriptomics, one only has to look at autoinflammatory/autoimmune (Ain/Aim) diseases where biomarkers lag well behind even the few available for cancer. There remains no good predictive/prognostic biomarker for lupus, or for that matter any autoimmune disease, and the promise initially predicted by examining type 1 interferon gene expression has been shown to not track disease activity within an individual over time [13]. For scleroderma, biopsies from affected regions have no substantive gene expression differences than samples from unaffected regions, and this profile does not change as the disease progresses [14]. Finally, a study looking at psoriasis where differentially expressed genes/proteins were compared to a diverse group of skin diseases (Mediterranean spotted fever, eschars, acne, and squamous cell carcinoma) identified only a single differentially expressed gene/protein [15].

What needs to be recognized is that for all the advances in technology, morphologic assessment remains the best method to diagnose the vast majority of diseases, and the most widely used biomarkers are tissue based. For cancer diagnosis, the current gold standard for cancer diagnosis is tissue biopsy followed by histopathology. For the diagnosis of the major autoimmune diseases, morphologic assessment whether by clinical assessment, imaging, or histopathology is generally required (Table 12.1), and those based on clinical or *in vivo* imaging methods have been historically validated by histologic assessment. As for the development of biomarkers, a listing of the sampling source of the 20 FDA-approved cancer biomarkers from 2011 to 2016 demonstrates only one, the Access Hybritech p2PSA assay used a circulating protein in the blood. The other three FDA-approved cancer biomarkers over this period that used body fluids (Cobas® EGFR Mutation Test v2, BRACAnalysis CDx, and PROGENSA® PCA3 Assay) used circulating tumor cells, peripheral

Table 12.1 The role of morphology in the diagnosis of major autoimmune diseases.

Disease	Disease diagnosis	Progress/response to therapy
Lupus nephritis	Combination of **clinical** features/lab based assays; **biopsy** needed to confirm lupus nephritis	**Imaging**, lab assays + **biopsy** (less common)
Scleroderma	Combination of **clinical** features/lab based assays	Rodnan score (**clinical**)
Rheumatoid arthritis	**Clinical/imaging** features (synovitis)/serologic changes	**Clinical/imaging** features (synovitis)/serologic changes
Idiopathic pulmonary fibrosis	**Imaging**/serologic + **biopsy** confirmation	**Imaging**
Psoriasis	**Clinical** features	**Clinical** features + **biopsy** (less common)
Inflammatory bowel disease	**Endoscopy + biopsy** confirmation	**Endoscopy + biopsy**, calprotectin assay

For autoimmune/autoinflammatory diseases, the structural/morphologic features of the disease (bolded) remain critical for diagnosis. Not all (i.e. psoriasis) require histopathology for diagnosis; however, even for psoriasis, the diagnostic clinical features have been validated by histopathology.

blood mononuclear cells for BRAC genomic, and prostatic epithelial cells in the urine after manual prostatic massage, respectively. All are tissue based. Of the approved cancer biomarkers derived from formalin fixed biopsy samples, seven used DNA for cancer mutation analysis, three used mRNA and were *in situ* hybridization-based, and six that defined protein expression used immunohistochemistry. This clearly indicates that with the exception of germline mutations, the vast majority of diagnoses and biomarkers will be derived from tissues, and blood or other body fluids are generally an inferior source for new biomarkers. This should serve as a cautionary note to those looking for biomarkers in diseases where biopsies are not routinely taken (i.e. most autoimmune/autoinflammatory diseases) that their development will at a minimum require correlation with histologic assessment, and if *in vivo* imaging is used as an alternative it too will have to be validated with histopathology (Table 12.2).

This emphasis away from direct tissue-based biomarkers tends to go against the trend in drug and biomarker development. At the same time, there are problems with using microscopic morphology to serve as biomarkers. First, with the arguable exception of cancer where excision/sampling is a standard of care, the invasive nature of a biopsy precludes it from being used as a "biomarker" in routine practice. Another major problem with histologic assessment is that a biopsy site cannot be easily "rebiospsied." Once a biopsy is removed, the progression at that site is lost forever. Biomarkers almost by definition need to be non- or minimally invasive and should be able to follow disease progression. Finally, in an era where genomes can be sequenced in days, histopathology can take a week to get a diagnostic result, too long for modern medicine. Arguably, it is these issues with morphologic assessment that have resulted in the search for molecular, enzymatic, and *in vivo* imaging methods to replace the traditional biopsy. There is also considerable emphasis to develop methods such as liquid biopsies in which cell-free circulating tumor DNA, RNA secreted in exosomes, or genomic/transcriptomic analysis of circulating tumor cells are used in place of a biopsy [16]. More recently, "histology agnostic trials" referring to clinical trials that are based solely on identifying targets or molecular aberrations with no morphologic assessment [17].

Although these methods pass the test of innovation, based on prior history, they are no more likely to succeed than methods that preceded them. At issue is that these methods run against the basic biology of disease. A physiologic abnormality that results in a disease is so downstream from genetics and transcriptomics that analysis using standard "bind and grind" methods often give data too dilute, or temporally removed, to parse out a good therapeutic target or a biomarker. This argument is far more compelling for inflammatory than neoplastic diseases but even for tumors, there are fewer and fewer advocates for the concept that cancer can be understood primarily by its genetics. The same applies for gene expression analyses. RNA is closer to most disease

Table 12.2 FDA-approved cancer biomarkers, 2011–2016.

Approval Year	Test	Type of sample	Molecule analyzed
2016	Cobas® EGFR Mutation Test v2	Blood/tissue (ctDNA)	ctDNA
	PD-L1 IHC 28-8 pharmDx	Tissue (tumor)	Protein (IHC)
	Ventana PD-L1 (SP142) Assay	Tissue (tumor)	Protein (IHC)
2015	Ventana ALK (D5F3) CDx Assay – P140025	Tissue (tumor)	Protein (IHC)
	PD-L1 IHC 28-8 pharmDx	Tissue (tumor)	Protein (IHC)
	Cobas KRAS Mutation Test	Tissue (tumor)	DNA
2014	BRACAnalysis CDx	Blood	DNA
	Therascreen® KRAS RGQ PCR Kit	Tissue (tumor)	DNA
2013	Therascreen EGFR RGQ PCR	Tissue (tumor)	DNA
	Cobas EGFR Mutation Test	Tissue (tumor)	DNA
	THxID™ BRAF Kit	Tissue (tumor)	DNA
2012	Ventana HER2 IHC System	Tissue (tumor)	Protein (IHC)
	PROGENSA® PCA3 Assay	Urine/tissue (tumor)	mRNA
	Bond™ Oracle™ HER2 Assay	Tissue (tumor)	Protein (IHC)
	Therascreen KRAS RGQ PCR Kit	Tissue (tumor)	DNA
	Access® Hybritech p2PSA	Blood	Protein
2011	HER2 CISH pharmDx	Tissue (tumor)	mRNA (ISH)
	Vysis ALK Break Apart FISH Probe Kit	Tissue (tumor)	mRNA (ISH)
	Cobas 4800 BRAF V600 Mutation Test	Tissue (tumor)	DNA
	Inform HER2 Dual ISH DNA Probe Cocktail	Tissue (tumor)	mRNA(ISH)

The table lists the FDA-approved biomarkers for a five-year period from 2011 to 2016. Almost all are based on analysis from tissue.
Source: https://www.fda.gov/MedicalDevices/ProductsandMedicalProcedures/DeviceApprovalsandClearances/Recently-ApprovedDevices.

processes, but it still is a poor surrogate for interpretation of the function of their derived proteins and its impact on morphology. Only 40% of the variation in protein concentration can be explained by knowing mRNA abundancies [18]. The transcriptomics also do not capture the vast post translational modifications that occur. There is the hope that improved proteomics methods will solve this problem but an argument can be put forward that inferring a disease

Table 12.3 Analysis of major autoimmune diseases by gene expression studies.

Disease	Citations under "gene expression" and "disease" listed (2007–2017)
Idiopathic pulmonary fibrosis	549
Inflammatory bowel disease	3205
Lupus	2339
Psoriasis	1428
Rheumatoid arthritis	3910
Scleroderma	337
Sjogren's syndrome	434

The table lists the number of articles when the term "gene expression" was coupled with one of the major autoimmune/autoinflammatory diseases in Medline. The table emphasizes that a large number of gene expression studies have been performed and more studies done in a similar manner may be redundant.

by the compositional proteins is similar to predicting the appearance of a building solely by knowledge of the building materials being used. What this means is that "binding and grinding" tissues with the hope of finding a new target has now approached scientific naïveté. For almost every organ and every disease in that tissue, hundreds if not thousands of studies have looked at gene expression and there is less and less to be gained by larger cohorts or subsampling large quantities of tissues. Finally, these methods largely ignore two other molecular classes: lipids and carbohydrates. A search for articles in which the words "gene expression" and one of the major autoimmune diseases (idiopathic pulmonary fibrosis, inflammatory bowel disease, lupus, psoriasis, rheumatoid arthritis, scleroderma, and Sjogren's syndrome) were used concurrently confirms this observation (Table 12.3).

Diseases are complex, composed of cells, and regulated by the intra- and extracellular milieu they produce, a milieu that is constantly changing. In short, they have both temporal and spatial aspects that cannot be ignored. This can be best understood if one looks at a disease as a battlefield, say the Battle of First Manassas (better known as the Battle of Bull Run) (Figure 12.3). A map demonstrates the complexity of what is occurring in association with the topography. If one were to do a metabolomics study over the entire battlefield (the equivalent of "binding and grinding"), there would undoubtedly be the identification of gun smoke. Modern science would consider gun smoke as a surrogate biomarker for an active battlefield and a chemist would go about finding a molecule that would inhibit gun smoke by inhibiting gunpowder. Let's assume that when applied to the Battle of Bull Run, the synthesized gunpowder antagonist would have its desired effect: gun smoke would be inhibited. However, this would not stop the battle for long. New weapons – knives, bows

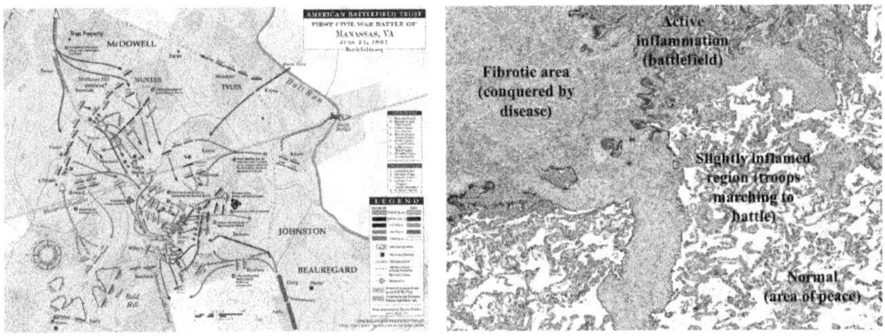

Figure 12.3 A comparison of a battlefield with a disease (idiopathic pulmonary fibrosis). The point to be made is that binding and grinding can result in data that is diluted with regards to pertinent information. Understanding the spatial and temporal aspects of both a land-based battle and a tissue-based disease is important to define outcome. Source: https://www.civilwar.org/learn/maps/first-manassas-july-21-1861. *(See insert for color representation of this figure.)*

and arrows, fisticuffs – would replace guns. More importantly, assuming the chemist works in the North, the objective is to not replace weapons but to have the North prevail.

If one replaces the map of Manassas with an inflammatory disease such as idiopathic pulmonary fibrosis, it becomes apparent that the analogy is not that far off. In the same photomicrograph, there are early, active, and late resolving changes, changes that are lost by "bind and grind" studies. Another aspect of the battlefield analogy needs to be mentioned: the goal is to stop the battle from being fought in the first place.

Most current therapies, be they for autoimmune/autoinflammatory diseases or cancer, result in an initial positive response but tend to lose effectiveness over time. For cancer, a good example is the mitogen-activated protein kinase kinase (MEK) inhibitors for melanoma [19]. For autoimmune/autoinflammatory (AIm/AIn) diseases, the best example is the anti-tumor necrosis factor antibodies where only 31% of patients are in remission with inflammatory bowel disease a year after therapy is initiated. [20]. In the case of AIm/AIn diseases the target toward which the therapy is directed is often equivalently expressed in both involved and uninvolved tissues, an indication that disease initiation is most likely upstream. Curiously, there is a unique successful exception to the treatment of AIm/AIn diseases – psoriasis. Inhibition of IL-17 by emerging therapies can result in a psoriasis area and severity index of 90 (meaning the percent of patients in whom 90% of their lesions had cleared) by 12 weeks of 70–85% [21]. What appears to be emerging from these data are that alterations in IL-17 are an early initiator for psoriasis, and this is the reason why this treatment is so effective [22]. For AIm/AIn diseases, this suggests that the most effective treatments target the earliest stage of disease progression, and defining the molecular profiles at this stage will require correlation with the spatial and temporal aspects of disease best performed by histopathology.

For cancer, identifying early stages of disease usually means diagnosis when the tumor is smaller and more localized; thus, the paradigm is different from AIm/AIn diseases. However, as the emphasis for newer therapies is advancing toward immunotherapies against checkpoint inhibitors than inhibitors of cancer growth itself, it will be interesting to learn if smaller tumors that have unrealized, but yet strong metastatic potential, respond better to immunotherapy or if treatments at this stage are less toxic. Regardless, the role of correlating morphology is becoming more important as the identification of tumor infiltrating lymphocytes and quantitation of markers for checkpoint inhibitors such as PD-1 and PDL1 by immunohistochemistry is becoming essential to predict response to immunotherapy.

In the context of molecular analyses, one final aspect of disease needs to be mentioned: every molecule targeted for therapy or for a biomarker has a useful purpose in maintaining homeostasis. There are no bad molecules. In addition, diseases associated with up- or downregulation of targets often can be considered as caricatures of their beneficial effect on tissues. Furthermore, it is important to understand what role targets for drugs or biomarkers play in maintaining homeostasis as well as the role they play in other diseases. For example, the effect of IL-17, so important in the development of psoriasis, needs to be understood on the basis of how it affects wound healing and morphologic simulant of psoriasis: a callus. Often this feature of evaluation of gene expression is downplayed as molecular analyses typically compare normal (or uninvolved sites) with morphologically abnormal samples. Note that for the development of both therapies and biomarkers, the need is not the identification of more targets but targets that are more specific. What this highlights is that developing a therapy or biomarker is difficult, that there is no stand-alone technology that can be used to identify what is a useable biomarker, and that ignoring the structural and temporal features of disease in biomarker analysis often is at the researcher's peril.

Suggestions for Improving Drug and Biomarker Development

Suggestion 1: Gene Expression Studies Analysis Should Be Correlated with Spatial and Temporal Aspects of the Disease

There is a growing appreciation that simply looking for gene/protein expression is no longer going to give new information and that molecular/morphologic correlates represent an unexplored and logically more fruitful way to pursue unidentified targets for therapies or biomarkers. Thus, site-directed transcriptomics is being used more and more [23]. The most widely used method is to view histologic sections, identify regions of interest, and then use laser microdissection. This entails making a subsequent section, usually on a specially designed slide in which the tissue is placed on a laser-sensitive substrate that allows for cutting and sampling tissues down to the cellular level using a microscope designed for that purpose. Although highly specific, laser capture microscopy is very labor intense as multiple slides often need to be sampled to get sufficient cellular material for RNA analysis. Other ways to sample tissues include simply scraping regions of interest off a slide using a small (25 gauge) needle or to take small (0.5 mm) cores from regions of interest. The latter method is much less exact but is extremely efficient and is done with the understanding that if interesting targets are identified, they can be confirmed using either laser microdissection, *in situ* hybridization, or if good antibodies are available, immunohistochemistry. RNA is then harvested from the tissues and analyzed using qPCR, microarrays, or, more commonly, RNA-Seq. The value of these methods cannot be understated. Using whole slide imaging, a pre- and post-coring image can be obtained so a digital record can be maintained of what was harvested. In addition, this allows for gene expression from regions in different stages of development. Newer methods have also been reported including methods to identify multiple targets on slides [24].

Suggestion 2: Think Carefully About the Value Gained by Compromising Morphology for RNA Quality

The common belief is that any research in which gene expression is being analyzed must be performed with RNA of the highest quality. Often there is little thought about compromising tissue morphology for the sake of intact RNA. The author has no belief that this chapter will change that bias, but will be content if the researcher simply realizes that there is a cost to pay in getting less than optimal morphology for the sake of transcriptomics. Formalin is far from a perfect fixative. It became a standard for tissue preservation largely because immersed tissues did not need refrigeration (not a problem today) and the other fixative of the 1900s, ethanol, could be consumed. In addition, it is carcinogenic and toxic, and causes cross linking of proteins and nucleic acids to each other; most nucleic acids are degraded to lengths of 200–300 base pairs. Furthermore, there are alternative fixatives that are nucleic acid preserving and result in equivalent morphology [25]. Still, it is the fixative standard and not using formalin-fixed paraffin embedded (FFPE) samples is at the exclusion of the vast majority of the world's archived tissue specimens that have good morphology. Although a decade ago any thought of using FFPE samples for genomics would have been unthinkable, with deeper sequencing and bioinformatics software, it is possible to piece together complete transcripts using small, 100–300 base pair amplicons. However, not all methods of RNA harvesting will result in successful analyses and when using more degraded RNA, one needs to use ribosomal depletion methods rather than polyA enrichment.

If ribosomal depletion methods are used, and the formalin blocks are relatively new (<5 years old), it is possible to get RNA of sufficient quality to use next generation sequencing methods such as RNA-Seq. A review of the literature is beyond the scope of this manuscript, and articles can be identified in which the use of archival samples is considered highly problematic [26]. At the same time, there is an emerging body of literature that suggests mRNA harvested from FFPE blocks is perfectly adequate for genomic analysis and is comparable to frozen tissues [27]. Furthermore, the development and validation of a robust and accurate methodology to perform whole-genome copy-number profiling of single nuclei obtained from FFPE clinical tumor samples has been described [28] demonstrating what expression analysis does and should not compromise morphology for RNA quality. What is on the horizon is the development of spatial transcriptomics in which single cells or small regions of morphometrically defined tissue can be analyzed using RNA-Seq methods.

Suggestion 3: Anatomic Pathology Methods Need to Become Standardized

If drug and biomarker development need to be better aligned with molecular biology, then morphologic assessment of tissues also needs to become more rigorous. This is not the author's opinion, but that of the FDA. In May 2016, the FDA released a guidance: "Considerations for the use of Histopathology and Its Associated Methodologies to Support Biomarker Qualification" [29]. It was released "to assist submitters of a biomarker for qualification that conduct nonclinical studies for which histopathology is used as a reference or truth standard." The document is unambiguous in the request that for diseases in which there are morphologic alterations (and very few diseases that are currently the targets for drug or biomarker development do not fall under this umbrella) the biomarker data need to be correlated with histopathology. The guidance is meant to inform preclinical studies in animals, but preclinical studies using human tissues are in no way exempted from the importance of using morphology to validate biomarkers either preclinically or during clinical trials. This does not just apply to correlating fluid-based biomarkers as FDA has long recognized that the use of *in vivo* imaging also requires this histologic correlation: "In regards to the indication of detecting amyloid in the brain (by an *in vivo* radiopharmaceutical), the committee overwhelmingly agreed that histopathological correlation should be the 'standard of truth' in Phase III clinical studies" (Summary Minutes of the Peripheral and Central Nervous System Drugs FDA Advisory Committee Meeting, 23 October 2008). What this means is regardless of the "measurable substance" that serves as the basis of a biomarker, the vast majority must be funneled through the slow, and as currently being practiced, largely subjective discipline of histologic review.

FDA also asks that the evaluation of tissue be done almost with the rigor of a good laboratory practice (GLP) study. It is interesting to note that in the Draft of this Guidance, FDA actually called for studies being done in accordance with good clinical practice (GCP) or GLP standards, but in the final document it is clear that although the methods do not have to be GCP/GLP, they certainly have to be "GLP-like":

> An important factor in planning confirmatory studies is a prospectively written protocol that all steps leading to and including the evaluation of the histopathology. All aspects of the necropsy, including timing, handling of the specimen and tissues, fixation, sample storage, transportation (if relevant), and slide preparation and evaluation, should be described clearly.

Finally, FDA asks that the histopathology be quantitative:

> Biomarker qualification studies should evaluate the quantitative correlation between the magnitude of biomarker change and the degree of morphologic change.

Using methods such as:

> … stereology, histochemistry, immunohistochemistry, in situ hybridization, electron microscopy with recognition that whole slide imaging can be used in this assessment.

There is a reason for this Guidance. Most anatomic pathology laboratories function with little standardization and a disconcerting lack of consistency in making a diagnosis. This can best be summarized by the position by the American Physiologic Association (APA) that histopathology is a theoretical "bottom feeder" when compared to other methods that employ reference standards. This puts histopathology in quandary – FDA generally views it as the "gold standard" to qualify biomarkers and yet the APA views it as not sufficiently robust to be used as a biomarker itself [30, 31]. An interesting point to realize is that without specifically stating so in its context, this Guidance (which to the author's knowledge has yet to be used as a "guide" in pharma) was meant to be applied only to preclinical studies on laboratory animals. This means that if the Guidance was taken seriously, the histologic assessment of laboratory rodents would be more rigorous than is currently being applied to humans.

For pre-analytical considerations, different histopathology laboratories will vary on the fixative, length of fixation, processing methods, sectioning thickness, and staining methods. Different methods result in unavoidably different appearance of the slides produced. Although familiarity with in-house methods should improve consistency among pathologists working in the same facility, there is very little inter-laboratory consistency and very little attempt for such. This difference becomes magnified in academic research laboratories where morphologic identification of gene expression by *in situ* hybridization, or protein expression by immunohistochemistry, is still largely done manually and not by automated immunostainers, the molecular equivalent of still doing Sanger sequencing on polyacrylamide gels.

In 2014, Begley and Ellis tried to replicate published results of 53 landmark studies from academic laboratories and were only able to confirm the findings in six of these studies (11.3%) [32]. A subsequent paper by Bradbury and Plückthun [33], which included Begley as a co-author, stated that poorly characterized and ill-defined antibodies were in large part to blame for this failure. Although Bradbury and Plückthun then go on to recommend that the use of polyclonal and monoclonal antibodies developed by hybridomas be replaced by antibodies produced as recombinant sequences, they fail to recognize that the studies that Begley and Ellis tried to replicate were published, and published in high impact journals. This indicates that the accuracy of antibody characterization is not a major concern by the scientists who submitted the manuscripts or the peers who reviewed them. This is disconcerting because it is often the demonstration of protein expression that validates most upstream molecular analyses. This poor characterization of antibodies applies not only to Western blots but also to enzyme-linked immunosorbent assays and the use of antibodies for immunohistochemistry that have an obvious impact on histologic evaluations. This is why, if employed, the guidance recently issued by FDA is so important. Although it is meant for preclinical animal studies, its impact will reverberate throughout anatomic pathology, resulting in more inter-laboratory consistency in time to fixation, length of time in a fixative, thickness of the tissue being placed in a fixative, trimming methods, tissue processing, post-processing embedding, section thickness, size of tissue examined, and number of slides to be examined to be representative as a whole.

Suggestion 4: Anatomic Pathology Needs to Change from a Descriptive to a Quantitative Discipline

The standardization of anatomic pathology cannot occur unless Suggestion 4 is realized. The practice of anatomic pathology has changed little since the time of Rudolph Virchow in the 1800s. Even microscopes are largely the same with the exception of adding binocular lens and electricity for illumination. What determines a quality section, or for that matter a quality diagnosis, is largely subjective. Thus, the critical question is can anatomic pathology be made more accurate and more informative?

The author believes the answer is a resounding "yes," but not as anatomic pathology is currently practiced. Rather it must transition from descriptive to quantitative. This transition is progressing slowly and is not widely embraced by all anatomic pathologists, but it is inevitable. The information on a histologic slide is arguably the greatest untapped resource for understanding disease and developing biomarkers. However, to capture these data, several interrelated steps need to be implemented which will require a quantum shift in thinking by many pathologists.

The first of these is digitization (almost better thought of as "pixilation") which transforms glass slides into a format where they can be quantified. Quantitation simply cannot be performed with efficiency or rigor by eye on analogue images. The use of whole slide images (WSI) routinely is standard practice in discovery laboratories in biopharma, but its use is not a standard due to regulatory/cultural, technical, and economic problems. Still, progress has been made. In May 2017, the FDA approved a whole slide scanner whose images are of acceptable quality and consistency that allow for a pathologist to make primary reads on them. This instrument requires both a scanner and an approved monitor. This is a big step. Prior to this, the United States was the only industrialized country in the world that did not allow primary reads of digitized histologic sections. The FDA does allow digitized images of IHC-stained breast tissue for the detection of HER-2 protein, but only after a primary diagnosis was made using traditional methods – glass slides and a microscope by a pathologist. One should note that there is considerable cultural resistance by anatomic pathologist to using WSI for making a diagnosis. As a pathologist who has used WSI extensively for over a decade, I can attest that I still greatly prefer looking through a microscope and recognize there is a rather steep learning curve getting used to looking at a microscopic image on a computer monitor.

Although there are cultural reasons for not using WSI, the biggest barriers to its widespread use are economic. Implementation of WSI is expensive. First, there is the acquisition and maintenance of the scanner; next, there is a need for a person to organize the images, enter them in a database, clean the slides, scan the images, and then remove and file them. Pathologists need high-end computers and monitors to view digitized images. Finally, storage of images that are sufficiently backed up is costly as is storage of the glass slides from which they were derived.

Additionally, there will be a coming debate as digitization becomes more commonplace about the cost-effectiveness of maintaining both analogue and digital versions of the same data. An argument can be made that while glass slides will fade over time, digital images are "ageless". However, when only small percentages of glass slides are ever reviewed after a diagnosis is made, does it make sense to store digitized images that will never be looked at again? As the technology evolves, one would assume that after images are scanned and quality defined, the glass slides will be destroyed. Advancing to that point is major change in pathology operations and may take years before only one medium is maintained.

The most important and yet largely ignored economic issue regarding implementation of WSI is reimbursement. The Centers for Medicare and Medicare Services and private payers are not interested in the convenience offered by making primary reads on WSI (if there is such a convenience), but they may be interested if it decreases the time required for a pathologist to interpret a lesion. A case can be made that if WSI allows for pathologists to read more slides in a shorter period of time, the professional component of reimbursement should be decreased relative to the technical component. What is the economic benefit of global use of WSI to the individual pathologist? More importantly, does the use of WSI as a replacement for a light microscope offer any benefit to the patient? Until these questions are answered, WSI will remain largely a "nice to have" technology. This is why it is unlikely that digital assessment of slides in and of itself will be widely used for primary reads even if the regulatory, culture, and economic issues are addressed.

To make WSI "reimbursable," it must be used as an instrument that serves as the doorway to quantitation. There is no exaggeration in the statement that digitization has transformed bioimaging in all fields except bright-field microscopy. One has only to compare the growth of radiology departments over the past two decades to see that this is true. In fact, the term "radiology department" is largely anachronistic because X-rays essentially have been replaced by CAT scans and imaging methods of positron-emission tomography (PET), single-photon-emission computed tomography (SPECT), magnetic resonance imaging (MRI), ultrasound, etc. and are now routine. This transformation has occurred not because digitization is more convenient, efficient, or more economical but because these technologies have improved the standard of care by giving more detailed information, thereby allowing the reviewer to make a better decision. This is the reality check needed for WSI: just digitally viewing the same image one would see through a microscope does little or nothing to improve the standard of care. This can only be done through quantification. In addition, the value of quantification using WSI will remain questionable if it is no better than the "gold standard" provided by a well-trained anatomic pathologist looking through a microscope.

The concept behind quantitation of WSI is simple. Each pixel in a digital image has at least 24-bit color depth using an RGB format. This means that each pixel contains one of 256 red, green, and blue intensities, respectively. Considering the range of red, green, and blue intensities means there can be >16 000 000 different colors. This is a bit of overkill because it is estimated that a trained human eye can only distinguish about 100 000 different colors. The digital image is to the computer no more than a large matrix and these matrices can be manipulated. There are hundreds of formulas and analytical methods one can use for this manipulation to allow for segmentation of digital images and then quantitation. There are also a number of commercially available software programs that apply these with a relatively user-friendly, graphical user interface.

The ability to utilize these matrices and formulas has progressed dramatically over the past 15 years when analysis was largely performed manually. The evolution has progressed from supervised (the user of the software tells the computer what to identify) to unsupervised (the computer tells the user textural or chromatic patterns that identify structures) to machine learning, a form of analysis where the computer trains itself. These methods are the basis of facial and signature recognition. A reason most machine learning is still nascent in its application to histopathology is that the information on a histologic section is far more complex than a human face or recognizing handwriting. This is why convolutional neural networks are needed for the evaluation of histologic sections. This refers to computer-generated "neurons" or nodes that are arranged in interconnected layers, analogous to how the brain functions. This connectivity allows the program to learn at multiple levels of image representation and to identify the nonlinear methods of analysis that allow for accurate and reproducible classification and predictability of changes in histologic sections [34–36].

Broad institution of quantitative analysis of histologic sections will have two major effects on anatomic pathology. First, it will allow for the standardization of the histotechnology mentioned above. With traditional visual methods of assessment quality control is subjective and for most pathologists, quite forgiving. Cutting artifacts, pallor, or overstaining are often perceived to be changes a pathologist can "look through" when evaluating a histologic section. Applying computer-based methods for assessment can distinguish differences in staining (including separation of hematoxylin and eosin to determine if one or both are over/understained, perform a quality control check to define areas out of focus on the scanned image, and identify cutting artifacts such as tears, folds and chatter, etc.). In short, pathology laboratories can define not only intra-laboratory variability but also inter-laboratory variability. Perhaps more important is the emergence of algorithms that can digitally normalize hematoxylin and eosin (H&E) and other stains on WSI images. Thus, H&E stained sections from two laboratories presumably using somewhat different staining methods can be changed to be similar in color space.

The second and most important change will be improving the standard of care offered by morphologic assessment, and its impact will be compounded as quantitative morphologic assessment is coupled with molecular analyses. There is unarguably a need for more accurate way to evaluate histologic sections. Seldom is there greater than 75% concurrence among pathologists interpreting the same slide. A recent study on breast cancer found that agreement among pathologists (including three pathologists who were considered true subject matter experts) was approximately 75%. What needs to be considered is that benign or highly malignant breast lesions are seldom misdiagnosed. Where the error rate becomes a factor are those lesions that straddle the border between benign and malignant [37]. As stated by the author of the breast cancer study cited above,

> "This is troubling when you know that women with DCIS [ductular carcinoma *in situ*] diagnosis are having mastectomies or lumpectomy and radiation therapy. And there was even less agreement on diagnoses for atypia – equivalent to the chance of heads or tails on a flip of a coin," http://hsnewsbeat.uw.edu/story/study-breast-biopsies-shows-rate-pathologists%E2%80%99-discord

Because histopathology is generally viewed as the "gold standard" for diagnosis, it is the basis of many therapeutic decisions. Thus, any inaccuracy is not good for the patient. Under diagnosis (a Type II or false negative error) will result in a needed therapy not being administered, whereas over diagnosis (a Type I or false positive error) will result in a therapy not needed being administered. The use of computer-assisted diagnosis appears to be the best way to avoid these errors. This is not to say that the computer will ever replace the pathologist in making a diagnosis, but rather a program will analyze the images and present to the pathologist a heat map of regions of concern and/or present a data table with a "confidence score." Although quantitative morphometry is not yet to the stage where it can assist a pathologist in making these diagnoses, it is not that far removed. However, it needs to be remembered that tying histopathology to advance drug or biomarker development will require more than a making diagnosis and validation of its associated prognosis. It will require that the tissue actually be analyzed for its unique morphologic features and those features be correlated with the molecular phenotype.

That there is lot of information being missed in the current "diagnostic-centric" analysis of histologic specimens is best defined by a paper by Beck et al. [38]. Largely by using more conventional image analysis methods on H&E stained sections, they were able to identify a feature set of multiple classifiers that defined breast cancer epithelium and its associate stroma. Based on this analysis, they reported that analysis of H&E samples was more informative with regard to prognosis than pathology grade, estrogen receptor status, tumor size, and lymph node status. They also were able to define three previously unrecognized stromal features that were significantly associated with survival and this association was stronger than that of the epithelial component of the analysis [38]. Although a single example, the study suggests that if new prognostic/predictive patterns can be recognized in breast cancer, a disease from which 1.6 million biopsies are obtained annually, how many other patterns are missed in other cancers and inflammatory diseases? Furthermore, how can these patterns be used to inform molecular analyses?

Conclusions

The development of drugs and biomarkers is fraught with difficulty. With the rising cost of both, methods need to be developed that will improve the ability to design and evaluate compounds that have a higher chance of success, developing assays that have a better chance of predicting disease behavior, and utilizing the best drugs to maximize effect. A critical analysis indicates that molecular methods, though powerful in and of themselves, have failed to accomplish this role. A major reason is that molecular correlation of the temporal and spatial aspects of disease are largely ignored. Developing drugs and biomarkers is too challenging not to use all the scientific tools available. Unfortunately, the discipline that provides the information regarding spatial and temporal information, anatomic pathology, has not changed substantively in over a century as to how it analyzes tissues and transmits this information. This needs to change. To accomplish this, anatomic pathology needs to evolve from a primarily descriptive and diagnostic discipline to one that is far more quantitative and analytical. This "multiplexing" of disciplines and technologies offers the potential of taking drug and biomarker development to a new and more successful plane.

References

1 Vaidya, V.S. and Bonventre, J.V. (eds.) (2010). *Biomarkers: In Medicine, Drug Discovery, and Environmental Health*. Wiley.
2 Colburn, W.A. (2001). Biomarkers and surrogate endpoints: preferred definitions and conceptual framework. Biomarkers Definitions Working Group. *Clin. Pharmacol. Ther.* 69: 89–95.
3 Goossens, N., Nakagawa, S., Sun, X., and Hishida, Y. (2015). Cancer biomarker discovery and validation. *Transl. Cancer Res.* 4 (3): 256–269.
4 Diamandis, E.P. (2014). Present and future of cancer biomarkers. *Clin. Chem. Lab. Med.* 52 (6): 791–794.
5 Ioannidis, J.P.A. and Panagiotou, O.A. (2011). Comparison of effect sizes associated with biomarkers reported in highly cited individual articles and in subsequent meta-analyses. *JAMA* 305 (21): 2200–2210.
6 Scannell, J.W., Blanckley, A., Boldon, H., and Warrington, B. (2012). Diagnosing the decline in pharmaceutical R&D efficiency. *Nat. Rev. Drug Discovery* 11 (3): 191–200.
7 DiMasi, J.A., Grabowski, H.G., and Hansen, R.W. (2016). Innovation in the pharmaceutical industry: new estimates of R&D costs. *J. Health Econ.* 47: 20–33.
8 Townsend, M.J. and Arron, J.R. (2016). Reducing the risk of failure: biomarker-guided trial design. *Nat. Rev. Drug Discovery* 15 (8): 517–518.

9 Thomas, D.M., James, P.A., and Ballinger, M.L. (2015). Clinical implications of genomics for cancer risk genetics. *Lancet Oncol.* 16 (6): e303–e308.

10 Crow, T.J. (2011). The missing genes: what happened to the heritability of psychiatric disorders? *Mol. Psychiatry* 16 (4): 362–364.

11 Leo, J. (Winter, 2016). The search for schizophrenia genes. *Issues Sci. Technol.* 32: 2.

12 Wishart, D.S. (2015). Is cancer a genetic disease or a metabolic disease? *EBioMedicine* 2 (6): 478–479.

13 Flint, S.M., Jovanovic, V., Teo, B.W. et al. (2016). Leucocyte subset-specific type 1 interferon signatures in SLE and other immune-mediated diseases. *RMD Open* 2 (1): e000183.

14 Assassi, S., Radstake, T.R., Mayes, M.D., and Martin, J. (2013). Genetics of scleroderma: implications for personalized medicine? *BMC Med.* 11 (1): 9–17.

15 Swindell, W.R., Remmer, H.A., Sarkar, M.K. et al. (2015). Proteogenomic analysis of psoriasis reveals discordant and concordant changes in mRNA and protein abundance. *Genome Med.* 7 (1): 86–92.

16 Karachaliou, N., Sosa, A.E., Molina, M.A. et al. (2017). Possible application of circulating free tumor DNA in non-small cell lung cancer patients. *J. Thorac. Dis.* 9 (Suppl 13): S1364–S1372.

17 Lacombe, D., Burock, S., Bogaerts, J. et al. (2014). The dream and reality of histology agnostic cancer clinical trials. *Mol. Oncol.* 8: 1057–1063.

18 Vogel, C. and Marcotte, E.M. (2012). Insights into the regulation of protein abundance from proteomic and transcriptomic analyses. *Nat. Rev. Genet.* 13: 227–232.

19 Poulikakos, P.I. and Solit, D.B. (2011). Resistance to MEK inhibitors: should we co-target upstream. *Sci. Signal.* 4 (166): e16.

20 Pérez-De-Lis, M., Retamozo, S., Flores-Chávez, A. et al. (2017). Autoimmune diseases induced by biological agents. A review of 12,731 cases (BIOGEAS Registry). *Expert Opin. Drug Saf.* 16 (11): 1255–1271.

21 Lebwohl, M., Strober, B., Menter, A. et al. (2015). Phase 3 studies comparing brodalumab with ustekinumab in psoriasis. *N. Engl. J. Med.* 373 (14): 1318–1328.

22 Gaspari, A.A. and Tyring, S. (2015). New and emerging biologic therapies for moderate-to-severe plaque psoriasis: mechanistic rationales and recent clinical data for IL-17 and IL-23 inhibitors. *Dermatol. Ther.* 28 (4): 179–193.

23 Mignardi, M., Ishaq, O., Qian, X., and Wahlby, C. (2017). Bridging histology and bioinformatics - computational analysis of spatially resolved transcriptomics. *Proc. IEEE* 105 (3): 530–541.

24 Ståhl, P.L., Salmen, F., Vickovic, S. et al. (2016). Visualization and analysis of gene expression in tissue sections by spatial transcriptomics. *Science* 353 (6294): 78–82.

25 Cox, M.L., Schray, C.L., Luster, C.N. et al. (2006). Assessment of fixatives, fixation, and tissue processing on morphology and RNA integrity. *Exp. Mol. Pathol.* 80 (2): 183–191.

26 Esteve-Codina, A., Arpi, O., Martinez-Garcia, M. et al. (2017). A comparison of RNA-Seq results from paired formalin-fixed paraffin-embedded and fresh-frozen glioblastoma tissue samples. *PLoS One* 12 (1): e0170632.

27 Hester, S.D., Bhat, V., Chorley, B.N. et al. (2016). Editor's highlight: dose-response analysis of RNA-Seq profiles in archival formalin-fixed paraffin-embedded samples. *Toxicol. Sci.* 154 (2): 202–213.

28 Martelotto, L.G., Baslan, T., Kendall, J. et al. (2017). Whole-genome single-cell copy number profiling from formalin-fixed paraffin-embedded samples. *Nat. Res.* 23 (3): 376–385.

29 Food and Drug Administration, United States 2016. Considerations for the Use of Histopathology and Its Associated Methodologies to Support Biomarker Qualification, https://www.fda.gov/media/82768/download (accessed 05 May 2019).

30 Cummings, J., Raynaud, F., Jones, L. et al. (2010). Fit-for-purpose biomarker method validation for application in clinical trials of anticancer drugs. *Br. J. Cancer* 103 (9): 1313–1317.

31 Booth, B., Arnold, M.E., DeSilva, B. et al. (2015). Workshop report: crystal city V – quantitative bioanalytical method validation and implementation: the 2013 revised FDA guidance. *AAPS J.* 17 (2): 277–288.

32 Begley, C.G. and Ellis, L.M. (2012). Drug development: raise standards for preclinical cancer research. *Nature* 483 (7391): 531–533.

33 Bradbury, A. and Plückthun, A. (2015). Reproducibility: standardize antibodies used in research. *Nature* 518: 27–29.

34 Gurcan, M.N., Boucheron, L., Can, A. et al. (2009). Histopathological image analysis: a review. *IEEE Rev. Biomed. Eng.* 2: 147–171.

35 Cruz-Roa, A., Gilmore, H., Basavanhally, A. et al. (2017). Accurate and reproducible invasive breast cancer detection in whole-slide images: a deep learning approach for quantifying tumor extent. *Sci. Rep.* 7 (46450): 1–14.

36 Litjens, G., Sanchez, C.I., Timofeeva, N. et al. (2016). Deep learning as a tool for increased accuracy and efficiency of histopathological diagnosis. *Sci. Rep.* 6: 26286.

37 Elmore, J.G., Longton, G.M., Carney, P.A. et al. (2015). Diagnostic concordance among pathologists interpreting breast biopsy specimens. *JAMA* 313 (11): 1122–1132.

38 Beck, A.H., Sangoi, A.R., Leung, S. et al. (2011). Systematic analysis of breast cancer morphology uncovers stromal features associated with survival. *Sci. Transl. Med.* (108): 108–113.

13

Development of Serum Calcium and Phosphorus as Safety Biomarkers for Drug-Induced Systemic Mineralization: Case Study with the MEK Inhibitor PD0325901[1]

Alan P. Brown

Novartis Institutes for Biomedical Research, Cambridge, MA, USA

Introduction

The mitogen-activated protein kinase (MAPK) signal transduction pathways control key cellular processes such as growth, differentiation, and proliferation, and provide a means for transmission of signals from the cell surface to the nucleus. As a part of the RAS–RAF–MEK–MAPK pathway, MEK (MAP kinase) phosphorylates the MAPK proteins ERK1 and ERK2 (extracellular signal-regulated kinases) as a means for intracellular signaling [1]. Although MEK has not been identified as having oncogenic properties, this kinase serves as a focal point in the signal transduction pathway of known oncogenes (e.g. RAS and RAF) [2]. MEK exists downstream of various receptor tyrosine kinases (such as the epidermal growth factor receptor) which have been demonstrated to be important in neoplasia [3]. RAS activation occurs first, followed by recruitment of RAF (A-RAF, B-RAF, or RAF-1) proteins to the cell membrane through binding to RAS, with subsequent activation of RAF. RAF phosphorylates MEK1 and MEK2 (which are highly homologous) on multiple serine residues in the activation process. MEK1 and MEK2 phosphorylate tyrosine or threonine residues on ERK proteins in the signal transduction process, with phosphorylated ERK activating various transcription factors [4]. Aberrant activation of this pathway has been observed in a diverse group of solid tumors, along with leukemia, and is believed to play a key role in tumorigenesis [5, 6].

Based on a significant amount of preclinical data, development of small-molecule inhibitors of MEK appears to be a rational approach for treatment of various malignancies [7–9]. This strategy was validated with the US FDA approval of the first MEK inhibitor, trametinib, in 2013 for melanoma with B-RAF V600E- or V600K-mutations [10]. A second MEK inhibitor,

[1] All research was conducted at Pfizer Global Research and Development, Ann Arbor, MI, USA

Biomarkers in Drug Discovery and Development: A Handbook of Practice, Application, and Strategy, Second Edition. Edited by Ramin Rahbari, Jonathan Van Niewaal, and Michael R. Bleavins.

cobimetinib, was approved for combination therapy with vemurafenib (BRAF inhibitor) for the treatment of melanoma by the US FDA in 2015. The first MEK inhibitor to enter clinical trials was CI-1040 (also known as PD184352), which inhibits MEK1/2 in a non-ATP competitive manner by binding into a hydrophobic pocket, thereby inducing conformational changes in unphosphorylated MEK and locking the kinase in a closed but catalytically inactive form [11, 12]. CI-1040 was intended for oral administration, but the level of antitumor activity in a multicenter Phase II study in patients with various solid tumors was not sufficient to warrant further development of this drug [13, 14]. CI-1040 exhibited low oral bioavailability and high metabolism, which were primary factors resulting in insufficient plasma drug levels for antitumor activity [12].

PD0325901 (Figure 13.1; chemical name of *N*-((*R*)-2,3-dihydroxypropoxy)-3,4-difluoro-2-(2-fluoro-4-iodo-phenylamino)benzamide) is a highly potent and specific non-ATP competitive inhibitor of MEK (K_i of 1 nM against activated MEK1 and MEK2 *in vitro*), and demonstrated anticancer activity against a broad spectrum of human tumors in murine models at ≥25 mg/kg [15]. Preclinical studies indicate that PD0325901 has the potential to impair growth of human tumors that rely on the MEK/MAPK pathway for growth and survival. PD0325901 inhibits the phosphorylation of MAPK proteins (ERK1 and ERK2) as a biochemical mechanism of action, and assays were developed to evaluate inhibition of protein phosphorylation in normal and neoplastic tissues [16]. This compound had greatly improved pharmacologic and pharmacokinetic properties compared with CI-1040 (i.e. greater potency for MEK inhibition, higher bioavailability, and increased metabolic stability) and provided an opportunity for investigating the therapeutic potential for treating cancer with an orally active MEK inhibitor [13, 15]. PD0325901 was selected for development as a clinical candidate due to its superior preclinical profile compared to CI-1040 [14]. Toxicology studies were conducted to

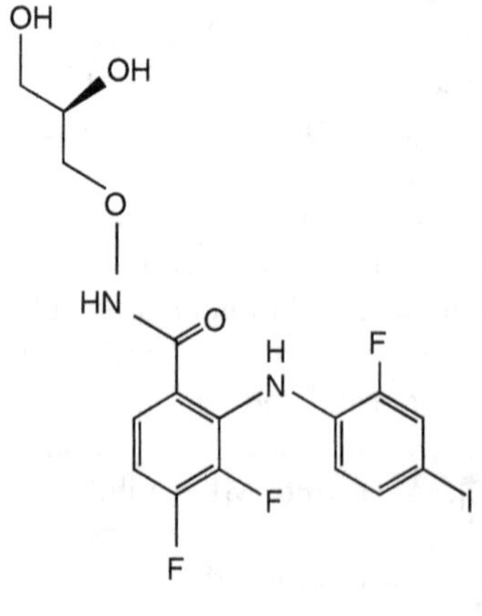

Figure 13.1 Chemical structure of PD0325901.

support Phase I and II clinical trials in cancer patients with various solid tumors (advanced breast cancer, colon cancer, melanoma, non-small cell lung cancer) utilizing oral administration of the drug.

Toxicology Studies

The nonclinical safety of PD0325901 was evaluated in Sprague–Dawley rats given single oral or intravenous (IV) doses, in beagle dogs given oral and IV escalating doses, and in cynomolgus monkeys given escalating oral doses to assess acute toxicity and assist in dose selection for subsequent studies. The potential effects of PD0325901 on central nervous system, cardiovascular, and pulmonary function were evaluated in single-dose safety pharmacology studies. Two-week dose-range finder (non-pivotal) oral toxicity studies were conducted in rats, dogs, and monkeys to assist in dose and species selection for the pivotal one-month oral toxicology studies. In addition, an investigative oral toxicity study was conducted in female Balb/c mice. The dog was selected as the non-rodent species for the pivotal toxicology study because of the following data. Metabolites of PD0325901 identified in human liver microsomal incubations were also present following incubation with dog liver microsomes, plasma protein binding of PD0325901 was similar in dogs and humans (>99%), and oral bioavailability in dogs was high (>90%). Finally, injury to the mucosa of the gastrointestinal tract occurred at lower doses and exposures in dogs than in monkeys (based on dose range–finding studies), indicating the dog as the more sensitive non-rodent species. Pivotal one-month oral toxicity studies, including one-month reversal phases, were conducted in beagle dogs and Sprague–Dawley rats to support submission of an investigational new drug (IND) application to the US Food and Drug Administration (FDA). A list of toxicology studies of PD0325901 conducted prior to initiation of human testing is presented in Table 13.1.

Upon completion of the first two-week dose range–finding study in rats, a significant and unique toxicity was observed that involved mineralization of vasculature (Figure 13.2) and various soft tissues (i.e. ectopic or systemic mineralization) as determined by routine light-microscopic evaluation. In a follow-up study in rats, dysregulation of serum calcium and phosphorus homeostasis and systemic mineralization occurred in a time- and dose-dependent manner. This toxicity was not observed in dogs or monkeys, despite systemic exposures to PD0325901 more than 10-fold higher than those associated with mineralization in rats and pharmacologic inhibition of phosphorylated MAPK in canine or monkey tissue (demonstrating biochemical activity of PD0325901 at the target protein, i.e. MEK). Various investigative studies were conducted to examine the time course and potential mechanism of systemic mineralization in rats and to identify biomarkers that could be used to monitor for this effect in clinical

Table 13.1 Summary of toxicology studies conducted with PD0325901.[a]

Acute and escalating dose
Single dose in rats
Single dose in rats, IV[b]
Escalating dose in dogs
Escalating dose in dogs, IV
Escalating dose in monkeys
Safety pharmacology
Neurofunctional evaluation in rats
Neurofunctional evaluation in rats, IV
Cardiovascular effects in monkeys
Pulmonary effects in rats
Purkinje fiber assay
HERG assay
Non-pivotal repeated-dose studies
2-Week dose-range finder in rats
Exploratory 2-week study in rats
2-Week dose-range finder in dogs
2-Week dose-range finder in monkeys
Pivotal repeated-dose studies
One month in rats (plus one-month reversal phase)
One month in dogs (plus one-month reversal phase)
Pivotal genetic toxicity studies
Bacterial mutagenicity
Structural chromosome aberration in human lymphocytes
In vivo micronucleus in rats
Special toxicity studies
Pharmacodynamic and toxicokinetic in rats, oral and IV
Time course and biomarker development in rats
Serum chemistry reversibility study in rats
Investigative study in mice
Enantiomer (*R* and *S*) study in rats
PD0325901 in combination with pamidronate or Renagel in rats

a) All animal studies were conducted by oral gavage unless otherwise indicated.
b) IV, intravenous (bolus).

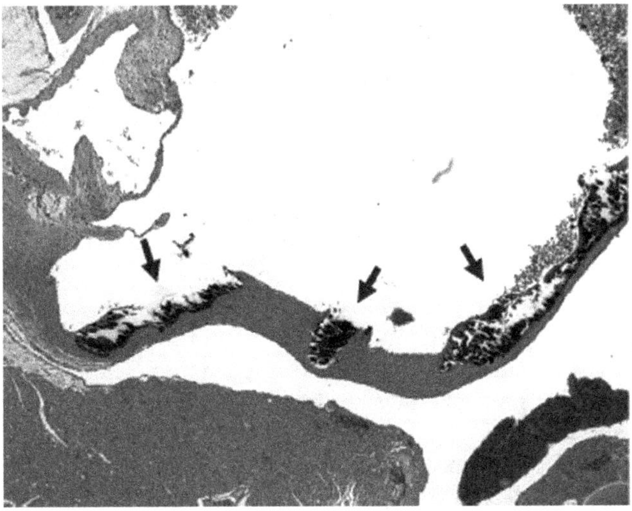

Figure 13.2 Mineralization of the aorta in a male rat administered PD0325901 at 3 mg/kg in a dose range–finding study. Arrows indicate mineral in the aorta wall. Hematoxylin and eosin–stained tissue section. (*See insert for color reproduction of this figure.*)

trials. Next are described the key studies conducted to investigate this toxicity, the results obtained, and how the nonclinical data were utilized to evaluate the safety risk of the compound, select a safe starting dose for a Phase I trial, and provide measures to ensure patient safety during clinical evaluation of PD0325901 in cancer patients.

At the beginning of the toxicology program for PD0325901, a two-week oral dose range–finding study was conducted in male and female rats in which daily doses of 3, 10, and 30 mg/kg (18, 60, and 180 mg/m^2, respectively) were administered. Mortality occurred in males at ≥3 mg/kg and females at ≥10 mg/kg, with toxicity occurring to a greater extent in males at all dose levels. Increased serum levels of phosphorus (13–69%) and decreased serum total protein (12–33%) and albumin (28–58%) were seen at all doses. Light-microscopic evaluation of formalin-fixed and hematoxylin- and eosin-stained tissues was performed. Mineralization occurred in the aorta (Figure 13.2) and coronary, renal, mesenteric, gastric, and pulmonary vasculature of males at ≥3 mg/kg and in females at ≥10 mg/kg. Parenchymal mineralization with associated degeneration occurred in the gastric mucosa and muscularis, intestines (muscularis, mucosa, submucosa), lung, liver, renal cortical tubules, and/or myocardium at the same doses. Use of the Von Kossa histology stain indicated the presence of calcium in the mineralized lesions. Vascular/parenchymal mineralization and degeneration were generally dose related in incidence and severity. PD0325901 produced increased thickness (hypertrophy) of the femoral growth plate (physis) in both sexes at all doses and degeneration and necrosis of the femoral metaphysis in males at ≥3 mg/kg and females at 30 mg/kg. In addition, skin ulceration, hepatocellular necrosis, decreased crypt goblet cells, reduced hematopoetic elements, and ulcers of cecum and duodenum were observed. Systemic mineralization of the vasculature and soft tissues was the most toxicologically significant finding of this study. At this time, it was not known whether the hyperphosphatemia was due to decreased renal clearance [17] and/or related to the mineralization. However, hyperphosphatemia and elevated serum calcium–phosphorus (Ca × P) product can result in vascular and/or soft tissue mineralization [18–20]. In addition, morphologic findings similar to those seen in this study are observed in various animal species (e.g. dogs, horses, pigs, rats) with vitamin D toxicosis and altered calcium homeostasis [21–24]. Tissue mineralization was observed in the aorta, various arteries, myocardium, gastric mucosa, and renal tubules, along with other soft tissues in these animals.

Following these findings, an exploratory two-week oral toxicity study was conducted in male and female rats to further investigate the toxicities observed in the initial two-week dose-range finder. The objectives of this study were to identify a minimal or no-observed adverse-effect level (NOAEL) and to provide toxicity, toxicokinetic, and pharmacodynamic data to aid in dose selection for future studies. In addition, an attempt was made to assess whether alterations in phosphorus and calcium homeostasis occur and whether changes can be monitored as potential biomarkers of toxicity. Doses tested in this study were 0.3, 1, or 3 mg/kg (1.8, 6, or 18 mg/m^2, respectively) and animals were dosed for up to 14 days. Cohorts of animals (5/sex/group) were necropsied on Days 4 and 15, and hematology, serum biochemistry, plasma intact parathyroid hormone (PTH), and urinary parameters were evaluated. Urinalysis included measurement of calcium, phosphorus, and creatinine levels. Select tissues were examined microscopically, and samples of liver and lung were evaluated for total and phosphorylated MAPK (pMAPK) levels by Western blot analysis to evaluate for pharmacologic activity of PD0325901 (method described in Ref. [16]). Satellite animals were included for plasma drug-level analyses on Day 8. In this study, systemic mineralization occurred at ≥0.3 mg/kg in a dose-dependent fashion, was first observed on Day 4, and was more severe in males. By Day 15, mineralization was generally more pronounced and widespread. Skeletal changes included hypertrophy of the physeal zone in males at ≥1 mg/kg and at 3 mg/kg in females, and necrosis of bony trabeculae and marrow elements with fibroplasia, fibro-osseous proliferation, and/or localized hypocellularity at ≥1 mg/kg in males and 3 mg/kg in females. The minimal plasma PD0325901 AUC(0–24) values associated with toxicity were 121–399 (ng h)/ml, which were well below exposure levels associated with antitumor efficacy in murine models (AUC of 1180–1880 (ng h)/ml). Pharmacologic inhibition of tissue pMAPK occurred at ≥1 mg/kg and was not observed in the absence of toxicity. The gastric fundic

Table 13.2 Mean clinical chemistry changes in male rats administered PD0325901 for up to two weeks.

	Day	Control	PD0325901		
			0.3 mg/kg	1 mg/kg	3 mg/kg
Serum phosphorus (mg/dl)	4	12.90	13.08	14.48	16.24[a]
	15	11.30	11.56	12.88	13.62[a]
Serum calcium (mg/dl)	4	10.58	10.36	10.10	10.16
	15	10.38	10.36	10.52	10.36
Serum albumin (g/dl)	4	2.74	2.56	2.10[a]	2.04[a]
	15	2.56	2.54	2.36[a]	1.98[a]
Plasma PTH (pg/ml)[b]	4	492	297	114[a]	155[a]
	15	1099	268	457	115[a]

a) $p' < 0.01$ vs. control; $n = 5$/group.
b) Intact parathyroid hormone.

mucosa appeared to be the most sensitive tissue for evaluating systemic mineralization, which probably resulted from alterations in serum calcium and phosphorus homeostasis. This was based on the following observations. On Day 4, serum phosphorus levels were increased 12–26%, and albumin was decreased 17–26% at ≥1 mg/kg (Table 13.2, male data only). In addition, PTH levels were decreased in a dose-dependent fashion (60–77%) at ≥1 mg/kg. On Day 15, phosphorus levels were increased 21% in males at 3 mg/kg, and albumin was decreased 8–32% at ≥0.3 mg/kg. PTH levels were decreased 77–89% at 3 mg/kg. Changes in urinary excretion of calcium and phosphorus were observed in both sexes at ≥1 mg/kg and included increased excretion of phosphorus on Day 15. Although increases in excretion of calcium were observed on Day 4 in females, males exhibited decreases in urinary calcium. In this study, PD0325901 administration resulted in significantly decreased levels of serum albumin without changes in serum (total) calcium levels [25–27]. This indicates that free, non-protein-bound calcium levels were increased. Hyperphosphatemia and hypercalcemia result in an increased Ca × P product, which is associated with induction of vascular mineralization [19, 20]. The changes observed in urinary excretion of calcium and phosphorus probably reflected the alterations in serum levels.

After completion of the two studies in rats described above, it was concluded that PD0325901 produces significant multi-organ toxicities in rats with no margin between plasma drug levels associated with antitumor efficacy, pharmacologic inhibition of pMAPK (as an index of MEK inhibition), and toxicity in rats. Systemic mineralization was considered the preclinical toxicity of greatest concern, due to the severity of the changes observed and expectation of irreversibility, and the data suggested that it was related to a dysregulation in serum phosphorus and calcium homeostasis. Furthermore, skeletal lesions were seen in the rat studies that were similar to those reported with vitamin D toxicity and may also have been related to the calcium–phosphorus dysregulation. In concurrent toxicology studies in dogs and monkeys, neither systemic mineralization nor skeletal changes were observed, despite higher plasma drug exposures, lethal doses, or pharmacologic inhibition of MEK. Therefore, the following questions were posed regarding PD0325901-induced systemic mineralization: (i) What is a potential mechanism? (ii) Is this toxicity relevant to humans or rat-specific?, and (iii) Can this toxicity be monitored clinically?

The ability of an anticancer agent that modulates various signal transduction pathways to produce dysregulation in serum calcium homeostasis is not unprecedented. 8-Chloro-cAMP is an experimental compound that has been shown to modulate various protein kinase signal transduction pathways involved in neoplasia. In preclinical models, this compound produced growth inhibition and increased differentiation in cancer cells [28]. In a clinical trial, 8-chloro-cAMP was administered to patients with advanced cancer via intravenous infusion and resulted in dose-limiting toxicity of reversible hypercalcemia, as serum calcium was increased by up to approximately 40% [29]. This drug produced a PTH–like effect in these patients, resulting in increased synthesis of 1,25-dihydroxyvitamin D (up to 14 times baseline value) as a mechanism for the hypercalcemia. Intravenous administration of 8-chloro-cAMP to beagle dogs also resulted in hypercalcemia (serum calcium increased 37–46%), indicating similar actions across species [30]. Experience with this compound was important with respect to designing investigative studies with PD0325901 in which the hormonal control of serum calcium and phosphorus were evaluated.

An investigative study was designed in rats to examine the time course for tissue mineralization in target organs and to determine whether clinical pathology changes occur prior to, or concurrent with, lesion development [31]. These clinical pathology parameters may therefore serve as biomarkers for systemic mineralization. Male rats (15/group) were used due to their increased sensitivity for this toxicity compared with females. Oral doses tested were 1, 3, or 10 mg/kg (6, 18, or 60 mg/m^2, respectively). Five animals per group were necropsied on Days 2, 3, or 4 following 1, 2, or 3 days of treatment, respectively. Clinical laboratory tests were conducted at necropsy that included serum osteocalcin, urinalysis, and plasma intact PTH, calcitonin, and 1,25-dihydroxyvitamin D. Lung samples were evaluated for inhibition of pMAPK, and microscopic evaluations of the aorta, distal femur with proximal tibia, heart, and stomach were conducted for all animals.

Administration of PD0325901 resulted in inhibition of pMAPK in lung at all doses, demonstrating pharmacologic activity of the drug. On Day 2, mineralization of gastric fundic mucosa and multifocal areas of necrosis of

Table 13.3 Mean serum phosphorus and plasma 1,25-dihydroxyvitamin D in male rats administered PD0325901 for up to three days of dosing.

			PD0325901		
	Day	Control	1 mg/kg	3 mg/kg	10 mg/kg
Serum phosphorus (mg/dl)	2	12.06	16.10[a]	17.22[a]	16.84[a] M[b]
	3	11.48	12.96[a]	15.62[a] M	19.02[a] M
	4	11.34	13.18[a] M	15.40[a] M	21.70[a] M
1,25-Dihydroxyvitamin D (pg/ml)	2	309	856[a]	1328[a]	2360[a] M
	3	257	396	776[a] M	1390[a] M
	4	191	236 M	604[a] M	1190[a] M

a) $p' < 0.01$ vs. control; $n = 5$/group.
b) M, systemic mineralization observed.

Table 13.4 Mean serum calcium and albumin in male rats administered PD0325901 for up to three days of dosing.

			PD0325901		
	Day	Control	1 mg/kg	3 mg/kg	10 mg/kg
Serum calcium (mg/dl)	2	10.42	11.04	11.00	10.66 M[a]
	3	9.60	10.58	10.64 M	10.58 M
	4	10.44	10.44 M	10.58 M	7.24[b] M
Serum albumin (g/dl)	2	3.08	2.92	2.82[c]	2.66[b] M
	3	2.88	2.68	2.62 M	2.34[b] M
	4	2.90	2.34[b] M	2.34[b] M	1.98[b] M

a) M, systemic mineralization observed.
b) $p < 0.01$ vs. control; $n = 5$/group.
c) $p < 0.05$ vs. control.

the ossifying zone of the physis were present only at 10 mg/kg. Necrosis of the metaphysis was present at ≥3 mg/kg. Serum phosphorus levels increased 33–43% and 1,25-dihydroxyvitamin D increased two- to sevenfold at all doses (Table 13.3). Osteocalcin increased 14–18%, and serum albumin decreased 8–14% at ≥3 mg/kg (Table 13.4). Osteocalcin is a major noncollagenous protein of bone matrix and synthesized by osteoblasts [32]. Changes in serum osteocalcin can reflect alterations in bone turnover (resorption/formation). Serum osteocalcin appears to reflect the excess of synthesized protein not incorporated into bone matrix, or released protein during bone resorption [33]. The increases in osteocalcin seen in this study may have been reflective of bone necrosis.

On Day 3, mineralization of gastric fundic mucosa, gastric and cardiac arteries, aorta, and heart were present in all rats at 10 mg/kg. Myocardial necrosis was also seen at 10 mg/kg. Mineralization of gastric fundic mucosa was present in all rats at 3 mg/kg, and focal, minimal myocyte necrosis was present in one rat at 3 mg/kg. Thickening of the physeal zone of hypertrophying cartilage and necrosis within the physeal zone of ossification and in the metaphyseal region in femur and tibia were seen in all animals at 10 mg/kg. Necrosis within the metaphyseal region was also present at 3 mg/kg. Serum phosphorus increased 13–66% at all doses and 1,25-dihydroxyvitamin D increased two-to fourfold at ≥3 mg/kg. Osteocalcin increased 12–28% at ≥3 mg/kg and serum albumin was decreased (7–19%) at all doses. Urine calcium increased fivefold at 10 mg/kg, resulting in a fivefold increase in urine calcium/creatinine ratio. This increase may have represented an attempt to achieve mineral homeostasis in response to the hypercalcemia. In addition, hypercalciuria can occur with vitamin D intoxication [34].

On Day 4, mineralization of gastric fundic mucosa, gastric muscularis, gastric and cardiac arteries, aorta, and heart were present in the majority of animals at ≥3 mg/kg. Myocardial necrosis with accompanying neutrophilic inflammation was also seen in all rats at 10 mg/kg and in one animal at 3 mg/kg. Mineralization of gastric fundic mucosa was present at 1 mg/kg. Thickening of the physeal zone of hypertrophying cartilage and necrosis within the physeal zone of ossification and/or in the metaphyseal region in femur and tibia were present at ≥3 mg/kg. At 1 mg/kg, thickening of the physeal zone of hypertrophying cartilage and metaphyseal necrosis were observed. Serum phosphorus increased 16–91% at all doses, and 1,25-dihydroxyvitamin D increased two-to fivefold at ≥3 mg/kg. Osteocalcin increased 14–24% at ≥3 mg/kg, and serum albumin decreased 19–32% at all doses. At 10 mg/kg, serum calcium was decreased 31% (possibly resulting from the hypercalciuria on Day 3), and calcitonin was decreased by 71%. Calcitonin is secreted by the thyroid gland and acts to lower serum calcium levels by inhibiting bone resorption [27]. The decrease in calcitonin may have resulted from feedback inhibition due to low serum calcium levels at 10 mg/kg on Day 4. Urine creatinine, calcium, and phosphorus were increased at 10 mg/kg. This resulted in decreases of 41% and 21% in the calcium/creatinine and phosphorus/creatinine ratios, respectively.

This four-day investigative study in rats resulted in several very important conclusions which were critical for supporting continued development of PD0325901. In the study, PD0325901 at ≥1 mg/kg resulted in systemic mineralization and skeletal changes in a dose- and time-dependent fashion. These changes were seen after a single dose at 10 mg/kg and after three doses at 1 mg/kg. Elevations in serum phosphorus and plasma 1,25-dihydroxyvitamin D occurred prior to tissue mineralization. Although serum albumin was decreased throughout the study, calcium remained unchanged, consistent with an increase in non-protein-bound calcium. This study set the stage for the proposal of using serum phosphorus and calcium measurements as clinical laboratory tests or biomarkers for PD0325901-induced systemic mineralization. Whereas measurement of plasma 1,25-dihydroxyvitamin D is technically complex and costly, evaluation of serum calcium and phosphorus is rapid and performed routinely in the clinical laboratory with historical reference ranges readily available. Although the data obtained with urinalysis were consistent with dysregulation of calcium and phosphorus homeostasis, concerns existed as to whether specific and reproducible urinalysis parameters could be developed for monitoring the safety of PD0325901. Based on the data obtained thus far, hyperphosphatemia appeared to be the primary factor for eliciting tissue mineralization, and serum phosphorus was proposed as the key analyte for monitoring.

An investigative study was conducted in male rats to assess the reversibility of serum chemistry changes following a single oral dose of PD0325901 [31]. The hypothesis was that serum phosphorus levels would return to control levels in the absence of drug administration. Male rats (10/group) received single oral doses at 1, 3, or 10 mg/kg, with controls receiving vehicle alone. Blood was collected on Days 2, 3, 5, and 8 for serum chemistry analysis. Hyperphosphatemia (serum phosphorus increased up to 58%) and minimal increases in calcium occurred at all doses on Days 2 and 3. Albumin was decreased at 10 mg/kg. These changes were completely reversible within a week. This study demonstrated that increases in serum phosphorus and calcium induced by PD0325901 are reversible following cessation of dosing. Although a single dose of 10 mg/kg produces systemic mineralization in rats, withdrawal of dosing results in normalization of serum calcium and phosphorus levels, indicating that the homeostatic mechanisms controlling these electrolytes remain intact. The results of this study were not unexpected. Oral administration to dogs of the vitamin D analogs dihydrotachysterol and Hytakerol (dihydroxyvitamin D_2-II) results in hypercalcemia that is reversible following termination of dosing [35]. Reversal of hypercalcemia and hypercalciuria has been demonstrated in humans following cessation of dosing of various forms of vitamin D (calciferol, dihydrotachysterol, 1-α-hydroxycholecalciferol, or 1-α,25-dihydroxycholecalciferol) [36].

Another investigative study was conducted in male rats to determine whether pamidronate (a bisphosphonate) or Renagel (sevelamer HCl; a phosphorus binder) would inhibit tissue mineralization induced by PD0325901 by inhibiting hyperphosphatemia. Bisphosphonates inhibit bone resorption and in so doing modulate serum calcium and phosphorus levels. Renagel is a non-absorbable resin that contains polymers of allylamine hydrochloride, which forms ionic and hydrogen bonds with phosphate in the gut, thereby inhibiting dietary phosphate absorption. Rats received daily oral doses of PD0325901 at 3 mg/kg for 14 days with or without co-treatment with pamidronate or Renagel. Pamidronate was given twice intravenously at 1.5 mg/kg one day prior to PD0325901 dosing and on Day 6. Renagel was given daily as 5% of the diet beginning one day prior to PD0325901 dosing. Treatment groups consisted of oral vehicle alone, PD0325901 alone, pamidronate alone, Renagel alone, PD0325901 + pamidronate, and PD0325901 + Renagel. PD0325901 plasma AUC(0–24) values were 11.6, 9.17, and 4.34 (μg h)/ml in the PD0325901 alone, PD0325901 + pamidronate, and PD0325901 + Renagel groups, respectively. Administration of PD0325901 alone resulted in hyperphosphatemia on Days 3 and 15, which was inhibited by co-treatment with pamidronate or Renagel on Day 3 only. PD0325901 alone resulted in systemic mineralization and skeletal changes consistent with changes seen in previous rat studies. Co-administration with either pamidronate or Renagel protected against systemic mineralization on Day 3 only. Bone lesions were decreased with the co-treatments. Inhibition of toxicity with Renagel may have been due in part to decreased systemic drug exposure. However, the inhibition of toxicity with pamidronate supports the role of a calcium–phosphorus dysregulation in PD0325901-induced systemic mineralization, because the inhibition of systemic mineralization observed on Day 3 coincided with attenuation in the rise in serum phosphorus in these animals.

A two-week oral dose range–finding study was conducted in dogs in which doses tested were 0.2, 0.5, and 1.5 mg/kg (4, 10, and 30 mg/m^2, respectively). Also, a two-week oral dose range–finding study was conducted in cynomolgus monkeys at doses of 0.5, 3, and 10 mg/kg (6, 36, and 120 mg/m^2, respectively). In addition to standard toxicology and toxicokinetic endpoints, determination of inhibition of tissue and peripheral blood mononuclear cell pMAPK was performed to assess pharmacologic activity of PD0325901 in both studies. PTH and 1,25-dihydroxyvitamin D were evaluated in the monkey study. In both studies, mortality occurred at ≥0.5 mg/kg (dogs) and at 10 mg/kg (monkeys) due to injury to the gastrointestinal tract mucosa, inhibition of pMAPK occurred at all doses, and systemic mineralization was not observed in either study. Increases in serum phosphorus were seen in moribund animals and/or associated with renal hypo-perfusion (resulting from emesis, diarrhea, and dehydration). These elevations in phosphorus were considered secondary to renal effects and were not associated with changes in serum calcium. Toxicologically significant increases in serum phosphorus or calcium were not evident at nonlethal doses in dogs or monkeys. In the two-week monkey study, a dose-related increase in 1,25-dihydroxyvitamin D was observed on Day 2 only (after a single dose) at ≥3 mg/kg. This increase did not occur on Days 7 or 15, and was not associated with changes in serum phosphorus or calcium, nor systemic mineralization. Therefore, there did not appear to be toxicologic significance to the Day 2 increase in 1,25-dihydroxyvitamin D in monkeys.

Discussion

Mineralization of vasculature and various soft tissues (systemic mineralization) was observed in toxicology studies in rats in a time- and dose-dependent manner. This change was consistent with the presence of calcium–phosphorus deposition within the vascular wall and parenchyma of tissues such as the stomach, kidney, aorta, and heart. The stomach appeared to be the most sensitive tissue, since mineralization of gastric fundic mucosa occurred prior to the onset of mineralization in other tissues. Gastric mineralization was also observed in rats administered the structurally related potent MEK inhibitors PD198306 and PD254552, whereas non-potent structural analogs (PD318894 and PD320125-2, respectively) did not produce this toxicity, demonstrating a relationship to MEK inhibition (Table 13.5; [37]). Male rats were consistently

Table 13.5 Toxicity to skin and gastrointestinal tract are associated with MEK inhibition.

Compound	IC_{50} (nM)[a)]	Plasma C_{ave} (μg/mL)[b)]	Lung pMAPK (% Inhibition)[c)]	Microscopic observations[d)]
PD198306	43	30 ± 17.4	93	Skin: epidermal ulcers, crusts, acanthosis Colon: decreased crypt goblet cells, thinning of mucosa Stomach: mineralization, degeneration of glandular mucosa Liver: hypertrophy of centrilobular hepatocytes
PD318894	3 000	53 ± 14.4	14	None
PD254552	4.2	42 ± 6.8	96	Skin: epidermal ulcers, crusts, acanthosis Cecum: epithelial hyperplasia, inflammation Stomach: mineralization, degeneration of glandular mucosa Liver: hypertrophy, vacuolation of hepatocytes, necrosis
PD320125-2	31 900	83 ± 7.3	11	None

Female Wistar rats received daily oral doses (525 μmol/kg) of the potent MEK inhibitors PD198306 and PD254552 or their non-potent analogs PD318894 and PD320125-2, respectively, for up to 21 days (histopathology; $N = 5$/group) or for a single dose ($N = 3$/group for plasma drug level and lung pMAPK measurements).

a) IC_{50} for pMAPK inhibition in murine C26 tumor cells.
b) C_{ave} ± standard deviation for 1, 3 and 6 hours post-dose.
c) Mean inhibition of pMAPK in lung, normalized to vehicle control rats.
d) Microscopic evaluations were conducted on a limited tissue list (Brown et al. [37]).

more sensitive to systemic mineralization than female rats despite similar systemic exposure to PD0325901. In the pivotal one-month toxicity study in rats, the no-effect level for systemic mineralization was 0.1 mg/kg (0.6 mg/m^2) in males and 0.3 mg/kg (1.8 mg/m^2) in females, which were associated with PD0325901 steady-state plasma AUC(0–24) values of 231 and 805 (ng h)/ml, respectively. Systemic mineralization was not observed in dogs or monkeys, despite pharmacologic inhibition of tissue pMAPK levels (>70%), administration of lethal doses, and exposures greater than 10-fold of those that induced mineralization in rats (10 600 (ng h)/ml in dogs and up to 15 000 (ng h)/ml in monkeys). Systemic mineralization was not observed in mice despite administration of PD0325901 at doses up to 50 mg/kg (150 mg/m^2).

Systemic mineralization observed in rats following administration of PD0325901 was consistent with vitamin D toxicity and dysregulation in serum calcium and phosphorus homeostasis [21, 38–41]. A proposed hypothesis for the mechanism of this toxicity is depicted in Figure 13.3. Elevated serum phosphorus levels (hyperphosphatemia) and decreased serum albumin were observed consistently in rats administered PD0325901. Although serum albumin levels are decreased in rats treated with PD0325901, calcium values typically remain unchanged or slightly elevated in these animals, indicating that free, non-protein-bound calcium is increased [25–27]. Decreased PTH levels were observed in the rat studies. PTH plays a central role in the

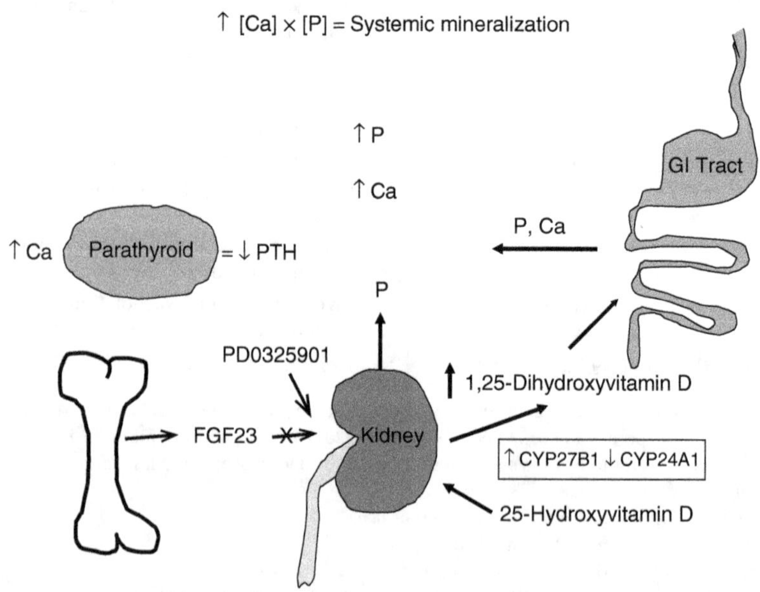

Figure 13.3 Hypothesis for the mechanism for systemic mineralization in the rat following PD0325901 administration. (*See insert for color reproduction of this figure.*)

hormonal control of serum calcium and phosphorus. PTH is produced by the parathyroid gland and induces conversion of 25-hydroxyvitamin D (which is produced in the liver) to 1,25-dihydroxyvitamin D (calcitriol) in the kidney. 1,25-Dihydroxyvitamin D elicits increased absorption of calcium from the gastrointestinal tract. In addition, PTH mobilizes calcium and phosphorus from bone by increasing bone resorption, increases renal absorption of calcium, and increases renal excretion of phosphorus (in order to regulate serum phosphorus levels). Elevations in serum calcium typically elicit decreased PTH levels as a result of the normal control (negative feedback loop) of this endocrine system [27]. The decreases in PTH observed in the rats were believed to be due to the elevations in serum calcium (hypercalcemia). Hyperphosphatemia in the presence of normo- or hypercalcemia can result in an increased Ca × P product, which is associated with systemic mineralization [19]. Hyperphosphatemia was also observed in rats administered PD176067, which is a reversible and selective inhibitor of fibroblast growth factor (FGF) receptor tyrosine kinase. In these animals, vascular and soft tissue mineralization also occurs (aorta and other arteries, gastric fundic mucosa, myocardium, renal tubules), probably due to increased Ca × P product [42]. The role of FGF in the mechanism of action of systemic mineralization is described below.

Administration of PD0325901 to rats resulted in significantly increased levels of plasma 1,25-dihydroxyvitamin D which is the most potent form of vitamin D and the primary metabolite responsible for regulating serum calcium and phosphorus. Vitamin D is converted to 25-hydroxyvitamin D in the liver and then 1-hydroxylated by CYP27B1to 1,25-dihydroxyvitamin D in renal proximal tubules. 1,25-Dihydroxyvitamin D is subsequently metabolized by 24-hydroxylase (CYP24A1) to an inactive form. 1,25-Dihydroxyvitamin D acts by increasing absorption of calcium and phosphorus from the gastrointestinal tract, and can increase calcium and phosphorus reabsorption by renal tubules. Hyperphosphatemia and increased plasma 1,25-dihydroxyvitamin D levels in rats occurred one to two days prior to the detection of tissue mineralization at doses $\leq$3 mg/kg (18 mg/m^2).

Fibroblast growth factor 23 (FGF23) is a circulating hormone produced in bone by osteocytes and osteoblasts that plays a significant role in modulating serum phosphate and 1,25-dihydroxyvitamin D. FGF23 binds to FGF receptors in the kidney, leading to decreased expression of CYP27B1 and induction of CYP24A1, thereby decreasing the levels of 1,25-dihydroxyvitamin D. In addition, FGF23-mediated signaling decreases the expression of sodium-phosphate co-transporters in renal tubules, resulting in decreased phosphate reabsorption and increased urinary excretion of phosphate, with the net effect of decreased serum phosphate [43, 44]. FGF23 (–/–) null mice display increased serum levels of phosphate, calcium, and 1,25-dihydroxyvitamin D, and decreased serum PTH, similar to what is observed in rats administered PD0325901 [45]. Mechanistic studies have shown that PD0325901 can disrupt FGF23 signaling in the kidney (via MEK inhibition), resulting in increased levels of serum phosphate and 1,25-dihydroxyvitamin D, and subsequent systemic mineralization [46–48]. PD0325901 can increase 1,25-dihydroxyvitamin D in rats due to induction of CYP27B1 and downregulation of CYP24A1 in the kidney, effects opposite to that of FGF23 [47].

Administration of PD0325901 to rats resulted in bone lesions that included necrosis of the metaphysis and the ossifying zone of the physis and thickening of the zone of hypertrophying cartilage of the physis. The expansion of chondrocytes in the physis may be a response to the metaphyseal necrosis and loss of osteoprogenitor cells. These changes are characterized by localized injury to bone that appear to be due to local ischemia and/or necrosis. Skeletal vascular changes may be present in these animals, resulting in disruption of endochondral ossification. Skeletal lesions, including bone necrosis, can result from vitamin D intoxication in non-rodents, including horses and pigs [23, 24, 49]. The skeletal lesions observed in rats administered PD0325901 are similar to those reported with vitamin D toxicity, which provides additional evidence that toxicity occurred via induction of 1,25-dihydroxyvitamin D. Bone lesions similar to those observed in rats were not seen in dogs, monkeys, or mice administered PD0325901.

In summary, PD0325901-induced systemic mineralization in the rat results from a dysregulation in serum phosphorus and calcium homeostasis. This dysregulation appears to result from toxicologically significant elevations in plasma 1,25-dihydroxyvitamin D following drug administration. Increased levels of 1,25-dihydroxyvitamin D appear to be due to MEK inhibition in the kidney, leading to disruption of FGF23 signaling pathways. Based on the toxicology data, rats are uniquely sensitive to this toxicity. A summary of the primary target organ toxicities observed in the preclinical studies is presented in Table 13.6. Toxicity to the skin (epidermal lesions) and gastrointestinal tract (primarily ulcers/erosions in the mucosa) were observed across species and

Table 13.6 Primary target organ toxicities observed in preclinical studies.

	Species		
Organ system	**Rat**	**Dog**	**Monkey**
Gastrointestinal tract	×[a)]	×	×
Skin	×	×	×
Systemic mineralization[b)]	×	—	—
Bone	×	—	—
Liver	×	—	—
Gallbladder	n/a	—	×

a) Toxicity observed.
b) Includes vascular (aorta, arteries) and soft tissue mineralization (e.g. stomach, heart, kidneys).

may have resulted from inhibition of MEK-related signal transduction pathways in these tissues (Table 13.5; [37]). Gastrointestinal tract toxicity was dose-limiting in dogs and monkeys and was anticipated to be the dose-limiting toxicity of PD0325901 in the clinic. Therefore, gastrointestinal tract toxicity may preclude the development of other potential adverse events in humans, including potential dysregulation in serum phosphorus or calcium. It was not known whether systemic mineralization is relevant to humans. However, if PD0325901 does induce a dysregulation in serum calcium–phosphorus metabolism in humans, monitoring serum levels would provide an early indication of effects and guide modifications to dosing regimens.

Safety Biomarkers

To ensure patient safety in the Phase I clinical trial with PD0325901, procedures were incorporated into the trial design to monitor for potential dysregulation in serum calcium–phosphorus homeostasis. Measurements of serum calcium, phosphorus, creatinine, albumin, and blood urea nitrogen were performed frequently during the initial treatment cycle (21 days of dosing in a 28-day cycle), with periodic measurement in subsequent cycles, and the serum Ca × P product was calculated. The serum Ca × P product has been determined to be a clinically useful value as a means for evaluating risk for tissue and/or vascular mineralization with the recommendation that the value not exceed 70 based on clinical use of vitamin D analogs such as Rocaltrol and Hectorol [19, 50, 51]. Serum calcium and phosphorus are readily measured in a clinical setting with well-established reference ranges available. The trial included a protocol-specific dose-limiting toxicity for a Ca × P product >70, which required a confirmatory measurement and dose interruption for that patient. In addition, serum vitamin D, PTH, alkaline phosphatase (total and bone), osteocalcin, and urinary C- and N-terminal peptide of collagen 1 (markers of bone resorption) were included for periodic measurement. Criteria for exclusion of candidate patients from the clinical trial included a history of malignancy-associated hypercalcemia, extensive bone metastasis, parathyroid disorder, hyperphosphatemia and renal insufficiency, serum calcium or phosphorus levels >1× the upper limit of normal, and/or concomitant use of calcium supplements and vitamin D in amounts exceeding normal daily allowances.

Clinical Doses and Responses

A common algorithm for calculating a starting dose in clinical trials with oncology drugs is to use one-tenth of the dose that causes severe toxicity (or death) in 10% of the rodents (STD_{10}) on a mg/m^2 basis, provided that this starting dose (i.e. 1/10 the STD_{10}) does not cause serious, irreversible toxicity in a non-rodent species (in this case, the dog) [52]. If irreversible toxicities are induced at the proposed starting dose in non-rodents, or if the non-rodent (i.e. the dog) is known to be the more appropriate animal model, the starting dose would generally be one-sixth of the highest dose tested in the non-rodent (the dog) that does not cause severe, irreversible toxicity.

Calculation of the initial Phase I starting dose of PD0325901 was based on the pivotal one-month toxicology studies in rats and dogs. Doses tested in the one-month rat study were 0.1, 0.3, and 1 mg/kg (0.6, 1.8, and 6 mg/m^2, respectively), and doses in the one-month dog study were 0.05, 0.1, and 0.3 mg/kg (1, 2, and 6 mg/m^2, respectively). Both studies included animals assigned to a one-month reversal phase, in the absence of dosing, to assess reversibility of any observed toxicities. In addition to standard toxicology and toxicokinetic parameters, these studies included frequent evaluation of serum chemistries and measurement of vitamin D, osteocalcin, PTH, and inhibition of tissue pMAPK levels.

In the one-month rat study, no drug-related deaths occurred and systemic mineralization occurred in multiple tissues in both sexes at 1 mg/kg. Hypocellularity of the metaphyseal region of distal femur and/or proximal tibia occurred in males at 1 mg/kg. Toxicologic findings at lower doses included skin sores (at $\geq$0.1 mg/kg) and mineralization of gastric mucosa in one male at 0.3 mg/kg. The findings at $\leq$0.3 mg/kg were not considered to represent serious toxicity. In previous two-week dose range–finding studies in rats, death occurred at 3 mg/kg (18 mg/m^2), indicating this to be the minimal lethal dose in rats. Based on these results, the STD_{10} in rats was determined to be 1 mg/kg (6 mg/m^2).

In the one-month dog study, doses up to 0.3 mg/kg (6 mg/m^2) were well tolerated with minimal clinical toxicity. Primary drug-related toxicity was limited to skin sores in two animals at 0.3 mg/kg. One-tenth the STD_{10} in rats is 0.6 mg/m^2, which is well below a minimally toxic dose (6 mg/m^2) in dogs. These data indicate an acceptable Phase I starting dose to be 0.6 mg/m^2, which is equivalent to 1 mg in a 60-kg person. The relationships between the primary toxicities in rats and dogs with dose and exposure are presented in Figure 13.4.

In the Phase I trial of PD0325901 in cancer patients (melanoma, breast, colon, and non-small cell lung cancer), oral doses were escalated from 1 to 30 mg twice daily (BID). Each treatment cycle consisted of 28 days, and three schedules of administration were evaluated: (i) three weeks of dosing with one week off, (ii) dosing every day, and (iii) five days of dosing with two days off per week [53–56]. Doses $\geq$2 mg BID suppressed tumor pMAPK (indicating biochemical activity of the drug) and rash (erythematous and maculopapular in nature, mainly on the face, upper body, and arms) was frequently dose limiting. There were no notable effects on serum Ca $\times$ P product, and the most common toxicities included rash, fatigue, diarrhea,

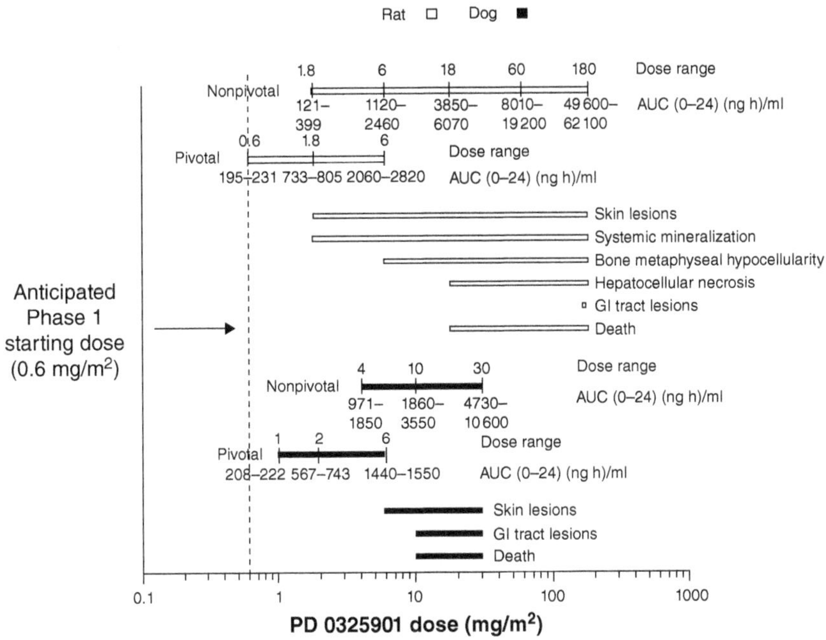

Figure 13.4 Relationships between dose and exposure with the primary toxicities of PD0325901 in rats and dogs. Exposure is expressed as PD0325901 plasma AUC(0–24) in (ng h)/ml and dose in mg/m^2. Results from non-pivotal (dose range–finding) studies and the pivotal one-month toxicity studies are presented.

nausea, visual disturbances/eye disorders, and vomiting. Acute neurotoxicity (including visual disturbances, balance, and gait disorders) was common in patients receiving ≥15 mg BID (all schedules), and several patients developed optic nerve ischemia, optic neuropathy, or retinal vein occlusion [56, 57]. In the Phase I trial, there were three partial responses (melanoma) and stable disease in 23 patients (primarily melanoma) [56, 57]. In a pilot Phase II study of PD0325901 in heavily pretreated patients with non-small cell lung cancer, 15 mg BID was given on various schedules over a 28-day cycle [58]. The main toxicities were reversible visual disturbances, diarrhea, nausea, vomiting, rash, and fatigue. Hallucinations were also reported. Although mean trough plasma concentrations of PD0325901 at 15 mg BID were ≥100 ng/ml, which are greater than the IC_{50} values for tumor xenograft mouse models (range of 16.5–53.5 ng/ml), there were no objective responses [58, 59]. Clinical trial data with PD0325901, along with other MEK inhibitors, have indicated that in addition to skin rash, diarrhea, and fatigue, these compounds can produce adverse neurological, ocular (including central serous retinopathy and retinal vein occlusion), and musculoskeletal effects [9, 60]. Ocular toxicities appear to be a class effect of MEK inhibitors [61]. Interestingly, PD0325901 has been shown to inhibit pERK in rat and mouse brain tissue following oral administration, suggesting that a pharmacodynamic effect can occur in the central nervous system [62]. In subsequent studies, intravitreal injection of PD0325901 in Dutch Belted rabbits produced retinal vein occlusion with retinal vasculature leakage and hemorrhage, followed by retinal detachment and degeneration, whereas ocular toxicity was not produced by the less-potent MEK inhibitor CI-1040 [63, 64].

Conclusions

Tissue mineralization produced in rats administered the MEK inhibitor PD0325901 provides a case study of how a unique and serious toxicity observed in preclinical safety testing was effectively managed to allow progression of an experimental drug into human clinical trials. A number of key factors were critical for allowing continued development of this compound to occur, rather than early termination. PD0325901 represented a novel and targeted therapeutic agent for the treatment of various solid tumors, thereby allowing a high risk–benefit ratio to exist due to the significant unmet medical need posed by cancer. Phase I oncology trials typically occur in cancer patients with limited treatment options. Therefore, the barriers to entry for novel anticancer agents in the clinic are generally lower than for Phase I trials involving healthy volunteers and therapies for non-life-threatening indications.

Early in the toxicology program with PD0325901, lesions observed in rats were similar to those seen with vitamin D toxicity, and serum chemistry data indicated changes in phosphorus and calcium. This information provided the basis for the hypotheses to be proposed regarding the mechanism for vascular and soft tissue mineralization. Because mineralization occurred in rats administered PD0325901 rather than only in dogs or monkeys, an animal model suitable for multiple investigative studies was readily available. Despite the apparent species specificity for this toxicity, it would not be appropriate to discount the risks toward humans because of a "rat-specific" finding. Rather, it was important to generate experimental data that characterized the toxicity which provided a plausible mechanism as a basis for risk management. Studies conducted with PD0325901 examined the dose–response and exposure–response relationships for toxicity and pharmacologic inhibition of MEK, the time course for lesion development, whether the changes observed were reversible or not, and whether associations could be made between clinical laboratory changes and anatomic lesions. We were able to identify biomarkers for tissue mineralization that were specifically related to the mechanism, were readily available in a clinical setting, were noninvasive, and had acceptable assay variability. It is important that biomarkers proposed for monitoring for drug toxicity are scientifically robust and obtainable and meet expectations of regulatory agencies. Finally, the data generated during the preclinical safety evaluation of PD0325901 were used to design the Phase I–II clinical trial to ensure patient safety. This included selection of a safe starting dose for Phase I, criteria for excluding patients from the trial, and clinical laboratory tests to be included as safety biomarkers for calcium–phosphorus dysregulation and tissue mineralization. In conclusion, robust data analyses, scientific hypothesis testing, and the ability to conduct investigative work were key factors in developing a safety biomarker for a serious preclinical toxicity, thereby allowing clinical investigation of a novel drug to occur.

Acknowledgments

Numerous people at the Pfizer Global Research and Development (PGRD), Ann Arbor, Michigan, Laboratories were involved in the studies performed with PD0325901, including the Departments of Cancer Pharmacology and Pharmacokinetics, Dynamics and Metabolism. In particular, the author would like to acknowledge the men and women of Drug Safety Research and Development, PGRD, Ann Arbor, who conducted the toxicology studies with this compound and made significant contributions in the disciplines of anatomic pathology and clinical laboratory testing during evaluation of this compound.

References

1 Sebolt-Leopold, J.S. (2000). Development of anticancer drugs targeting the MAP kinase pathway. *Oncogene* 19: 6594–6599.
2 Mansour, S.J., Matten, W.T., Hermann, A.S. et al. (1994). Transformation of mammalian cells by constitutively active MAP kinase. *Science* 265: 966–970.
3 Jost, M., Huggett, T.M., Kari, C. et al. (2001). Epidermal growth factor receptor–dependent control of keratinocyte survival and Bcl-XL expression through a MEK-dependent pathway. *J. Biol. Chem.* 276 (9): 6320–6326.
4 Friday, B.B. and Adjei, A.A. (2008). Advances in targeting the Ras/Raf/MEK/Erk mitogen-activated protein kinase cascade with MEK inhibitors for cancer therapy. *Clin. Cancer Res.* 14 (2): 342–346.
5 Hoshino, R., Chatani, Y., Yamori, T. et al. (1999). Constitutive activation of the 41-/43-kDa mitogen-activated protein kinase pathway in human tumors. *Oncogene* 18: 813–822.
6 Milella, M., Kornblau, S.M., Estrov, Z. et al. (2001). Therapeutic targeting of the MEK/MAPK signal transduction module in acute myeloid leukemia. *J. Clin. Invest.* 108 (6): 851–859.

7 Sebolt-Leopold, J.S., Dudley, D.T., Herrera, R. et al. (1999). Blockade of the MAP kinase pathway suppresses growth of colon tumors in vivo. *Nat. Med.* 5 (7): 810–816.

8 Dent, P. and Grant, S. (2001). Pharmacologic interruption of the mitogen-activated extracellular-regulated kinase/mitogen-activated protein kinase signal transduction pathway: potential role in promoting cytotoxic drug action. *Clin. Cancer Res.* 7: 775–783.

9 Zhao, Y. and Adjei, A.A. (2014). The clinical development of MEK inhibitors. *Nat. Rev. Clin. Oncol.* 11: 385–400.

10 Flaherty, K.T., Robert, C., Hersey, P. et al. (2012). Improved survival with MEK inhibition in BRAF-mutated melanoma. *N. Engl. J. Med.* 367 (2): 107–114.

11 LoRusso, P.M., Adjei, A.A., Varterasian, M. et al. (2005). Phase I and pharmacodynamic study of the oral MEK inhibitor CI-1040 in patients with advanced malignancies. *J. Clin. Oncol.* 23 (23): 5281–5293.

12 Sebolt-Leopold, J.S. (2008). Advances in the development of cancer therapeutics directed against the RAS-mitogen-activated protein kinase pathway. *Clin. Cancer Res.* 14 (12): 3651–3656.

13 Rinehart, J., Adjei, A.A., LoRusso, P.M. et al. (2004). Multicenter phase II study of the oral MEK inhibitor, CI-1040, in patients with advanced non-small-cell lung, breast, colon, and pancreatic cancer. *J. Clin. Oncol.* 22 (22): 4456–4462.

14 Wang, D., Boerner, S.A., Winkler, J.D., and LoRusso, P.M. (2007). Clinical experience of MEK inhibitors in cancer therapy. *Biochim. Biophys. Acta* 1773: 1248–1255.

15 Sebolt-Leopold, J.S., Merriman, R., and Omer, C. (2004). The biological profile of PD0325901: a second generation analog of CI-1040 with improved pharmaceutical potential. *Proc. Am. Assoc. Cancer Res.* 45: 925. (abstract 4003).

16 Brown, A.P., Carlson, T.C.G., Loi, C.M., and Graziano, M.J. (2007). Pharmacodynamic and toxicokinetic evaluation of the novel MEK inhibitor, PD0325901, in the rat following oral and intravenous administration. *Cancer Chemother. Pharmacol.* 59: 671–679.

17 York, M.J. and Evans, G.O. (1996). Electrolyte and fluid balance. In: *Animal Clinical Chemistry: A Primer for Toxicologists* (ed. G.O. Evans), 163–176. New York: Taylor & Francis.

18 Spaulding, S.W. and Walser, M. (1970). Treatment of experimental hypercalcemia with oral phosphate. *J. Clin. Endocrinol.* 31: 531–538.

19 Block, G.A. (2000). Prevalence and clinical consequences of elevated Ca × P product on hemodialysis patients. *Clin. Nephrol.* 54 (4): 318–324.

20 Giachelli, C.M., Jono, S., Shioi, A. et al. (2001). Vascular calcification and inorganic phosphate. *Am. J. Kidney Dis.* 38(4, Suppl 1): S34–S37.

21 Grant, R.A., Gillman, T., and Hathorn, M. (1963). Prolonged chemical and histochemical changes associated with widespread calcification of soft tissues following brief calciferol intoxication. *Br. J. Exp. Pathol.* 44 (2): 220–232.

22 Spangler, W.L., Gribble, D.H., and Lee, T.C. (1979). Vitamin D intoxication and the pathogenesis of vitamin D nephropathy in the dog. *Am. J. Vet. Res.* 40: 73–83.

23 Harrington, D.D. and Page, E.H. (1983). Acute vitamin D3 toxicosis in horses: case reports and experimental studies of the comparative toxicity of vitamins D2 and D3. *J. Am. Vet. Med. Assoc.* 182 (12): 1358–1369.

24 Long, G.G. (1984). Acute toxicosis in swine associated with excessive dietary intake of vitamin D. *J. Am. Vet. Med. Assoc.* 184 (2): 164–170.

25 Payne, R.B., Carver, M.E., and Morgan, D.B. (1979). Interpretation of serum total calcium: effects of adjustment for albumin concentration on frequency of abnormal values and on detection of change in the individual. *J. Clin. Pathol.* 32: 56–60.

26 Meuten, D.J., Chew, D.J., Capen, C.C., and Kociba, G.J. (1982). Relationship of serum total calcium to albumin and total protein in dogs. *J. Am. Vet. Med. Assoc.* 180: 63–67.

27 Rosol, T.J. and Capen, C.C. (1997). Calcium-regulating hormones and diseases of abnormal mineral (calcium, phosphorus, magnesium) metabolism. In: *Clinical Biochemistry of Domestic Animals*, 5e, 619–702. San Diego, CA: Elsevier.

28 Ally, S., Clair, T., Katsaros, D. et al. (1989). Inhibition of growth and modulation of gene expression in human lung carcinoma in athymic mice by site-selective 8-Cl-cyclic adenosine monophosphate. *Cancer Res.* 49: 5650–5655.

29 Saunders, M.P., Salisbury, A.J., O'Byrne, K.J. et al. (1997). A novel cyclic adenosine monophosphate analog induces hypercalcemia via production of 1,25-dihydroxyvitamin D in patients with solid tumors. *J. Clin. Endocrinol. Metab.* 82 (12): 4044–4048.

30 Brown, A.P., Morrissey, R.L., Smith, A.C. et al. (2000). Comparison of 8-chloroadenosine (NSC-354258) and 8-chloro-cyclic-AMP (NSC-614491) toxicity in dogs. *Proc. Am. Assoc. Cancer Res.* 41: 491. (abstract 3132).

31 Brown, A.P., Courtney, C., Carlson, T., and Graziano, M. (2005). Administration of a MEK inhibitor results in tissue mineralization in the rat due to dysregulation of phosphorus and calcium homeostasis. *Toxicologist* 84 (S-1): 108. (abstract 529).

32 Fu, J.Y. and Muller, D. (1999). Simple, rapid enzyme-linked immunosorbent assay (ELISA) for the determination of rat osteocalcin. *Calcif. Tissue Int.* 64: 229–233.

33 Ferreira, A. and Drueke, T.B. (2000). Biological markers in the diagnosis of the different forms of renal osteodystrophy. *Am. J. Med. Sci.* 320 (2): 85–89.

34 Knutson, J.C., LeVan, L.W., Valliere, C.R., and Bishop, C.W. (1997). Pharmacokinetics and systemic effect on calcium homeostasis of 1α,25-dihydroxyvitamin D2 in rats. *Biochem. Pharmacol.* 53: 829–837.
35 Chen, P.S., Terepka, A.R., and Overslaugh, C. (1962). Hypercalcemic and hyperphosphatemic actions of dihydrotachysterol, vitamin D2 and Hytakerol (AT-10) in rats and dogs. *Endocrinology* 70: 815–821.
36 Kanis, J.A. and Russell, R.G.G. (1977). Rate of reversal of hypercalcaemia and hypercalciuria induced by vitamin D and its 1α-hydroxylated derivatives. *Br. Med. J.* 1: 78–81.
37 Brown, A.P., Reindel, J.F., Grantham, L. et al. (2006). Pharmacologic inhibitors of the MEK-MAP kinase pathway are associated with toxicity to the skin, stomach, intestines, and liver. *Proc. Am. Assoc. Cancer Res.* 47: 308. (abstract 1307).
38 Rosenblum, I.Y., Black, H.E., and Ferrell, J.F. (1977). The effects of various diphosphonates on a rat model of cardiac calciphylaxis. *Calcif. Tissue Res.* 23: 151–159.
39 Kamio, A., Taguchi, T., Shiraishi, M. et al. (1979). Vitamin D sclerosis in rats. *Acta Pathol. Jpn.* 29 (4): 545–562.
40 Mortensen, J.T., Lichtenberg, J., and Binderup, L. (1996). Toxicity of 1,25-dihydroxyvitamin D3, tacalcitol, and calcipotriol after topical treatment in rats. *J. Invest. Dermatol. Symp. Proc.* 1: 60–63.
41 Morrow, C. (2001). Cholecalciferol poisoning. *Vet. Med.* 905–911.
42 Brown, A.P., Courtney, C.L., King, L.M. et al. (2005b). Cartilage dysplasia and tissue mineralization in the rat following administration of a FGF receptor tyrosine kinase inhibitor. *Toxicol. Pathol.* 33 (4): 449–455.
43 Prie, D. and Friedlander, G. (2010). Reciprocal control of 1,25-dihydroxyvitamin D and FGF23 formation involving the FGF23/klotho system. *Clin. J. Am. Soc. Nephrol.* 5: 1717–1722.
44 Lederer, E. (2014). Regulation of serum phosphate. *J. Physiol.* 592: 3985–3995.
45 Shimada, T., Kakitani, M., Yamazaki, Y. et al. (2004). Targeted ablation of Fgf23 demonstrates an essential physiological role of FGF23 in phosphate and vitamin D metabolism. *J. Clin. Invest.* 113: 561–568.
46 Ranch, D., Zhang, M.Y.H., Portale, A.A., and Perwad, F. (2011). Fibroblast growth factor 23 regulates renal 1,25-dihydroxyvitamin D and phosphate metabolism via the MAP kinase signaling pathway in Hyp mice. *J. Bone Miner. Res.* 26 (8): 1883–1890.
47 Diaz, D., Allamneni, K., Tarrant, J.M. et al. (2012). Phosphorous dysregulation induced by MEK small molecule inhibitors in the rat involves blockade of FGF-23 signaling in the kidney. *Toxicol. Sci.* 125 (1): 187–195.
48 Yanochko, G.M., Vitsky, A., Heyen, J.R. et al. (2013). Pan-FGFR inhibition leads to blockade of FGF23 signaling, soft tissue mineralization, and cardiovascular dysfunction. *Toxicol. Sci.* 135 (2): 451–464.

49 Haschek, W.M., Krook, L., Kallfelz, F.A., and Pond, W.G. (1978). Vitamin D toxicity, initial site and mode of action. *Cornell Vet.* 68 (3): 324–364.
50 Roche Laboratories (1998). Package insert, Rocaltrol® (calcitriol) capsules and oral solution. Nov. 20.
51 Bone Care International, Inc. (1999). Package insert, Hectorol™ (doxercalciferol) capsules. June 9.
52 DeGeorge, J.J., Ahn, C.H., Andrews, P.A. et al. (1998). Regulatory considerations for preclinical development of anticancer drugs. *Cancer Chemother. Pharmacol.* 41: 173–185.
53 LoRusso, P., Krishnamurthi, S., Rinehart, J.R. et al. (2005). A phase 1–2 clinical study of a second generation oral MEK inhibitor, PD0325901 in patients with advanced cancer. 2005 ASCO Annual Meeting Proceedings. *J. Clin. Oncol.* 23 (16S): 3011–3011.
54 Menon, S.S., Whitfield, L.R., Sadis, S. et al. (2005). Pharmacokinetics (PK) and pharmacodynamics (PD) of PD0325901, a second generation MEK inhibitor after multiple oral doses of PD0325901 to advanced cancer patients. 2005 ASCO Annual Meeting Proceedings. *J. Clin. Oncol.* 23 (16S): 3066–3066.
55 Tan, W., DePrimo, S., Krishnamurthi, S.S. et al. (2007). Pharmacokinetic (PK) and pharmacodynamic (PD) results of a phase I study of PD-0325901, a second generation oral MEK inhibitor, in patients with advanced cancer. Presented at the AACR-NCI-EORTC International Conference on Molecular Targets and Cancer Therapy, abstract B109.
56 LoRusso, P.M., Krishnamurthi, S.S., Rinehart, J.J. et al. (2007). Clinical aspects of a phase I study of PD-0325901, a selective oral MEK inhibitor, in patients with advanced cancer. Presented at the AACR-NCI-EORTC International Conference on Molecular Targets and Cancer Therapy, abstract B113.
57 LoRusso, P.M., Krishnamurthi, S.S., Rinehart, J.J. et al. (2010). Phase I pharmacokinetic and pharmacodynamic study of the oral MAPK/ERK kinase inhibitor PD-0325901 in patients with advanced cancers. *Clin. Cancer Res.* 16 (6): 1924–1937.
58 Haura, E.B., Larson, T.G., Stella, P.J., et al. (2007). A pilot phase II study of PD-0325901, an oral MEK inhibitor, in previously treated patients with advanced non-small cell lung cancer. Presented at the AACR-NCI-EORTC International Conference on Molecular Targets and Cancer Therapy, abstract B110.
59 Haura, E.B., Ricart, A.D., Larson, T.G. et al. (2010). A phase II study of PD-0325901, an oral MEK inhibitor, in previously treated patients with advanced non-small cell lung cancer. *Clin. Cancer Res.* 16 (8): 2450–2457.
60 Boasberg, P.D., Redfern, C.H., Daniels, G.A. et al. (2011). Pilot study of PD-0325901 in previously treated patients with advanced melanoma, breast cancer, and colon cancer. *Cancer Chemother. Pharmacol.* 68: 547–552.

61 Duncan, K.E., Chang, L.Y., and Patronas, M. (2015). MEK inhibitors: a new class of chemotherapeutic agents with ocular toxicity. *Eye* 29: 1003–1012.
62 Iverson, C., Larson, G., Lai, C. et al. (2009). RDEA119/BAY 869766 : a potent, selective, allosteric inhibitor of MEK1/2 for the treatment of cancer. *Cancer Res.* 69 (17): 6839–6847.
63 Huang, W., Yang, A.H., Matsumoto, D. et al. (2009). PD0325901, a mitogen-activated protein kinase kinase inhibitor, produces ocular toxicity in a rabbit animal model of retinal vein occlusion. *J. Ocul. Pharmacol. Ther.* 25 (6): 519–530.
64 Smith, A., Pawar, M., Van Dort, M.E. et al. (2018). Ocular toxicity profile of ST-162 and ST-168 as novel bifunctional MEK/PI3K inhibitors. *J. Ocul. Pharmacol. Ther.* 34 (6): 477–485.

14

New Markers of Kidney Injury

Sven A. Beushausen

Zoetic Pharmaceuticals, Amherst, New York

Introduction

The current biomarker standards for assessing acute kidney injury (AKI), caused by disease or as a consequence of drug-induced toxicity, include blood urea nitrogen (BUN) and serum creatinine (SC). Retention of either marker in the blood is indicative of impedance in the glomerular filtration rate (GFR), which if left untreated could escalate to serious kidney injury through loss of function and ultimately, death. Although the colorimetric assays developed for SC and BUN are relatively quick, seconds as compared to hours for antibody-based analysis platforms like enzyme-linked immunosorbent assay (ELISA), they are poor *predictors* of kidney injury because they both suffer from a lack of sensitivity and specificity. For example, SC concentration is greatly influenced by other nonrenal factors, including gender, age, muscle mass, race, drugs, and protein intake [1]. Consequently, increases in BUN and SC levels *report* out injury only after serious kidney damage has occurred. These shortcomings limit their clinical utility to patients who are at risk of developing drug-induced AKI or where instances of AKI have already been established and require frequent monitoring, and time to treatment is critical.

Renal failure or AKI is often a direct consequence of disease, can result from complications associated with disease or postsurgical trauma like sepsis, or is produced by drug-induced nephrotoxicity. Drug-induced renal injury is of great concern to physicians. Knowledge of toxicities associated with U.S. Food and Drug Administration (FDA)–approved compounds helps to guide product selection in an effort to manage risk and maximize patient safety. Drug-induced nephrotoxicity is of even greater concern to the pharmaceutical industry, where patient safety is the principal driver in the need to discover safer and more efficacious drugs. Because BUN and SC are insensitive predictors of early kidney injury, many instances of subtle renal damage caused by

Biomarkers in Drug Discovery and Development: A Handbook of Practice, Application, and Strategy, Second Edition. Edited by Ramin Rahbari, Jonathan Van Niewaal, and Michael R. Bleavins.

drugs may go unrecognized. Consequently, true estimates for drug-induced nephrotoxicity are likely to be far lower than previously realized. For example, studies have indicated that the incidence of acute tubular necrosis or acute interstitial nephritis due to medication has been estimated to be as high as 18.6% [2]. In addition, renal injury attributed to treatment with aminoglycosides has been reported to approach 36% [3, 4]. Not surprisingly, many common drugs have been associated with renal injury that cause site-specific damage (Table 14.1).

Fortunately, most instances of drug-induced nephrotoxicity are reversible if discovered early and medication is discontinued. Collectively, the combined shortcomings of BUN and SC as *predictors* of nephrotoxicity and the propensity for many classes of medicines to cause drug-induced nephrotoxicity underscore the urgent need for the development and qualification of more sensitive and specific biomarkers. The benefits such tools will provide include predictive value and earlier diagnosis of drug-induced kidney injury before changes in renal function or clinical manifestations of AKI are evident. More importantly, biomarkers of nephrotoxicity with increased sensitivity and specificity will be invaluable to drug development both preclinically and clinically. Preclinically, new biomarkers will aid in the development of safer drugs having fewer liabilities with an ultimate goal to considerably lower or even possibly eliminate drug-induced nephrotoxicity. Clinically, the biomarkers will be used to monitor potential nephrotoxic effects due to therapeutic intervention or the potential for new drugs to cause renal toxicity in Phase I to III clinical trials.

New Preclinical Biomarkers of Nephrotoxiciy

In recent years, two consortia led by the nonprofit organizations ILSI-HESI (International Life Sciences Institute, Health and Environmental Sciences Institute, http://www.hesiglobal.org) and C-Path (Critical Path Institute, http://www.c-path.org) aligned with leaders in academia, industry, and the FDA with a mission to evaluate the potential utility of newly identified biomarkers of nephrotoxicity for use in preclinical safety studies and to develop a process for the acceptance of the new biomarkers in support of safety data accompanying new regulatory submissions. Several criteria for the evaluation and development of new biomarkers of nephrotoxicity were considered, including the following:

- A preference should be given for noninvasive sample collection.
- New biomarkers developed for preclinical use should optimally be translated to the clinic.
- Assays for new biomarkers should be robust and kits readily available for testing.

Table 14.1 Common medications associated with acute renal injury[a].

Pathoetiology	Medication	Clinical Findings	Treatment
Prerenal injury	Diuretics, NSAIDs, ACE inhibitors, ciclosporin, tacrolimus, radiocontrast media, interleukin-2, vasodilators (hydralazine, calcium-channel blockers, minoxidil, diazoxide)	Benign urine sediment, FENa < 1%, UOsm > 500	Suspend or discontinue medication, volume replacement as clinically indicated
Intrinsic renal injury Vascular effects Thrombotic microangiopathy	Ciclosporin, tacrolimus, mitomycin C, conjugated estrogens, quinine, 5-fluorouracil, ticlopidine, clopidogrel, interferon, valaciclovir, gemcitabine, bleomycin	Fever, microangiopathic, hemolytic anemia, thrombocytopenia	Discontinue medication, supportive care, plasmapheresis if indicated
Cholesterol emboli	Heparin, warfarin, streptokinase	Fever, microangiopathic, hemolytic anemia, thrombocytopenia	Discontinue medication, supportive care, plasmapheresis if indicated
Tubular toxicity	Aminoglycosides, radiocontrast media, cisplatin, nedaplatin, methoxyflurane, outdated tetracycline, amphotericin B, cephaloridine, streptozocin, tacrolimus, carbamazepine, mithramycin, quinolones, foscarnet, pentamidine, intravenous gammaglobulin, fosfamide, zoledronate, cidofovir, adefovir, tenofovir, mannitol, dextran, hydroxyethyl starch	FENa > 2%, UOsm < 350 urinary sediment with granular casts, tubular epithelial cells	Drug discontinuation, supportive care
Rhabdomyolysis	Lovastatin, ethanol, codeine, barbiturates, diazepam	Elevated CPK, ATN urine sediment	Drug discontinuation, supportive care
Severe hemolysis	Quinine, quinidine, sulfonamides, hydralazine, triamterene, nitrofurantoin, mephenytoin	High LDH, decreased hemoglobin	Drug discontinuation, supportive care

(Continued)

Table 14.1 (Continued)

Pathoetiology	Medication	Clinical Findings	Treatment
Immune-mediated interstitial inflammation	Penicillin, methicillin ampicillin, rifampin, sulfonamides, thiazides, cimetidine, phenytoin, allopurinol, cephalosporins, cytosine arabinoside, furosemide, interferon, NSAIDs, ciprofloxacin, clarithromycin, telithromycin, rofecoxib, pantoprazole, omeprazole, atazanavir	Fever, rash, eosinophilia, urine sediment showing pyuria, white cell casts, eosinophiluria	Discontinue medication, supportive care
Glomerulopathy	Gold, penicillamine, captopril, NSAIDs, lithium, mefenamate, fenoprofen, mercury, interferon-α, pamidronate, fenclofenac, tolmetin, foscarnet	Edema, moderate to severe proteinuria, red blood cells, red blood cell casts possible	Discontinue medication, supportive care
Obstruction Intratubular: (crystalluria and/or renal lithiasis)	Aciclovir, methotrexate, sulfanilamide, triamterene, indinavir, foscarnet, ganciclovir	Sediment can be benign with severe obstruction, ATN might be observed	Discontinue medication, supportive care
Ureteral (secondary to retroperitoneal fibrosis)	Methysergide, ergotamine, dihydroergotamine, methyldopa, pindolol, hydralazine, atenolol	Benign urine sediment, hydronephrosis on ultrasound	Discontinue medication, decompress ureteral obstruction by intrarenal stenting or percutaneous nephrostomy

a) ACE, angiotensin-converting enzyme; ATN, acute tubular necrosis; CPK, creatinine phosphokinase; FENa, fractional excretion of sodium; LDH, lactate dehydrogenase; NSAIDs, nonsteroidal anti-inflammatory drugs; UOsm, urine osmolality.
Source: Adapted from Choudhury and Ziauddin 2005 [2].

- Assays should be multiplexed to minimize cost and expedite sample analysis.
- Biomarkers should ideally predict or report out site-specific injury.
- Biomarkers must be more sensitive and specific of kidney injury than existing standards.
- Biomarkers should be predictive (prodromal) of kidney injury in the absence of histopathology.

Table 14.2 Biomarkers of renal injury by region of specificity, onset, platform, and application.

Biomarker	Injury related to	Onset	Platforms	Application
β_2-Microglobulin	Proximal tubular injury	Early	Luminex, ELISA	Mouse, rat, human, chicken, turkey
Clusterin	Tubular epithelial cells	Early	Luminex, ELISA	Mouse, rat, dog, monkey, human
Cystatin-C	Tubular dysfunction	Late	Luminex, ELISA	Mouse, rat, human
GSTα	Proximal tubular injury	Early	Luminex, ELISA	Mouse, rat, human
GST Yb1	Distal tubule	Early	Luminex, ELISA	
KIM-1	General kidney injury and disease	Early	Luminex, ELISA	Zebrafish, mouse, rat, dog, monkey, human
Microalbumin	Proximal tubular injury	Early	Luminex, ELISA	Mouse, rat, dog, monkey, human
Osteopontin	Tubulointerstitial fibrosis	Late	Luminex, ELISA	Mouse, rat, monkey, human
NGAL	Proximal tubular injury	Early	Luminex, ELISA	Mouse, rat, human
RPA-1	Renal papilla and collecting ducts	Early	ELISA	Rat

Source: Adapted from Vaidya et al, 2008 [5].

The preference for noninvasive sample collection made urine the obvious choice of biofluid. Urine has proven to be a fertile substrate for the discovery of promising new biomarkers for the early detection of nephrotoxicity [5]. A number of these markers were selected for further development and qualification by the ILSI-HESI and C-Path Nephrotoxicity Working Groups in both preclinical and clinical settings, with the exception of Renal papillary antigen 1 (RPA-1) and GST Yb1 (Biotrin), which are markers developed specifically for the analysis of kidney effects in rats (Table 14.2). The utility and limitations of each marker used in the context of early and site-specific detection are discussed below.

β_2-Microglobulin

Human β_2-microglobulin (β_2M) was isolated and characterized in 1968 [6]. β_2M was identified as a small 11 815-Da protein found on the surface of human cells expressing the major histocompatibility class I molecule [7]. β_2M is shed into the circulation as a monomer, from which it is normally filtered by the glomerulus and subsequently reabsorbed and metabolized within proximal tubular cells [8]. Twenty-five years ago, serum β_2M was advocated for use as an index of renal function because of an observed proportional increase in serum β_2M levels in response to decreased renal function [9]. It has since been abandoned as a serum biomarker due to a number of factors complicating the interpretation of the findings. More recently, increased levels of intact urinary β_2M have been directly linked to impairment of tubular uptake. Additional work in rats and humans has demonstrated that increased urinary levels of β_2M can be used as a marker for proximal tubular function when β_2M production and glomerular filtration are normal in a setting of minimal proteinuria [10–13].

Urinary β_2M has been shown to be superior to *N*-acetyl-β-glucosaminidase (NAG) as a marker in predicting prognosis in idiopathic membranous neuropathy [14]. In this context, β_2M can be used to monitor and avoid unnecessary immunosuppressive therapy following renal transplantation. β_2M can be used as an early predictor of proximal tubular injury in preclinical models of drug-induced nephrotoxicity. Although easily detected in urine, there are several factors that may limit its value as a biomarker. For example, β_2M is readily degraded by proteolytic enzymes at room temperature and also degrades rapidly in an acidic environment at or below $pH < 6.0$ [15]. Therefore, care must be taken to collect urine in an ice-cold, adequately buffered environment with the addition of stabilizers to preserve β_2M levels during the period of collection and in storage. It is unlikely that β_2M will be used as a stand-alone marker to predict or report proximal tubule injury preclinically or in the clinic. Rather, it is likely to be used in conjunction with other proximal tubule markers to support such a finding. A brief survey for commercially available antibodies used to detect β_2M indicates that most are species specific (http://www.abcam.com). Instances of cross-reactivity were noted for specific reagents between human and pig, chicken and turkey, and human and other primates. A single monoclonal reagent is reported to have cross-reactivity with bovine, chicken, rabbit, and mouse, and none were listed that specifically recognized dog β_2M. Because both commonly used preclinical species rat and dog β_2M proteins share only 69.7 and 66.7% amino acid identity with the human protein (http://www.expasy.org), it would be prudent to develop and characterize antibody reagents specific to each species and cross-reacting antisera to specific amino acid sequences shared by all three proteins.

Clusterin

Clusterin is a highly glycosylated and sulfated secreted glycoprotein first isolated from ram rete testes fluid in 1983 [16]. It was named clusterin because of its ability to elicit clustering of Sertoli cells *in vitro* [17]. Clusterin is found primarily in the epithelial cells of most organs. Tissues with the highest levels of clusterin include testis, epididymis, liver, stomach, and brain. Metabolic and cell-specific functions assigned to clusterin include sperm maturation, cell transformation, complement regulation, lipid transport, secretion, apoptosis, and metastasis [18]. Clusterin is also known by a number of synonyms as a consequence of having been identified simultaneously in many parallel lines of inquiry. Names include glycoprotein III (GPIII), sulfated glycoprotein-2 (SG-2), apolipoprotein J (apo J), testosterone-repressed message-2 (TRPM-2), complement associated protein SP-40, and complement cytolysis inhibitor protein (see Table 14.1). Clusterin has been cloned from a number of species, including the rat [19]. The human homolog is 449 amino acids in length, coding for a protein with a molecular weight of 52 495 Da [20]. However, due to extensive posttranslational modification, the protein migrates to an apparent molecular weight of 70–80 kDa following sodium dodecyl sulfate polyacrylamide gel electrophoresis (SDS-PAGE). Amino acid identity between species is moderate. Human clusterin shares 70.3, 76.6, 71.7, and 77 with the bovine, mouse, pig, and rat homologs, respectively (http://www.expasy.org).

Clusterin is a heterodimer comprised of an α and a β subunit, each having an apparent mass of 40 kDa by SDS-PAGE. The subunits result from the proteolytic cleavage of the translated polypeptide at amino acid positions 23 and 277. This eliminates the leader sequence and produces the mature 205-amino acid β subunit and the remaining 221-amino acid α subunit. The α and β subunits are held together by five sulfhydryl bonds afforded by cysteine residues clustered within each of the subunits [21]. In addition, each subunit has three N-linked carbohydrates that are also heavily sulfated, giving rise to the higher apparent molecular weight observed following SDS-PAGE. Considerable evidence has been provided suggesting that clusterin plays an important role in development. For example, clusterin mRNA expression has been observed at 12.5 days postgestation in mice, where it is present in all germ cell layers [22]. Furthermore, stage-specific variations of the transcript have been observed, as have changes in specific localization during development. Similarly, changes in the developmental expression of clusterin in kidney, lung, and nervous system have also been reported [23]. These observations suggest that clusterin might play a role in tissue remodeling.

In the developing murine kidney, clusterin is expressed in the tubular epithelium and later in development is diminished as tubular maturation progresses [24]. Interestingly, clusterin is observed in newly formed tubules but appears to be absent in glomeruli. Of interest to many investigators of renal function is the reemergence of clusterin observed following induction of a variety of kidney diseases and drug-induced renal injury. Clusterin induction has been observed following ureteral obstruction [25] and ischemia reperfusion injury [26]. Elevations in the levels of clusterin have also been observed in the peri-infarct region following subtotal nephrectomy [27] and in animal models of hereditary polycystic kidney disease [28]. Marked increases of clusterin released in urine have also been recorded in animal models of aminoglycoside-induced nephrotoxicity [29–31].

Authors have opined that clusterin either functions in a protective role by scavenging cell debris or may play a role in the process of tissue remodeling following cellular injury based on these observations. Collectively, the body of work linking elevated levels of urinary clusterin to kidney damage has suggested that measurement of urinary clusterin may be useful as a marker of renal tubular injury. Indeed, an early study comparing urinary levels of clusterin against NAG following chronic administration of gentamicin over a two-month period demonstrated that while the urinary levels of both proteins rose rapidly, peaked, and then declined, clusterin levels remained significantly higher than control values over the duration of the experiment. By contrast, NAG levels dropped to within control values within 10 days of treatment even though evidence of tubulointerstitial disease persisted [30]. More recent work examining the levels of urinary clusterin in the autosomal-dominant polycystic kidney disease (cy/+) rat model compared to the fawn hooded hypertensive (FHH) rat model of focal segmental glomerulosclerosis following bilateral renal ischemia demonstrated that clusterin levels correlated with the severity of tubular damage and suggested use as a marker for differentiating between tubular and glomerular damage [32]. Although the value of clusterin as an early marker of tubular epithelial injury has not been established clinically, preclinical findings suggest that it is an ideal candidate for translation to the clinic as an early marker of nephrotoxicity.

Cystatin-C

Cystatin C (Cys-C) is a 13-kDa nonglycosylated protein belonging to the superfamily of cysteine protease inhibitors [33]. Cys-C is produced by all nucleated cells and, unlike SC, is unaffected by muscle mass. Serum Cys-C was suggested to be closer to the "ideal" biomarker reporting GFR because although freely filtered by the glomerulus, it is not secreted. Instead, Cys-C is adsorbed by tubular epithelial cells, where it is catabolized and is not returned to the bloodstream, thus obviating the need to calculate urinary Cys-C to measure GFR [34]. Several studies have been designed to examine the usefulness of serum Cys-C as a measure or biomarker of GFR [35]. In one such study, serum Cys-C was shown to be a useful biomarker of acute renal failure and could be detected one to two days prior to the elevation in levels of SC, the accepted clinical diagnosis of AKI [36]. Although earlier in detection than SC, serum Cys-C levels were not predictive of kidney disease and, like SC, reported out kidney injury long after serious damage had occurred. In another study, investigators monitored and compared the levels of serum Cys-C and urinary Cys-C in patients following cardiothoracic surgery with and without complicating AKI [37]. The results clearly demonstrated that while plasma Cys-C was not a useful predictor of AKI, early and persistent increases in urinary Cys-C correlated with the development and severity of AKI. Another interesting but unexplained observation in this study was that women had significantly higher postoperative levels of urinary Cys-C than did men even though preoperative Cys-C levels were similar. These data have prompted groups like ILSI-HESI and C-Path to examine the utility of urinary Cys-C as a preclinical biomarker of drug-induced renal injury in the hope that elevated levels of Cys-C can be detected in urine prior to the emergence of overt tubular dysfunction.

Glutathione S-Transferases

The glutathione S-transferases (GSTs) form a family of homo-and heterodimeric detoxifying enzymes [38] identified originally as a group of soluble liver proteins that play a major role in the detoxification of electrophilic compounds [39]. They have since been shown to be products of gene superfamilies [40] and are classified into α, μ, π, and θ subfamilies based on sequence identity and other common properties [41]. Tissue distribution and levels of GST isoform expression have been determined by immunohistochemical localization [42], isoform-specific peptide antibody Western blotting, and mass spectrometry [40]. Analysis of GST subunit diversity and tissue distribution using peptide-specific antisera has shown GST μ isoforms to be the most widely distributed class of GSTs, with expression evident in brain, pituitary, heart, lung, adrenal gland, kidney, testis, liver, and pancreas, with the highest levels of GST μl observed in adrenals, testis, and liver. Isoforms of the GSTα subfamily, also known by the synonyms glutathione S-transferase-1, glutathione S-transferase Ya-1, GST Ya1, ligandin, GST 1a-1a, GST B, GST 1-1, and GST A1-1 (http://www.expasy.org/uniprot/P00502), are more limited in distribution, with highest levels of expression observed in hepatocytes and proximal tubular cells of the kidney [42].

Indeed, proximal tubular GSTα levels have been reported to approximate 2% of total cytosolic protein following exposure to xenobiotics or renal toxins [43]. In the Rowe study [40], GSTα was found to be rather evenly distributed between adrenals, kidney, and pancreas, with highest levels observed in liver, whereas isoforms of the GSTπ subclass were expressed in brain, pituitary, heart, liver, kidney, and adrenals, with highest levels of expression observed in kidney. The high levels of expression and differential distribution of GST isoforms made them attractive candidates as biomarkers that could be used to indicate site-specific drug-induced nephrotoxicity. For example, development of a radioimmunoassay to quantify leakage of ligandin (GSTα) into the urine as a measure of nephrotoxicity in the preclinical rat model was reported as early as 1979 [44]. Subsequent work described the development of a radioimmunoassay for the quantitation of GSTπ in the urine [45a] later used as an indicator of distal tubular damage in the human kidney [45b]. Additional work described the development of a multiplexed ELISA for the simultaneous quantitation of GSTα and GSTπ to discriminate between proximal and distal tubular injury, respectively [46].

In terms of sensitivity, a study examining the nephrotoxic effects of the sevoflurane degradation product, fluoromethyl-2,2-difluoro-1-(trifluoromethyl)vinyl ether, in rats showed urinary GSTα to be the most sensitive marker of mild proximal tubular damage compared to other urinary markers measured, including protein and glucose [47]. A second study in which four human volunteers were given sevoflurane demonstrated abnormalities in urinary glucose, albumin, GSTα, and GSTπ, while levels of BUN or SC were unaffected, suggesting that the GSTs were more sensitive markers of site-specific dug-induced nephrotoxicity [48]. Immunohistochemical staining of the rat kidney with antibodies to different GST isoforms has shown that GSTα subunits are expressed selectively in the proximal tubule, whereas GSTμ and π subunits are localized to the thin loop of Henle and proximal tubules, respectively [38]. An examination of the distribution of the rat GSTμ equivalent, GSTYb1, in the kidney indicates that it is localized to the distal tubules. Simultaneous measurement of urinary GSTα and GSTYb1 has been used to discriminate between drug-induced proximal and distal tubular injury (cited by Kilty et al. [49]). The high levels of GSTs in the kidney and site-specific localization of different GST classes in addition to increased sensitivity in detecting drug-induced nephrotoxicity in humans make them ideal candidates for the development and testing of preclinical markers that predict or report early signs of nephrotoxicity to support preclinical safety studies and subsequent compound development.

Kidney Injury Molecule 1

Rat kidney injury molecule 1 (KIM-1) was discovered as part of an effort to identify genes implicated in kidney injury and repair [50] using the polymerase chain reaction (PCR) subtractive hybridization technique of representational difference analysis originally developed to look at differences in genomic DNA [51] but adapted to examine differences in mRNA expression [52]. Complementary DNA generated from poly(A+) mRNA purified from normal and 48-hour postischemic rat kidneys was amplified to generate driver and tester amplicons, respectively. The amplicons were used as templates to drive the subtractive hybridization process to generate designated differential products, three of which were ultimately gel purified and subcloned into the pUC18 cloning vector. Two of these constructs were used to screen λZapII cDNA libraries constructed from 48-hour postischemic rat kidneys. Isolation and purification of positively hybridizing plaques resulted in the recovery of a 2.5-kb clone that contained sequence information on all three designated differential products. A basic local alignment search tool (BLAST) search of the National Center of Biotechnology Information (NCBI) database revealed that the rat KIM-1 sequence had limited (59.8%) amino acid homology to HAVcr-1, identified earlier as the monkey gene coding for the hepatitis A virus receptor protein [53]. The human homolog of KIM-1 was isolated by low-stringency screening of a human embryonic liver λgt10 cDNA library using the same probe that yielded the rat clones [50]. The plaque of one of two clones purified from this exercise was shown to code for a 334-amino acid protein sharing 43.8% identity and 59.1% similarity to the rat KIM-1 protein. Comparison to the human HAVcr protein [54] revealed 85.3% identity demonstrating a clear relationship between the two proteins.

Subsequent work has demonstrated that KIM-1 and HAVcr are synonyms for the same protein, also known as T-cell immunoglobulin and mucin domain–containing protein 1 (TIMD-1) and TIM-1. The TIMD proteins are all predicted to be type I membrane proteins that share a characteristic immunoglobulin V, mucin, transmembrane, and cytoplasmic domain structure [55]. It is not clear what the function of KIM-1 (TIMD-1) is, but it is believed that TIMD-1 is involved in the preferential stimulation of Th2 cells within the immune system [56]. In the rat, KIM-1 mRNA expression is highest in liver and barely detected in kidney [50]. KIM-1 mRNA and protein expression are dramatically upregulated following ischemic injury. Immunohistochemical examination of kidney sections using a rat-specific KIM-1 antibody showed that KIM-1 is localized to regenerating proximal tubule epithelial cells. KIM-1 was proposed as a novel biomarker for human renal proximal tubule injury in a study that demonstrated that KIM-1 could be detected in the urine of patients with biopsy-proven acute tubular necrosis [57]. Human KIM-1 occurs as two splice variants that are identical with respect to the extracellular domains but differ at the carboxy termini and are differentially distributed throughout tissues [58]. Splice-variant KIM-1b is 25 amino acids longer than the originally identified KIM-1a and is found predominantly in human kidney.

Interestingly, cell lines expressing endogenous KIM-1 or recombinant KIM-1b constitutively shed KIM-1 into the culture medium, and shedding of KIM-1 could be inhibited with metalloprotease inhibitors, suggesting a mechanism for KIM-1 release into the urine following the regeneration of proximal tubule epithelial cells as a consequence of renal injury. Evidence supporting KIM-1's potential as a biomarker for general kidney injury and repair was clearly demonstrated in another paper describing the early detection of urinary KIM-1 protein in a rat model of drug-induced renal injury. In this study, increases in KIM-1 were observed before significant increases in SC levels could be detected following injury with folic acid and prior to measurable levels of SC in the case of cisplatin-treated rats [59]. In later, more comprehensive studies examining the sensitivity and specificity of KIM-1 as an early biomarker of mechanically [60] or drug-induced renal injury [61], KIM-1 was detected earlier than any of the routinely used biomarkers of renal injury, including BUN, SC, urinary NAG, glycosuria, and proteinuria. Certainly, the weight of evidence described above supports the notion that KIM-1 is an excellent biomarker of AKI and drug-induced renal injury. The increasing availability of antibody-based reagents and platforms to rat and human KIM-1 proteins offers convenient and much needed tools for preclinical safety assessment of drug-induced renal toxicity and for aid in diagnosing or monitoring mild to severe renal injury in the clinic. Further work is required to determine if KIM-1 is a useful marker for long-term injury and whether it can be used in combination with other makers to determine site-specific kidney injury.

Microalbumin

The examination of proteins excreted into urine provides useful information about renal function (reviewed in [62]). Tamm–Horsfall proteins that originate from renal tubular cells comprise the largest fraction of protein excreted in normal urine. The appearance of low-molecular-weight urinary proteins normally filtered through the basement membrane of the glomerulus, including insulin, parathormone, lysozyme, trypsinogen, and β_2-microglobulin, indicates some form of tubular damage [63]. The detection of higher-molecular-weight (40- to 150-kDa) urinary proteins not normally filtered by the glomerulus, including albumin, transferrin, IgG, caeruloplasmin, α1-acid glycoprotein, and high-density lipoprotein (HDL), indicate compromised glomerular function [64]. Albumin is by far the most abundant protein constituent of proteinuria. Although gross increases in urinary albumin measured by the traditional dip-stick method with a reference interval of 150–300 mg/ml have been used to indicate impairment of renal function, there are many instances of subclinical increases of urinary albumin within the defined reference interval that are predictive of disease [65–67].

The term *microalbuminuria* was coined to define this phenomenon, where such increases had value in predicting the onset of nephropathy in insulin-dependent diabetes mellitus [68]. The accepted reference interval defined for microalbuminuria is between 30 and 300 mg in 24 hours [69, 70]. Because microalbuminuria appears to be a sensitive indicator of renal injury, there is a growing interest in the nephrotoxicity biomarker community to evaluate this marker in the context of an early biomarker predictive of drug-induced renal injury. Although microalbuminuria has traditionally been used in preclinical drug development to assess glomerular function, there is growing evidence to suggest that albuminuria is a consequence of impairment of the proximal tubule retrieval pathway [71]. Evidence that microalbuminuria might provide value in diagnosing drug-induced nephrotoxicity was reported in four of 18 patients receiving cisplatin, ifosamide, and methotrexate to treat osteosarcoma [72]. Because microalbuminuria can be influenced by other factors unrelated to nephrotoxicity, including vigorous exercise, hematuria, urinary tract infection, and dehydration [5], it may have greater predictive value for renal injury in the context of a panel of markers with increased sensitivity and site specificity. Indeed, further evaluation of microglobulin as an early biomarker of site-specific or general nephrotoxicity is required before qualification for preclinical and clinical use.

Osteopontin

Osteopontin (OPN) is a 44-kDa highly phosphorylated secreted glycoprotein originally isolated from bone [73]. It is an extremely acidic protein with an isoelectric point of 4.37 (http://www.expasy.org/uniprot/P10451), made even more acidic through phosphorylation on a cluster of up to 28 serine residues [74]. OPN is widely distributed among different tissues, including kidney, lung, liver, bladder, pancreas, and breast [75] as well as macrophages [76], activated T-cells [77], smooth muscle cells [78], and endothelial cells [79]. Evidence has been provided demonstrating that OPN functions as a calcium oxalate crystal formation inhibitor in cultured murine kidney cortical cells [80]. Immunohistochemical and *in situ* hybridization examination of the expression and distribution of OPN protein and mRNA in the rat kidney clearly demonstrated that levels are highest in the descending thin loop of Henle and cells of the papillary surface epithelium [81]. Uroprontin, first described as a relative of OPN, was among the first examples of OPN isolation from human urine [82].

Although normally expressed in kidney, OPN expression can be induced under a variety of experimental pathologic conditions [83, 84], including tubulointerstitial nephritis [85], cyclosporine-induced neuropathy [86], hydronephrosis as a consequence of unilateral ureteral ligation [87], renal ischemia [88], nephropathy induced by cisplatin, and crescentric glomerulonephritis [89]. Upregulation of OPN has been reported in a number of animal models of renal injury, including drug-induced nephrotoxicity by puromycin, cylcoaporine, streptozotocin, phenylephrine, and gentamicin (reviewed in [90a]). In the rat, gentamicin-induced acute tubular necrosis model OPN levels were highest in regenerating proximal and distal tubules, leading the authors to conclude that OPN is related to the proliferation and regeneration of tubular epithelial cells following tubular damage [90b]. Although OPN has been proposed as a selective biomarker of breast cancer [91] and a useful clinical biomarker for the diagnosis of colon cancer [92], OPN shows great promise and requires further evaluation as a clinical biomarker for renal injury. Certainly, the high levels of OPN expression following chemically or physically induced renal damage coupled with the recent availability of antibody-based reagents to examine the levels of mouse, rat, and human urinary OPN provide ample opportunity to evaluate OPN as an early marker of AKI in the clinic and a predictive marker of drug-induced nephrotoxicity preclinically. Further planned work by the ILSI-HESI and C-Path groups hopes to broaden our understanding regarding the utility of OPN in either capacity as an early predictor of renal injury.

Neutrophil Gelatinase–Associated Lipocalin

Neutrophil gelatinase–associated lipocalin (NGAL) was first identified as the small molecular-weight glycoprotein component of human gelatinase affinity purified from the supernatant of phorbol myristate acetate–stimulated human neutrophils. Human gelatinase purifies as a 135-kDa complex comprised of the 92-kDa gelatinase protein and the smaller 25-kDa NGAL [93]. NGAL has subsequently been shown to exist primarily in monomeric or dimeric form free of gelatinase. A BLAST search of the 178-amino acid NGAL protein yielded a high degree of similarity to the rat α2 microglobulin-related protein and mouse protein 24p3, suggesting that NGAL is a member of the lipocalin family. Lipocalins are characterized by the ability to bind small lipophilic substances and are thought to function as modulators of inflammation [94]. More recent work has shown that NGAL, also known as siderocalin, complexes with iron and iron-binding protein to promote or accelerate recovery from proximal tubular damage (reviewed in [95]). RNA dot blot analysis of 50 human tissues revealed that NGAL expression is highest in trachea and bone tissue, moderately expressed in stomach and lung with low levels of transcript expression in the remaining tissues examined, including kidney [94]. Because NGAL is a reasonably stable small-molecular-weight protein, it is readily excreted from the kidney and can be detected in urine.

NGAL was first proposed as a novel urinary biomarker for the early prediction of acute renal injury in rat and mouse models of acute renal failure induced by bilateral ischemia [96]. Increases in the levels of urinary NGAL were detected in the first hour of postischemic urine collection and shown to be related to dose and length of exposure to ischemia. In this study, the authors reported NGAL to be more sensitive than either NAG or β_2M, underscoring its usefulness as an early predictor of acute renal injury. Furthermore, the authors proposed NGAL to be an earlier marker predictive of acute renal injury than KIM-1, since the latter reports injury within 24 hours of renal injury compared to 1 hour for NGAL. Marked upregulation of NGAL expression was observed in proximal tubule cells within three hours of ischemia-induced damage, suggesting that NGAL might be involved in postdamage reepithelialization. Additional work demonstrated that NGAL expression was induced following mild ischemia in cultured human proximal tubule cells. This paper also addressed the utility of NGAL as an early predictor of drug-induced renal injury by detecting increased levels of NGAL in the urine of cisplatin-treated mice.

Adaptation of the NGAL assay to address utility and relevance in a clinical setting showed that both urinary and serum levels of NGAL were sensitive, specific, and highly predictive biomarkers of acute renal injury following cardiac surgery in children [97]. In this particular study, multivariate analysis showed urinary NGAL to be the most powerful predictor in children that developed acute renal injury. Measurable increases in urinary NGAL concentrations were recorded within two hours of cardiac bypass surgery, whereas increases in SC levels were not observed until one to three days postsurgery. Other examples demonstrating the value of NGAL as a predictive biomarker of early renal injury include association of NGAL with severity of renal disease in proteineuric patients [98] and NGAL as an early predictor of renal disease resulting from contrast-induced nephropathy [99]. NGAL has been one of the most thoroughly studied new biomarkers predictive of AKI as a consequence of disease or surgical intervention and, to a lesser extent, drug-induced renal injury.

Sensitive and reliable antibody-based kits have been developed for a number of platforms in both humans and rodents (Table 14.2), and there is considerable interest in examining both the specificity and sensitivity of NGAL for acceptance as a fit-for-purpose predictive biomarker of drug-induced renal injury to support regulatory submissions. Certainly, because NGAL is such an early marker of renal injury, it will have to be assessed as a stand-alone marker of renal injury as well as in the context of a larger panel of markers that may help define site specific and degree of kidney injury.

Renal Papillary Antigen 1

RPA-1 is an uncharacterized antigen that is highly expressed in the collecting ducts of the rat papilla and can be detected at high levels in rat urine following exposure to compounds that induce renal papillary necrosis [100]. RPA-1 was identified by an IgG_1 monoclonal antibody, designated Pap X 5C10, that was generated in mice immunized with pooled fractions of a cleared lysate of homogenized rat papillary tissue following crude diethyl-aminoethyl (DEAE) anion-exchange chromatography. Immunohistochemical analysis of rat papillae shows that RPA-I is localized to the epithelial cells lining the luminal side of the collecting ducts and to a lesser extent in cortical collecting ducts. A second publication described the adaptation of three rat papilla–specific monoclonal antibodies, including Pap X 5C10 (PapA1), to an ELISA assay to examine antigen excretion in rat urine following drug-induced injury to the papillae using 2-bromoethanamine, propyleneimine, indomethicin, or ipsapirone [101]. Of the three antibodies evaluated, PapA1 was the only antigen released into the urine of rats following exposure to each of the toxicants. The authors concluded that changes in the rat renal papilla caused by xenobiotics could be detected early by urinary analysis and monitored during follow-up studies. This study also clearly demonstrated that the Pap X 5C10, PapA1, RPA-1 antigen had the potential for use as a site-specific biomarker predictive of renal papillary necrosis. Indeed, the Pap X 5C10 monoclonal antibody was adapted for commercial use as an RPA-1 ELISA kit marketed specifically to predict or monitor site-specific renal injury in the rat [49].

The specificity and sensitivity of the rat reagent has generated a great deal of interest in developing an equivalent reagent for the detection of human papillary injury. Identification of the RPA-1 antigen remains elusive. Early biochemical characterization of the antigen identified it as a large-molecular-weight protein (150–200 kDa) that could be separated into two molecular-weight species with isoelectric points of 7.2 and 7.3, respectively [100]. However, purification and subsequent protein identification of the antigen were extremely challenging. A recent attempt at the biochemical purification and identification of the RPA-1 antigen has been equally frustrating, with investigators providing some evidence that the antigen may be a large glycoprotein and suggesting that the carbohydrate moiety is the specific epitope recognized by the Pap X 5C10 monoclonal antibody (S. Price and G. Betton, Personal communication, unpublished data). This would be consistent with and help to explain why the rat reagent does not cross-react with a human antigen in the collecting ducts, as protein glycosylation of related proteins often differs dramatically between species, thereby precluding the likelihood of presenting identical epitopes. Nevertheless, continued efforts toward identifying a human RPA-1 antigen will provide investigators with a sorely needed clinical marker for the early detection of drug-induced renal papillary injury.

Summary

A considerable amount of effort has gone into identifying, characterizing, and developing new biomarkers of renal toxicity having greater sensitivity and specificity than the traditional markers, BUN and SC. The issue of sensitivity is a critical one, as the ideal biomarker would detect renal injury before damage is clinically evident or cannot be reversed. Such prodromal biomarkers would provide great predictive value to the pharmaceutical industry in preclinical drug development, where compounds could be modified or development terminated early in response to the nephrotoxicity observed. Even greater value could be realized in the clinic, where early signs of kidney injury resulting from surgical or therapeutic intervention could be addressed immediately before serious damage to the patient has occurred.

Several of the candidate markers described above, including β_2M, GSTα, microalbumin, KIM-1, and NGAL, have demonstrated great promise as early predictors of nephrotoxicity. Continued investigation should provide ample data from which to make a determination regarding the utility of these markers for preclinical, and ultimately, clinical use. The issue of biomarker specificity is also of great value because it provides information regarding where and to what extent injury is occurring. For example, increases in levels of SC and BUN inform us that serious kidney injury has occurred but does not reveal the precise nature of that injury, whereas the appearance of increased levels of β_2M, GSTα, microalbumin, and NGAL indicates some degree of proximal tubule injury. Similarly, RPA-1 reports injury to the papilla, clusterin indicates damage to tubular epithelial cells, and GSTYb1 is specific to distal tubular damage. Low-level increases of early markers warn the investigator or clinician that subclinical damage to the kidney is occurring and provides the necessary time to alter or terminate a course of treatment or development.

Monitoring toxicity is an important aspect of achieving a positive clinical outcome and increased safety in drug development. Incorporation of many or all of these markers into a panel tied to an appropriate platform allows for the simultaneous assessment of site-specific kidney injury with some understanding of the degree of damage. Several detection kits are commercially available for many of these new biomarkers of nephrotoxicity. For example, Biotrin International provides ELISA kits for the analysis of urinary GSTs and RPA-1, while Rules Based Medicine and Meso Scale Discovery offer panels of kidney biomarkers multiplexed onto antibody-based fluorescence or chemiluminescent platforms, respectively. As interest in new biomarkers of kidney injury continues to develop, so will the technology that supports them. Presently, all of the commercial reagents and kits supporting kidney biomarker detection are antibody based. The greatest single limitation of such platforms is how well the reagents perform with respect to target identification, nonspecific protein interaction, and species cross reactivity. Although kits are standardized and come complete with internal controls, kit-to-kit and lab-to-lab variability can be high.

Another technology being developed for the purpose of quantifying biomarkers in complex mixtures such as biofluids is mass spectrometry–based multiple reaction monitoring. This technology requires the synthesis and qualification of small peptides specific to a protein biomarker that can be included in a sample as an internal standard to which endogenous peptide can be compared and quantified. This platform is extremely sensitive (femtomolar detection sensitivity), requires very little sample volume, and offers the highest degree of specificity with very short analysis times. Limitations of the platform are related to the selection of an appropriate peptide and the expense of assay development and qualification for use. For example, peptides need to be designed that are isoform specific, being able to discriminate between two similar but not identical proteins. Peptide design is also somewhat empirical with respect to finding peptides that will "fly" in the instrument and produce a robust signal at the detector. The choice of peptides available in a particular target may be limiting given these design restrictions. Consequently, not all proteins may be amenable to this approach.

In conclusion, continued improvement in technology platforms combined with the availability of reagents to detect new biomarkers of nephrotoxicity provides both the clinician and the investigator with a variety of tools to predict and monitor early or AKI. This will be of tremendous value toward saving lives in the clinic and developing safer, more efficacious drugs without nephrotoxic side effects.

References

1 Bjornsson, T.D. (1979). Use of serum creatinine concentrations to determine renal function. *Clin. Pharmacokinet.* 4: 200–222.

2 Choudhury, D. and Ziauddin, A. (2005). Drug-associated renal dysfunction and injury. *Nat. Clin. Pract. Nephrol.* 2: 80–91.

3 Kleinknecht, D., Landais, P., and Goldfarb, B. (1987). Drug-associated renal failure: a prospective collaborative study of 81 biopsied patients. *Adv. Exp. Med. Biol.* 212: 125–128.

4 Kaloyanides, G.J., Bosmans, J.-L., and ME, D.B. (2001). Antibiotic and immunosuppression-related renal failure. In: *Diseases of the Kidney and Urogenital Tract* (ed. R.W. Schrier), 1137–1174. Philadelphia PA: Lippincott Williams & Wilkinson.

5 Vaidya, V.S., Ferguson, M.A., and Bonventre, J.V. (2008). Biomarkers of acute kidney injury. *Annu. Rev. Pharmacol. Toxicol.* 48: 463–493.

6 Berggard, I. and Bearn, A.G. (1968). Isolation and properties of a low molecular weight β2-globulin occurring in human biological fluid. *J. Biol. Chem.* 213: 4095–4103.

7 Harris, H.W. and Gill, T.J. III (1986). Expression of class I transplantation antigens. *Transplantation* 42: 109–117.

8 Bernier, G.M. and Conrad, M.E. (1969). Catabolism of β2-microglobulin by the rat kidney. *Am. J. Physiol.* 217: 1350–1362.

9 Shea, P.H., Mahler, J.F., and Horak, E. (1981). Prediction of glomerular filtration rate by serum creatinine and β2-microglobulin. *Nephron* 29: 30–35.

10 Eddy, A.A., McCullich, L., Liu, E., and Adams, J. (1991). A relationship between proteinuria and acute tubulointerstitial disease in rats with experimental nephritic syndrome. *Am. J. Pathol.* 138: 1111–1123.

11 Holm, J., Hemmingsen, L., and Nielsen, N.V. (1993). Low-molecular-mass proteinuria as a marker of proximal renal tubular dysfunction in normo- and microalbuminuric non-insulin-dependent subjects. *Clin. Chem.* 39: 517–519.

12 Kabanda, A., Jadoul, M., Lauwerys, R. et al. (1995). Low molecular weight proteinuria in Chinese herbs nephropathy. *Kidney Int.* 48: 1571–1576.

13 Kabanda, A., Vandercam, B., Bernard, A. et al. (1996). Low molecular weight proteinuria in human immunodeficiency virus–infected patients. *Am. J. Kidney Dis.* 27: 803–808.

14 Hofstra, J.M., Deegans, J.K., Willems, H.L., and Wetzels, F.M. (2008). Beta-2-microglobulin is superior to *N*-acetyl-beta-glucosaminindase in predicting prognosis in idiopathic membranous neuropathy. *Nephrol. Dial. Transplant* 23: 2546–2551.

15 Davey, P.G. and Gosling, P. (1982). Beta-2-microglobulin instability in pathological urine. *Clin. Chem.* 28: 1330–1333.

16 Blashuck, O., Burdzy, K., and Fritz, I.B. (1983). Purification and characterization of cell-aggregating factor (clusterin), the major glycoprotein in ram rete testis fluid. *J. Biol. Chem.* 12: 7714–7720.

17 Fritz, I.B., Burdzy, K., Setchell, B., and Blashuck, O. (1983). Ram rete testes fluid contains a protein (clusterin) which influences cell–cell interactions in vitro. *Biol. Reprod.* 28: 1173–1188.

18 Rosenberg, M.E. and Silkensen, J. (1995). Clusterin: physiologic and pathophysiologic considerations. *Int. J. Biochem. Cell Biol.* 27: 633–645.

19 Collard, M.W. and Griswold, M.D. (1987). Biosynthesis and molecular cloning of sulfated glycoprotein 2 secreted by rat Sertoli cells. *Biochemistry* 26: 3297–3303.

20 Kirszbaum, L., Sharpe, J.A., Murphy, B. et al. (1989). Molecular cloning and characterization of the novel, human complement-associated protein, SP40,40: a link between the complement and reproductive systems. *EMBO J.* 8: 711–718.

21 Kirszbaum, L., Bozas, S.E., and Walker, I.D. (1992). SP-40,40, a protein involved in the control of the complement pathway, possesses a unique array of disulfide bridges. *FEBS Lett.* 297: 70–76.

22 French, L.E., Chonn, A., Ducrest, D. et al. (1993). Murine clusterin: molecular cloning and mRNA localization of a gene associated with epithelial differentiation processes during embryogenesis. *J. Cell. Biol.* 122: 1119–1130.

23 O'Bryan, M.K., Cheema, S.S., Bartlett, P.F. et al. (1993). Clusterin levels increase during neuronal development. *J. Neurobiol.* 24: 6617–6623.

24 Harding, M.A., Chadwick, L.J., Gattone, V.H. II, and Calvet, J.P. (1991). The SGP-2 gene is developmentally regulated in the mouse kidney and abnormally expressed in collecting duct cysts in polycystic kidney disease. *Dev. Biol.* 146: 483–490.

25 Pearse, M.J., O'Bryan, M., Fisicaro, N. et al. (1992). Differential expression of clusterin in inducible models of apoptosis. *Int. Immunol.* 4: 1225–1231.

26 Witzgall, R., Brown, D., Schwarz, C., and Bonventre, J.V. (1994). Localization of proliferating cell nuclear antigen, vimentin, c-Fos, and clusterin in the post-ischemic kidney: evidence for a heterogeneous genetic response

among nephron segments, and a large pool of mitotically active and dedifferentiated cells. *J. Clin. Invest.* 93: 2175–2188.

27 Correa-Rotter, R., Hostetter, T.M., Manivel, J.C. et al. (1992). Intrarenal distribution of clusterin following reduction of renal mass. *Kidney Int.* 41: 938–950.

28 Cowley, B.D. Jr. and Rupp, J.C. (1995). Abnormal expression of epidermal growth factor and sulfated glycoprotein SGP-2 messenger RNA in a rat model of autosomal dominant polycystic kidney disease. *J. Am. Soc. Nephrol.* 6: 1679–1681.

29 Aulitzky, W.K., Schlegel, P.N., Wu, D. et al. (1992). Measurement of urinary clusterin as an index of nephrotoxicity. *Proc. Soc. Exp. Biol. Med.* 199: 93–96.

30 Eti, S., Cheng, S.Y., Marshall, A., and Reidenberg, M.M. (1993). Urinary clusterin in chronic nephrotoxicity in the rat. *Proc. Soc. Exp. Biol. Med.* 202: 487–490.

31 Rosenberg, M.E. and Silkensen, J. (1995). Clusterin and the kidney. *Exp. Nephrol.* 3: 9–14.

32 Hidaka, S., Kranzlin, B., Gretz, N., and Witzgall, R. (2002). Urinary clusterin levels in the rat correlate with the severity of tubular damage and may help to differentiate between glomerular and tubular injuries. *Cell Tissue Res.* 310: 289–296.

33 Abrahamson, M., Olafsson, I., Palsdottir, A. et al. (1990). Structure and expression of the human cystatin C gene. *Biochem. J.* 268: 287–294.

34 Grubb, A. (1992). Diagnostic value of cystatin C and protein HC in biological fluids. *Clin. Nephrol.* 38: S20–S27.

35 Laterza, O., Price, C.P., and Scott, M. (2002). Cystatin C: an improved estimator of glomerular function? *Clin. Chem.* 48: 699–707.

36 Herget-Rosenthal, S., Marggraf, G., Husing, J. et al. (2004). Early detection of acute renal failure by serum cystatin C. *Kidney Int.* 66: 1115–1122.

37 Koyner, J.L., Bennett, M.R., Worcester, E.M. et al. (2008). Urinary cystatin C as an early biomarker of acute kidney injury following adult cardiothoracic surgery. *Kidney Int.* 74: 1059–1069.

38 Rozell, B., Hansson, H.-A., Guthenberg, M. et al. (1993). Glutathione transferases of classes α, μ, and π show selective expression in different regions of rat kidney. *Xenobiotica* 23: 835–849.

39 Smith, G.J., Ohl, V.S., and Litwack, G. (1977). Ligandin, the glutathione S-transferases, and chemically induced hepatocarcinogenesis: a review. *Cancer Res.* 37: 8–14.

40 Rowe, J.D., Nieves, E., and Listowsky, I. (1997). Subunit diversity and tissue distribution of human glutathione S-transferases: interpretations based on electrospray ionization-MS and peptide sequence–specific antisera. *Biochem. J.* 325: 481–486.

41 Mannervik, B., Awasthi, Y.C., Board, P.G. et al. (1992). Nomenclature for human glutathione transferases. *Biochem. J.* 282: 305–306.

42 Sundberg, A.G., Nilsson, R., Appelkvist, E.L., and Dallner, G. (1993). Immunohistochemical localization of alpha and pi class glutathione transferases in normal human tissues. *Pharmacol. Toxicol.* 72: 321–331.

43 Beckett, G.J. and Hayes, J.D. (1993). Glutathione S-transferases: biomedical applications. *Adv. Clin. Chem.* 30: 281–380.

44 Bass, N.M., Kirsch, R.E., Tuff, S.A. et al. (1979). Radioimmunoassay measurement of urinary ligandin excretion in nephrotoxin-treated rats. *Clin. Sci.* 56: 419–426.

45 (a) Sundberg, A.G., Appelkvist, E.L., Backman, L., and Dallner, G. (1994). Quantitation of glutathione transferase-pi in the urine by radioimmunoassay. *Nephron* 66: 162–169. (b) Sundberg, A.G., Appelkvist, E.L., Backman, L., and Dallner, G. (1994). Urinary pi-class glutathione transferase as an indicator of tubular damage in the human kidney. *Nephron* 67: 308–316.

46 Sundberg, A.G., Nilsson, R., Appelkvist, E.L., and Dallner, G. (1995). ELISA procedures for the quantitation of glutathione transferases in the urine. *Kidney Int.* 48: 570–575.

47 Kharasch, E.D., Thorning, D., Garton, K. et al. (1997). Role of renal cysteine conjugate b-lyase in the mechanism of compound A nephrotoxicity in rats. *Anesthesiology* 86: 160–171.

48 Eger, E.I. II, Koblin, D.D., Bowland, T. et al. (1997). Nephrotoxicity of sevofluorane versus desulfrane anesthesia in volunteers. *Anesth. Analg.* 84: 160–168.

49 Kilty, C.G., Keenan, J., and Shaw, M. (2007). Histologically defined biomarkers in toxicology. *Expert Opin. Drug Saf.* 6: 207–215.

50 Ichimura, T., Bonventre, J.V., Bailly, V. et al. (1998). Kidney injury molecule-1 (KIM-1), a putative epithelial cell adhesion molecule containing a novel immunoglobulin domain, is up-regulated in renal cells after injury. *J. Biol. Chem.* 273: 4135–4142.

51 Lisitsyn, N., Lisitsyn, N., and Wigler, M. (1993). Cloning the differences between two complex genomes. *Science* 259: 946–951.

52 Hubank, M. and Schatz, D.G. (1994). Identifying differences in mRNA expression by representational difference analysis of cDNA. *Nucleic Acids Res.* 22: 5640–5648.

53 Kaplan, G., Totsuka, A., Thompson, P. et al. (1996). Identification of a surface glycoprotein on African green monkey kidney cells as a receptor for hepatitis A virus. *EMBO J.* 15: 4282–4296.

54 Feigelstock, D., Thompson, P., Mattoo, P. et al. (1998). The human homolog of HAVcr-1 codes for a hepatitis A virus cellular receptor. *J. Virol.* 72: 6621–6628.

55 Kuchroo, V.K., Umetsu, D.T., DeKruyff, R.H., and Freeman, G.J. (2003). The TIM gene family: emerging roles in immunity and disease. *Nat. Rev. Immunol.* 3: 454–462.

56 Meyers, J.H., Chakravarti, S., Schlesinger, D. et al. (2005). TIM-4 is the ligand for TIM-1 and the TIM-1-TIM-4 interaction regulates T cell proliferation. *Nat. Immunol.* 6: 455–464.

57 Won, K.H., Bailly, V., Aabichandani, R. et al. (2002). Kidney injury molecule-1(KIM-1): a novel biomarker for human renal proximal tubule injury. *Kidney Int.* 62: 237–244.

58 Bailly, V., Zhang, Z., Meier, W. et al. (2002). Shedding of kidney injury molecule-1, a putative adhesion protein involved in renal regeneration. *J. Biol. Chem.* 277: 39739–39748.

59 Ichimura, T., Hung, C.C., Yang, S.A. et al. (2004). Kidney injury molecule-1: a tissue and urinary biomarker for nephrotoxicant-induced renal injury. *Am. J. Renal Physiol.* 286: F552–F563.

60 Vaidya, V.S., Ramirez, V., Ichimura, T. et al. (2006). Urinary kidney injury molecule-1: a sensitive quantitative biomarker for early detection of kidney tubular injury. *Am. J. Renal Physiol.* 290: F517–F529.

61 Zhou, Y., Vaidya, V.S., Brown, R.P. et al. (2008). Comparison of kidney injury molecule-1 and other nephrotoxicity biomarkers in urine and kidney following acute exposure to gentamicin, mercury and chromium. *Toxicol. Sci.* 101: 159–170.

62 Lydakis, C. and Lip, G.Y.H. (1998). Microalbuminuria and cardiovascular risk. *Q. J. Med.* 91: 381–391.

63 Kaysen, J.A., Myers, B.D., Cowser, D.G. et al. (1985). Mechanisms and consequences of proteinuria. *Lab Invest.* 54: 479–498.

64 Noth, R., Krolweski, A., Kaysen, G. et al. (1989). Diabetic nephropathy: hemodynamic basis and implications for disease management. *Ann. Int. Med.* 110: 795–813.

65 Viberti, G.C. (1989). Recent advances in understanding mechanisms and natural history of diabetic disease. *Diabetes Care* 11: 3–9.

66 Morgensen, C.E. (1987). Microalbuminuria as a predictor of clinical diabetic nephropathy. *Kidney Int.* 31: 673–689.

67 Parving, H.H., Hommel, E., Mathiesen, E. et al. (1988). Prevalence of microalbuminuria, arterial hypertension, retinopathy, neuropathy, in patients with insulin-dependent diabetes. *Br. Med. J.* 296: 156–160.

68 Viberti, G.C., Hill, R.D., Jarrett, R.J. et al. (1982). Microalbuminuria as a predictor of clinical nephropathy in insulin-dependent diabetes mellitus. *Lancet* 319: 1430–1432.

69 Morgensen, C.K. and Schmitz, O. (1988). The diabetic kidney: from hyperfiltration and microalbuminuria to end-stage renal failure. *Med. Clin. North Am.* 72: 466–492.

70 Rowe, D.J.F., Dawnay, A., and Watts, G.F. (1990). Microalbuminuria in diabetes mellitus: review and recommendations for measurement of albumin in urine. *Ann. Clin. Biochem.* 27: 297–312.

71 Russo, L.M., Sandoval, R.M., McKee, M. et al. (2007). The normal kidney filters nephritic levels of albumin retrieved by proximal tubule cells: retrieval is disrupted in nephritic states. *Kidney Int.* 71: 504–513.

72 Koch Nogueira, P.C., Hadj-Assa, A., Schell, M. et al. (1998). Long-term nephrotoxicity of cisplatin, ifosamide, and methotrexate in osteosarcoma. *Pediatr. Nephrol.* 12: 572–575.

73 Prince, C.W., Oosawa, T., Butler, W.T. et al. (1987). *J. Biol. Chem.* 262: 2900–2907.

74 Sorensen, E.S., Hojrup, P., and Petersen, T.E. (1995). Posttranslational modifications of bovine osteopontin: identification of twenty-eight phosphorylation and three O-glycosylation sites. *Protein Sci.* 4: 2040–2049.

75 Brown, L.F., Berse, B., Van de Water, L. et al. (1982). Expression and distribution of osteopontin in human tissues: widespread association with luminal epithelial surfaces. *Mol. Cell Biol.* 3: 1169–1180.

76 Singh, P.R., Patarca, R., Schwartz, J. et al. (1990). Definition of a specific interaction between the early T lymphocyte activation 1 (Eta-1) protein and murine macrophages in vitro and its effect upon macrophages in vivo. *J. Exp. Med.* 171: 1931–1942.

77 Weber, G.F. and Cantor, H. (1996). The immunology of eta-1/osteopontin. *Cytokine Growth Factor Rev.* 7: 241–248.

78 Giachelli, C., Bae, N., Lombardi, D. et al. (1991). The molecular cloning and characterization of 2B7, a rat mRNA which distinguishes smooth muscle cell phenotypes in vitro and is identical to osteopontin (secreted phosphoprotein I, 2a). *Biochem. Biophys. Res. Commun.* 177: 867–873.

79 Liaw, L., Lindner, V., Schwartz, S.M. et al. (1995). Osteopontin and beta 3 integrin are coordinately expressed in regenerating endothelium in vivo and stimulate ARG-GLY-ASP-dependent endothelial migration in vitro. *Circ. Res.* 77: 665–672.

80 Worcester, E.M., Blumenthal, S.S., Beshensky, A.M., and Lewand, D.L. (1992). The calcium oxalate crystal growth inhibitor protein produced by mouse kidney cortical cells in culture is osteopontin. *J. Bone. Miner. Res.* 7: 1029–1036.

81 Kleinman, J.G., Beshenky, A., Worcester, E.M., and Brown, D. (1995). Expression of osteopontin, a urinary inhibitor of stone mineral crystal growth, in rat kidney. *Kidney Int.* 47: 1585–1596.

82 Shiraga, H., Min, W., Vandusen, W.J. et al. (1992). Inhibition of calcium oxalate growth in vitro by uropontin: another member of the aspartic acid-rich protein superfamily. *Proc. Natl. Acad. Sci. U.S.A.* 89: 426–430.

83 Wuthrich, R.P. (1998). The complex role of osteopontin in renal disease. *Nephrol. Dial. Transplant* 13: 2448–2450.

84 Rittling, S.R. and Denhardt, D.T. (1999). Osteopontin function in pathology: lessons from osteopontin-deficient mice. *Exp. Nephrol.* 7: 103–113.

85 Giachelli, C.M., Pichler, R., Lombardi, D. et al. (1994). Osteopontin expression in angiotensin II–induced tubulointerstitial nephritis. *Kidney Int.* 45: 515–524.

86 Pichler, R.H., Franseschini, N., Young, B.A. et al. (1995). Pathogenesis of cyclosporine nephropathy: roles of angiotensin II and osteopontin. *J. Am. Soc. Nephrol.* 6: 1186–1196.

87 Diamond, J.R., Kees-Folts, D., Ricardo, S.D. et al. (1995). Early and persistent up-regulated expression of renal cortical osteopontin in experimental hydronephrosis. *Am. J. Pathol.* 146: 1455–1466.

88 Kleinman, J.G., Worcester, E.M., Beshensky, A.M. et al. (1995). Upregulation of osteopontin expression by ischemia in rat kidney. *Ann. NY Acad. Sci.* 760: 321–323.

89 Yu, X.Q., Nikolic-Paterson, D.J., Mu, W. et al. (1998). A functional role for osteopontin expression in experimental crescentic glomerulonephritis in the rat. *Proc. Assoc. Am. Physicians* 110: 50–64.

90 (a) Xie, Y., Sakatsume, M., Nishi, S. et al. (2001). Expression, roles, receptor, and regulation of osteopontin in the kidney. *Kidney Int.* 60: 1645–1657. (b) Xie, Y., Nishi, S., Iguchi, S. et al. (2001). Expression of osteopontin in gentamicin induced acute tubular necrosis and its recovery process. *Kidney Int.* 59: 959–974.

91 Mirza, M., Shaunessy, E., Hurley, J.K. et al. (2008). Osteopontin-C is a selective marker for breast cancer. *Int. J. Cancer* 122: 889–897.

92 Agrawal, D., Chen, T., Irby, R. et al. (2002). Osteopontin identified as lead marker of colon cancer progression, using pooled sample expression profiling. *J. Natl. Cancer Inst.* 94: 513–521.

93 Kjeldsen, L., Johnson, A.H., Sengelov, H., and Borregaard, N. (1993). Isolation and primary structure of NGAL, a novel protein associated with human neutrophil gelatinase. *J. Biol. Chem.* 268: 10425–10432.

94 Cowland, J.B. and Borregaard, N. (1997). Molecular characterization and pattern of tissue expression of the gene for neutrophil gelatinase–associated lipocalin from humans. *Genomics* 45: 17–23.

95 De Broe, M. (2006). Neutrophil gelatinase–associated lipocalin in acute renal failure. *Kidney Int.* 69: 647–648.

96 Mishra, J., Ma, Q., Prada, A. et al. (2003). Identification of neutrophil gelatinase–associated protein as a novel early urinary biomarker of renal ischemic injury. *J. Am. Soc. Nephrol.* 14: 2534–2543.

97 Mishra, J., Dent, C., Tarabishi, R. et al. (2005). Neutrophil gelatinase–associated lipocalin (NGAL) as a biomarker for acute renal injury after cardiac surgery. *Lancet* 365: 1231–1238.

98 Bolignano, D., Coppolino, G., Campo, S. et al. (2007). Urinary neutrophil gelatinase–associated lipocalin (NGAL) is associated with severity of renal disease in proteinuric patients. *Nephrol. Dial. Transplant* 23: 414–416.
99 Hirsch, R., Dent, C., Pfriem, H. et al. (2007). HNGAL as an early predictive biomarker of contrast-induced nephropathy in children. *Pediatr. Nephrol.* 22: 2089–2095.
100 Falkenberg, F.W., Hildebrand, H., Lutte, L. et al. (1996). Urinary antigens as markers of papillary toxicity: I. Identification and characterization of rat kidney papillary antigens with monoclonal antibodies. *Arch. Toxicol.* 71: 80–92.
101 Hildebrand, H., Rinke, M., Schluter, G. et al. (1999). Urinary antigens as markers of papillary toxicity: II. Application of monoclonal antibodies for the determination of papillary antigens in rat urine. *Arch. Toxicol.* 73: 233–245.

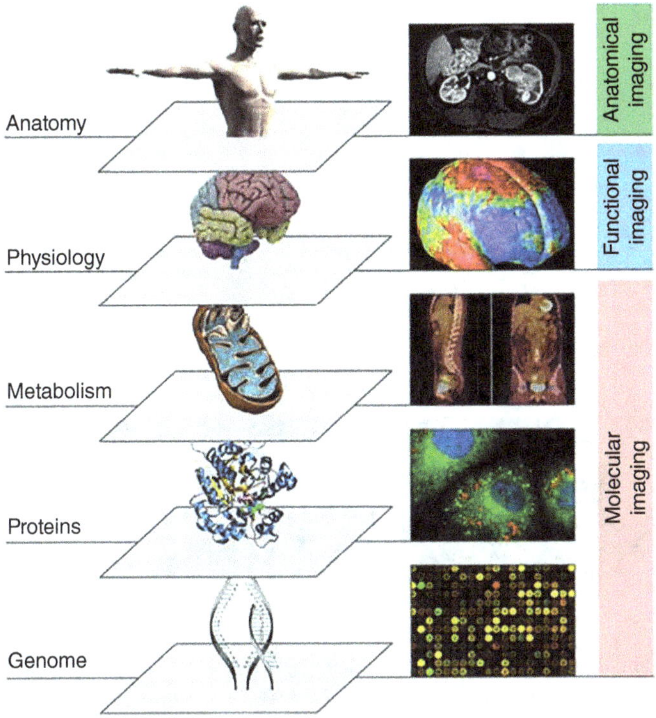

Figure 5.2 Quantitative imaging biomarkers can provide important information at multiple levels of biological "resolution" from anatomical and functional levels of organ systems down to the level of cellular and molecular processes. Source: Lambin et al. 2012 [89]. Reprinted with permission of Elsevier).

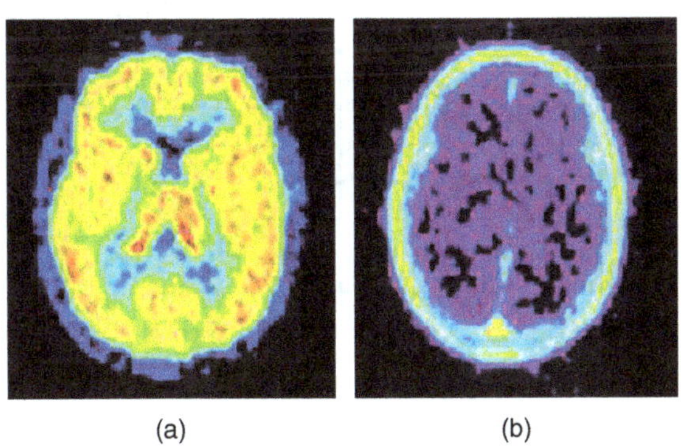

Figure 6.2 PET examples of a drug with high brain uptake (a) and one with limited brain access (b).

Biomarkers in Drug Discovery and Development: A Handbook of Practice, Application, and Strategy, Second Edition. Edited by Ramin Rahbari, Jonathan Van Niewaal, and Michael R. Bleavins.

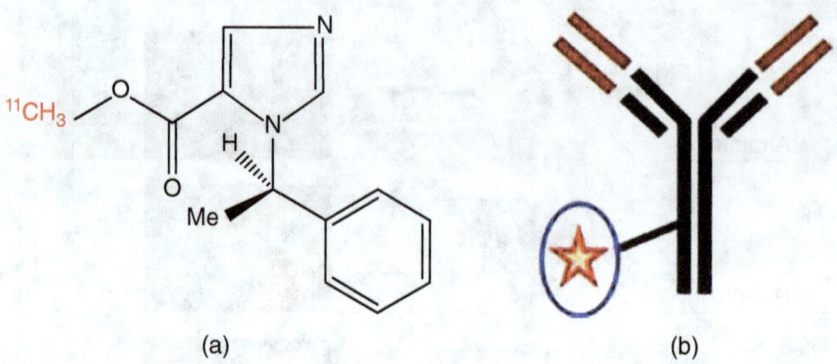

Figure 6.3 Illustration of a radiolabeled small organic molecule (a) and an antibody (b).

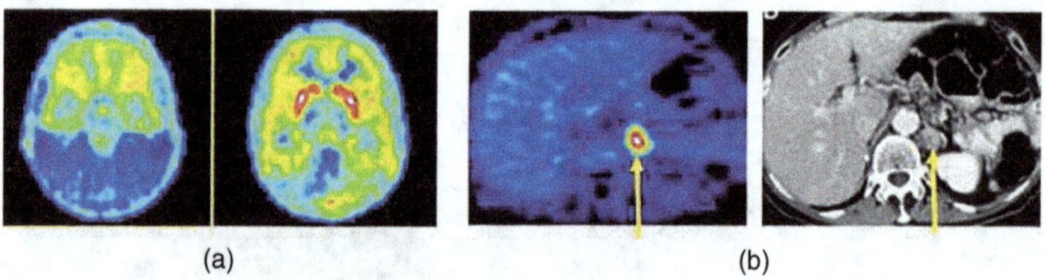

Figure 6.4 Illustration of a neuroreceptor ligand (a) with high binding to striatum and low in cerebellum, and a ligand binding to an enzyme expressed in adrenocortical tumor (b).

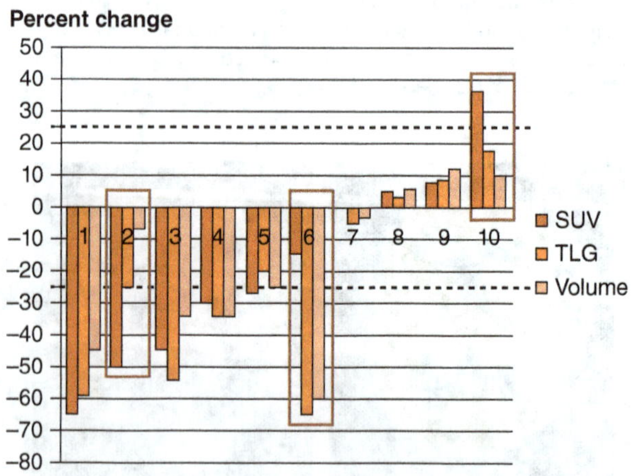

Figure 6.5 FDG-PET may supply results on three parameters: maximum FDG uptake as standardized uptake value (SUV), total lesion glycolysis as TLG, and tumor volume. A "waterfall plot" indicates the proportions of patients (1–10)/lesions with different degrees of response.

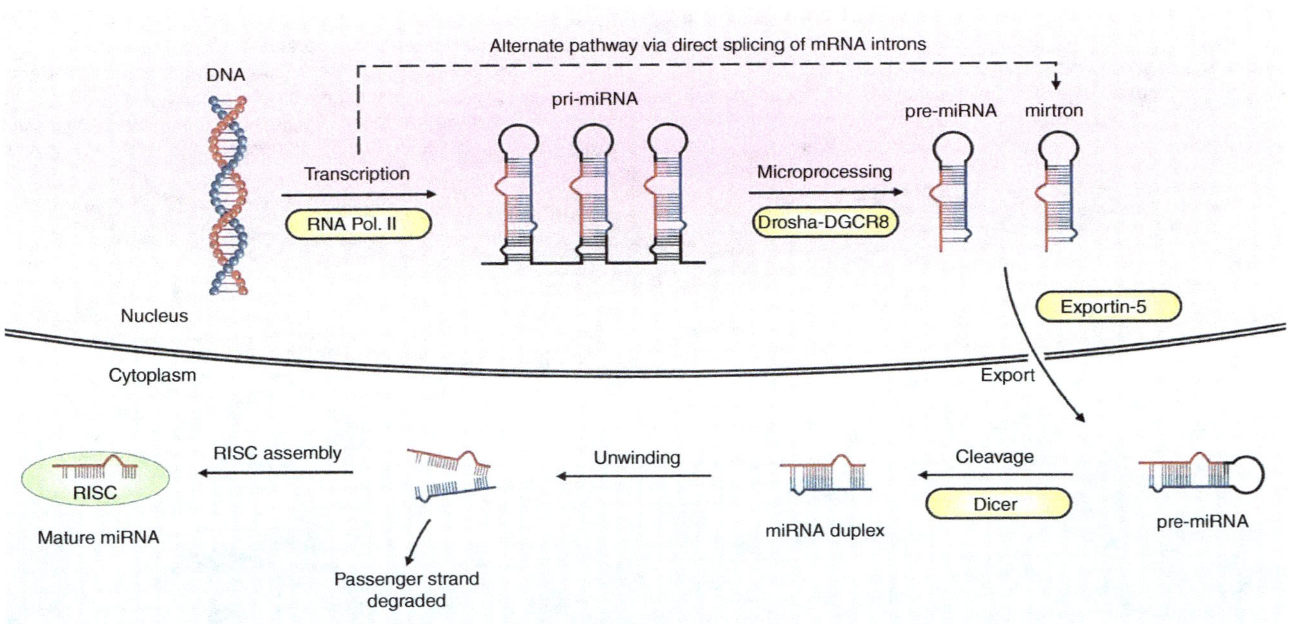

Figure 7.1 Biogenesis of miRNA. miRNAs are transcribed in the nucleus by RNA polymerase II into the primary transcript (pri-miRNA). The pri-miRNA transcript is cleaved to pre-miRNA by the *microprocessor complex*, which is made up of the Drosha endonuclease and DGCR8 cofactor. The pre-miRNA is exported via exportin 5 from the nucleus to the cytoplasm where the transcript is further modified by Dicer endonuclease to yield an RNA duplex. Following the release of the "passenger strand," the now mature, single-stranded miRNA is loaded on to Ago2 protein to form the RNA-induced silencing complex (RISC). This miRNA as a part of the RISC can specifically bind the 3′-untranslated region of messenger RNA to downregulate gene expression via translational repression or induction of mRNA degradation. Source: Adapted with permission of Romaine et al. 2015 [1]. https://heart.bmj.com/content/101/12/921. Licensed under CC-BY-4.0 http://creativecommons.org/licenses/by/4.0/.

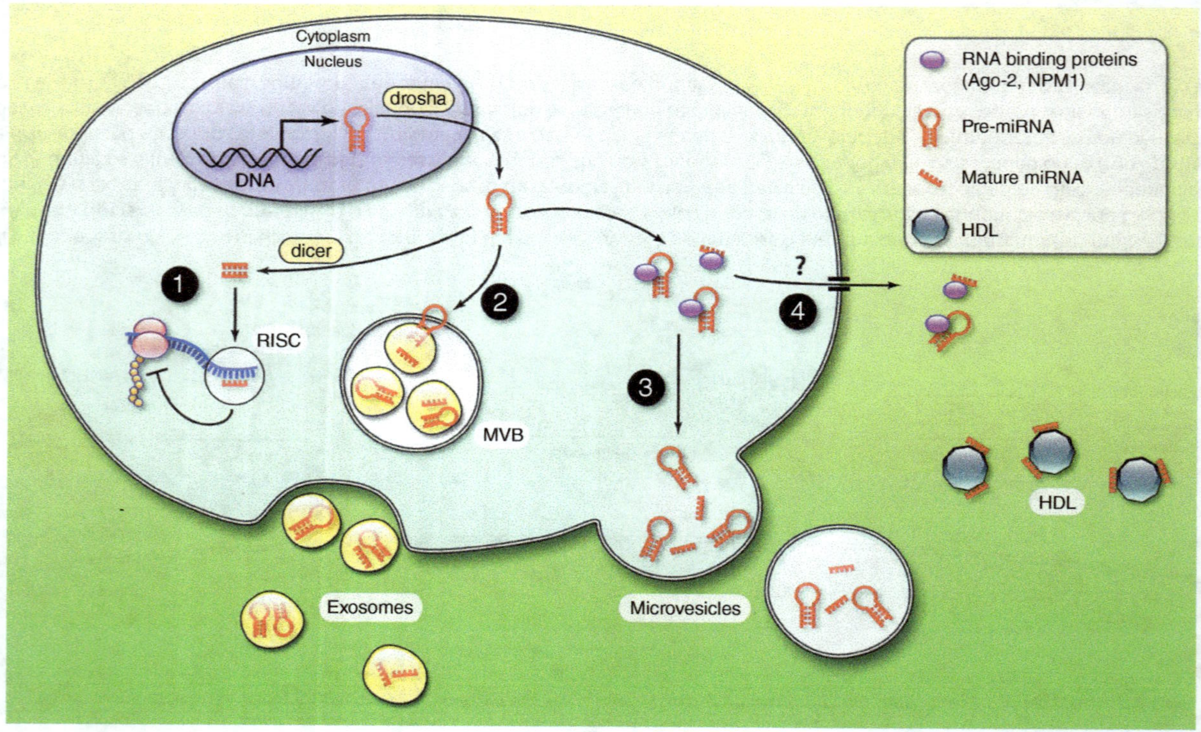

Figure 7.2 Transport of extracellular miRNA. For intracellular miRNA, one strand of this double-stranded mature miRNA duplex is loaded into the RNA-induced silencing complex (RISC) (1). Separately, miRNAs can be incorporated into exosomes, which originate from the endosome and are released from cells when multivesicular bodies (MVBs) fuse with the plasma membrane (2). Cytoplasmic miRNAs can also be released by microvesicles, which are released from the cell through plasma membrane blebbing (3). Circulating miRNAs can be bound to RNA-binding proteins such as Ago2 (4) or associated with high-density lipoproteins (HDL). The mechanisms by which specific miRNAs are packaged and released still remains largely uncharacterized. Source: Creemers et al. 2012 [18]. Adapted with permission of Wolters Kluwer Health, Inc.

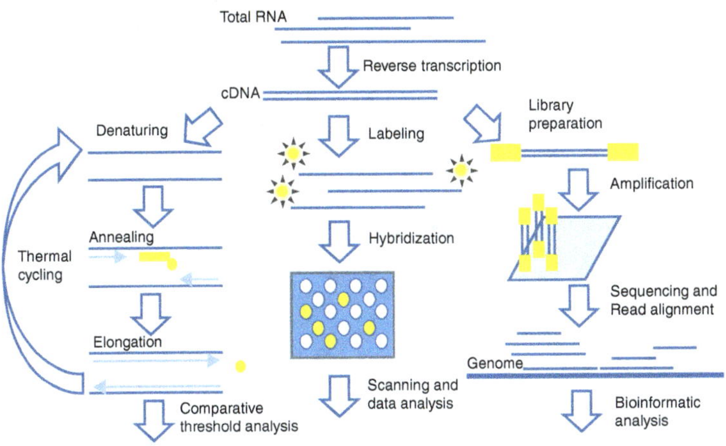

	RT-qPCR	Microarray	RNA-seq
Strengths of the technology	- High yield and high specificity - Commercially available at relatively low cost	- Multiple miRNAs identified In parallel - Useful for healthy versus disease comparison	- High-throughput large-scale data generation - Identifies novel miRNAs and distinguishes between single nucleotide variants and separate isomiRs
Limitations of the technology	- Target gene expression must be normalized to control genes	- Target gene expression must be normalized to healthy controls	- Relatively high cost - Requires tremendous analytical support - High rate of false positives and negatives
miRNA-specific challenges of quantification	- Lack of a standardized endogenous control miRNA - Short length - GC content	- Low abundance relative us cellular RNA - Low specificity - Cannot detect novel miRNAs	- Inefficient library preparation by random hexamers due to the short length of the template transcript
Technological **improvements**	- Combined polyadenylation and reverse transcription - S-poly(T) RT primer - Digital PCR provides absolute quantification	- Increase the number of complementary oligos included on each CGH chip	- Blocking techniques to remove high abundance miRNA - Complementary long padlock proteins - Synthetic and protein-based nanopore sequencing

Figure 7.3 Techniques for miRNA quantification. Technologies for the measurement of c-miRNAs were adapted from methods used to identify and measure mRNAs, including RT-qPCR, microarray hybridization, and RNA sequencing. These three major techniques share similarities in their initial steps: RNA extraction and reverse transcription reaction to convert RNA to cDNA. Therefore, these techniques are similarly limited by the inherent characteristics of miRNAs, short length, and high GC content. RT-qPCR uses thermal cycling during which denaturing, annealing of the DNA-specific primers, and subsequent elongation of these template strands take place. Comparative threshold analysis compares the cycle number at which the fluorescence measurement passes a fixed threshold value for the c-miRNA of interest vs. a positive synthetic miRNA control. RT-qPCR is commercially available, relatively cost-effective, and highly specific but is limited to one transcript quantified per reaction well and by its inability to quantify absolute abundance. Microarray applies labeled sample cDNA to a chip of known oligonucleotide fragments and then employs a scanner to quantify fluorescence to compare specific transcript expression between treated and control samples. This technique allows for the rapid relative quantification of several c-miRNAs in parallel but cannot provide absolute measurements and has relatively low specificity. Lastly, RNA sequencing performs single-nucleotide sequencing in thousands of cDNA transcripts at one time. This high-throughput NGS technique can distinguish between isomiRs and identify novel c-miRNAs in a short amount of time but is limited by the relatively high cost, requirement for advanced biostatistical analysis, and high rate of false positives and negatives, among other challenges. CGH, comparative genomic hybridization.

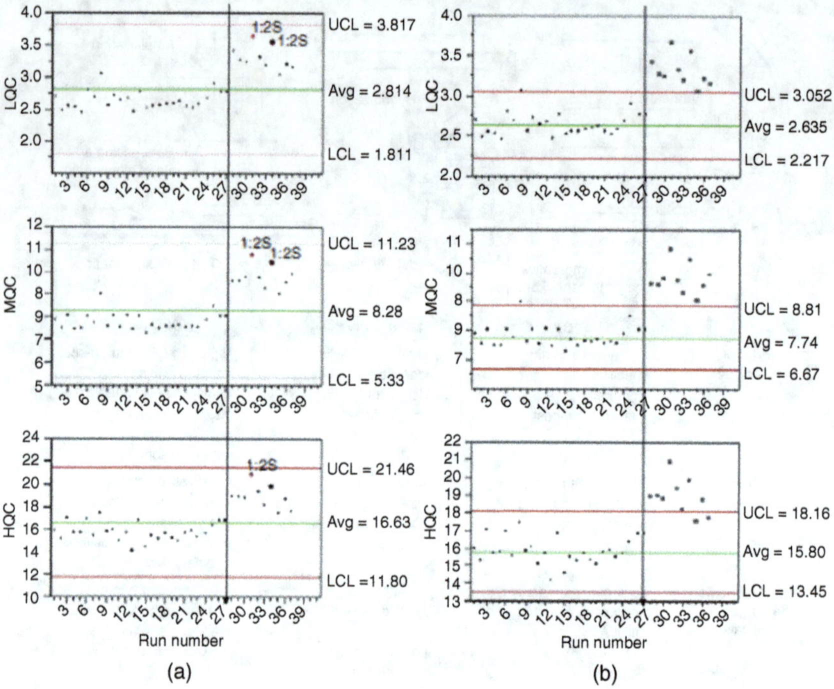

Figure 9.4 Use of sample controls for trend analysis on variability. Low, middle, and high sample controls of a biomarker were monitored in Levey Jennings control charts. There was a noticeable shift in trend in all the sample control levels after run 27. The average, upper, and lower control limits of analytical runs up to run 27 (b) were compared to the overall parameters (a).

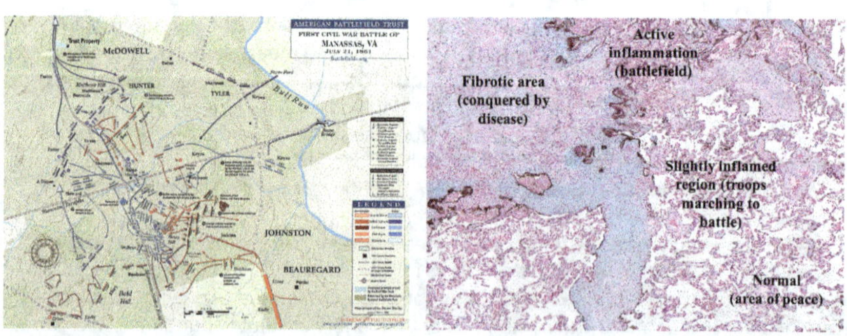

Figure 12.3 A comparison of a battlefield with a disease (idiopathic pulmonary fibrosis). The point to be made is that binding and grinding can result in data that is diluted with regards to pertinent information. Understanding the spatial and temporal aspects of both a land-based battle and a tissue-based disease is important to define outcome. Source: https://www.civilwar.org/learn/maps/first-manassas-july-21-1861.

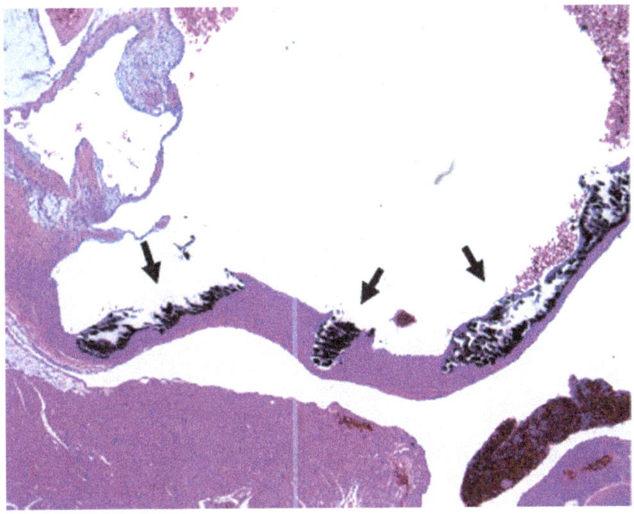

Figure 13.2 Mineralization of the aorta in a male rat administered PD0325901 at 3 mg/kg in a dose range–finding study. Arrows indicate mineral in the aorta wall. Hematoxylin and eosin–stained tissue section.

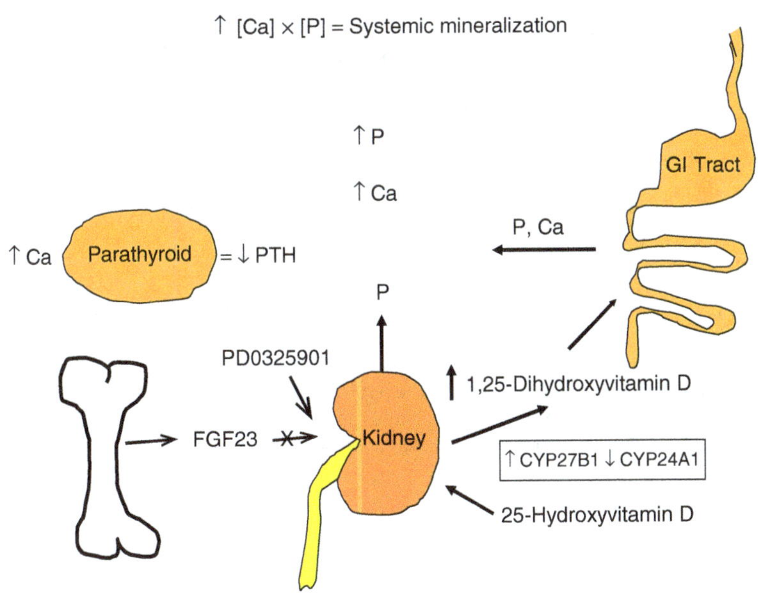

Figure 13.3 Hypothesis for the mechanism for systemic mineralization in the rat following PD0325901 administration.

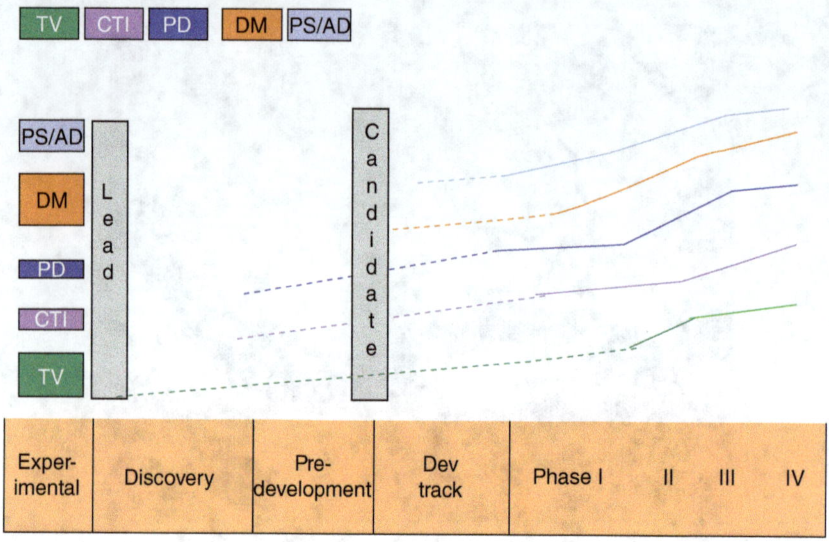

Figure 16.3 Building translational medicine via biomarker research.

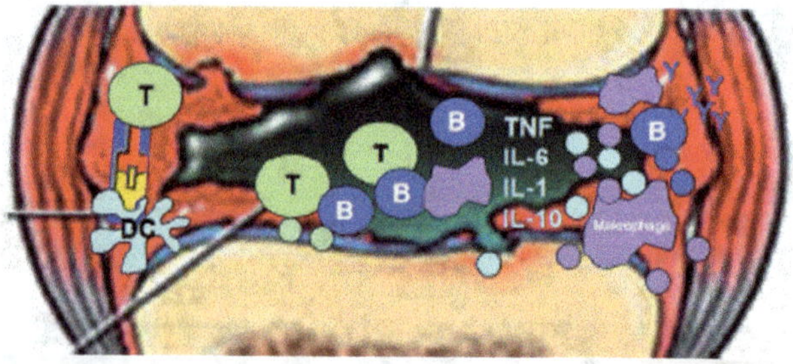

Figure 18.2 Inflammatory cells in the RA joint. A number of immune cells have infiltrated the joint, and local production of inflammatory mediators, including cytokines and antibodies, occurs. The synovial fluid, which functions as a cushion during joint movements, is in a healthy joint acellular. Illustrated to the left is the presentation of an antigen by the dendritic cell to a T-cell; the yellow connector is a MHC class II molecule. Cytokines, released from the immune cells, function as signaling molecules between cells.

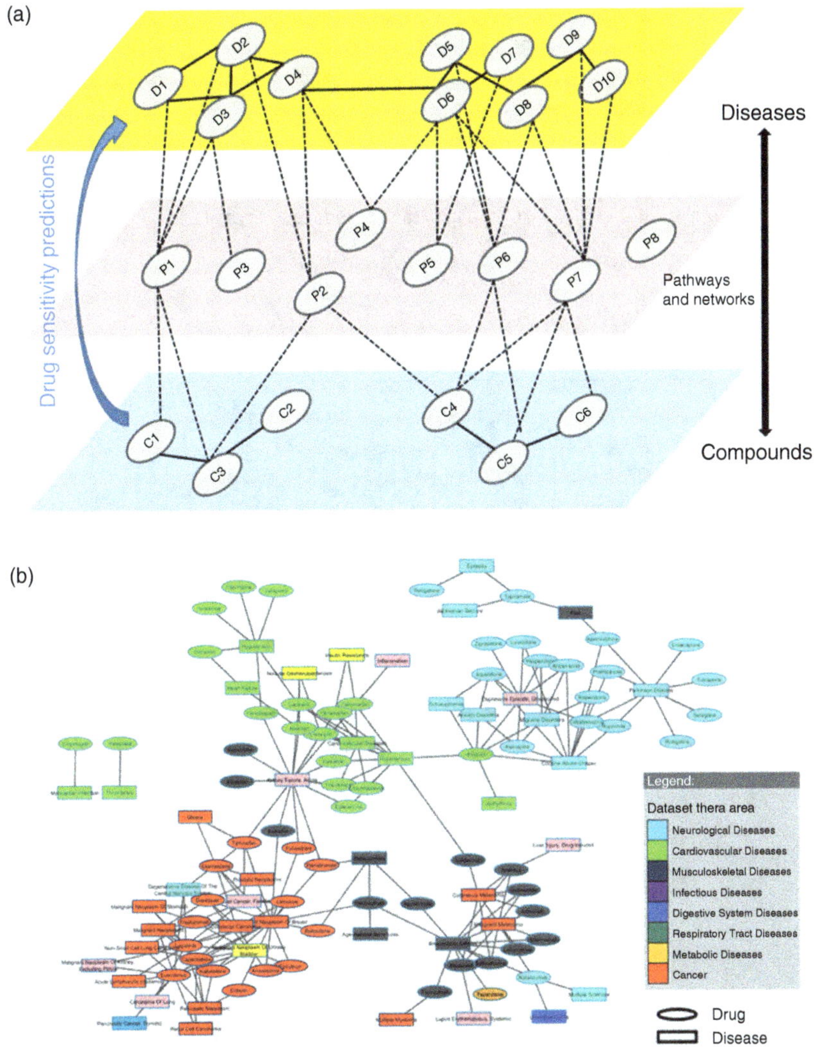

Figure 24.1 Diseases and/or drug maps for drug purposing and repurposing. (a) Information spaces and networks for disease, drugs, and disease drugs. (b) A disease–drug map defined by significant overlapping on pairs of disease/drug associated genes. The associations (i.e. similarities) between diseases or between drugs or between diseases and drugs were measured by taking $-\log_{10}$ of *p*-values from Fisher's exact test given two gene sets.

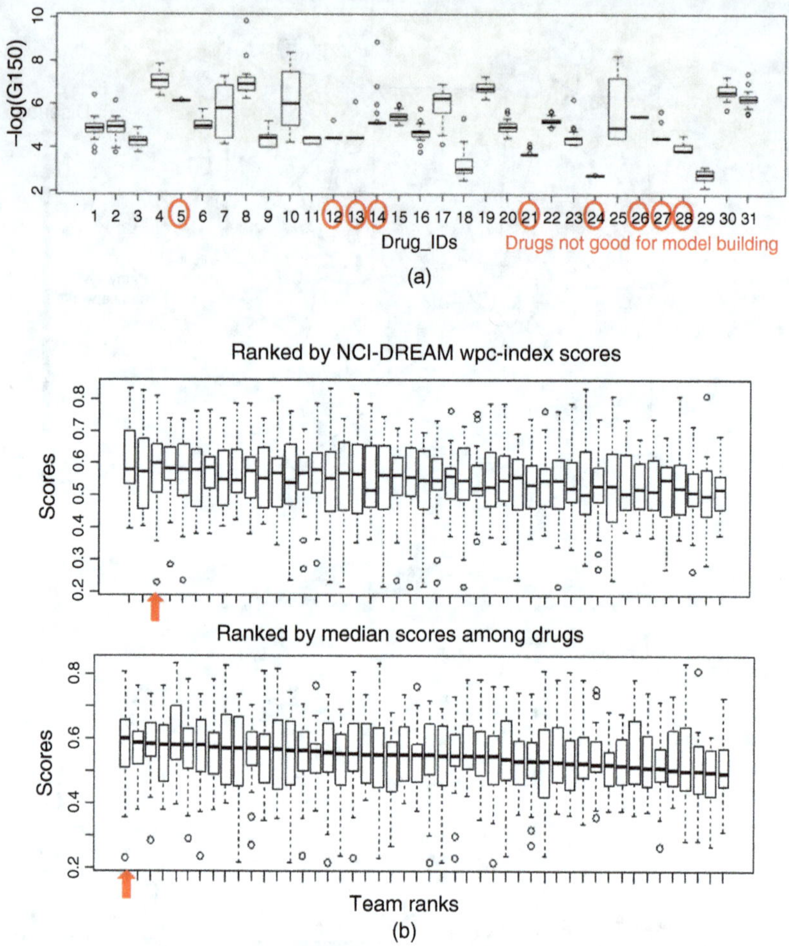

Figure 24.2 Method and data source evaluation using NCI-DREAM drug sensitivity challenge data. (a) Drug sensitivity data (IC_{50}) QC that removed 9 out of 31 drugs. (b) Our models (highlighted with red arrows) built from gene expression–only input data were highly competitive when compared to NCI-DREAM teams' models.

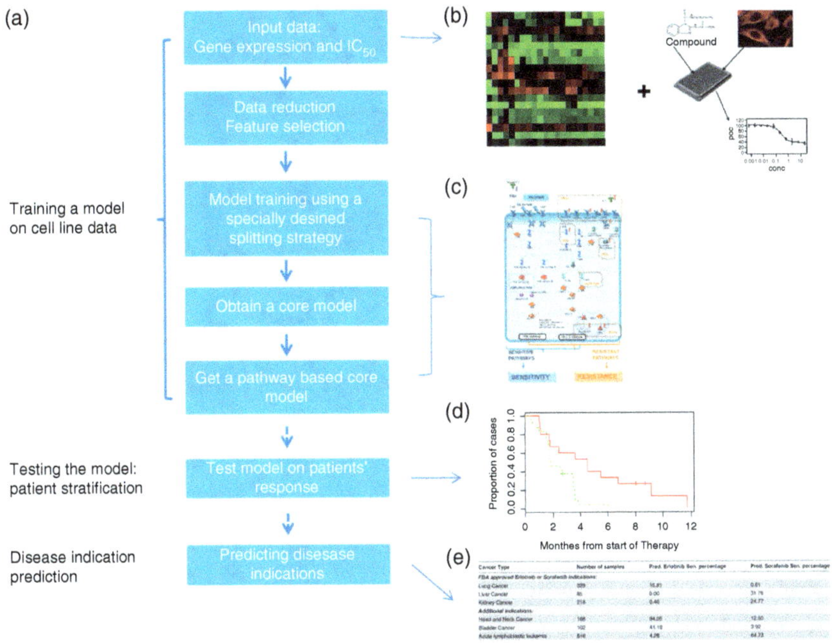

Figure 24.3 A case study on computational supports of translational biomarker research for patient stratification and disease indication selection. (a) Flowchart on the model building, testing, and application. (b) Baseline gene expression and drug sensitivity screen as model inputs. (c) Causal network to depict functional relations between sensitivity-specific and resistance-specific signature genes. (d) Survival analysis on biomarker identified treatment sensitive/resistant subgroups. (e) Combined with translational data repository effort with globally normalized GEO public domain data, the drug sensitivity predictive models were used to predict each drug's sensitive cancer indications.

Part V

Translating from Preclinical to Clinical and Back

15

Biomarkers from Bench to Bedside and Back – Back-Translation of Clinical Studies to Preclinical Models

Damian O'Connell[1], *Zaki Shaikhibrahim*[2], *Frank Kramer*[3], *and Matthias Ocker*[2,4]

[1] *Experimental Drug Development Centre A*STAR, Singapore*
[2] *Bayer AG, Berlin, Germany*
[3] *Bayer AG, Wuppertal, Germany*
[4] *Charite University Medicine, Berlin, Germany*

Introduction

Today's medicine is dominated by attempts to tailor treatments to individual patients and buzzwords like personalized medicine or precision medicine have been coined as a consequence. The identification of individual profiles for patients who will have the best benefit from a novel treatment has emerged as an important tool and goal for researchers, pharmaceutical development, and regulatory approval. The pharmaceutical industry now seems poised for rapid growth in this area because of advances in the field, including more sophisticated diagnostic technologies and a better appreciation of disease heterogeneity. This targeted approach is believed to carry the promise of reducing compound failures during the drug development process and providing the most appropriate treatment to patients while concomitantly reducing drug-associated risks.

The promise of personalized medicine is driving systematic changes in how research and development (R&D) is conducted within pharmaceutical companies, particularly in the integration and use of biomarkers and genomic information. Previously, pharma companies were hesitant about personalized medicine given the perceived threat of diminishing commercial value of a drug due to restricted patient numbers potentially being eligible for the new drug. Yet, today a majority of companies advancing new drug products embrace personalized medicine as part of their R&D strategy. Biomarkers are an essential part of pharmaceutical drug development because they offer a faster alternative to the conventional drug development approaches: earlier decision-making on ability to impact a target, the promise of safer drugs given

Biomarkers in Drug Discovery and Development: A Handbook of Practice, Application, and Strategy, Second Edition. Edited by Ramin Rahbari, Jonathan Van Niewaal, and Michael R. Bleavins.

to those patients most likely to benefit, increases in the number of novel new drugs delivered to patients, and reduced times to regulatory approval [1]. Biomarkers thus play a crucial role in achieving the goal of a more successful and straightforward drug development process that brings compelling clinical benefit to patients. Overall, attrition rates during drug development are still high and failure is attributed mostly to lack of efficacy in early development phases [2–4]. In this setting, candidate molecules with an existing biomarker program (e.g. activity/efficacy biomarkers, biomarkers determining patient's future risk, markers identifying a genetic linkage such as distinct driver mutations) still possess a higher likelihood for success than those without these features [5]. The highest chance for success thus is seen for drugs that provide maximum confidence in the translation of pharmacokinetic drug exposure and pharmacology (PK-PD modeling) and having relevant test systems established (biomarkers) [6]. A recent survey [7] has further emphasized the criticality of biomarkers in underpinning the probability of successful development. Phase III transition success rates in programs utilizing selection biomarkers in the last decade were 76.5% (n = 132) compared to only 55.0% (n = 1254) for non-biomarker trials. As many rare diseases (usually defined by a prevalence rate of 7.5 or less per 10 000 inhabitants) are identified by specific genetic mutations, it is not surprising that success rates in rare disease indications closely match the success rates observed in clinical trials that utilized selection biomarkers. Phase transition success rates for rare disease candidates and candidates utilizing selection biomarkers were very similar for every clinical stage of development. Transition success rates for each classification, rare disease programs, and programs with selection biomarkers, respectively, were as follows: Phase I, 76.0% and 76.7%; Phase II, 50.6% and 46.7%; Phase III, 73.6% and 76.5%; and new drug application/biologics license application (NDA/BLA): 89.2% and 94.5%. Both of these specific classifications significantly outpaced the success rates seen for chronic, high prevalence disease drug development at 58.7% in Phase I, 27.7% in Phase II, 61.6% in Phase III, and 87.2% in the NDA/BLA stage [7].

Several approaches have been initiated to predict the translatability of drug development, most of which has focused on biomarkers. A biomarker scoring system has been proposed that includes the availability, quantity, and quality of animal models and human data and clinical relevance of potential markers [8]. In this scoring system, a grading of up to five points per category was used. While 0 points in any of 10 categories (like number of species investigated, availability of *in vitro*/*in vivo* and human data, accessibility of technology, etc.) were sufficient to stop the model or the compound, an exceptional score of 6 (to be given if a feature brings the marker in the proximity of a surrogate) in any parameter was accepted as a single proof-of-concept (PoC) parameter. In general, a score of 20–30 was considered as a suitable biomarker for proof of mechanism (PoM), 31–40 as proof of principle (PoP), and a score of 41–50 was adequate to describe a desirable PoP parameter, with a score of >50 (maximum score is 55 in this model) as close to PoC.

This model was further refined by introducing a weighting factor for the different categories which was independent of the test compound [9, 10]. The model's forward looking utility was recently demonstrated in a retrospective analysis of several drug development candidates [11]. Interestingly, the anti-CTLA-4 antibody ipilimumab, which is currently widely used in immuno-oncology approaches for various solid tumors, only reached a medium score for successful translatability, while the introduction of a pivotal biomarker raised the score for the epidermal growth factor receptor (EGFR) inhibitor gefitinib from weak to high translatability, underlining the importance of adequate biomarkers and companion diagnostics (like EGFR mutation status for gefitinib) during drug development. In addition, the study revealed that the translatability is higher for oncology compounds due to the availability of animal models including patient-derived xenografts and validated biomarkers compared to central nervous system (CNS), cardiovascular, or metabolic drug targets.

Preclinical models that faithfully recapitulated clinical scenarios in which the compounds will ultimately be applied are the Holy Grail. Though models like this are established (e.g. the KRAS-p53-PDX1 model recapitulation pancreatic intraepithelial neoplasia and carcinogenesis or anaplastic lymphoma kinase [ALK]-translocated lung cancer models), they are not widely available for forward translation from bench to bedside. A similar approach is missing for the back-translation from bedside to bench. Except for rare cases of monogenetic diseases such as BCR-ABL fused chronic myeloid leukemia (CML) [12] and early onset hypertension due to autosomal dominant mutations of the epithelial sodium channel (ENaC) in Liddle's syndrome [13], human diseases are usually complex interactions between directly affected cells, surrounding stroma, and the immune system. These multidimensional interactions are difficult to capture in existing preclinical models and may involve chronic effects only manifested after a prolonged period. Often even distant organs are affected by a disease (e.g. the kidney in heart failure [HF] or the heart in chronic kidney disease). This limits the applicability of preclinical models toward prediction of clinical success and translatability, but does even more to impact the selection and design of appropriate animal models *per se*.

In oncology, back-translation is further complicated by the still unanswered questions about which tissues (primary vs. metastatic lesions, synchronous vs. metachronous lesions) best reflect the pathophysiology and are of highest importance and value to predict treatment efficacy. Although primary patient tumor specimens can be subjected to xenograft mouse models for drug and biomarker testing, it is still unclear if all metastases show the same behavior and genetic or functional profile [14–16]. These patient-derived xenograft models already helped to overcome limitations of cell line–based models like long-term adaptation to cell culture conditions, lack of cellular and genetic heterogeneity, and lack of stromal microenvironment, but are still subject to missing immune system control in murine immunocompromised models. The latter will become even more important with the now widespread application of immune checkpoint modulators where a humanized immune system is essential for the anticancer effect. Otherwise, surrogate development using murine specific antibodies and targets would even further increase the development costs and timelines and might have a negative impact on the translatability of these preclinical data.

In cardiovascular diseases, it is challenging to establish preclinical disease models, which reflect the targeted pathomechanism and the complexity of common human comorbidities. A means to overcome the hurdle of back-translation in oncology is to take serial biopsies from a patient before and under treatment.

Despite clinical limitations (e.g. patient status to allow repeated biopsies, accessibility of biopsy site), there also are technical limitations such as low tumor cell numbers compared to stroma and low signal intensity that limit this approach [17]. These limitations are reflected by the still low prevalence of paired biopsies before and after or during treatment within an individual study participant in clinical studies, and further improvement is urgently needed to accurately assess biomarker value in predicting outcomes [18].

Despite the increasing clinical and experimental understanding of these connections, there still is a lack of translation in both directions. In this chapter, we review examples from oncology, cardiovascular disease, and metabolic liver disease to highlight strengths and weakness of the currently available approaches to overcome the hurdles described above. As recently argued by Lötsch and Geisslinger [19] by preferring to start research in a bed to bench manner, as is favored by pure translational approaches, one still favors deductive reasoning despite the fact that "The proper study of mankind is man," which was already stated by Alexander Pope in the eighteenth century.

Current Immuno-Oncology Approaches – One Size Fits All? The Case of PD-L1

The programmed cell death ligand (PD-L1) is located on the surface of tumor and immune cells and plays a critical role in the suppression of an anti-tumor immune response. By binding to its cognate receptor (PD-1) on T-lymphocytes, an anti-tumor immune response is efficiently suppressed by cancer cells. Recently, cancer immunotherapy has focused on inhibiting the PD-1/PD-L1 signaling [20]. The introduction of several specific antibodies targeting this interface has led to a paradigm shift in cancer treatment and induced durable responses in previously untreatable advanced solid tumors like malignant melanoma [21, 22]. However, a question that is still unclear is whether utilizing the expression of PD-L1 as a biomarker for patient selection can predict response rates, progression-free survival, and overall survival [20]. In some indications, response to anti-PD-L1 treatment was independent of PD-L1 expression on tumor cells, on infiltrating immune cells, or on the viral status (e.g. Merkel-cell carcinoma and squamous non-small cell lung cancer [NSCLC]) [23, 24]. In contrast, the response to cancer immunotherapies showed a clear correlation with these parameters in melanoma or genitourinary cancers [25]. Interestingly, a proportion of PD-L1 negative patients still benefit from therapy in these indications. Furthermore, regardless of using a cut-off of either 1% or 5% for a positive PD-L1 expression, the correlation with response rates remained unaffected [20].

This uncertainty on the value of PD-L1 expression as a predictive biomarker for cancer immunotherapy from the currently existing clinical data is further confounded by the existence of multiple immunohistochemistry companion diagnostic tests using various clones, antibody classes, cellular targets, and assessment methods (Table 15.1) [26, 27].

Therefore, PD-L1 expression is currently not considered to be a robust biomarker for cancer immunotherapy [28]. It is crucial to back-translate these

Table 15.1 PD-L1 assays used for patient selection.

Profile	Pembrolizumab	Nivolumab	Duruvalumab	Atezolizumab
Monoclonal antibody description	Humanized IgG4	Human IgG4	Human Fc-modified IgG1	Human Fc-modified IgG1
	22C3 mouse	28-8 rabbit	SP263 rabbit	SP142
	DAKO	DAKO	Ventana	Ventana
Target	PD-1	PD-1	PD-L1	PD-L1
Approved indication	Melanoma	Melanoma, NSCLC		Bladder, NSCLC
Expression	Tumor cells and stroma	Tumor cells	Tumor cells	Tumor-infiltrating immune cells
				Tumor cells
Positive cut-off point	≥1%	≥1% (NSCLC)	≥25%	≥5% to <10% tumor cells
		≥5% (renal)		Tumor-infiltrating immune cells
				Immunohistochemistry score 3 + ≥10% of any cell type

Source: Adapted from Hansen and Sui 2016 [26].

Table 15.2 Most commonly used experimental mouse model systems in cancer immunology research: strengths and weaknesses.

Experimental Model	Strengths	Weaknesses	Clinical validation
Transplantable tumors	• Simplicity of experimentation • Rapid tumor formation • Quick evaluation of therapeutic responses	• Limited recapitulation of tumor microenvironment	• Checkpoint blockade (CTLA4- or PD1-specific monoclonal antibodies)
Genetically engineered tumor models	• Tumors harbor genetic lesions found in human cancers • Improved modeling of tumor progression • Tumor microenvironment, tumor-induced tolerance, and immunosuppression show similarities to human cancers	• Long period of time for tumor development • Extended evaluation of therapeutic response • Increased cost and complexity of generating and maintaining transgenic mice	• Prognostic importance of intra-tumoral lymphocyte infiltrates and CTLA4-specific monoclonal antibodies
Humanized mice	• Human immune populations studied directly and might be able to capture the heterogeneity of patient responses	• Logistically difficult to obtain matched donor and tumor samples • Incomplete reconstitution of human immune response • Impact of mouse-derived factors on human immune cells not fully defined • Expensive	• Chimeric antigen receptor T cell technology

Source: Adapted from Dranoff 2012 [30].

clinical findings including expression level cut-offs to preclinical *in vitro* and *in vivo* models to better understand the role of this marker.

Given the above-mentioned controversy regarding the value of PD-L1 expression in predicting response to anti-PD-1/PD-L1 immunotherapy, combined with the fact that there are many cancer patients who failed to respond to this class of treatment, there is urgent need to identify biomarkers that are able to predict response to PD-1/PD-L1 checkpoint blockades [29]. Therefore, it is important, in addition to PD-L1 expression, to determine whether tumor-infiltrating immune cells, as well as molecules in the tumor microenvironment, could predict which patients may or may not benefit from anti-PD-1/PD-L1 immunotherapy [29]. Such relevant cells and molecules in the tumor microenvironment that are directly implicated with the immune response include tumor-infiltrating immune cells, the BCL-2-interacting mediator of cell death (Bim), and cytokines like interferon gamma [29]. Furthermore, genetic analysis such as determining the mutational landscape and mismatch-repair deficiency have proven to be valuable new approaches [29]. Mismatch-repair deficiency is found in colorectal cancer patients, and it was suggested that such deficiency may render the tumors more responsive to PD-1 blockade than mismatch-repair proficient tumors [29]. The mutational landscape in which high mutational burden has been correlated with clinical benefit includes the case of melanoma patients who were treated with CTLA-4 blockade agents, and this treatment could also be a valuable approach [29]. In addition, as somatic mutations could encode immunogenic neo-antigens which are crucial for tumor immunity, response to immune checkpoint blockade could be predicted based on the somatic mutational level [29].

Lastly, given that the experimental mouse tumor models play an important role into our understanding of host antitumor immune responses, Table 15.2 summarizes our current knowledge of the three most commonly used experimental mouse model systems in cancer immunology research (transplantable tumors, genetically engineered tumor models, and humanized mice) in terms of their strengths and weaknesses, as well as their level of clinical validation [30].

Cardiovascular Diseases – Heart Failure

The complexity of cardiovascular disease demands better back-translation from clinical finding into preclinical models for the development of relevant and effective therapies. In the United States, heart diseases are the leading causes of death [31]. Heart Failure (HF) is an especially serious health condition affecting approximately 5.1 million people in the United States [32]. Almost 1 million hospitalizations per year in the United States attribute HF as the primary discharge diagnosis, and approximately 25% of patients hospitalized with HF are rehospitalized within 30 days after discharge. Though multiple therapy options are available, mortality rates remain above 50% at five years from initial diagnosis. More than 200 000 patients die per year from HF. The disease has a high socioeconomic impact causing overall costs of approximately $30.7 billion per year and 12–15 million outpatient office visits [33]. Overall, HF is associated with a prognosis comparable to many forms of cancer with a comparable five-year survival rate [34].

HF is a complex syndrome driven by multiple pathological mechanisms and characterized by structural and/or functional impairment of the heart. These changes comprise limitations in filling and ejection of blood from the ventricle, which leads to fatigue, dyspnea, and systemic fluid retention. The latter manifests itself in pulmonary congestion and peripheral edema. A large number of conditions such as persistent hypertension, coronary artery disease (CAD), myocardial infarction, cardiac arrhythmias, infections, diabetes, some chemotherapies, and normal aging may cause HF. The development and progression of HF are driven by a complex interplay of effects such as pressure-volume overload, impaired microcirculation, inflammation, oxidative stress, cell apoptosis, tissue necrosis, remodeling, and metabolic dysregulation of the myocardium.

Two different forms of HF are distinguished based on the ejection fraction (EF) of the left ventricle (LV): (i) heart failure with preserved ejection fraction (HFpEF) and (ii) heart failure with reduced ejection fraction (HFrEF). In recent years, the incidence of HFpEF has been increasing [35]. HFpEF is most prevalent in older women with a history of hypertension. Additional risk factors for the development and progression of HFpEF are diabetes mellitus, hyperlipidemia, obesity, CAD, and atrial fibrillation (AF) [36, 37]. Once hospitalized, prognosis for HFpEF and HFrEF patients are comparable with up to 15% mortality and 35% rehospitalization within the first two to three months [38].

Due to the complexity of the HF, many pharmacotherapies targeting different mechanisms have been developed and established as standard of care (SoC) in HFrEF. These comprise, among others, angiotensin-converting enzyme inhibitors (ACEIs), angiotensin receptor blockers (ARBs), beta-blockers (BBs), mineral corticoid receptor antagonists (MRAs), diuretics, and calcium channel blockers. Application of these therapies has led to a significant improvement in patient care and a decline in deaths from cardiovascular (CV) diseases [39]. In contrast to that for HFrEF, no life-saving therapy has yet been approved for HFpEF, though many of the established HFrEF therapies (MRAs, ACEIs, ARBs, and BBs) have been evaluated clinically [40–45]. The lack of life-extending pharmacotherapies for HFpEF, combined with the current hospitalization and mortality rate in HFrEF patients, demonstrates the pressing demand for new, safe, and more efficacious therapies.

Role of Biomarkers in Heart Failure Drug Development

To enable successful HF drug development, a comprehensive definition of underlying causes/mechanisms in patients is essential, which can provide the basis for patient selection in clinical studies. Furthermore, the ability to reliably monitor safety and detect early signs of efficacy during a clinical study are key aspects. Biomarkers serve as a tool for both. First, to stratify patients for a clinical study investigating a certain mode of action (MoA; e.g. identify patients most likely to respond or at increased risk for a certain outcome) and second, to detect undesired side effects and early signs of efficacy.

In contrast to oncology, where biomarkers are better accepted and characterized to support patient selection, in cardiovascular indications biomarkers generally play a less prominent role. This is nicely illustrated by the fact that more than 20 companion diagnostics (CDx) kits are U.S. Food and Drug Administration (FDA) approved for oncological indications, whereas in CV disease no CDx assay has been approved to date [46]. A number of analytes in blood and urine routinely are used as biomarkers in animal HF models to support profiling of early drug candidates. A few biomarkers such as the pressure-volume overload indicator B-type natriuretic peptide (BNP), its longer half-lived N-terminal pro-peptide (Nt-proBNP), the myocyte stress marker troponin, and renal injury markers such as cystatin c also are routinely used in clinical studies in humans. Nevertheless, insufficient work has been performed to back-translate patient stratification approaches from the clinic into the animal disease model and to provide new investigative tools. Most recently, in clinical studies investigating safety, optimal dose, and efficacy of the novel soluble guanylate cyclase (sGC) stimulator vericiguat, or the combination of the Angiotensin receptor antagonist valsartan and the neprilysin inhibitor (NI) sacubitril, in HF a certain cut-off plasma concentration for BNP and NtproBNP was defined as inclusion criteria [47, 48]. However, at least to the author's knowledge, no attempt to test this approach in preclinical disease models was undertaken. It has not been tested systematically whether animals with high NtproBNP or BNP at onset of therapy respond better than those with moderately increased levels of the blood biomarkers. In multiple studies, BNP/NtproBNP has been used as a response to therapy biomarker in preclinical models. For example, treatment with the partial adenosine-1 (A1) receptor agonist capadenoson led to a significant decrease of NtproBNP in dogs with micro-embolization–induced HF, which was paralleled by an increase in EF and a decrease of interstitial cardiac fibrosis [49]. However, whether those dogs with the highest levels of natriuretic peptides responded best was not reported in this study. Furthermore, very recently, it has been shown that treatment with a novel mitochondria-targeting peptide elamipretide (MTP-131) improves LV systolic function and decreases NTproBNP levels, but again no analysis was performed to investigate whether levels of natriuretic peptides at the start of treatment correlated with response to therapy [50]. To get as much information as possible from animal disease models regarding which patient population(s) most probably will benefit from a treatment with a new MoA, it would be desirable to evaluate whether dogs or rats with certain biomarker profiles respond better to a new therapy versus other therapeutic interventions. The back-translation of natriuretic peptide–based patient stratification into animal models is very attractive and feasible as just recently new clinical data were generated in patients treated with new drug candidates like MRAs, sGC stimulators, angiotensin receptor-neprilysin inhibitors (ARNIs), and others. These human data could now be used to fine-tune existing animal models to improve the development process of upcoming drug candidates. Biomarkers like BNP/NtproBNP are very promising tools in this process, and their role will become more important in an environment where a more personalized cardiovascular medicine is requested by society, physicians, health authorities, and payers.

The Role of Medical Devices as Tools to Back-Translate the Human Situation into Preclinical Disease Models

With the advent of digital technologies in the health care sector, more and more functional parameters and vital signs can be captured by implanted and wearable devices. Many of these devices are capable of continuously monitoring patients and submitting the acquired information to treating physicians or clinical trial study data bases. Especially, the use of data derived from devices such as implanted cardiac defibrillators (ICDs), cardiac re-synchronization therapy (CRT) devices, pacemakers, subcutaneously implanted monitors, or cutaneously worn patches bears the potential to allow better surveillance of a patient's disease status under real-life conditions than what can be achieved by point in time assessments at the clinic or at the doctor's office. Therefore, it is reasonable to extent the use of the term "biomarker" to data derived from medical devices.

Interestingly, what has become available in recent years to monitor patients is already available for a longer time for use in a laboratory setting for animal disease models. There are several technical solutions available which allow monitoring of multiple parameters in animal disease models. These are important to characterize the disease status and assess therapeutic response in human patients. For example, the effect of drug candidates on blood pressure, heart rate, myocardial contractility, and electro-cardiographic (ECG) parameters (QRS complex and QTc interval) of large animals like dogs and monkeys can be measured after implantation of the respective sensors into subcutaneous tissue, the femoral artery, and the LV [51]. Using constant monitoring, precious information about pharmacokinetic and pharmacodynamic effects of new drug candidates can be collected in conscious free-roaming animals. This is very important as anesthesia, which is required when comparable data are acquired using traditional invasive catheter techniques in humans, has a strong influence on cardiac function and vital signs. Using implanted sensors combined with telemetric features allows the generation of data in animals which can be more realistically compared to real-life data acquired by implantable or wearable cardiac monitors in human patients.

Beyond monitoring, data derived from implanted or wearable devices can be used to identify patients with an increased risk to reach a certain undesirable clinical endpoint, such as HF or hospitalization. A relevant example of a device-derived digital biomarker which could be translated from animal models to patients, and back to the preclinical setting, is intra-thoracic impedance (ITI). ITI is an attractive tool to monitor health status of HF patients. The inability of the LV to pump a sufficient amount of blood leads to fluid accumulation in the lung tissue of HF patients. Increasing tissue fluid content leads to signs and symptoms of pulmonary congestion. Fluid accumulation is a silent process which starts long before it culminates in readily recognized symptoms and HF hospitalization. Therefore, continuous measurement of ITI, which decreases with increased tissue fluid content, could become an elegant tool for the early detection of increased patient risk to decompensate. ITI can easily be measured by implanted pacemakers or CRT devices. It has been shown in a study with 33 New York Heart Association (NYHA) class III and IV HF patients that tissue fluid accumulation began 15.3 ± 10.6 days before the onset of symptoms and 18.3 ± 10.1 days before patients were hospitalized due to HF [52].

Early identification of when a patient is prone to decompensate opens the opportunity to optimize therapy early and prevent HF hospitalization. This reduces both disease burden to the patient and saves resources in the health care system. Finally, history of hospitalizations predicts risk for future hospitalization and mortality and thereby determines prognosis of HF patients, and therefore prevention of hospitalization is a key goal of HF therapy. Early detection of indicators for pending decompensation may also help to select patients at high risk for HF hospitalization for clinical studies. It has been shown in a study with 326 HF patients that the number of impedance threshold crossing events (TCEs) in an initial observation period of 4 months was independently correlated with hospitalization rate in the subsequent period of approximately 3–10 months [53]. These examples nicely illustrate that medical device–derived digital biomarkers, such as ITI, could be used to enrich clinical studies with patients at high risk for HF hospitalization. In addition, the combination of different device readouts into an overall risk score is even more promising as can be seen in the recently published study, in which impedance, night heart rate, AF burden, heart rate during AF, and time of pacemaker activity were used to predict patient's individual risk [54]. ITI could be measured in dog, pig, or sheep HF models by implanting the aforementioned medical devices, initially produced for human application, into these large animals. Rademaker et al. [55] showed in an ovine HF model that ITI decreased in proportion to HF severity. Further investigations should address whether ITI in animal disease models at baseline can determine response to HF therapies and whether longitudinal change of ITI reflects beneficial effects of diuresis, as was shown in human HF patients [53].

Taken together, there are protein biomarkers such as BNP/NtproBNP, and digital biomarkers like ITI, which could be measured in HF disease models in animals as well as in human patients. Theoretically, this allows: (i) assessment of parameters in preclinical models, which will later be captured in clinical studies in human, (ii) fine-tuning of disease models to better mimic the human disease context, and (iii) generation of real-life data in conscious animals by using tele-monitoring devices. So far, few attempts have been undertaken to back-translate biomarker findings from large clinical studies into the preclinical arena to make the present disease models more predictive for the human situation. However, with the digital transformation of the health care sector and the growing number of functional measurements which can be performed in a real-life setting in patients, it is expected that this trend will have a positive influence on the design of disease models used to profile new drug candidates in the future.

Metabolic Liver Diseases

Chronic metabolic liver diseases are characterized by fatty degeneration of hepatocytes (non-alcoholic fatty liver disease, NAFLD), chronic inflammation (non-alcoholic steatohepatitis, NASH), and tissue remodeling that eventually leads to the development of fibrosis, end-stage liver disease with impaired liver function and cirrhosis, hepatocellular carcinoma, and liver failure [56]. Liver fibrosis is among the top 10 causes of death worldwide, and its incidence and prevalence are predicted to continuously rise in the coming years [57] due to increasing number of metabolic disorders like diabetes and obesity, still high prevalence of viral hepatitis, and ongoing high alcohol consumption, which renders metabolic liver diseases an urgent unmet medical need.

The pathophysiology of end-stage liver disease and fibrosis is complex. The final step in disease pathogenesis is the activation of myofibroblast-like hepatic stellate cells (other cell types like mesenchymal stem cells or portal fibroblast contribute to a lesser extent), which leads to collagen and other extracellular matrix protein deposition in the Space of Disse that finally induces the morphological and functional consequences of chronic liver tissue damage [58]. The initiation of this process is also quite well understood for biliary, toxic, or drug-induced liver injury and, to some extent, for chronic viral hepatitis B and C. Several animal models (e.g. bile duct ligation, humanized mice, thioacetamide, or concanavalin A injections) have been developed and extensively characterized. Under these pathophysiologic conditions, several cellular (e.g. restorative macrophages) and soluble factors (e.g. transforming growth factor, interleukins, tumor necrosis factor, other chemokines, etc.) also form a network of pro- and anti-inflammatory conditions and pathways [59]. Etiology is even less understood for the metabolic liver diseases like NAFLD and NASH than fibrosis.

In addition to factors already mentioned, reactive oxygen species, pro-inflammatory cytokines released from adipose tissue, and disturbances in hepatic cholesterol and lipid metabolism are implicated in the disease progression.

The influence of dietary factors (e.g. fructose intake), endocrine conditions like insulin resistance, and impaired intestinal barrier function with activation of toll-like receptors on Kupffer cells have been previously described [59, 60]. Genetically modified animal models like the Mdr2 (Abc4$^{-/-}$) knockout mouse or the ApoE-LDLR double knockout mouse have been established to overcome the limitations of dietary models available thus far [61, 62] by mimicking human fatty liver disease and metabolic conditions like atherosclerosis.

These examples highlight the complexity of metabolic and other end-stage liver diseases leading to fibrosis and cirrhosis. The available animal models are thus only reflecting selected pathophysiologic conditions yet are still widely used for specific aspects of drug development [59]. The translatability of these models is limited which is further confounded by the difficult assessment of the disease condition and level of clinical understanding. Although clinical meaningful endpoints from humans like the NAFLD activity score or portal vein pressure are also used in these models, these determinations are technically challenging in rodents. Therefore, larger animal models using cynomolgus monkeys or minipigs were recently established and allowed the serial assessment of the disease with noninvasive imaging techniques and/or sequential biopsies [63, 64].

The complexity of several etiologies and numerous pathogenetic factors also limits the translatability to human studies. It is recommended that novel anti-fibrotic drugs demonstrate efficacy in more than one model to reflect the complex and divergent pathophysiologic conditions. Furthermore, disease progression is usually slow, and clear endpoints, as well as generally accepted biomarkers leading to regulatory approval, are still missing [65]. While liver biopsy remains the gold standard to assess the disease activity and stage, it is an invasive measure that is clinically often limited by patient's performance status, especially under decompensated liver function, and biopsies show significant sampling error and interobserver diagnostic variability [66–68]. Serum-based algorithms like Fibrotest, aspartate aminotransferase (AST)-platelet ratio index, fibroindex, enhanced liver fibrosis (ELF) score, and many others are currently being explored in humans, but are often validated and of predictive value only in distinct and specific disease settings [59, 69]. Clinical translatability of biomarkers is further limited by an ongoing discussion on identifying appropriate patient subpopulations and meaningful endpoints. It is currently suggested to include patients with intermediate fibrosis scores, and high risk for rapid disease progression, to obtain clinically meaningful results with reasonable study duration as these patients are considered to have high disease kinetics in both directions (fibrosis resolution progression to cirrhosis) which allows quick readout of study endpoints [65, 70–72]. It is noteworthy that these surrogate parameters may be acceptable to regulatory authorities if the chance of predicting clinical benefit to patients is demonstrated.

Discussion

As outlined above, there is plenty of evidence regarding the "forward" translation from *in vitro* experiments and preclinical models into clinical studies and applications. Yet, the overall success rate of drug development remains disappointing with an overall success rate of 5–20% depending on the therapeutic area [2]. Despite numerous technological advances and the introduction of various *omics* approaches to research, the attrition rate remains high and relatively unchanged over the past years. Several examples have shown that a stronger focus on "back" translation from clinical experience could overcome the risk of failure in early drug development, lead to faster approvals, and provide better identification of the patient population(s) likely to benefit.

Palbociclib is a selective cyclin-dependent kinase (CDK) 4/6 inhibitor developed for the treatment of metastatic breast cancer. A clear preclinical rationale existed even before the initiation of clinical trials that indicated interfering with the Cyclin D-CDK4/6 complex should interrupt uncontrolled cell cycle progression in breast cancer [73, 74]. Several phase I studies were initiated and showed a manageable safety profile, but only limited signs of clinical efficacy [75, 76]. Despite the preclinical rationale, these studies did not enroll biomarker-selected patients. Instead the studies used a conventional "all-comer" trial design that allowed anyone with certain tumor types to be enrolled, and no on-target pharmacodynamic biomarkers were assessed or used to stratify patient responses. This led to a decreased interest in continuing the clinical development program through pharmaceutical company–sponsored studies, and only smaller investigator sponsored studies were pursued. In a translational approach, palbociclib was profiled in a larger cell-line panel *in vitro*. Estrogen receptor positive (ER+) breast cancer cell lines were identified as being most sensitive to the compound [77], and ER+ status was identified as a predictive biomarker. Phase II and phase III data confirmed this hypothesis by showing a significant improvement in progression-free survival in this specific patient population [78, 79]. Palbociclib received accelerated approval by FDA for the treatment of postmenopausal ER+ and Her2-negative metastatic breast cancer [80].

How back-translation contributes to value generation and reduction of attrition is also exemplified when considering the development of second generation tyrosine kinase inhibitors. While this compound class consistently has induced good initial clinical responses, resistance to such inhibitors is commonly and rapidly observed in patients. The targeted kinase still represents an attractive target and is considered an oncogenic driver under these circumstances. Understanding the molecular mechanism and exact genetic characterization of such resistance development can open the possibility for novel drugs with improved profiles. The development of next generation ALK inhibitors will be used as an example here.

Alterations (point mutations, gene fusions) in the ALK gene were initially described in anaplastic large-cell non-Hodgkin lymphoma and later also shown in neuroblastoma and NSCLC [81–83]. The discovery of these alterations coincided with the early clinical development of crizotinib, which was originally designed as a kinase inhibitor against the hepatocyte growth factor receptor tyrosine kinase c-Met. During the dose-escalation phase of the compound's first-in-human study, ALK rearrangements in NSCLC were described and preclinical data with a specific mouse model proposed outstanding efficacy in this patient population [84]. Clinical trials enrolling only NSCLC patients with ALK alterations showed a significant improvement in progression-free survival and overall response rate from 7.0 to 10.9 months and from 45% to 75% in a first-line setting compared to standard chemotherapy and from 3.0 to 7.7 months and from 20% to 65% in a second-line setting, respectively [85, 86]. Despite good initial responses, many patients develop recurrent or resistant disease due to the appearance of new brain metastases or further genetic mutations within the ALK gene possibly due to *de novo* point mutations, gene amplifications, or activation of bypass pathways or due to a growth advantage for clones already harboring such alterations [87].

The development of brain metastases under crizotinib treatment was associated with its pharmacokinetic properties that limited penetration of the blood-brain barrier due to ABCB1-mediated drug efflux [88]. This hypothesis was confirmed in an ABCB1-knockout mouse model, and subsequently inhibitors with improved pharmacokinetic and pharmacodynamic properties were designed [89]. Among them, ceritinib and alectinib also inhibit ALK activity in cell lines harboring mutations that mediate crizotinib resistance and both show improved intracranial activity. Interestingly, alectinib is not transported via ABCB1 and therefore achieves higher CNS levels leading to higher response rates in NSCLC patients with brain metastases [90, 91]. Both second generation ALK inhibitors were granted accelerated approval for the treatment of ALK+ NSCLC [92, 93].

There are also examples for the back-translation of clinical observations into refined disease models in cardiovascular indications. One example is drug R&D in pulmonary arterial hypertension (PAH), a fatal syndrome in which remodeling of the pulmonary vasculature induces an increase of pulmonary vascular resistance (PVR) and pulmonary arterial pressure (PAP). This leads to pressure volume overload of the right ventricle (RV) of the heart and ultimately to RV dysfunction and failure. Mechanisms leading to onset and progression of PAH were not well understood in the past and still are not completely unraveled [94]. Animal models used in the attempt to develop new treatments in the past minimally reflected the pathologic mechanisms activated in human patients. One example of a routinely used animal model that only partially reflects the clinical situation is the monocrotaline rat model [95]. Only when the human disease was better understood by applying immunohistochemical (IHC) biomarkers to characterize molecular and morphological changes in patients' lung tissue samples, could new animal models like the Sugen 5416/hypoxia rodent be developed, which mimics several additional mechanisms of the human disease and produces a similar phenotype [96]. Today, this model is one of the gold standards in PAH drug R&D and the assessment of IHC biomarkers is one of the efficacy parameters in this preclinical model. Without back-translating clinical biomarker findings, this model would not have been developed.

These examples highlight how back-translation can easily be achieved through investigative approaches based on clinical characterization. Important prerequisites are to carefully analyze the novel clinical data, to understand the pathophysiology behind these results, and to continue with the selection and refinement of preclinical *in vitro* and *in vivo* models. This primarily requires a change in mindset and willingness to better understand these interrelations, with the results enabling a faster, cheaper, and better drug development process. New therapeutic agents can bring significant benefit to patients and generate sufficient revenue so that the additional investment into drug development can be sustained.

References

1 (a) Goodsaid, F. and Frueh, F. (2007). Biomarker qualification pilot process at the US Food and Drug Administration. *AAPS J.* 9 (1): E105–E108. (b) Merriam-Webster (1988). *Webster's Ninth New Collegiate Dictionary.* Springfield, MA: Merriam-Webster, Inc.

2 Kola, I. and Landis, J. (2004). Can the pharmaceutical industry reduce attrition rates? *Nat. Rev. Drug Discovery* 3 (8): 711–715.

3 DiMasi, J.A., Reichert, J.M., Feldman, L. et al. (2013). Clinical approval success rates for investigational cancer drugs. *Clin. Pharmacol. Ther.* 94 (3): 329–335.

4 Paul, S.M., Mytelka, D.S., Dunwiddie, C.T. et al. (2010). How to improve R&D productivity: the pharmaceutical industry's grand challenge. *Nat. Rev. Drug Discovery* 9 (3): 203–214.

5 Cook, D., Brown, D., Alexander, R. et al. (2014). Lessons learned from the fate of AstraZeneca's drug pipeline: a five-dimensional framework. *Nat. Rev. Drug Discovery* 13 (6): 419–431.

6 Morgan, P., Van Der Graaf, P.H., Arrowsmith, J. et al. (2012). Can the flow of medicines be improved? Fundamental pharmacokinetic and pharmacological principles toward improving phase II survival. *Drug Discovery Today* 17 (9–10): 419–424.

7 Thomas, D.W., Burns, J., Audette, J. et al. (2016). Clinical Development Success Rates 2006–2015: BIO Industry Analysis. https://www.bio.org/sites/default/files/Clinical%20Development%20Success%20Rates%202006-2015

%20-%20BIO,%20Biomedtracker,%20Amplion%202016.pdf (accessed 05 May 2019).

8 Wehling, M. (2006). Translational medicine: can it really facilitate the transition of research "from bench to bedside"? *Eur. J. Clin. Pharmacol.* 62 (2): 91–95.

9 Wehling, M. (2009). Assessing the translatability of drug projects: what needs to be scored to predict success? *Nat. Rev. Drug Discovery* 8 (7): 541–546.

10 Wendler, A. and Wehling, M. (2010). The translatability of animal models for clinical development: biomarkers and disease models. *Curr. Opin. Pharmacol.* 10 (5): 601–606.

11 Wendler, A. and Wehling, M. (2012). Translatability scoring in drug development: eight case studies. *J. Transl. Med.* 10: 39.

12 Peng, C. and Li, S. (2016). Chronic myeloid leukemia (CML) mouse model in translational research. *Methods Mol. Biol.* 1438: 225–243.

13 Hanukoglu, I. and Hanukoglu, A. (2016). Epithelial sodium channel (ENaC) family: phylogeny, structure-function, tissue distribution, and associated inherited diseases. *Gene* 579 (2): 95–132.

14 Gerlinger, M., Rowan, A.J., Horswell, S. et al. (2012). Intratumor heterogeneity and branched evolution revealed by multiregion sequencing. *N. Engl. J. Med.* 366 (10): 883–892.

15 Klein, C.A. (2013). Selection and adaptation during metastatic cancer progression. *Nature* 501 (7467): 365–372.

16 Klein, C.A. (2009). Parallel progression of primary tumours and metastases. *Nat. Rev. Cancer* 9 (4): 302–312.

17 Arnedos, M., Vicier, C., Loi, S. et al. (2015). Precision medicine for metastatic breast cancer—limitations and solutions. *Nat. Rev. Clin. Oncol.* 12 (12): 693–704.

18 Sweis, R.F., Drazer, M.W., and Ratain, M.J. (2016). Analysis of impact of post-treatment biopsies in phase I clinical trials. *J. Clin. Oncol.* 34 (4): 369–374.

19 Lotsch, J. and Geisslinger, G. (2010). Bedside-to-bench pharmacology: a complementary concept to translational pharmacology. *Clin. Pharmacol. Ther.* 87 (6): 647–649.

20 Gandini, S., Massi, D., and Mandala, M. (2016). PD-L1 expression in cancer patients receiving anti PD-1/PD-L1 antibodies: a systematic review and meta-analysis. *Crit. Rev. Oncol. Hematol.* 100: 88–98.

21 Robert, C., Long, G.V., Brady, B. et al. (2015). Nivolumab in previously untreated melanoma without BRAF mutation. *N. Engl. J. Med.* 372 (4): 320–330.

22 Postow, M.A., Chesney, J., Pavlick, A.C. et al. (2015). Nivolumab and ipilimumab versus ipilimumab in untreated melanoma. *N. Engl. J. Med.* 372 (21): 2006–2017.

23 Nghiem, P.T., Bhatia, S., Lipson, E.J. et al. (2016). PD-1 blockade with pembrolizumab in advanced Merkel-cell carcinoma. *N. Engl. J. Med.* 374 (26): 2542–2552.

24 Brahmer, J., Reckamp, K.L., Baas, P. et al. (2015). Nivolumab versus docetaxel in advanced squamous-cell non-small-cell lung cancer. *N. Engl. J. Med.* 373 (2): 123–135.

25 Carbognin, L., Pilotto, S., Milella, M. et al. (2015). Differential activity of nivolumab, pembrolizumab and MPDL3280A according to the tumor expression of programmed death-ligand-1 (PD-L1): sensitivity analysis of trials in melanoma, lung and genitourinary cancers. *PLoS One* 10 (6): e0130142.

26 Hansen, A.R. and Siu, L.L. (2016). PD-L1 testing in cancer: challenges in companion diagnostic development. *JAMA Oncol.* 2 (1): 15–16.

27 Topalian, S.L., Drake, C.G., and Pardoll, D.M. (2015). Immune checkpoint blockade: a common denominator approach to cancer therapy. *Cancer Cell* 27 (4): 450–461.

28 Aguiar, P.N. Jr., Santoro, I.L., Tadokoro, H. et al. (2016). The role of PD-L1 expression as a predictive biomarker in advanced non-small-cell lung cancer: a network meta-analysis. *Immunotherapy* 8 (4): 479–488.

29 Meng, X., Huang, Z., Teng, F. et al. (2015). Predictive biomarkers in PD-1/PD-L1 checkpoint blockade immunotherapy. *Cancer Treat. Rev.* 41 (10): 868–876.

30 Dranoff, G. (2012). Experimental mouse tumour models: what can be learnt about human cancer immunology? *Nat. Rev. Immunol.* 12 (1): 61–66.

31 Center for Disease Control and Prevention NCHS (2015). https://www.cdc.gov/nchs/data/hus/2015/019.pdf (accessed 05 May 2019).

32 Go, A.S., Mozaffarian, D., Roger, V.L. et al. (2013). Heart disease and stroke statistics—2013 update: a report from the American Heart Association. *Circulation* 127 (1): e6–e245.

33 Mozaffarian, D., Benjamin, E.J., Go, A.S. et al. (2016). Heart disease and stroke statistics—2016 update: a report from the american heart association. *Circulation* 133 (4): e38–e360.

34 Stewart, S., MacIntyre, K., Hole, D.J. et al. (2001). More 'malignant' than cancer? Five-year survival following a first admission for heart failure. *Eur. J. Heart Fail.* 3 (3): 315–322.

35 Steinberg, B.A., Zhao, X., Heidenreich, P.A. et al. (2012). Trends in patients hospitalized with heart failure and preserved left ventricular ejection fraction: prevalence, therapies, and outcomes. *Circulation* 126 (1): 65–75.

36 Owan, T.E., Hodge, D.O., Herges, R.M. et al. (2006). Trends in prevalence and outcome of heart failure with preserved ejection fraction. *N. Engl. J. Med.* 355 (3): 251–259.

37 Lee, D.S., Gona, P., Vasan, R.S. et al. (2009). Relation of disease pathogenesis and risk factors to heart failure with preserved or reduced ejection fraction: insights from the framingham heart study of the national heart, lung, and blood institute. *Circulation* 119 (24): 3070–3077.

38 Gheorghiade, M., Vaduganathan, M., Fonarow, G.C. et al. (2013). Rehospitalization for heart failure: problems and perspectives. *J. Am. Coll. Cardiol.* 61 (4): 391–403.

39 Nabel, E.G. and Braunwald, E. (2012). A tale of coronary artery disease and myocardial infarction. *N. Engl. J. Med.* 366 (1): 54–63.

40 McMurray, J.J. and O'Connor, C. (2014). Lessons from the TOPCAT trial. *N. Engl. J. Med.* 370 (15): 1453–1454.

41 Deswal, A., Richardson, P., Bozkurt, B. et al. (2011). Results of the randomized aldosterone antagonism in heart failure with preserved ejection fraction trial (RAAM-PEF). *J. Card. Fail.* 17 (8): 634–642.

42 Yusuf, S., Pfeffer, M.A., Swedberg, K. et al. (2003). Effects of candesartan in patients with chronic heart failure and preserved left-ventricular ejection fraction: the CHARM-preserved trial. *Lancet* 362 (9386): 777–781.

43 McMurray, J.J., Carson, P.E., Komajda, M. et al. (2008). Heart failure with preserved ejection fraction: clinical characteristics of 4133 patients enrolled in the I-PRESERVE trial. *Eur. J. Heart Fail.* 10 (2): 149–156.

44 van Veldhuisen, D.J., Cohen-Solal, A., Bohm, M. et al. (2009). Beta-blockade with nebivolol in elderly heart failure patients with impaired and preserved left ventricular ejection fraction: data from SENIORS (Study of Effects of Nebivolol Intervention on Outcomes and Rehospitalization in Seniors With Heart Failure). *J. Am. Coll. Cardiol.* 53 (23): 2150–2158.

45 Cleland, J.G., Taylor, J., Freemantle, N. et al. (2012). Relationship between plasma concentrations of N-terminal pro brain natriuretic peptide and the characteristics and outcome of patients with a clinical diagnosis of diastolic heart failure: a report from the PEP-CHF study. *Eur. J. Heart Fail.* 14 (5): 487–494.

46 FDA (2019). http://www.fda.gov/MedicalDevices/ProductsandMedicalProcedures/InVitroDiagnostics/ucm301431.htm (accessed 05 May 2019).

47 Pieske, B., Butler, J., Filippatos, G. et al. (2014). Rationale and design of the SOluble guanylate Cyclase stimulatoR in heArT failurE Studies (SOCRATES). *Eur. J. Heart Fail.* 16 (9): 1026–1038.

48 McMurray, J.J., Packer, M., Desai, A.S. et al. (2013). Dual angiotensin receptor and neprilysin inhibition as an alternative to angiotensin-converting enzyme inhibition in patients with chronic systolic heart failure: rationale for and design of the Prospective comparison of ARNI with ACEI to determine impact on global mortality and morbidity in heart failure trial (PARADIGM-HF). *Eur. J. Heart Fail.* 15 (9): 1062–1073.

49 Sabbah, H.N., Gupta, R.C., Kohli, S. et al. (2013). Chronic therapy with a partial adenosine A1-receptor agonist improves left ventricular function and remodeling in dogs with advanced heart failure. *Circ. Heart Fail.* 6 (3): 563–571.

50 Sabbah, H.N., Gupta, R.C., Kohli, S. et al. (2016). Chronic therapy with elamipretide (MTP-131), a novel mitochondria-targeting peptide, improves left ventricular and mitochondrial function in dogs with advanced heart failure. *Circ. Heart Fail.* 9 (2): e002206.

51 Caruso, A., Frances, N., Meille, C. et al. (2014). Translational PK/PD modeling for cardiovascular safety assessment of drug candidates: methods and examples in drug development. *J. Pharmacol. Toxicol. Methods* 70 (1): 73–85.

52 Yu, C.M., Wang, L., Chau, E. et al. (2005). Intrathoracic impedance monitoring in patients with heart failure: correlation with fluid status and feasibility of early warning preceding hospitalization. *Circulation* 112 (6): 841–848.

53 Small, R.S., Wickemeyer, W., Germany, R. et al. (2009). Changes in intrathoracic impedance are associated with subsequent risk of hospitalizations for acute decompensated heart failure: clinical utility of implanted device monitoring without a patient alert. *J. Card. Fail.* 15 (6): 475–481.

54 Small, R.S., Whellan, D.J., Boyle, A. et al. (2014). Implantable device diagnostics on day of discharge identify heart failure patients at increased risk for early readmission for heart failure. *Eur. J. Heart Fail.* 16 (4): 419–425.

55 Rademaker, M.T., Charles, C.J., Melton, I.C. et al. (2011). Monitoring of heart failure: comparison of left atrial pressure with intrathoracic impedance and natriuretic peptide measurements in an experimental model of ovine heart failure. *Clin. Sci. (London)* 120 (5): 207–217.

56 Schuppan, D. and Afdhal, N.H. (2008). Liver cirrhosis. *Lancet* 371 (9615): 838–851.

57 Lim, Y.S. and Kim, W.R. (2008). The global impact of hepatic fibrosis and end-stage liver disease. *Clin. Liver Dis.* 12 (4): 733–746.

58 Friedman, S.L. (2008). Mechanisms of hepatic fibrogenesis. *Gastroenterology* 134 (6): 1655–1669.

59 Hauff, P., Gottwald, U., and Ocker, M. (2015). Early to Phase II drugs currently under investigation for the treatment of liver fibrosis. *Expert Opin. Invest. Drugs* 24 (3): 309–327.

60 Schuppan, D. and Schattenberg, J.M. (2013). Non-alcoholic steatohepatitis: pathogenesis and novel therapeutic approaches. *J. Gastroenterol. Hepatol.* 28 (Suppl 1): 68–76.

61 Kampschulte, M., Stockl, C., Langheinrich, A.C. et al. (2014). Western diet in ApoE-LDLR double-deficient mouse model of atherosclerosis leads to hepatic steatosis, fibrosis, and tumorigenesis. *Lab. Invest.* 94 (11): 1273–1282.

62 Popov, Y., Patsenker, E., Fickert, P. et al. (2005). Mdr2 (*Abcb4*)-/- mice spontaneously develop severe biliary fibrosis via massive dysregulation of pro- and antifibrogenic genes. *J. Hepatol.* 43 (6): 1045–1054.

63 Ding, K., Liu, M.R., Li, J. et al. (2014). Establishment of a liver fibrosis model in cynomolgus monkeys. *Exp. Toxicol. Pathol.* 66 (5–6): 257–261.

64 Zhang, J.J., Meng, X.K., Dong, C. et al. (2009). Development of a new animal model of liver cirrhosis in swine. *Eur. Surg. Res.* 42 (1): 35–39.

65 Schuppan, D. and Pinzani, M. (2012). Anti-fibrotic therapy: lost in translation? *J. Hepatol.* 56 (Suppl 1): S66–S74.

66 Regev, A., Berho, M., Jeffers, L.J. et al. (2002). Sampling error and intraobserver variation in liver biopsy in patients with chronic HCV infection. *Am. J. Gastroenterol.* 97 (10): 2614–2618.

67 Zhou, K. and Lu, L.G. (2009). Assessment of fibrosis in chronic liver diseases. *J. Dig. Dis.* 10 (1): 7–14.

68 Lee, Y.A., Wallace, M.C., and Friedman, S.L. (2015). Pathobiology of liver fibrosis: a translational success story. *Gut* 64 (5): 830–841.

69 Castera, L. (2015). Noninvasive assessment of liver fibrosis. *Dig. Dis.* 33 (4): 498–503.

70 Schuppan, D. and Kim, Y.O. (2013). Evolving therapies for liver fibrosis. *J. Clin. Invest.* 123 (5): 1887–1901.

71 Sanyal, A.J., Friedman, S.L., McCullough, A.J. et al. (2015). Challenges and opportunities in drug and biomarker development for nonalcoholic steatohepatitis: findings and recommendations from an American Association for the Study of Liver Diseases-U.S. Food and Drug Administration Joint Workshop. *Hepatology* 61 (4): 1392–1405.

72 Sanyal, A.J., Brunt, E.M., Kleiner, D.E. et al. (2011). Endpoints and clinical trial design for nonalcoholic steatohepatitis. *Hepatology* 54 (1): 344–353.

73 Yu, Q., Geng, Y., and Sicinski, P. (2001). Specific protection against breast cancers by cyclin D1 ablation. *Nature* 411 (6841): 1017–1021.

74 Landis, M.W., Pawlyk, B.S., Li, T. et al. (2006). Cyclin D1-dependent kinase activity in murine development and mammary tumorigenesis. *Cancer Cell* 9 (1): 13–22.

75 Schwartz, G.K., LoRusso, P.M., Dickson, M.A. et al. (2011). Phase I study of PD 0332991, a cyclin-dependent kinase inhibitor, administered in 3-week cycles (Schedule 2/1). *Br. J. Cancer* 104 (12): 1862–1868.

76 Flaherty, K.T., Lorusso, P.M., Demichele, A. et al. (2012). Phase I, dose-escalation trial of the oral cyclin-dependent kinase 4/6 inhibitor PD 0332991, administered using a 21-day schedule in patients with advanced cancer. *Clin. Cancer Res.* 18 (2): 568–576.

77 Finn, R.S., Dering, J., Conklin, D. et al. (2009). PD 0332991, a selective cyclin D kinase 4/6 inhibitor, preferentially inhibits proliferation of luminal estrogen receptor-positive human breast cancer cell lines in vitro. *Breast Cancer Res.* 11 (5): R77.

78 Finn, R.S., Crown, J.P., Lang, I. et al. (2015). The cyclin-dependent kinase 4/6 inhibitor palbociclib in combination with letrozole versus letrozole alone as first-line treatment of oestrogen receptor-positive, HER2-negative, advanced breast cancer (PALOMA-1/TRIO-18): a randomised phase 2 study. *Lancet Oncol.* 16 (1): 25–35.

79 Turner, N.C., Ro, J., Andre, F. et al. (2015). Palbociclib in hormone-receptor-positive advanced breast cancer. *N. Engl. J. Med.* 373 (3): 209–219.

80 Beaver, J.A., Amiri-Kordestani, L., Charlab, R. et al. (2015). FDA approval: palbociclib for the treatment of postmenopausal patients with estrogen receptor-positive, HER2-negative metastatic breast cancer. *Clin. Cancer Res.* 21 (21): 4760–4766.

81 Morris, S.W., Kirstein, M.N., Valentine, M.B. et al. (1994). Fusion of a kinase gene, ALK, to a nucleolar protein gene, NPM, in non-Hodgkin's lymphoma. *Science* 263 (5151): 1281–1284.

82 Soda, M., Choi, Y.L., Enomoto, M. et al. (2007). Identification of the transforming EML4-ALK fusion gene in non-small-cell lung cancer. *Nature* 448 (7153): 561–566.

83 Chen, Y., Takita, J., Choi, Y.L. et al. (2008). Oncogenic mutations of ALK kinase in neuroblastoma. *Nature* 455 (7215): 971–974.

84 Soda, M., Takada, S., Takeuchi, K. et al. (2008). A mouse model for EML4-ALK-positive lung cancer. *Proc. Natl. Acad. Sci. U.S.A.* 105 (50): 19893–19897.

85 Shaw, A.T., Kim, D.W., Nakagawa, K. et al. (2013). Crizotinib versus chemotherapy in advanced ALK-positive lung cancer. *N. Engl. J. Med.* 368 (25): 2385–2394.

86 Solomon, B.J., Mok, T., Kim, D.W. et al. (2014). First-line crizotinib versus chemotherapy in ALK-positive lung cancer. *N. Engl. J. Med.* 371 (23): 2167–2177.

87 Katayama, R., Lovly, C.M., and Shaw, A.T. (2015). Therapeutic targeting of anaplastic lymphoma kinase in lung cancer: a paradigm for precision cancer medicine. *Clin. Cancer Res.* 21 (10): 2227–2235.

88 Costa, D.B., Kobayashi, S., Pandya, S.S. et al. (2011). CSF concentration of the anaplastic lymphoma kinase inhibitor crizotinib. *J. Clin. Oncol.* 29 (15): e443–e445.

89 Tang, S.C., Nguyen, L.N., Sparidans, R.W. et al. (2014). Increased oral availability and brain accumulation of the ALK inhibitor crizotinib by coadministration of the P-glycoprotein (ABCB1) and breast cancer resistance protein (ABCG2) inhibitor elacridar. *Int. J. Cancer* 134 (6): 1484–1494.

90 Gadgeel, S.M., Gandhi, L., Riely, G.J. et al. (2014). Safety and activity of alectinib against systemic disease and brain metastases in patients with crizotinib-resistant ALK-rearranged non-small-cell lung cancer (AF-002JG):

results from the dose-finding portion of a phase 1/2 study. *Lancet Oncol.* 15 (10): 1119–1128.

91 Kodama, T., Hasegawa, M., Takanashi, K. et al. (2014). Antitumor activity of the selective ALK inhibitor alectinib in models of intracranial metastases. *Cancer Chemother. Pharmacol.* 74 (5): 1023–1028.

92 Khozin, S., Blumenthal, G.M., Zhang, L. et al. (2015). FDA approval: ceritinib for the treatment of metastatic anaplastic lymphoma kinase-positive non-small cell lung cancer. *Clin. Cancer Res.* 21 (11): 2436–2439.

93 AACR (2016). Alectinib approved for ALK+ lung cancer. *Cancer Discovery* 6 (2): 115.

94 Velayati, A., Valerio, M.G., Shen, M. et al. (2016). Update on pulmonary arterial hypertension pharmacotherapy. *Postgrad. Med.* 128 (5): 460–473.

95 Gomez-Arroyo, J.G., Farkas, L., Alhussaini, A.A. et al. (2012). The monocrotaline model of pulmonary hypertension in perspective. *Am. J. Physiol. Lung Cell. Mol. Physiol.* 302 (4): L363–L369.

96 Vitali, S.H., Hansmann, G., Rose, C. et al. (2014). The Sugen 5416/hypoxia mouse model of pulmonary hypertension revisited: long-term follow-up. *Pulm. Circ.* 4 (4): 619–629.

16

Translational Medicine – A Paradigm Shift in Modern Drug Discovery and Development: The Role of Biomarkers

Giora Z. Feuerstein[1], Salvatore Alesci[2], Frank L. Walsh[3], J. Lynn Rutkowski[4], and Robert R. Ruffolo Jr.[5]

[1] *United States Department of Defense, Defense Threat Reduction Agency, Fort Belvoir, VA, USA*
[2] *Takeda Pharmaceuticals, Cambridge, MA, USA*
[3] *Wyeth Research, Collegeville, PA, USA*
[4] *Ossianix, Philadelphia, PA, USA*
[5] *Ruffolo Consulting, Spring City, PA, USA*

Drug Targets: Historical Perspectives

Drugs are natural or designed substances used deliberately to produce pharmacological effects in humans or animals. Drugs have been part of human civilizations for millennia. However, until the very recent modern era, drugs have been introduced to humans by empiricism and largely by serendipitous events such as encounters with natural products in search of food or by avoiding hazardous plants and animal products. The emergence of the scientific era in drug discovery evolved alongside the emergence of physical and chemical sciences at large, first as knowledge to distill, isolate, and enrich the desired substance from its natural environment, followed by deliberate attempts to modify natural substances to better serve the human needs and desires.

Scientific evolution throughout the past two centuries enabled identification of biologically active substances in humans (e.g. hormones) which were manipulated chemically to improve (potency, duration of action, and exposure), or to mitigate or abrogate undesirable actions. The cumulative knowledge of human, animal, and plant biology and chemistry provided the scientific foundation and technical capabilities to alter natural substances purposely in order to improve them. Such evolution marked the era of *forward pharmacology*. The era of forward pharmacology is about drug design that emanates from primary knowledge of the action of the biological target that has clear biological action.

The exponential progress in molecular biology since the mid-twentieth century, culminating in deciphering the complete human genome in the year 2000, brought the dawn of *pharmacogenomics* and the *reverse pharmacology*

Biomarkers in Drug Discovery and Development: A Handbook of Practice, Application, and Strategy, Second Edition. Edited by Ramin Rahbari, Jonathan Van Niewaal, and Michael R. Bleavins.

era. The reverse pharmacology era is defined by the need to clarify the biology and medical perspectives of the target so as to qualify it as a drugable and pharmaceutically exploitable for drug discovery and development. The pharmacogenomic era provides vast opportunities for selection of new molecular targets from a gamut of approximately 30 000 primary genes, over 100 000 proteins, and multiples of their translational and metabolomic products. Thus, the permutations in respect to opportunities for pharmacological interventions are unprecedented, vast, and most promising for innovative medicines.

The pharmacogenomics era as a source for drug targets also poses unprecedented hurdles in selection, validation, and translation into effective and safe drugs. New technologies continue to drive efficiency and robustness of mining the genomic drug discovery opportunities, but physiological and integrated biology knowledge is lagging. In this perspective, translational medicine and biomarkers research have taken center stage in validation of the molecular target for pharmaceutical exploitation.

In this chapter, we offer a utilitarian approach to biomarkers and target selection and validation that is driven by the translational medicine prospect of the target to become a successful drug target. We hereby offer classification and analytical process aimed to assess risk, innovation, feasibility, and predictability of success of translating novel targets into successful drugs. We provide clear definitions of the type of biomarkers that are the core of translational medicine and biomarkers research in modern pharmaceutical companies.

Biomarkers: Utilitarian Classification

Biomarkers are the stepping-stones for modern drug discovery and development [1–4]. *Biomarkers* are defined as biological substances or biophysical parameters that can be monitored objectively and reproducibly and used to predict drug effect or outcome. This broad definition is, however, of little utility to the pharmaceutical process since it carries no qualification for the significance and use of the biomarker. The following classes and definitions of biomarkers are therefore offered:

1. *Target validation*: biomarkers that assess the relevance and the potential for a given target to become the subject of manipulation that will modify the disease to provide clear therapeutic benefits while securing a sufficient therapeutic index of safety and tolerability.
2. *Compound-target interaction biomarkers*: biomarkers that define the discrete parameters of the compound (or biological) interaction with the molecular target. Such parameters include binding of the compound to the target, its residency time on the target, the specific site of interaction with the target, and the physical or chemical consequences to the target induced by the compound (or biological).

3. *Pharmacodynamic biomarkers*: biomarkers that predict the consequence(s) of compound (biological) interaction with the target. The pharmacodynamic biomarkers include events that are desired therapeutically and adverse events based on mechanism of action. Pharmacodynamic biomarkers can report on discrete molecular events that are proximal to the biochemical pathway that is modified by the manipulated target or remote consequences such as *in vivo* or clinical outcomes (morbidity or mortality). Pharmacodynamic biomarkers are diverse and frequently not obvious. Advanced and sophisticated bioinformatics tools are required for tracking the divergence and convergence of signaling pathways triggered by compound interaction with the target. A subset of the pharmacodynamic biomarkers are consequences induced by the compound outside its intended mechanism of action. Such pharmacodynamic effects are often termed "off-target" effects, as they are not the direct consequence of the compound interaction with the target. Usually, such pharmacodynamic events are due to unforeseen lack of selectivity or metabolic transformations that yielded metabolites not present (or detected) in the animals used for safety and metabolic studies prior to launch of the compound into human trials or into human use. These issues are not dealt with in this chapter.
4. *Disease biomarkers*: biomarkers that correlate statistically with the disease phenotype (syndrome) for which therapeutics are developed. Correlation of levels (in the circulation, other fluids or tissue) or expression patterns (gene, protein) in peripheral blood cells or tissues should signify disease initiation, progression, regression, remission, or relapse. In addition, duration of aberrantly expressed biomarkers could also be associated with risk for disease, even if the level of the biomarker does not change over time. Since disease biomarkers are defined by their statistical correlation to features of the disease, it is imperative that the clinical phenotyping is clearly defined. Stratification of all possible phenotypic variables is clearly a prerequisite for accurate assessment of the discrete relationships of the biomarker to the disease. Gender, age, lifestyle, medications, and physiological and biochemical similarities are often not sufficiently inclusive, resulting in a plethora of disease biomarker claims that are often confusing and futile.
5. *Patient selection*: biomarkers that are used for selection of patients for clinical studies, specifically proof-of-concept studies or confirmation Phase III clinical trials that are required for drug registration. These biomarkers are important in helping to select patients likely to respond (or conversely, not respond) to a particular treatment or a drug's specific mechanism of action, and potentially predict those patients who may experience adverse effects. Such biomarkers are frequently genetic (single-nucleotide polymorphism, haplotypes) or pharmacogenomic biomarkers (gene expression), but could be any of the primary pharmacodynamic biomarkers. Biomarkers for patient selection are now mainstream in exploratory clinical trials in

oncology, where genotyping of tumors in view of establishing the key oncogenic "driver(s)" is critical for the prediction of potential therapeutic benefits of modern treatments with molecular targeting drugs. The success of the new era of molecular oncology (as compared to the cytotoxic era) will depend largely on the ability to define these oncogenic signaling pathways via biomarkers such as phosphorylated oncogenes, or the functional state due to mutations that cause gain or loss of function.

6. *Adaptive trial design*: The objectives of adaptive design trials are to establish an integrated process to plan, design, and implement clinical programs that leverage innovative designs and enable real-time learning. The method is based on simulation-guided clinical drug development. In a first step, the situation is being assessed, the path forward and decision criteria defined, and assumptions analyzed. Adaptive trials have become an enabler strategy, and they work to integrate competing positions and utilities into a single aligned approach and to force much clearer articulation and quantification on the path forward. Once this framework is established, a formal scenario analysis that compares the fingerprints of alternative designs through simulation is conducted. Designs that appear particularly attractive to the program are further subjected to more extensive simulation. Decision criteria steer away from doses that are either unsafe or non-efficacious and aim quickly to hone in onto the most attractive dose range. Response-adaptive dose-ranging studies deploy dynamic termination rules (i.e. as soon as a no effective-dose scenario is established, the study is recommended for termination). Bayesian approaches are ideally suited to enable ongoing learning and dynamic decision-making [5]. The integrator role of adaptive trials is particularly strong in establishing links between regulatory accepted "confirm"-type endpoints and translational medicine's efforts to develop biomarkers. Search for biomarkers that may enable early decision-making needs to be read out early to gain greater confidence in basing decisions on them. A biomarker can be of value even if it only allows a pruning decision. These considerations highlight the importance of borrowing strength from indirect observations and use mathematical modeling techniques to enhance learning about the research question. For example, in a dose-ranging study, it is assumed that there should be some relationship between the response of adjacent doses, and this assumption can be used to model an algorithm. Both safety and efficacy considerations can be built into this model: ideally, integration of all efforts, from disease modeling in discovery to pharmacokinetic/pharmacodynamic (PK/PD) modeling in early clinical development to safety/risk and business case modeling in late development [4–7].

The utility of this system is represented in Figures 16.1 and 16.2, which suggest a semiquantitative scoring system that helps assess the strength of the program overall and identification of the areas of weaknesses in each of

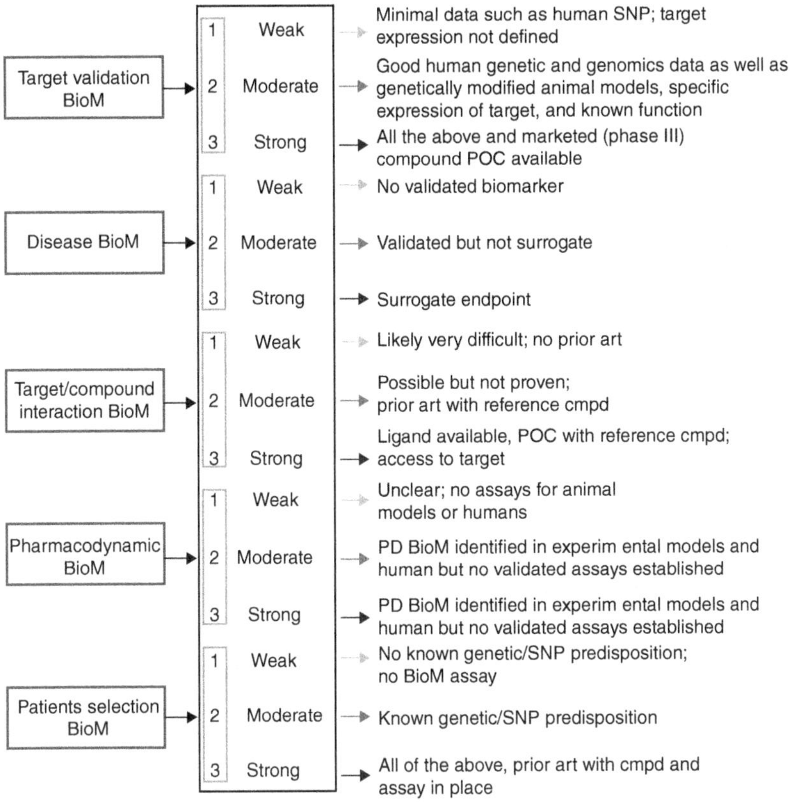

Figure 16.1 Criteria for biomarker scoring. SNP, single-nucleotide polymorphism.

the biomarkers needed along the compound (biological) progression path. Figure 16.3 illustrates the continuum of biomarkers research, validation, and implementation along the complete time line of drug discovery and development, including life-cycle management (Phase IV) and new indication (Phase V) when appropriate. Figure 16.3 illustrates the interfaces of translational medicine within the traditional drug discovery and development process, while Figure 16.4 represents the new model of "learn and confirm," where biomarkers figure prominently in driving the "learn" paradigm.

A program for which a STRONG score is established across all five biomarkers' specifications provides confidence of the likelihood for success from the biological and medical perspectives and is likely to result in a more promising development outcome. Similarly, it would be prudent to voice concerns regarding programs that score WEAK, especially if low scores are assigned to target validation, pharmacodynamic, and, in special cases, target-compound interaction (e.g. central nervous system target). This scoring system is complementary to other definitions of biomarkers based on certain

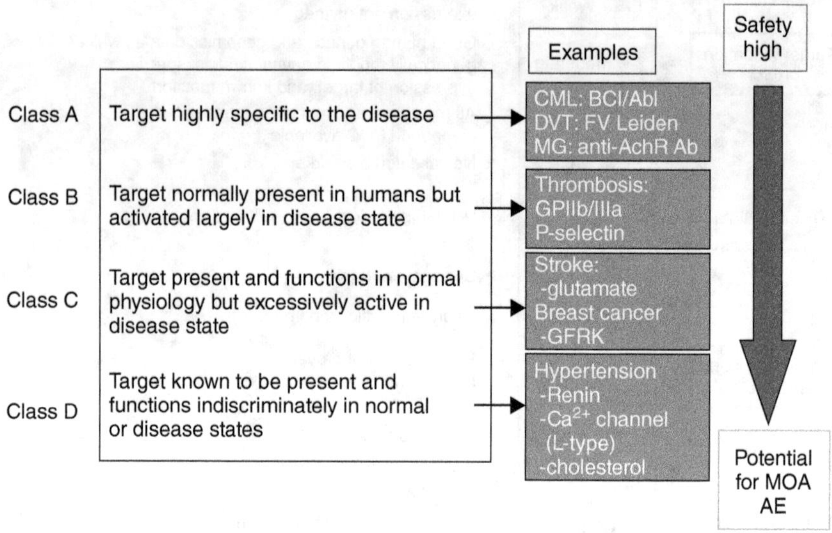

Figure 16.2 Type I biomarkers: target validation translational medicine perspectives. CML, chronic myelocytic leukemia; DVT, deep vein thrombosis; FV, factor V; MG, myasthenia gravis; AchR, acetylcholine receptor; GPIIb/IIIa, platelet integrin receptor; GFRK, growth factor receptor kinase; MOA, mechanism of action; AE, adverse effects.

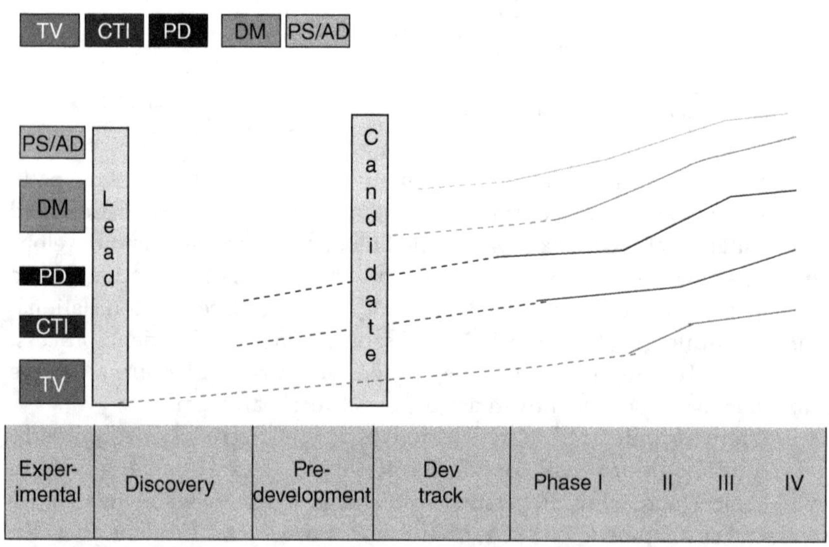

Figure 16.3 Building translational medicine via biomarker research. (*See insert for color representation of this figure.*)

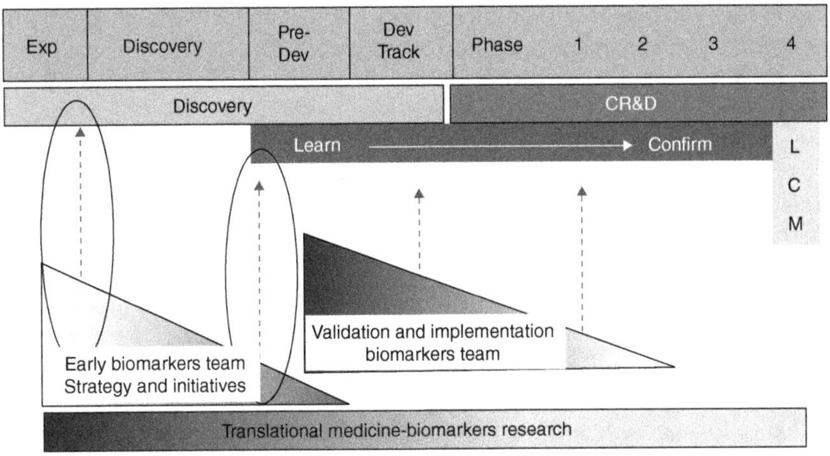

Figure 16.4 Translational medicine: biomarker implementation along the pipeline. Exp, exploratory phase; Pre-Dev, predevelopment track; CR&D, clinical research and development; LCM, life-cycle management.

needs. For example, surrogate biomarkers as defined by the US Food and Drug Administration (FDA) are markers that can be used for drug registration in lieu of more definitive clinical outcome data. Surrogate biomarkers are few and difficult to establish (e.g. blood pressure and cholesterol; Figure 16.2).

Principles of Target Selection

There are two essential guiding principles in the early selection process of molecular targets:

1. Modulating the target carries the prospect of unequivocal medical benefit (efficacy) to patients beyond a standard of care.
2. Benefits can be garnered while maintaining a sufficient level of safety that can be realized within the attainable compound exposure.

Such a mission is frequently unachievable, and hence establishing an acceptable *therapeutic index* is the practical goal for most drug development schemes. Commonly, a therapeutic index is established by calculation of the ratio of the maximum tolerated dose (MTD) and the minimum effective dose (MED) in animal efficacy and safety studies. In this light, targets selected for drug development can be classified with respect to risk assessment based on the following categories (Figure 16.2):

Class A: the target is only present and contributing to the disease process.

Class B: the target is present physiologically but in a nonactive form, but then is activated and contributes to the disease.
Class C: the target functions physiologically but in an augmented, uncontrolled fashion that contributes to the disease.
Class D: the target functions in normal states and indiscriminately in disease (e.g. no difference in target expression, functions, or distribution can be identified in disease as compared to the normal physiological state).

Class A: Disease-Specific Target

A disease-specific molecular target should be a molecule that operates only in the disease state and does not participate in physiological (normal) functions. Drug interaction with such targets should provide efficacy with the lowest chance for mechanism-based adverse effects when manipulated by drugs. Examples of such targets are genetic disorders, which result in either overactivity or loss of activity of the target. Such is the case in chronic myelogenous leukemia (CML), which results from aberrant recombination of DNA from chromosome 22 into chromosome 9 (Philadelphia chromosome), fusing the *Bcr* and *Abl* genes into an overacting tyrosine kinase, which drives oncogenic transformation. To cure the disease, potent and selective inhibitors of this aberrant kinase had to be discovered, a task that took over a decade to accomplish [8]. Such targets have the potential for a high safety profile. It is, however, important to note that this example may not necessarily represent the ultimate approach for this disease since the activity of the kinase (*Bcr/Abl*) is driven by the *Abl* kinase catalytic site, which is preserved in its physiological format. Thus, inhibitors of this target/kinase by drugs such as Gleevec may still carry the potential for interference in tissue and cells in which the *Abl* kinase is physiologically active.

Another example that is applicable to this category is a disease such as myasthenia gravis, where specific antibodies that block the acetylcholine receptors cause progressive muscle weakness. Specific neutralizing agents to these antibodies are likely to provide high efficacy in treating the disease, with the likelihood of fewer adverse effects [9] since such antibodies are not physiologically present in human beings. These examples are typical for type 1 class A target validation. The biomarkers that need to be established for this category should focus on validating the specificity of the target to the disease state.

Class B: Target Present Physiologically in a Nonactive Form but Is Activated and Contributes to the Disease

This class of targets has little or no discernible physiological activity in normal states, yet in certain pathophysiological situations, the target is presented, activated, and plays a role in a pathophysiological event. An example of such targets in the type 1 class B category is the P-selectin adhesion molecule. This adhesion molecule is normally cryptic within platelets and endothelial cells. Upon activation of these cells, P-selectin is presented on the surface of the cell and mediates adhesion interaction with its ligand, a mechanism believed to play a role in thrombosis and inflammation. Inhibitors of P-selectin binding to its ligand, the P-selectin glycoprotein ligand 1 (PSGL-1), are expected to provide clinical benefit with a lower likelihood of adverse events. To validate this situation, biomarkers that confirm the preferential role of the activated target in a pathophysiological process while maintaining little physiological function are essential.

However, one must be aware of potentially serious limitations to this approach where cryptic targets in the physiological state that are activated in pathological conditions and where inhibition of the target may not only provide for significant therapeutic benefit but where inhibition of such a target may also expose the patient to some other risk, such as loss of host defense from injury. Such is the case of the platelet adhesion integrin molecule, GPIIb/IIIa, which serves as the final common pathway for platelet aggregation. Interfering with activated GPIIb/IIIa binding to its ligand (e.g. fibrinogen) provides an effective and often life-saving therapy in patients with acute risk for thrombosis; however, chronic treatment with GPIIb/IIIa antagonists have not been particularly effective in providing benefits, due to the relatively high frequency of significant adverse effects due to bleeding since platelet adhesion to matrix protein is essential to seal bleeding sites in trauma and disease conditions. Thus, biomarkers for this class must establish the full physiological significance of the target in order to assess the therapeutic index of tolerability (benefits as well as risks).

Class C: Target Functions Physiologically but in an Augmented, Uncontrolled Fashion That Contributes to the Disease

This class of targets includes molecules that play an active role in normal physiological processes, some of which may be critical to health. Such is the neurotransmitter glutamate in the central nervous system, which is essential to cognition, memory, thought processes, and state of arousal. However, in ischemic stroke or asphyxia, glutamate release is uncontrolled and reaches to high levels over prolonged periods that are believed to be neurotoxic and likely contribute to neuronal death following a stroke. Inhibitors of glutamate release or antagonists of its action at various receptors are believed to carry the potential for effective treatment for stroke provided that the inhibition of excess release of the neurotransmitter can be achieved in a timely manner and only to an extent that preserves the physiological need of this transmitter, and over short periods (only that limited period where excess glutamate is neurotoxic). Such targets may be pharmaceutically exploitable when their manipulation is tuned carefully to the pathophysiological context.

Another example of a target in this category includes the human growth factor receptor kinase (hGFRK) inhibitor, Herceptin, which in certain cancers (e.g. breast cancer) is constitutively activated and participates in the oncogenic drive. Inhibition of the hGFRK, while clearly of therapeutic value in breast cancer, has also been associated with heart failure due to the physiological role of growth factor receptor kinase (GFRK) in the cardiac myocyte survival signaling pathway [10]. Thus, the biomarker challenge in modulation of class C targets of this nature is in identifying biomarkers that assess the needed "titration" for inhibition of the target activity only to an extent that is necessary for the maintenance of normal physiological function.

Class D: Target Maintains Physiological Functions in Normal and Disease States

This class of targets encompasses the largest group of molecular targets exploited so far by modern drugs. Many members of this class have yielded highly beneficial therapies. This class consists of molecular targets that are known to have important physiological functions, which cannot be differentiated within a disease context; that is, the target is not different in its expression levels (gene, protein) or signaling pathway in normal and disease states. A priori, such targets harbor the greatest risk for mechanism-based adverse effects, as there is no apparent reason to expect that modulation of the target in the disease state will spare the normal physiological function of the target.

Examples of such targets include the coagulation factor inhibitors (e.g. FIX, FXa, thrombin), which are critical to maintain a physiological level of homeostasis; hence, inhibition of these targets carries inherent bleeding liabilities. Similarly, antiarrhythmic drugs (e.g. amiodarone, lidocaine, dofetilide), while effective in treating life-threatening arrhythmias, all carry significant liability for mechanism-based pro-arrhythmic effects and the potential for sudden death. The biomarker challenges for this class are defining the fine balance needed between efficacy in the disease context and expected safety limitations. Biomarkers that define the acceptable therapeutic index are key to the successful utility of drugs that modulate such targets. However, targets in this class do not necessarily exhibit a narrow safety margin for clinically meaningful adverse effects. Significant examples are the L-type Ca^{2+} channel blockers. The L-type Ca^{2+} channel is an essential conduit of Ca^{2+} needed for "beat by beat" Ca^{2+} fluxes that secure precise rhythm and contractility of the heart, skeletal muscle, neuronal excitability, and hormone and neurotransmitter release. Yet L-type Ca^{2+} channel blockers are important and sufficiently safe drugs that are used to treat hypertension, angina, and cardiac arrhythmias with undisputable medical benefits. However, inherent to the L-type Ca^{2+} channel blockers in this class of targets are mechanism-based adverse effects associated with rhythm disturbances, hypotension, edema, and other liabilities.

Probably the best example for system specificity of physiological targets that provide major medical benefits with a high safety margin is the renin–angiotensin–aldosterone system (RAAS). The RAAS is an important blood pressure, blood volume, and blood flow regulatory system, yet its manipulation by several different pharmacological agents (rennin inhibitors, angiotensin I converting enzyme inhibitors, angiotensin II receptor antagonists) has yielded highly beneficial drugs that reduce risk of morbidity and mortality from hypertension, heart failure, and renal failure, despite the fact that the system does not demonstrate significant operational selectivity between normal and disease states (especially hypertension). However, mechanism-based hypotension and electrolyte disturbances can limit the therapeutic benefit of these drugs, and elicit significant adverse effects when the RAAS is excessively inhibited [11]. The biomarker challenge for these targets is to define the relative or preferential role of the target in its various physiological activities, where minor manipulation in one organ might provide sufficient therapeutic potential while providing a low likelihood for adverse effects that result from more substantial inhibition of the same target in other organs.

Summary

The analysis and classification offered in this chapter regarding biomarkers in drug discovery and development aim to highlight the need for careful study and analysis of the significance of the target selected for therapeutic intervention as the first crossroad for success or failure in the development of effective and safe drugs [12]. The analysis and utility of biomarkers along the process of drug discovery and development have become an integral part of the "learn and confirm" paradigm of drug discovery and development in leading pharmaceutical organizations. Such analyses are useful to guide the "learn phase" in search for biomarkers that can better assess the benefits and risks associated with manipulation of the molecular target.

The scope of this chapter does not allow for a detailed review of the "learn and confirm" paradigm, for which the readers are directed elsewhere [13, 14]. Various technological and strategic activities are needed to establish the biomarker strategies for the various targets described. The need to address these issues via biomarker research, validation, and implementation beginning at the very early stages of the drug discovery and development process is emphasized. In the pharmaceutical setting, it means beginning efforts to identify biomarkers for all five categories listed above. Such efforts could begin even before a tractable compound (biological) is in hand, a time where target validation is a clear focus of the program. As a compound becomes available, compound–target interaction, pharmacodynamic (efficacy and safety) biomarkers, and strategies for patient selection and adaptive design needs must be explored. At the onset of the first-in-human studies, all strategies, plans, and biomarker research should be as well worked out as possible. We believe that fundamental changes in the structure, function, and interfaces of pharmaceutical R&D are urgently needed to provide a key role for translational medicine and biomarkers research toward more successful discovery and development of innovative medicines.

References

1 Biomarker Definition Working Group (2001). Biomarkers and surrogate biomarkers endpoints: preferred definitions and conceptual framework. *Clin. Pharmacol. Ther.* 69: 89–95.

2 Trusheim, R., Berndt, E.R., and Douglas, F.L. (2007). Stratified medicine: strategic and economic implications of combining drugs and clinical biomarkers. *Nat. Rev. Drug Discovery* 6: 287–293.

3 Feuerstein, G.Z., Rutkowski, J.L., Walsh, F.L. et al. (2007). The role of translational medicine and biomarkers research in drug discovery and development. *Am. Drug Discovery* 2: 23–28.

4 FDA (2004). Challenge and opportunity on the critical path to new medical products. https://web.archive.org/web/20090512144844/www.fda.gov/oc/initiatives/criticalpath/whitepaper.pdf (accessed 26 April 2019).

5 Berry, D.A. (2006). Bayesian clinical trials. *Nat. Rev. Drug Discovery* 5: 27–36.

6 Gallo, P., Chuang-Stein, C., Dragalin, V. et al. (2006). Adaptive design in clinical drug development: an executive summary of the PhRMA Working Group. *J. Biopharm. Stat.* 16: 275–283.

7 Krams, M., Lees, K.R., Hacke, W. et al. (2003). Acute stroke therapy by inhibition of neutrophils (ASTIN): an adaptive dose–response study of UK-279,276 in acute ischemic stroke. *Stroke* 34: 2543–2548.

8 Kurzrock, R. (2007). Studies in target-based treatment. *Mol. Cancer Ther.* 6 (9): 2385.

9 Hampton, T. (2007). Trials assess myasthenia gravis therapies. *JAMA* 298 (1): 29–30.

10 Chien, K.R. (2006). Herceptin and the heart: a molecular modifier of cardiac failure. *N. Engl. J. Med.* 354: 789–790.

11 Hershey, J., Steiner, B., Fischli, W., and Feuerstein, G.Z. (2005). Renin inhibitors: an antihypertensive strategy on the verge of reality. *Drug Dev. Today* 2: 181–185.

12 Simmons, D. (2006). What makes a good anti-inflammatory drug target? *Drug Discovery Dev.* 5–6: 210–219.

13 Gombar, C. and Loh, E. (2007). Learn and confirm. *Drug Discovery Dev.* 10: 22–27.

14 Sheiner, L.B. (1997). Learning versus confirming in clinical drug development. *Clin. Pharmacol. Ther.* 61: 275–291.

17

Clinical Validation and Biomarker Translation

Ji-Young V. Kim, Raymond T. Ng, Robert Balshaw, Paul Keown, Robert McMaster, Bruce McManus, Karen Lam, and Scott J. Tebbutt

PROOF Centre of Excellence and Biomarkers in Transplantation Team, Vancouver, BC, Canada

Introduction

Throughout history, biological and pathogenic processes have been measured to monitor health and disease. The sweet taste of urine was recognized, since antiquity, as an indication of the disorder now known as diabetes. Midwives in the medieval period knew that if mothers detected too salty a taste from the forehead of a newborn, then the child was "bewitched" and would die shortly [1]. We now know these children had cystic fibrosis. Any such measures were forerunners of what are now popularly referred to as biomarkers. With the advent of clinical technology that allows detection of various types of molecules from biospecimens, modern efforts in biomarker discovery still aim to predict, prognose, and diagnose health and disease development and progression.

The National Institutes of Health (NIH) defines a biomarker as any characteristic that reflects a normal biological process, pathogenic process, or a pharmacologic response to an intervention [2]. These biomarkers may be constituents of molecular, physiological, or structural features of biological systems. The focus on using biomarkers in clinical decision-making and diagnosis has expanded, and biomarkers are playing an ever-increasing role in drug development processes and for regulatory decision-making.

One popular use of biomarkers is to mitigate the risks and costs associated with drug development. For example, costs and time can be reduced when biomarkers help to identify the optimal patient groups early in trials, monitor drug safety and serve as timely and convenient clinical study endpoints or outcomes [3]. Also, biomarkers can generate proof-of-concept or dose range information more rapidly than a downstream clinical endpoint [4]. Ultimately,

Biomarkers in Drug Discovery and Development: A Handbook of Practice, Application, and Strategy,
Second Edition. Edited by Ramin Rahbari, Jonathan Van Niewaal, and Michael R. Bleavins.

by reducing the investment to demonstrate a drug's safety and efficacy, biomarkers may greatly reduce the costs and risks of performing a clinical trial.

An emerging challenge now lies in selecting biomarkers that are clinically useful (i.e. positively affecting the clinical outcome of interest) and determining the most appropriate context for their clinical use.

Biomarker Discovery to Clinical Use

Biomarker development is usually initiated by a specific, clinically-unmet need and the health economics case built around solving that need. A clinically useful biomarker must favorably affect clinical outcomes such as decreasing toxicity or increasing survival [5]. Then, the decision to develop biomarker candidates is made based on its potential to contribute cost-effectively to disease management [6] (Figure 17.1).

The initial biomarker discovery is usually performed on human disease and healthy samples from a well-defined target population, using a suitable technology measurement platform. Because these platforms generally produce large amounts of complex data, high-quality data analytical and statistical approaches are required to examine and extract useful information. Upon generating a list of potential targets, it is critical to establish parameters for rational, statistically-sound, and evidence-based criteria for selecting or rejecting candidate biomarkers [8, 9]. When promising biomarkers are found, their reproducibility in an external, independent cohort should be tested [7].

The next phase, validation and qualification, begins with transferring the biomarkers onto a clinical platform with proper analytical validation of the assay. Analytical validation demonstrates the accuracy, precision, and reproducibility of the test: How well does the test measure what it claims to measure? How reliably? Once the biomarker test is established, its clinical validation and clinical utility must be demonstrated. Clinical validation is the documented process demonstrating the evidentiary link of a biomarker measurement with a clinical endpoint(s) [10]. Clinical utility measures the impact of incorporating the biomarker assay in clinical practice in terms of improved disease management and/or patient outcomes. These processes, whereby biomarkers are translated from discovery to clinical use, is customized according to biomarker type, use, variability, and prevalence. The goals for the validation and qualification is proper contextualization of the test for the management of the disease in question.

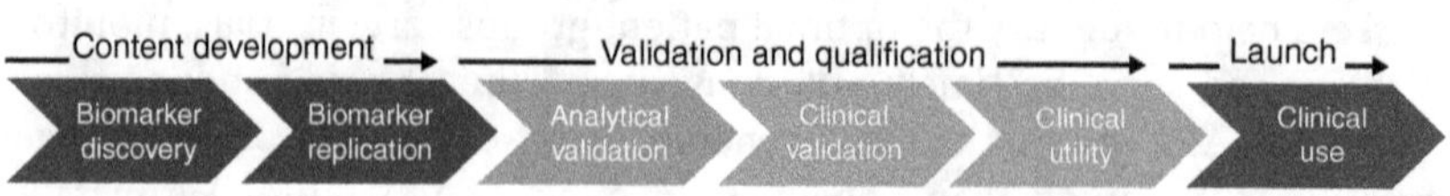

Figure 17.1 Biomarker development pipeline. Source: Adapted from Sin et al. 2015 [7].

In this chapter, we examine the current practice of biomarker validation and qualification and provide developing efforts in the regulatory sectors in establishing guidelines and standards that can facilitate more efficient biomarker discovery and development pipelines.

Biomarker Validation and Qualification: Current Practice

Validation is a process of assessing the biomarker and its measurement performance characteristics, and determining the range of conditions under which the biomarker will give reproducible and accurate data [3]. The evidentiary process of proving a linkage between the biomarker and a clinical end point is termed **Qualification** [3].

Development and validation of biomarkers are as difficult as the development and approval of a new drug; only 3–5% of the approximately 30–50% of biomarkers that are coupled to drug development programs reach the clinic [11–13]. When co-developing a drug and a biomarker, it is helpful to use the **fit-for-purpose** (FFP) approach to biomarker method validation (Box 17.1), in which methods can be refined throughout the development phases of the validation [14, 15]. Utilizing the FFP strategy, various questions can be asked at the beginning of any biomarker research and development project. For what reason is the biomarker(s) being identified? Is the clinical validation going to validate the biomarker for specificity, sensitivity, and reproducibility for diagnostic purposes, or to serve as a surrogate endpoint? What socioeconomic and other business-critical decisions will be made based on the biomarker data? These questions help determine the level of confidence required for the biomarker validation. For example, in the case of developing surrogate biomarkers to replace clinical endpoints, candidate biomarkers would evolve toward the point of surrogacy with greater regulatory scrutiny and evidentiary requirements as the research project advances through sequential phases of validation.

Box 17.1 Fit-for-Purpose Strategy

Despite the lack of universal guidelines or agreement on the specific requirements for biomarker assay development and validation, there is a general consensus in the biomarker research community that the foundation of validation efforts is to ensure that the biomarker(s), or the assay, is reliable for its intended use [16]. This principle is now commonly referred to as the fit-for-purpose validation strategy. This approach embraces the notion that depending on the intended purpose of the biomarker, the objectives and processes or types of validation can be different [15, 17]. Dose selection, early efficacy assessment or candidate selection, and surrogacy development are all examples of biomarker clinical purposes, each of which have differing

(Continued)

Box 17.1 (Continued)

levels of validation requirement. The risks and consequences may be different depending on the purpose, even when the same biomarker is used. According to the position paper by Lee and colleagues, the fit-for-purpose biomarker assay validation can be separated into four continuous iterative activities:

- the pre-validation process during which the definition of purpose and selection of the candidate assay occurs – perhaps the most critical activity
- the exploratory validation process where the goal is to assemble all the appropriate reagents and components and write the method validation plan and decide upon the final classification of the assay
- the advanced validation process, which characterizes the formal performance of the assay with regard to its intended use, leading to the evaluation of fit-for-purpose culminating in writing a standard operating procedure
- the in-study validation process that allows further assessment of fit-for-purpose and the robustness of the assay in the clinical context and enables identification of patient sampling issues, such as collection, storage, and stability.

The driver of this process is one of continual improvement, which may necessitate a series of iterations that can lead back to any one of the earlier stages [16].

The classification and validation of a biomarker are context-specific, and the validation criteria required are dependent on the intended use of the biomarker. Any tests that are to be used in isolation for decision-making require particularly stringent validation, whereas drug efficacy biomarkers that are typically used in groups or involve subsequent testing may require less stringency since the consequence of a false indication of efficacy is less harmful to the patient [18].

Biomarker Validation and Qualification: Regulatory Development and Case Studies

The *Guidance for Industry: Pharmacogenomic Data Submission*, published by the US Food and Drug Administration (FDA), has helped first introduce and define exploratory and valid biomarkers [19]. The FDA defines a valid pharmacogenomic biomarker as one that is measured in an analytical test system with well-established performance characteristics and for which there is an established scientific framework or body of evidence that elucidates the toxicological, pharmacological, or clinical significance of the test result. The FDA further classifies valid biomarkers as “probable” and “known,” describing the level of confidence that they attain through the validation process. **Probable valid** biomarkers may not yet be widely accepted or validated externally but appear to have predictive value for clinical outcomes, whereas **known valid** status is achieved by those that have been widely accepted by the scientific community.

Traditionally, the validity of clinical biomarkers has become established in a lengthy process through consensus and test of time [20]. As technological advances have drastically increased biomarker discovery rates, the FDA has put forth efforts in establishing basic practice that can facilitate the development of biomarkers. Great efforts are now being undertaken to accelerate the acceptance of biomarkers from exploratory to valid with a goal to streamline the translation of biomarkers from basic science and discovery to clinical use [21].

There are two other pathways through which biomarkers can be accepted by the FDA for use in therapeutic product development [22] (Table 17.1). First is

Table 17.1 Examples of companion diagnostics and qualified biomarkers approved by the FDA [23].

Type	Drug	Diagnostic	Indication	Methodology
CD	Exjade	Ferriscan	Thalassemia	MRI
CD	Erbitux	Cobas KRAS test	Colorectal Cancer	RT-PCR
CD	Vectibix	KRAS RGQ kit	Colorectal Cancer	RT-PCR
CD	Vectibix	EGFR PharmDx Kit	Colorectal Cancer	IHC
CD	Herceptin	INFORM HER2/NEU	Breast Cancer	FISH
CD	Herceptin	PATHWAY® ANTI-HER-2/NEU	Breast Cancer	IHC
CD	Herceptin	SPOT-Light® HER2 CISH Kit	Breast Cancer	CISH

Type	Submitter	Diagnostic	Qualified for specific context of use
QB	Mycoses Study Group	Platelia Aspergillus EIA	Serum/bronchoalveolar lavage fluid biomarker: Galactomannan
QB	Chronic Obstructive Pulmonary Disease Biomarker Qualification Consortium (CBQC)	Glycated Serum Protein Assay	Plasma biomarker: Fibrinogen
QB	Polycystic Kidney Disease Outcomes Consortium	NA	Imaging Biomarker: Total Kidney Volume (TKV)

CD, companion diagnostic; QB, qualified biomarker; MRI, magnetic resonance imaging; RT-PCR, real-time polymerase chain reaction; FISH, fluorescence *in situ* hybridization; IHC, *in situ* hybridization; CISH, chromogenic *in situ* hybridization.

co-development of the biomarker with drug development, which is filed under an investigational new drug (IND) application. A pharmaceutical developer can engage directly with FDA to reach agreement on the use of a particular biomarker in a given drug's development program. Accepted biomarkers can serve as surrogate endpoints or be used as criteria for restricting use in the population. This pathway can be efficient for drug developers, but because the biomarker's development takes place in a confidential setting, the FDA and drug sponsor are not subject to the broad input of the scientific community, and therefore only have limited contribution to ongoing agency policy development.

The second, fairly new, mechanism allows pharmaceutical developer, patient- or disease-specific foundation, health research organization, or consortium to request regulatory qualification of a biomarker for a particular context of use (COU) through the FDA's Biomarker Qualification Program (BQP) [22] (Table 17.1). The FDA defines qualification as "a conclusion that within the stated COU, a biomarker can be relied upon to have a specific interpretation and application in drug development and regulatory review" [24]. This mechanism offers flexibility in the ranges of areas where the biomarker can be used and a biomarker can have broader application across multiple therapeutic areas. This is particularly advantageous for biomarkers that require aggregation of disparate data sources to provide sufficient evidence of biomarker utility. Once positive qualification decisions are made, the FDA's Center for Drug Evaluation and Research (CDER) publicly communicates through regulatory guidance documents.

Pathways for Regulatory Approval of Diagnostics

FDA-approved biomarkers can be sold as in vitro diagnostics (IVD) for their specified purpose and intended use after obtaining premarket regulatory clearance from the FDA to sell them as **diagnostic kits** (a packaged product) or **companion diagnostics** [25]. In the United States, the FDA Center for Devices and Radiological Health (CDRH) regulates the premarketing process by placing IVDs into three "classes" based on their risks (Box 17.2). In this context, the risk refers to the impact of the results generated by the test on patients, particularly false-negative or false-positive results. Individual components of diagnostic tests can be marketed separately as analyte-specific reagents (ASRs). ASRs, individually, are regulated by the FDA but are exempt from premarket notification. This can allow early-market penetration and enhance early adoption of the technology [26].

Box 17.2 FDA IVD Classes [25]

Class I devices are of low risk and have the least regulatory controls. Some examples include dental floss, tongue depressors, arm slings, and handheld surgical instruments. Most Class I devices are exempt from the premarket notification 510(k).

Class II devices are considered higher risk devices and must prove the device's safety and effectiveness. For example, X-ray systems, gas analyzers, and surgical drapes are classified as Class II devices. Manufacturers introducing Class II medical devices must submit a 510(k) to the FDA. The 510(k) identifies the proposed medical device and compares it to an existing medical device to demonstrate that it is substantially equivalent and at least as safe and effective.

Class III devices are the highest risk devices and must be approved by the FDA before they are marketed. Class III medical devices that are high risk or novel devices for which no "predicate device" exist require clinical trials of the medical device. Examples of Class III IVDs include HercepTest, for detection of human epidermal growth factor receptor 2 positive (HER2+) breast cancer, and DxS' TheraScreen KRAS, for detection of mutated kirsten rat sarcoma viral oncogene homolog (KRAS) in metastatic colorectal cancer.

Alternatively, biomarker diagnostics can be filed as **laboratory-developed tests (LDTs)** and sold as the in-house performance of the test as a service or sold as a kit after obtaining premarket regulatory clearance from the FDA. The FDA does not typically review these "home brewed" tests, but they are subject to the test performance standards of Clinical Laboratory Improvement Amendments (CLIA). Under CLIA regulations, certified laboratories are required to adhere to standards of quality control, personnel qualifications, and documentation, as well as to analytically validated tests, but the clinical validation of the tests is not required. The level of scrutiny of CLIA inspections and certification requirements will depend on the complexity of the tests performed. One example of an LDT is Genomic Health's Oncotype DX, a test for the detection of 21 genes in a patient's breast tumor tissue that indicate both the likely benefit of chemotherapy to patients who have node-negative, estrogen receptor positive breast cancer, and the chance of cancer recurrence in early-stage breast cancer [27].

The FDA route has not been the path of choice for most diagnostic tests to date, as it is the costliest and most time-consuming option. Indeed, the majority of biomarker diagnostics are currently marketed as LTDs [28]. Such distribution may reflect the difficulty of obtaining adequately controlled clinical data to support the IVD. Bypassing the rigorous regulatory scrutiny to expedite the market placement of biomarker diagnostics may lower that particular hurdle, but there can be a major challenge in following this route. Clinical adoption is one of the biggest barriers to the success of a new test (and ultimately the drug), and meeting the high standards of a regulator like the FDA is the right step forward in convincing clinicians and payers of a test's validity and clinical utility. That being said, FDA-approved tests can cost laboratories significantly more than unapproved generic versions of these tests and therefore can also influence market uptake since an expensive patient selection test places higher hurdles for an IVD's adoption [28].

While certain LDTs have undoubtedly brought benefits to many patients, the FDA has raised concerns about the safety and the quality of LDTs and has recently put forward suggestions for the need for appropriate oversight [29]. The FDA has highlighted that the current oversight framework is inadequate and there is a need to address several serious concerns, including the lack of evidence supporting the clinical validity of tests and lack of adverse event reporting [30]. Most fundamentally, any clinical tests administered or ordered by health care providers should generate reliable results that can be used to make decisions that can potentially affect patients' health. However, lack of clinical validity of some LDTs was shown to result in a failure to detect life-threatening diseases and cause patients to elect suboptimal therapeutic options [30]. In addition, device adverse event reporting is not required for LDTs. This is an important tool both for manufacturers, to identify problems with their tests that was not anticipated, and for FDA, to identify potential public health issues and to take appropriate regulatory action. Currently, information on adverse events associated with LDTs is not systematically collected or reported. An in-depth report on the FDA's increasing awareness of harms caused by the lack of LDT regulation is found in The Public Health Evidence for FDA Oversight of Laboratory Developed Tests: 20 Case Studies [30].

Amur and colleagues suggest that it is frequently of benefit to seek discussions with the FDA proactively during biomarker development in order to find out which branch of the "FDA might be interested in a particular diagnostic topic, to understand the agency parameters and expectations for data submission, and to gain buy-in for an investigational or confirmatory trial" [22, 28]. Such discussion can take place during a pre-investigational device exemption (IDE) session or through the voluntary exploratory data submission (VXDS) process that can provide informal peer-review feedback to help shape the strategic thinking and prevent delays in reviews of future formal FDA submissions.

Biomarker Validation and Qualification: Challenges

As biomarker candidates continue to pour out of research laboratories, it has become increasingly evident that validation is a difficult and complex process.

There are a multitude of general and specific considerations and challenges with evaluation of the analytical performance of the test to ensure the biomarker test is reliable, reproducible, and accurate (analytical validation), ensure the biomarker is associated with the clinical outcome of concern (clinical validation), and determine that the biomarker is appropriate for the proposed use (clinical utility), all of which contribute to the clinical value of a biomarker. Specifically, in this section, we address challenges with biological variability; sample quality and accuracy and precision in handling and processing of the samples; and the FDA's evidentiary standards for determining clinical relevance and specific COU.

Biological Variability

Starting with the population selection process, variations at a biological level can lead to discernible differences in body fluid and tissue compositions and biomarker measurements [31]. As such, basic criteria such as sex, age, medical history, and severity of the underlying condition are all potential sources of variability. Further biological obstacles arise from the tissue availability and the complexity of the biomarkers of interest (e.g. DNA, RNA, and proteins). For example, among recognized positive or negative predictive biomarkers in oncology, genetic alterations such as mutations, amplifications, or translocations were found to be more concordant between primary tumors and associated metastases than protein level expressions [32]. This observation probably reflects the complexity of protein biochemistry, including posttranslational modifications and catabolism, which can hinder the development of robust and accurate analytical standards for certain types of biomarkers. Biological variability may be difficult to assess, but it is important to control for this factor as the overall noise of a sample is the sum of both analytical and biological variability [15].

Technical Factors

Even with the establishment of a relatively homogeneous cohort for external validation or clinical trials, results from the intended biomarker assays are valid only if sample integrity is maintained and reproducible from sample collection through analysis [14]. Major components of assay and technical validation include reference materials, quality controls, quality assurance, accuracy, precision, sensitivity, and specificity. The FDA outlines comprehensive guidance on analytical validation processes appropriate to biomarkers [33].

Reproducibility is the key to successful biomarker validation. It is important to establish standardized operating protocols (SOPs) that can minimize variability in collecting and processing biological fluids or tissues. Prior to the start of a validation phase, the team needs to decide on the sampling procedure, sample quality parameters, and sample processing and provide training to personnel performing specimen collection. Once the samples are collected, systematic monitoring of their quality over time should also be established. Random tests can be conducted to ensure the short-term, benchtop, or long-term stability of the samples to uphold the integrity of the biolibrary [34].

In practice, during external and clinical validations, it is likely that the independent collaborators utilize a different set of SOPs. Therefore, it is necessary to establish specific guidelines that would help minimize site-to-site variability. There are several ways to achieve this. Provision to the laboratories and centers of kits that contain all chemicals needed for the processing of the sample can potentially reduce the risk of batch effects. For example, assessment of accuracy and reproducibility of the diagnostic evaluation of HER2 using PathVysion HER2 DNA probe kit by the North Central Cancer Treatment Group (NCCTG), which is consisted of over one hundred CLIA-accredited laboratories, showed an overall concordance of 82% [35]. Such findings indicate that attempts to standardize biomarker measurement methods can ensure uniformity and reproducibility of data. Similarly, it may be feasible to collect a separate set of samples internally (to minimize sample collection variability) and send them to a collaborator's site for external validation using independent processing of samples.

The analytical validity of the test should be closely monitored along with the clinical validity of the biomarker throughout the biomarker development and that any deviation from the SOP is noted [6]. These features, including bias or deviation from protocols or batch processing, can then later be accommodated by statisticians for better modeling and thereby improve the chance of successful clinical translation of biomarkers.

Technological Advancements and Emergence of "Omics"

Technological improvements generally have a positive impact on clinical practice, but sometimes these improvements can result in confusion when making clinical decisions. For example, one study reported discrepancies in tumour mutation-status assessment by standard methods at local laboratories compared with more sensitive TaqMan mutation-specific probes for KRAS screening performed in a different laboratory [36]. In this study, tissue samples from 60 patients with metastatic colorectal cancer that were assessed locally were defined as wild-type KRAS at codon 12, when 10% of these samples were actually found to harbor KRAS mutation when reassessed at a central laboratory. Clinicians then have to arbitrarily decide whether or not the differences in sensitivity, considering analytical and biological variables, represent clinically meaningful information that should influence medical decisions. For instance, should the detection of a low-frequency KRAS mutation be a contraindication to anti-epidermal growth factor receptor (EGFR) antibody therapy in patients with metastatic colon cancer? [37]. These decisions can have important impact on patients and the therapeutic treatments they receive.

Recent technological advancements in high-throughput "omics" techniques, such as genomics, transcriptomics, proteomics, and metabolomics, have catalyzed discoveries of novel biomarkers. Genomics/transcriptomics focus on the sequencing and analysis of genes and mRNA transcripts using sequencing platforms (e.g. RNA-seq), whereas proteomic analysis characterizes the proteins encoded by the genome using tools like mass spectrometry and protein arrays. Now, instead of having to test individual molecules one by one for their biomarker potential, a large number of possible biomarkers can be screened simultaneously, enabling large-scale analysis of complex biological systems to identify candidate biomarkers and characterize relevant pathways.

The biomarker field is now looking at the predictive power of sets of multiple factors analyzed together, biosignatures, in hopes of finding better biomarkers that are able to confer much more information than the measurement of a single gene [38]. Such efforts have led discoveries of biosignatures in various disease types and a few of them have made it into clinics, including AlloMap and ImmuKnow (Box 17.3). While most omics type analyses have generated valuable findings, the complex workflow of computational approaches and the irreproducibility of some candidate markers in subsequent studies suggested that such analyses are, in general, insufficiently robust to guide therapeutic choices at present, and the intermediate values from such biomarker assay techniques are difficult to interpret [39]. Nevertheless, comprehensive broad-scale assessments of various biomolecule measurements have the potential to provide more extensive and/or more robust predictive data, and thus these high-throughput approaches hold promise in advancing the current state of clinical practice and knowledge of disease mechanisms.

Box 17.3 AlloMap and ImmuKnow

AlloMap (CareDx) is the first FDA-approved noninvasive diagnostic test intended to aid in identifying heart transplant patients with low probability of moderate to severe acute cellular rejection. This test is based on measuring 20 genes from the peripheral blood mononuclear cells (PMBC) and combining the expression patterns into a single score ranging from 0 to 40. The higher scores represent increased risk of acute rejection [40]. Since the FDA clearance in 2008, AlloMap has conducted a clinical utility study, where they demonstrated non-inferiority of AlloMap to the current golden standard, endomyocardial biopsy (EMB) [41]

(Continued)

Box 17.3 (Continued)

Another approach of monitoring allograft rejection has exploited the intracellular ATP (iATP) levels in CD4+ T-cells where high iATP levels indicate high immune activation (increased risk of rejection) and low levels indicate low immune activation (increased risk of infections and malignancies) [42, 43]. This work led to the development of a therapeutic response assay, ImmuKnow (Cyclex), and has been used to detect immune function after transplantation of various organs including heart, liver, lung, and kidney [44, 45].

While ImmuKnow and AlloMap have helped clinicians provide improved post-transplant care, they are still far from being perfect biomarker diagnostic tools. AlloMap cannot be used within six months after the transplantation, when the rates of acute rejection is at the highest and the low positive predictive value renders little value in identifying patients who truly have rejection (i.e. high AlloMap score does not necessarily mean that those patients are actually rejecting). ImmuKnow measures the status of a patient's general immune function and not the T-cell–mediated activity toward the transplant, which is the understood mechanism underlying acute rejection [46].

Clearly, new biomarker tests are needed in order to improve clinical care to patients who have undergone organ transplant. This will require a biomarker panel that has much improved performance in discriminating rejectors from non-rejectors, likely achieved by availability of large number of clinical samples, development of sophisticated algorithms, and willingness of the medical communities to support and adopt biomarker tests.

Biomarker	Sample	AUC	Sensitivity (%)	Specificity (%)	PPV (%)	NPV (%)
AlloMap[a)]	Blood (PBMCs, mRNA)	0.70	90	18	2	99
ImmuKnow[b)]	Serum (protein)	0.51	43	75	58	67

a) Threshold at 19 [40].
b) Across kidney, liver, heart, and lung post-transplant samples [45].

Evidentiary Standards

Qualification starts by defining the intended COU of the biomarker in drug development and then determining the data sources and level of evidence required for that COU [22]. The first step in qualification is then to show that a relationship exists between the biomarker and the clinical outcome. Once the relationship between the biomarker and clinical outcome has been established and reasonably well characterized, specifics about the intended use come into play.

While establishing COU seems intuitive and straightforward because most biomarker studies begin with some ideas of application, different evidentiary standards are needed for the various types of biomarkers and their varying COUs and these are not concretely established. For example, prognostic biomarkers can be used to stratify patients during clinical trials for enrichment of a specific subpopulation, whereas safety biomarkers may be used to permit more aggressive dosing. A prototypical framework for developing evidentiary standards for biomarkers was developed by Pharmaceutical Research and Manufacturers of America (PhRMA) [47]. They propose that the tolerability of the risk introduced by use of the biomarker should be the key factor in determining the weight of evidence. For example, the benefit of using a surrogate endpoint can be significant because it can allow for an earlier and often easier determination of a drug's efficacy. However, if the surrogate endpoint is not adequately predictive, its use may result in approval of a drug that is not effective. Thus, the level of evidence needed to support the confidence in a surrogate endpoint is likely to be high. Thus, it was proposed that the "specific purpose and COU" would determine tolerability of risk, which can determine threshold for type and level of evidence required to qualify any biomarker.

A similar suggestion was made by the Institute of Medicine (IOM) in 2008 when the FDA's Center for Food Safety and Applied Nutrition (CFSAN), in conjunction with the Center for Drug Evaluation and Research (CDER), sought out their inputs in developing clear evidentiary standards [48]. The IOM report recommended that the FDA should convene expert panels to evaluate biomarkers and biomarker tests without considering their COU but then reevaluate them with COUs in mind on a continual and a case-by-case basis. IOM concluded that it was important to "separate the parts of the evaluation of framework that have the goal of being objective and those for which subjective judgement is necessary." As such, all available data on analytical validation and qualification should be used as source of evidence for determining appropriate COU and the level of evidence needed for the biomarker qualification.

Considering the rapid emergence of novel biomarkers, their applications, and associated analytical methods, it is not surprising that there remains a lack of specific guidelines on evidentiary standards for all types of biomarkers for various COUs (Box 17.4). However, as more research and clinical development occurs, the regulatory experience and practice-based considerations will be able to contribute to forming better standardization on the qualification process.

Box 17.4 Determining the Optimal COU for Biomarker

While specific guidelines for statistical analyses to support qualification for the different COUs have not been developed or agreed upon by all interested parties, there are a few basic principles that have been used by the FDA to guide the statistical evaluation of biomarkers [22, 49].

The statistical rigor required for qualifying a prognostic biomarker should be less than what is required for a predictive biomarker, with most rigor demanded for a surrogate endpoint biomarker. Within each of these categories of use, required level of rigor depends on other aspects of the qualification process, such as the availability of additional biomarkers serving the same purpose, past experience with and/or validation of the biomarker's use in similar contexts, etc.

Adequacy of data sources for characterizing the relationship between the biomarker and other variables is an important factor to consider in determining the feasibility of the qualification process; in some cases, qualification may not be feasible until additional data or other types of data (e.g. from randomized controlled trials) are available. A statistical analysis plan describing data sources and details of the statistical analyses proposed to support qualification should be developed and locked down prior to the qualification process.

Conclusion

Precision medicine is at the frontier of improving patient health outcomes and reducing the overall cost of health care. The potential to improve patient care and management by allowing an individualized approach to diagnosis, drug development, and therapeutic regimen is truly encouraging. Biomarkers are the tools that we can use to exactly achieve such goals of improving diagnostic precision. Biomarkers can be correlational (associated with disease) and/or functional (identified mechanism of action related to disease), and their measurements can be used to infer risk, diagnosis, prognosis, and therapeutic response. In other words, patients with different sets of biomarker reactivity can mean different risks of developing a disease, different disease prognoses, or different responses to treatment; therefore, new standards of care will then be developed for these patient populations. One success story comes from lung adenocarcinoma; EGFR-positive non-small cell lung adenocarcinoma is currently its own disease subtype (not considered as a general lung adenocarcinoma) and is treated with specific type of chemotherapy [50]. Such improved diagnostic precision can allow clinicians to treat patients with the most optimal therapeutic options that are specific to each patient and their disease type, and this is just the beginning of what precision medicine biomarkers can help us accomplish.

Acknowledgment

We would like to thank David Lin, Andreas Scherer, and Shawna Flynn for their insightful contributions to the initial publication [51].

References

1 Quinton, P.M. (2007). Cystic fibrosis: lessons from the sweat gland. *Physiology* 22: 212–225.
2 Food and Drug Administration. The Public Health Evidence for FDA Oversight of Laboratory Developed Tests: 20 Case Studies. 2015.
3 Colburn, W.A. (2003). Biomarkers in drug discovery and development: from target identification through drug marketing. *J. Clin. Pharmacol.* 43 (4): 329–341.
4 Kuhlmann, J. and Wensing, G. (2006). The applications of biomarkers in early clinical drug development to improve decision-making processes. *Curr. Clin. Pharmacol.* 1 (2): 185–191.
5 Mandrekar, S.J., Grothey, A., Goetz, M.P., and Sargent, D.J. (2005). Clinical trial designs for prospective validation of biomarkers. *Am. J. PharmacoGenomics* 5 (5): 317–325.
6 Bast, R.C. Jr., Lilja, H., Urban, N. et al. (2005). Translational crossroads for biomarkers. *Clin. Cancer Res.* 11 (17): 6103–6108.
7 Sin, D.D., Hollander, Z., DeMarco, M.L. et al. (2015). Biomarker development for chronic obstructive pulmonary disease from discovery to clinical implementation. *Am. J. Respir. Crit. Care Med.* 192 (10): 1162–1170.
8 Listgarten, J. and Emili, A. (2005). Statistical and computational methods for comparative proteomic profiling using liquid chromatography–tandem mass spectrometry. *Mol. Cell. Proteomics* 4 (4): 419–434.
9 Hunt, S.M., Thomas, M.R., Sebastian, L.T. et al. (2005). Optimal replication and the importance of experimental design for gel-based quantitative proteomics. *J. Proteome Res.* 4 (3): 809–819.
10 Wagner, J., Williams, S., and Webster, C. (2007). Biomarkers and surrogate end points for fit-for-purpose development and regulatory evaluation of new drugs. *Clin. Pharmacol. Ther.* 81 (1): 104–107.
11 Arrowsmith, J. (2011). Trial watch: phase II failures: 2008–2010. *Nat. Rev. Drug Discovery* 10 (5): 328–329.
12 Arrowsmith, J. (2011). Trial watch: phase III and submission failures: 2007–2010. *Nat. Rev. Drug Discovery* 10 (2): 87–87.
13 DiMasi, J.A., Feldman, L., Seckler, A., and Wilson, A. (2010). Trends in risks associated with new drug development: success rates for investigational drugs. *Clin. Pharmacol. Ther.* 87 (3): 272.

14 Lee, J.W., Devanarayan, V., Barrett, Y.C. et al. (2006). Fit-for-purpose method development and validation for successful biomarker measurement. *Pharm. Res.* 23 (2): 312–328.

15 Lee, J.W., Figeys, D., and Vasilescu, J. (2006). Biomarker assay translation from discovery to clinical studies in cancer drug development: quantification of emerging protein biomarkers. *Adv. Cancer Res.* 96: 269–298.

16 Lee, J.W., Weiner, R.S., Sailstad, J.M. et al. (2005). Method validation and measurement of biomarkers in nonclinical and clinical samples in drug development: a conference report. *Pharm. Res.* 22 (4): 499–511.

17 Ilyin, S.E., Belkowski, S.M., and Plata-Salamán, C.R. (2004). Biomarker discovery and validation: technologies and integrative approaches. *Trends Biotechnol.* 22 (8): 411–416.

18 Jarnagin, K. (2006). ID and validation of biomarkers: a seven-fold path for defining quality and acceptable performance. *Genet. Eng. Biotechnol. News* 26 (12): 4–5.

19 Food and Drug Administration (2005). *Guidance for industry: Pharmacogenomic Data Submissions.* Rockville, Maryland: Food and Drug Administration.

20 Goodsaid, F. and Frueh, F. (2007). Biomarker qualification pilot process at the US food and drug administration. *AAPS J.* 9 (1): E105–E108.

21 Ball, J.R. and Micheel, C.M. (2010). *Evaluation of Biomarkers and Surrogate Endpoints in Chronic Disease.* National Academies Press.

22 Amur, S., LaVange, L., Zineh, I. et al. (2015). Biomarker qualification: toward a multiple stakeholder framework for biomarker development, regulatory acceptance, and utilization. *Clin. Pharmacol. Ther.* 98 (1): 34–46.

23 Food and Drug Administration (2019). List of cleared or approved companion diagnostic devices (in vitro and imaging tools). https://www.fda.gov/MedicalDevices/ProductsandMedicalProcedures/InVitroDiagnostics/ucm301431.htm. (accessed 01 April 2019).

24 US Food and Drug Administration (2015). *Guidance for Industry and FDA Staff: Qualification Process for Drug Development Tools.* US Food and Drug Administration.

25 US Food and Drug Administration (2011). Draft guidance for industry and food and drug administration staff-in vitro companion diagnostic devices. CDRH, CBER, CDER (eds.). US Food and Drug Administration https://www.fda.gov/media/81309/download.

26 Food and Drug Administration (2014). *Framework for Regulatory Oversight of Laboratory Developed Tests (LDTS).* Food and Drug Administration.

27 Paik, S., Tang, G., Shak, S. et al. (2006). Gene expression and benefit of chemotherapy in women with node-negative, estrogen receptor–positive breast cancer. *J. Clin. Oncol.* 24 (23): 3726–3734.

28 Groves, E. (2010). *From Biomarkers to Diagnostics–The Road to Success.* Whitepaper. Quintiles.

29 US Food and Drug Administration (2014). *Draft Guidance for Industry, Food and Drug Administration Staff, and Clinical Laboratories: Framework for Regulatory Oversight of Laboratory Developed Tests (LDTs).* Food and Drug Administration.

30 Food and Drug Administration (October 3, 2014). *FDA Notification and Medical Device Reporting for Laboratory Developed Tests (LDTs): Draft Guidance,* vol. 2015. Food and Drug Administration.

31 Moore, R.E., Kirwan, J., Doherty, M.K., and Whitfield, P.D. (2007). Biomarker discovery in animal health and disease: the application of post-genomic technologies. *Biomarker Insights* 2: 185.

32 Loupakis, F., Pollina, L., Stasi, I. et al. (2009). PTEN expression and KRAS mutations on primary tumors and metastases in the prediction of benefit from cetuximab plus irinotecan for patients with metastatic colorectal cancer. *J. Clin. Oncol.* 27 (16): 2622–2629.

33 US FDA: *Draft guidance for industry: Bioanalytical method validation. US Department of Health and Human Services, US FDA*. 2013.

34 Maruvada, P. and Srivastava, S. (2006). Joint National Cancer Institute-Food and Drug Administration workshop on research strategies, study designs, and statistical approaches to biomarker validation for cancer diagnosis and detection. *Cancer Epidemiol. Biomarkers Prev.* 15 (6): 1078–1082.

35 Perez, E.A., Suman, V.J., Davidson, N.E. et al. (2006). HER2 testing by local, central, and reference laboratories in specimens from the north central cancer treatment group N9831 intergroup adjuvant trial. *J. Clin. Oncol.* 24 (19): 3032–3038.

36 Andre, T., Blons, H., Mabro, M. et al. (2013). Panitumumab combined with irinotecan for patients with KRAS wild-type metastatic colorectal cancer refractory to standard chemotherapy: a GERCOR efficacy, tolerance, and translational molecular study. *Ann. Oncol.* 24 (2): 412–419.

37 De Gramont, A., Watson, S., Ellis, L.M. et al. (2015). Pragmatic issues in biomarker evaluation for targeted therapies in cancer. *Nat. Rev. Clin. Oncol.* 12 (4): 197–212.

38 Christians, U., Klawitter, J., and Klawitter, J. (2016). Biomarkers in transplantation – proteomics and metabolomics. *Ther. Drug Monit.* 38: S70–S74.

39 Van't Veer, L.J. and Bernards, R. (2008). Enabling personalized cancer medicine through analysis of gene-expression patterns. *Nature* 452 (7187): 564–570.

40 Crespo-Leiro, M.G., Stypmann, J., Schulz, U. et al. (2016). Clinical usefulness of gene-expression profile to rule out acute rejection after heart transplantation: CARGO II. *Eur. Heart J.* 37 (33): 2591–2601.

41 Pham, M.X., Teuteberg, J.J., Kfoury, A.G. et al. (2010). Gene-expression profiling for rejection surveillance after cardiac transplantation. *N. Engl. J. Med.* 362 (20): 1890–1900.

42 Roedder, S., Vitalone, M., Khatri, P., and Sarwal, M.M. (2011). Biomarkers in solid organ transplantation: establishing personalized transplantation medicine. *Genome Med.* 3 (6): 37.

43 Zeevi, A. and Lunz, J. (2013). Cylex ImmuKnow cell function assay. *Methods Mol. Biol.* 1034: 343–351. https://doi.org/10.1007/978-1-62703-493-7_21.

44 Fernández-Ruiz, M., Kumar, D., and Humar, A. (2014). Clinical immune-monitoring strategies for predicting infection risk in solid organ transplantation. *Clin. Transl. Immunol.* 3 (2): e12.

45 Ling, X., Xiong, J., Liang, W. et al. (2012). Can immune cell function assay identify patients at risk of infection or rejection? A meta-analysis. *Transplantation* 93 (7): 737–743.

46 Ingulli, E. (2010). Mechanism of cellular rejection in transplantation. *Pediatr. Nephrol.* 25 (1): 61–74.

47 Altar, C., Amakye, D., Bounos, D. et al. (2008). A prototypical process for creating evidentiary standards for biomarkers and diagnostics. *Clin. Pharmacol. Ther.* 83 (2): 368–371.

48 National Research Council (2010). *Evaluation of Biomarkers and Surrogate Endpoints in Chronic Disease.* National Research Council.

49 Gosho, M., Nagashima, K., and Sato, Y. (2012). Study designs and statistical analyses for biomarker research. *Sensors* 12 (7): 8966–8986.

50 Cancer Genome Atlas Research Network (2014). Comprehensive molecular profiling of lung adenocarcinoma. *Nature* 511 (7511): 543–550.

51 Lin, D., Scherer, A., Ng, R. et al. (2010). Clinical validation and biomarker translation. In: *Pharmaceutical Sciences Encyclopedia.* https://doi.org/10.1002/9780470571224.pse270.

18

Predicting and Assessing an Inflammatory Disease and Its Complications: Example from Rheumatoid Arthritis

Christina Trollmo[1] *and Lars Klareskog*[2]

[1] *Roche Pharmaceuticals, Stockholm, Sweden*
[2] *Karolinska Institute, Stockholm, Sweden*

Introduction

Chronic inflammatory diseases include a number of rheumatic, neurological, dermatological, and gastrointestinal diseases, which develop as a result of immune and inflammatory reactions. These perpetuating reactions ultimately cause the clinical symptoms, which are subsequently used to classify the symptoms as a "disease." Analyzing the disease course longitudinally reveals several distinct steps during disease progression, for which the presence of biomarkers is of importance, both for identification of disease status and for prediction of disease course and treatment options. However, biomarkers per se are not always available today. We have chosen here to discuss disease characteristics and potential biomarkers in one common chronic inflammatory disease, rheumatoid arthritis (RA), which affects approximately 0.5–1% of the population worldwide. In this chapter, we focus on the following factors:

1. onset of disease to discuss the questions in whom and why the disease occurs, and whether and how onset can be predicted from biomarkers that are present before the occurrence of clinical symptoms
2. progression of disease with respect to not only development of joint destruction but also complications such as extra-articular manifestations, cardiovascular (CV) events, and lymphoma development in this patient group
3. treatment of the individual patient and of specific symptoms and disease manifestations
4. selection of patients in clinical trials of new drugs

Biomarkers in Drug Discovery and Development: A Handbook of Practice, Application, and Strategy, Second Edition. Edited by Ramin Rahbari, Jonathan Van Niewaal, and Michael R. Bleavins.

Rheumatoid Arthritis Disease Process

RA is a disease defined by seven criteria, with four that should be fulfilled to make the diagnosis (Table 18.1). These criteria have been useful in harmonizing clinical trials and clinical practice. However, they are not based on what is now known about etiology or pathogenesis, and they are not too helpful in selecting treatment for the single patient. Hence, there are needs to redefine the diagnosis for RA and related diseases: first to define entities more related to distinct etiologies and pathogenetic mechanisms, then to use such new entities for stratification and selection of patients in clinical trials and clinical practice.

Basic features of the immune and inflammatory process in RA are, on the one hand, processes that can be identified in the peripheral circulation, initially rheumatoid factors (RFs) and, on the other hand, processes in the inflamed tissue, mainly the joints (Figure 18.1). RFs, identified almost 70 years ago, are part of the diagnostic criteria for RA. Being present in 50–60% on incident RA cases and increasing over time with active disease, these autoantibodies are also seen in many non-RA conditions and are thus not very specific for the disease.

RFs have never been shown to be pathogenic by themselves, neither in patients nor in experimental models. They are thus seen mainly as biomarkers of importance for diagnosis and prognosis of a more severe disease course, but not necessarily directly involved in disease pathogenesis. Joint inflammation in RA is focused on synovial inflammation, in many cases associated with cartilage destruction and concomitant erosions in bone. This inflammation has been studied in large detail over the years, demonstrating that its features are common to many other types of chronic inflammation in other tissues. No real RA pathognomonic features have yet been identified, the most unique feature identified so far being the way the inflammatory cells and molecules attack and destroy bone and cartilage.

Table 18.1 Classification criteria for rheumatoid arthritis [1].

1. Morning stiffness in and around joints lasting at least 1 h before maximal improvement
2. Soft tissue swelling (arthritis) of three or more joint areas observed by a physician
3. Swelling (arthritis) of the proximal interphalangeal, metacarpophalangeal, or wrist joints
4. Symmetric swelling (arthritis)
5. Rheumatoid nodules
6. Presence of rheumatoid factor
7. Radiographic erosions and/or periarticular osteopenia in hand and/or wrist joints

Criteria 1–4 must have been present for at least 6 wk. Rheumatoid arthritis is defined by the presence of four or more criteria, and no further qualifications (classic, definite, or probable) or list of exclusions are required. These criteria demonstrated 91–94% sensitivity and 89% specificity for RA compared with non-RA rheumatic disease control subjects.

Source: Arnett et al. 1988 [1]. Reproduced with permission of John Wiley & Sons.

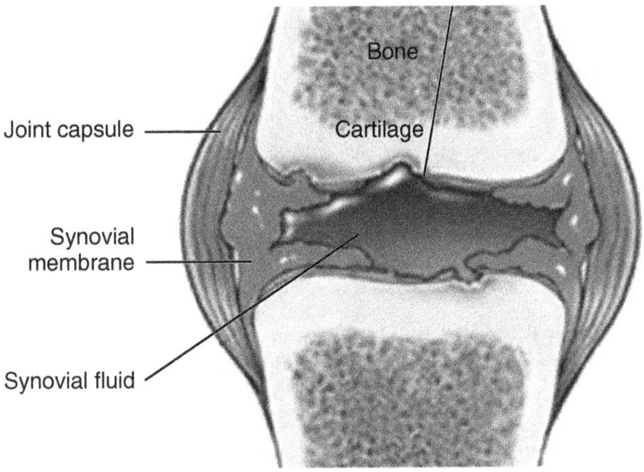

Figure 18.1 Inflamed RA joint. In healthy joints, a thin synovial membrane lines the joint capsule and the synovial fluid. In the RA joint, both the synovial membrane and the synovial fluid are infiltrated by inflammatory cells, leading to tender and swollen joints. The synovial membrane also "grows" over the cartilage, aiding in the process of cartilage and bone destruction.

Having the major features of RA, the synovial joint inflammation and presence of RF being typical but by no means unique for RA, there is an obvious need to define more specific features of the disease. The identification of such features would enable us to search for a better understanding of the pathogenesis of RA, a more accurate diagnosis based on biomarkers, and more specific treatments.

Studies on Etiology and Pathogenesis as a Basis for Development of Biomarkers for Diagnosis and Prognosis in RA

Any understanding of a complex, partly genetic disease is based on an understanding of how genes and environment interact in giving risk to immune reactions that contribute to the joint destruction and other inflammatory reactions in RA. In healthy subjects, the major role for the immune system and subsequent inflammatory reactions is to defend us against pathogens, but in RA the immune system has partly changed focus to attack our own tissues, primarily the joints, and is thus denoted as an autoimmune disease.

Genes

There is strong evidence to support a significant genetic component to the susceptibility of RA. Twin studies clearly demonstrate an overrepresentation

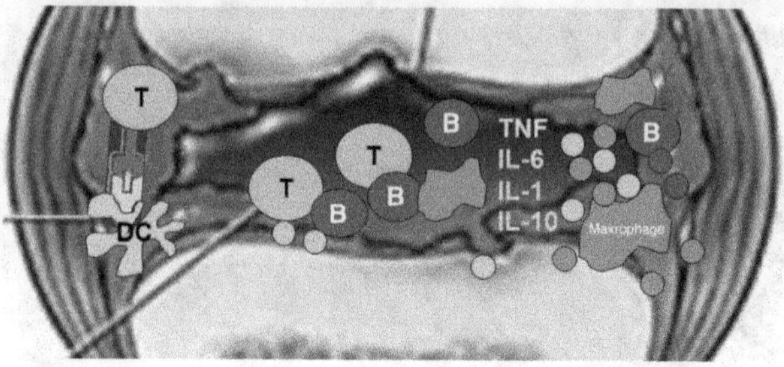

Figure 18.2 Inflammatory cells in the RA joint. A number of immune cells have infiltrated the joint, and local production of inflammatory mediators, including cytokines and antibodies, occurs. The synovial fluid, which functions as a cushion during joint movements, is in a healthy joint acellular. Illustrated to the left is the presentation of an antigen by the dendritic cell to a T-cell; the yellow connector is a MHC class II molecule. Cytokines, released from the immune cells, function as signaling molecules between cells. (*See insert for color representation of this figure.*)

of disease concordance in monozygotic twins (12–20%, depending on study) compared to dizygotic twins (4–5%) and the general population (0.5–1%). Analysis on the gene level demonstrates the strongest genetic association with genes within the HLA region, specifically the *HLA-DRB1* gene. Its gene products, the MHC class II molecules, were described in the 1970s to be present on cells in the inflamed joint, allowing antigens (part of proteins, generally pathogens, but in autoimmune diseases self-proteins) to be presented to the immune system and subsequently to trigger inflammatory reactions (Figure 18.2). Serologic and later genetic typing of the various MHC molecules revealed that a few allotypes of *HLA-DRB1* were overrepresented in RA patients. A closer analysis demonstrated even identical amino acid sequences in those regions of the MHC class II DRβ1 chain, which are in contact with the antigen and the T-cell receptor on the T-cell mediating the specific immune responses. This amino acid sequence is termed *shared epitope.* The nature of the specific immune reactions mediated by the MHC class II molecules has, however, been surprisingly difficult to define. Identifying the specific antigens would help to identify the autoimmune trigger and make it possible to interfere specifically to break the autoimmune reactions. In total, the HLA region contributes 30–50% of the genetic component for RA in Caucasians. It appears that other MHC class II genes and allotypes may contribute to RA in other ethnic groups, and much is still not known about the contributions of different genes and allotypes within the MHC concerning susceptibility and disease course in RA.

Recently, a second genetic risk allele has been identified in populations of European descent. This minor allele of a nonsynonymous single-nucleotide polymorphism (SNP) in the protein tyrosine phosphatase nonreceptor 22

Table 18.2 Contribution of genetic risk factors in rheumatoid arthritis.

Gene	Odds ratio
HLA	6.4
PTPN22	1.8
6q23	1.2
STAT4	1.2–1.4
TRAF1/C5	1.1–1.4

(*PTPN22*) gene confers the second largest genetic risk to the development of RA, with an odds ratio of about 1.8. *PTPN22* encodes the intracellular protein lymphoid tyrosine phosphatase, which plays a central role in immune responses by playing an integral part in signal transduction and T-cell receptor signaling pathway and inhibiting T-cell activation. This variant was first demonstrated for type 1 diabetes, but was soon confirmed in a number of autoimmune diseases, including RA, systemic lupus erythematosus (SLE), and Grave disease. However, other autoimmune diseases show no association with this SNP, suggesting subsets of autoimmune diseases to be defined accordingly.

The introduction of whole genome-wide association studies allowed, for the first time, good coverage of common variations in the human genome. Hundreds of thousands of SNPs in thousands of samples were genotyped and compared. Interestingly, studies focusing on RA again demonstrated the strongest effects for the two well-documented RA susceptibility genes *HLA-DRB1* and *PTPN22*. Other genes (Table 18.2) make a more modest contribution to susceptibility. Future studies have to identify remaining, probably smaller, genetic effects and how all the genetic effects interact with each other as well as with environmental factors in inducing and perpetuating the disease.

Environment

Information on environmental factors important for the development, perpetuation, or course of RA is surprisingly scarce. Smoking is the only conventional environmental factor that has been linked reproducibly to an increased risk of developing RA. Other exposures, such as silica dust and mineral oils, have been reported in a few studies. It has not yet been possible to verify frequently hypothesized stimuli such as microbial infections with the methods used to date. Smoking was initially considered as an unspecific risk factor, of interest mainly from a public health perspective. However, newer studies indicate that smoking is a specific trigger of RA, as will be discussed in more detail below.

Immunity

Studies on specific immune reactions in RA have been confined almost entirely to those involving autoantibodies. As described above, they were initially restricted to the measure of RFs, but more recently, antibodies specific for citrullinated proteins have been shown to be of great importance. This is discussed in detail below, since their presence covariates with genetic and environmental factors, providing an important tool in subgrouping patients into different entities of RA. In the process of joint inflammation and cartilage and bone destruction, many different cells and molecules of the immune system participate. Some of these are illustrated in Figure 18.2. However, even if some of these inflammatory processes are revealed in detail, a specific trigger has still not been identified.

In recent years, advances within the field of cytokine regulation and cytokine-directed therapy have largely dominated the research field of RA, illustrating how therapeutic progress is possible even though the role of adaptive immunity in the disease is not fully understood. Cytokines are soluble molecules that mediate the communication between cells of the immune system as well as with other cells of the body, such as the endothelium. Interestingly, the first cytokine that was targeted, tumor necrosis factor (TNF), belongs to the innate immune system. Blocking IL-1, another cytokine belonging to the innate immune system, has not proven to be as effective as TNF blockade for the majority of the RA patients. Clinical trials blocking a third cytokine in this family, IL-6, show promising results. This cytokine exerts effects within both the innate and the adaptive immune systems.

Temporarily eliminating B-cells, which are the producers of antibodies, has also proven to be a successful therapy. A third alternative, blocking the interaction between cells presenting antigens and T-cells, has resulted in an approved therapy. Together, these treatment-based data also demonstrate the significant role of the various parts of the innate as well as adaptive immune system for disease progression.

Citrulline Immunity

The presence of antibodies specifically identifying citrullinated antigens, specifically termed *antibodies to citrullinated protein antigens* (ACPAs), are a strong predictor of developing RA. Approximately 60% of all RA patients carry such antibodies. These antibodies are highly specific for the disease; that is, they are rare in the normal population (<2%) and are also quite rare in other inflammatory conditions. Almost all patients who carry ACPAs developed them before disease onset. ACPAs are directed against proteins, which, as the name implicates, contain the amino acid citrulline. Citrulline occurs only upon posttranslational modification of the amino acid arginine by enzymatic conversion. Even if these autoantibodies are more specific for RA than for RF, it is important to ask the question if their presence is an epi-phenomenon or is causally related to the disease. This helps in further dissection of the pathophysiology of the disease and in the development of new treatments. Here, animal models provide a tool to test the hypothesis of citrullination to be causative of the disease. One example is arthritis induced in the rat by immunization with collagen type II, a protein expressed almost exclusively in the joint. Immunization with the citrullinated form of collagen type II resulted in a more severe arthritis than immunization with the same non-citrullinated protein. In a mouse model, where transfer of antibodies directed toward collagen induces a mild arthritis, the addition of antibodies directed against citrullinated fibrinogen enhanced the arthritis. Thus, in experimental arthritis models, citrullination can change the immunogenicity of self-antigens, and some citrullinated proteins may contribute to arthritis development.

From the data above, we can conclude that immunity to citrullinated proteins may play a pathogenic role in a subset of RA patients; however, so far we have not identified the specific proteins that are targeted by the immune system. How do we then measure ACPAs? Test kits have been developed which contain cyclic citrullinated peptides (CCPs), where the amino acid citrulline is flanked by a variety of amino acids to mimic many different peptide sequences. Thus, the test measures anti-CCP antibodies, and this is what is generally reported in patient files. In summary, the presence of ACPAs in the circulation is more predictive for RA than high blood pressure is for cardiovascular disease and thus can be considered as a biomarker for a subset of RA patients.

Better Diagnosis by Means of Biomarkers and Genetics

By the late 1970s, the MHC class II genes were already identified as a major genetic risk factor for RA. However, only the presence of specific MHC class II alleles does not result in RA, since the disease-associated *HLA-DRB1* alleles are common also in healthy subjects. But combining genetic studies with ACPA analysis, the genetic association of RA to *HLA-DRB1* SE has recently been shown to be confined entirely to the ACPA-positive subset of RA. In contrast, ACPA-negative RA may be associated with an unrelated *HLA-DR* allele, *HLA-DR3*. A second major genetic risk factor for RA, the polymorphism in the *PTPN22* gene was also shown to be associated only with the ACPA-positive disease.

In contrast, other genetic risk factors, in particular variations in the interferon regulating factor 5 (IRF-5), as well as polymorphisms in a newly identified risk gene in the C-type lectin complex, were associated exclusively with ACPA-negative disease. Taken together, the descriptive studies of disease course and genetic linkages strongly indicate that ACPA reactivity splits RA into two major and clinically relevant subsets of disease and thus becomes an

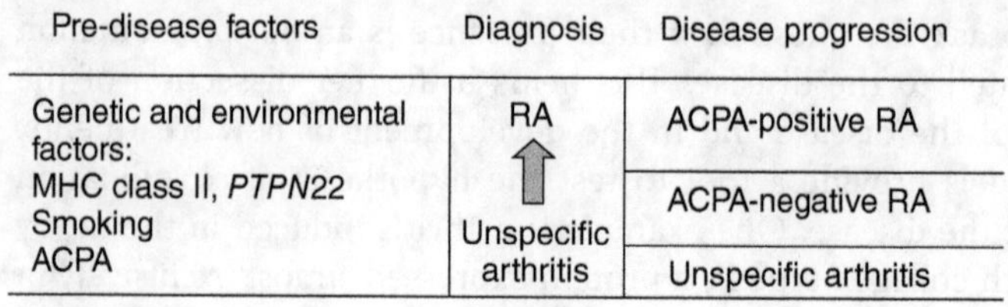

Figure 18.3 Rheumatoid arthritis is a multifactorial disease. A number of pre-disease factors have been identified, which are of importance for the later disease course but not enough to initiate clinical disease. However, the trigger for clinical disease onset has still not been identified. The presence or absence of anticitrulline protein antibodies (ACPAs) distinguishes two forms of RA. Some patients will have swollen and tender joints but not fulfill four of the seven criteria for RA and thus carry an unspecific arthritis.

important biomarker for one subgroup of RA. Practically, ACPA-positive and ACPA-negative RA should be treated as separate entities when studying the molecular pathophysiology of RA and probably also when selecting treatment for some treatments for the single patient (Figure 18.3).

Including smoking habits in these combined analyses provides an even clearer risk assessment for developing RA. The relative risk of developing ACPA-positive RA was shown to be over 20 times higher for smokers carrying two copies of the *HLA-DRB1* SE alleles than for nonsmokers with no SE alleles. On the other hand, no increased risk was discerned from smoking concerning development of ACPA-negative RA. This leads directly to the question of whether smoking can trigger an ACPA immunity. Recent studies suggest that this could be the case since cells from bronchoalveolar lavage from smokers, but not from nonsmokers, contain large amounts of citrullinated proteins, and smoking can activate macrophages and thus enhance their antigen-presenting capacity.

The disease onset of RA can be gradual, and thus patients can have swollen and tender joints without being given the diagnosis for RA, which is based on achieving four of seven specific criteria. These patients are classified as *undifferentiated arthritis* (UA), having an inflammatory arthritis for which no specific diagnosis can be made. An important question is if these patients should be treated in order to ameliorate disease progression to RA and retard radiographic joint damage or even prevent the development of RA in these patients. A recent double-blind, placebo-controlled, randomized clinical trial indicated that methotrexate (MTX) treatment as used for RA patients can postpone progression to RA and retard radiographic progression but not prevent the development of RA from UA. Subgroup analysis revealed that the beneficial outcome was most pronounced in patients with anti-CCP antibodies. In contrast, in the anti-CCP-negative subgroup, the effect of MTX on the development of RA, the radiographic progression, and even the signs and symptoms was not demonstrable. Thus, this study provides evidence of distinguishing between anti CCP-positive and CCP-negative arthritis, even before the diagnosis RA is met, and therefore also becomes an important biomarker for UA.

The identification of pre-disease biomarkers might in the future lead to treatments which can be given to healthy subjects before the clinical diagnosis of RA and so prevent disease onset. An example already applied in medicine is the treatment of high blood pressure to prevent cardiovascular disease.

Better Prognostic Tools with the Help of Biomarkers and Genetics

RA affects not only joints but also, possibly, inflammation in other organs. Manifestations include rheumatoid nodules, secondary Sjögren's syndrome, rheumatoid lung disease, pleuritis, pericarditis, vasculitis, neuropathy, Felty syndrome, and severe eye disease, and are commonly summarized as extra-articular manifestations. Different comorbidities also occur during the disease course. Cardiovascular morbidity and mortality occur at rates greater than what would be expected from the profile of established risk factors. The average risk of lymphoma is increased in RA; however, a more detailed analysis indicates that the risk is increased primarily in patients with the most severe disease (Figure 18.4).

Extra-articular Disease Manifestations

Extra-articular disease manifestations are more likely to occur in patients with more severe joint disease. Rheumatoid nodules often precede the onset of severe RA and thus can be considered a biomarker for severe disease. Also, their clustering with all different extra-articular manifestations indicates that

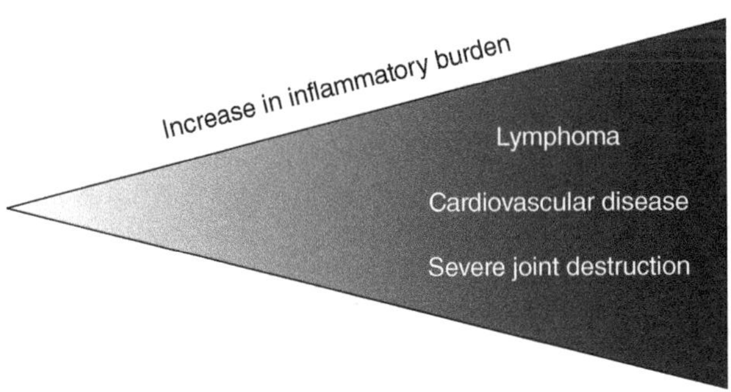

Figure 18.4 Inflammatory burden. With increased accumulated inflammatory burden, the risk for additional disease states and complications increases for patients with RA. It is thus of importance to control the inflammation at all time points, a difficult task when not all patients respond to the antirheumatic therapies available today.

vascular pathogenetic mechanisms are important in all types of extra-articular RA. This is of particular interest given the association of RA with cardiovascular comorbidity. The different manifestations tend to cluster in specific patterns and suggest shared disease mechanisms in these systemic manifestations. There are very likely several genetic and environmental factors affecting the development of these manifestations, of which some are described below.

An association of *HLA-C3* with vasculitis has been described. *HLA-C3* is the product of a specific MHC class I gene and thus with CD8+ cytotoxic T-cells and NK-cells, in contrast to MHC class II proteins, which interact with CD4+ T-helper cells. Smoking is an independent predictor of vasculitis and is probably involved in vascular damage by antigen modification as described above. The increased number of patients with specific autoantibodies such as RF and ANAs (antinuclear antibodies) supports a role for immune complexes in the pathogenesis of vasculitis, which in turn indicates B-cell abnormalities in the development of extra-articular RA.

In summary, extra-articular manifestation indicates a more severe disease; however, this risk group is not routinely analyzed early in the disease course. Here, specific biomarkers could aid an early aggressive treatment of this patient group in order to limit the progression of extra-articular manifestations.

Cardiovascular Disease in RA

Cardiovascular (CV) morbidity and mortality in RA occur at rates greater than what would be expected from the profile of established CV risk factors. Diabetes mellitus, hypercholesterolemia, hypertension, cigarette smoking, and obesity are generally powerful classifiers of CV risk and when comparing the profiles of these risk factors in people with and without RA, they are similar. Thus, there must be RA-specific factors explaining the increased risk. Dividing the patients in younger and older patients demonstrates that such factors are more common in the younger age group. In older patients, the proportion of atherosclerosis can be explained by established CV risk factors as described above. In contrast, in the younger age group, systemic inflammation seems to exert its atherosclerotic effects early. High-resolution carotid ultrasound is used to measure carotid intima media thickness (IMT). These measures may thus be used as a biomarker of subclinical atherosclerosis. Interaction calculations suggest that the ESR's effect on IMT varies according to the number of CV risk factors; thus, higher ESR values, a measure for inflammation, are associated with greater IMT only in the presence of CV risk factors. In patients without CV risk factors, the ESR is not associated significantly with IMT.

Taking genetic and environmental components of CV disease into consideration in RA patients demonstrates that the presence of two SE alleles predicts death from cardiovascular disease in these patients, and this is independent from autoantibody status. However, the greatest risk of death from CV disease is associated with an interaction among smoking, SE alleles, and anti-CCP antibodies. Interestingly, no association of the *PTPN22 gene* with mortality was detected.

In summary, both established CV risk factors and RA manifestations account for a significant proportion of atherosclerosis in RA. Factors related to RA may have a greater influence on the extent of atherosclerosis in young patients. The presence of established risk factors may be necessary for systemic inflammation to promote atherosclerosis; however, more studies are needed to understand these interactions.

Lymphoma in RA

Among RA patients, there is in general a doubled risk for lymphoma. However, a closer analysis of disease severity, as measured by degree of joint destruction, ESR, number of affected joints, and accumulated inflammatory load, revealed a more pronounced relative risk in patients with the most severe disease and little or no increase in those with mild to moderate disease. This raises the question of whether lymphomas arise as a consequence of treatment or because of a lack thereof. Larger cohort studies have thus far not confirmed any treatment-related excess of lymphoma in general. However, it is premature to make firm conclusions about the true lymphoma risk in patients treated with TNF antagonists. In summary, the data so far indicate that the inflammatory burden should be kept as low as possible to possibly reduce the risk for lymphoma development.

Cartilage and Bone Destruction

During the inflammation in the joint, cartilage, and underlying bone are broken down. Matrix metalloproteinases (MMPs) are enzymes active in this process, which results in the release of a number of collagenous and noncollagenous cartilage and bone-derived molecules. Serum levels of MMP-3 and urinary levels of C-telopepetide of collagen type II (CTX II) have been identified as two independent baseline factors that predict radiographic progression. Upon successful MTX treatment, MMP-3 levels decrease. Also, serum cartilage oligomeric matrix protein (COMP) levels are predictive of radiographic damage. Proof-of-concept studies demonstrate that targeting RANK-mediated osteoclastogenesis prevents inflammatory bone loss, and clinical application has only just begun by blocking RANKL.

Replacing radiographic measures with measures of biochemical molecules in serum and/or urinary samples might in the future become the preferred way to measure joint destruction specifically. Traditionally, high inflammatory burden and measures of CRP have been included in the prediction of joint destruction, but the concomitant measures of inflammation and joint destruction indicate that the inflammatory and destructive processes are uncoupled and therefore need to be identified and treated separately.

Biomarkers Proving Leads Concerning Effects of Treatment of Rheumatoid Arthritis

The drugs used in the treatment of RA are generally divided into three groups: nonsteroidal anti-inflammatory drugs (NSAIDs) to relieve pain, traditional disease-modifying antirheumatic drugs (DMARDs), and biologics. There are today some general guidelines for the treatment of RA; however, biomarkers guiding the most adequate selection for the single patient are missing.

In the majority of RA patients, treatment with MTX or other DMARDs only partly reduces clinical and radiological progression of the disease. For many patients, therefore, the introduction of the biological drugs provided a major breakthrough. By blocking the proinflammatory cytokine TNF, joint progression could be halted much more efficiently than before in a large group of patients. More recent drugs eliminating circulating B-cells or blocking co-stimulation of T-cells also indicate a slowdown in joint destruction. Measuring levels of TNF in the inflamed tissue has been suggested to be a biomarker for treatment response to TNF blocking agents. Unfortunately, it has not yet been possible to describe such a marker for serum samples.

With the identification of the association of smoking, SE alleles, and anti-CCP antibodies with the greatest risk of death from CVD, it will be possible to be more active in treatment to prevent cardiovascular disease in these patients at high risk for CV disease. In the treatment of lymphoma, which in the majority of cases are B-cell derived, a specific treatment option exists for RA patients. Rituximab was established in 1998 as a drug to treat patients with B-cell lymphomas, and the indication was extended in 2006 to treat patients with moderate to severe RA, independent of lymphoma associations. Thus, in these cases, both the lymphoma and the RA are treated with the same drug.

Concluding Remarks

In summary, RA is a heterogeneous disease with regard to severity of the joint inflammation and other inflammatory manifestations. To prevent disease onset or progression of disease in the single patient, distinct biomarkers are needed, and only a few have yet been identified. The fact that the presence or absence of ACPAs divides the disease into two distinct entities will very likely influence future treatments and clinical trials. Clinical trials, if not already focusing on a subgroup of patients, need also to be adequately powered to allow subgrouping patients according to genetic setup, environmental exposures, and different nonsynovial inflammations in order to find biomarkers for individualized therapies. Today's and tomorrow's biological therapies targeting different parts of the immune and inflammatory pathways will also increase our understanding of disease etiology and progression and help identify new drugs.

Further Reading

Bowes, J. and Barton, A. (2008). Recent advances in the genetics of RA susceptibility. *Rheumatology* https://doi.org/10.1093/rheumatology/ken005.

Ekström-Smedeby, K., Baecklund, E., and Askling, J. (2006). Malignant lymphomas in autoimmunity and inflammation: a review of risks, risk factors, and lymphoma characteristics. *Cancer Epidemiol. Biomarkers Prev.* 15: 2069–2077.

Farragher, T.M., Goodson, N.J., Naseem, H. et al. (2008). Association of the *HLA-DRB1* gene with premature death, particularly from cardiovascular disease, in patients with rheumatoid arthritis and inflammatory polyarthritis. *Arthritis Rheum.* 58: 359–369.

Klareskog, L., Rönnelid, J., Lundberg, K. et al. (2008). Immunity to citrullinated proteins in rheumatoid arthritis. *Annu. Rev. Immunol.* 26: 651–675.

Romas, E. and Gillespie, T. (2006). Inflammation-induced bone loss: can it be prevented? *Rheum. Dis. Clin. North Am.* 32: 759–773.

Turesson, C., Schaid, D.J., Weyand, C.M. et al. (2006). Association of *HLA-C3* and smoking with vasculitis in patients with rheumatoid arthritis. *Arthritis Rheum.* 54: 2776–2783.

Van der Helm-van Mil, A.H.M., Huizinga, T.W.J., De Vries, R.R.P., and Toes, R.E.M. (2007). Emerging patterns of risk factor make-up enable subclassification of rheumatoid arthritis. *Arthritis Rheum.* 56: 1728–1735.

Van Dongen, H., van Aken, J., Lard, L.R. et al. (2007). Efficacy of methotrexate treatment in patients with probable rheumatoid arthritis: a double-blind, randomized, placebo-controlled trial. *Arthritis Rheum.* 56: 1424–1432.

Young-Min, S., Cawston, T., Marshall, N. et al. (2007). Biomarkers predict radiographic progression in early rheumatoid arthritis and perform well compared with traditional markers. *Arthritis Rheum.* 56: 3236–3247.

Reference

1 Arnett, F.C., Edworthy, S.M., Bloch, D.A. et al. (American Rheumatism Association) (1988). The American Rheumatism Association 1987 revised criteria for the classification of rheumatoid arthritis. *Arthritis Rheum.* 31: 315–324.

19

Validating *In Vitro* Toxicity Biomarkers Against Clinical Endpoints

Calvert Louden[1] and Ruth A. Roberts[2]

[1] *Johnson & Johnson Pharmaceutical, Raritan, NJ, USA*
[2] *Apconix, Alderley Edge, Cheshire, UK*

Introduction

Biomarkers have been used for many years to predict and track biological changes in tissue and organ systems. The conceptual framework for their use was captured in the output of the Biomarkers Definitions Working Group [1], which provided a valuable working definition and some challenges to the scientific community. Here, a *biomarker* was defined as "a characteristic that is objectively measured and evaluated as an indicator of normal biological processes, pathogenic processes or pharmacological responses to a therapeutic intervention" [1]. The working group noted that biomarkers may have the greatest value in early efficacy studies and safety evaluations with applications that could include prediction and monitoring of clinical response to an intervention [1].

Biomarkers in Drug Development

Biomarkers have been used for many years as indicators of biological change. In the course of drug development, biomarkers enjoy utility both as predictors of possible efficacy and as warning signs of potential toxicity (Figure 19.1). For efficacy endpoints, biomarkers are used to address the pharmacodynamic question "What is the drug doing to the patient?" commensurate with desired outcomes such as receptor modification or blood glucose changes. For safety endpoints, biomarkers may be used as indicators of unwanted endpoints such as tissue damage or adverse hematology.

Although these two types of biomarkers serve diverse purposes, the parameters dictating their identification, development, and utility are very similar. However, there is one key aspect in which safety biomarkers differ; safety

Biomarkers in Drug Discovery and Development: A Handbook of Practice, Application, and Strategy, Second Edition. Edited by Ramin Rahbari, Jonathan Van Niewaal, and Michael R. Bleavins.

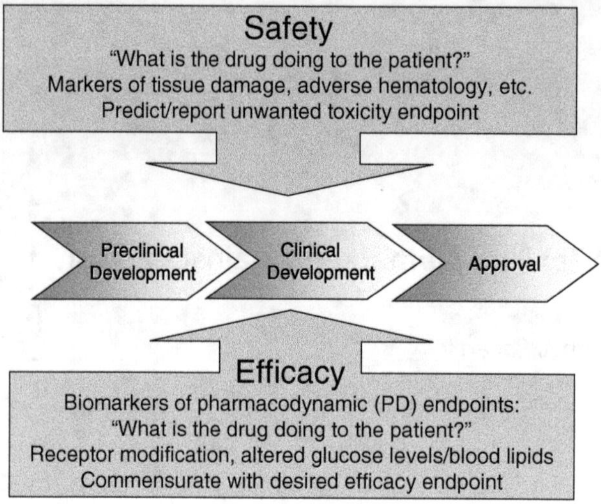

Figure 19.1 Biomarkers in clinical medicine. Biomarkers of safety and efficacy can be used together as part of the overall data set to assist in decision-making.

biomarkers are most useful when they predict rather than report damage. For example, an efficacy biomarker can provide rapid confirmation that the drug has hit its target, perhaps by confirmation of growth factor receptor inhibition in a skin biopsy. However, a safety biomarker of liver damage would ideally precede the damage (either in dose or time) to be of maximum use to the clinician. Despite this, biomarkers that report rather than predict can still be useful if they prevent further damage or permit cessation of treatment and subsequent recovery.

Toxicity Biomarkers *In Vitro*

The driving force for better clinical biomarkers of potential adverse events is to monitor patient safety, and many such biomarkers will be discovered and developed within the context of human medicine. However, from a nonclinical perspective, the discovery and development of toxicological biomarkers facilitate the translation to the clinic of knowledge that will avoid or minimize the risk of occurrence of a potential unwanted outcome identified in animal studies. Thus, it is key to identify preclinical biomarkers of the adverse response of animals to potential new drugs and to evaluate fully their relevance and applicability to humans.

Potential biomarkers can be identified *in vitro*, and their translation to *in vivo*, either in animal or humans, evaluated at a later date. This concept is developed later in this review with specific examples and case histories. One argument against this approach is that findings in animals *in vivo* or in animal-derived tissue *in vitro* may not be relevant for human safety; however, the job of the clinician is to manage the best possible therapy while minimizing the risk of adverse reactions. Thus, such data would form part of an integrated assessment of safety and efficacy that can inform preclinical and clinical decisions.

Clinically useful biomarkers of toxicity can be derived from appropriately designed preclinical experiments, and such biomarkers can significantly assist safe progression into the clinic with an approach designed to minimize patient risk. These biomarkers can be derived from *in vitro* or *in vivo* mode-of-action studies in preclinical species, then translated to potential clinical use or, as we present here, could already be established in clinical practice. This translational science approach requires close collaboration between the clinical and the preclinical scientists.

Characteristics of Biomarkers

An important consideration in the use of biomarkers is the practicality of subsequent analysis, which is likely to be far more successful and robust if based on an uncomplicated physiological or biochemical assay that can be carried out with minimal training and preexisting equipment (Figure 19.2). Nonetheless, the need for specialist techniques or expertise should not exclude a potential biomarker, especially where it contributes to advancing therapies for a serious or unmet medical need. Other practical considerations include the

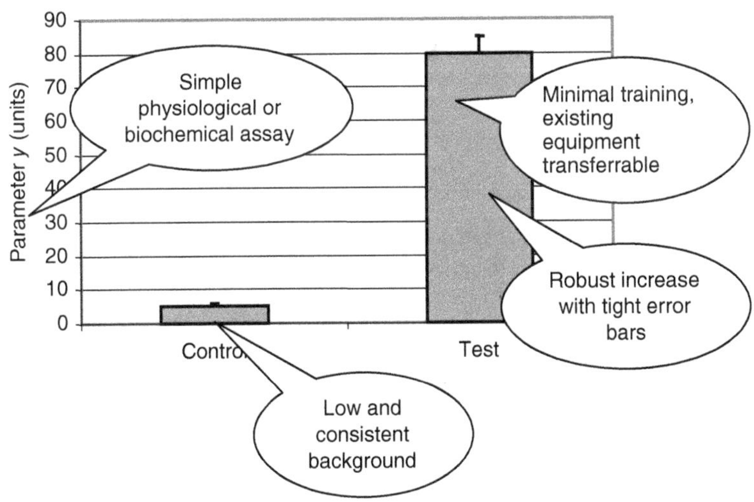

Figure 19.2 Technical aspects of biomarker utility. Biomarkers must be sufficiently robust biologically and technically to transfer across patient groups and across international boundaries. This is illustrated by the different variables that need to be overcome or accommodated in the assay.

need for assays to be transferrable across international boundaries, supported by low background and variability of the measured parameter coupled with a robust increase or decrease signifying the biological endpoint of interest. Indeed, lower and upper limits for normal values would be better if minimally influenced by age, gender, or race within the human population. Similarly, a biomarker of drug-induced toxicity would be better if it were largely unaffected by disease states expected in the patient population of interest or, indeed, any co-medication that might be present or recently administered. However, this is often not the case, and the best that can be expected is some understanding of these relationships to mitigate against difficult or erroneous interpretation. In addition, an ideal biomarker should exhibit sufficient sensitivity (not too many false negatives) and specificity (not too many false positives) to underpin clinical decisions.

As well as the technical aspects of biomarker assays, there are additional practical considerations to enhance appropriateness and usability. First, a biomarker needs to be accessible and ideally to be noninvasive. If this is not possible, a biomarker could be measurable in blood and other body fluids obtained with minimal additional impact on the subject. If this cannot be achieved, biomarkers may be detectable with increasing impact on the patient via a peripheral or central biopsy. Finally, biomarker research for the advancement of medical knowledge can be conducted after autopsy. Examples of noninvasive methods would include standard physiological measures of cardiovascular parameters such as blood pressure, heart rate, and electrocardiogram, or could be ophthalmological. Fluids could include feces, saliva, and urine, in addition to standard blood sampling. Skin and hair follicles would be examples of peripheral biopsy; muscle and liver needle samples provide the most common route of more invasive biopsy.

Biomarkers of Safety

One key aspect where biomarkers of safety differ from those of efficacy is that they preferably predict rather than report the biological effect they are flagging. Thus, the biomarker should ideally precede that biological effect either in dose (i.e. be detectable at doses below where the damage occurs) or in time (i.e. be detectable perhaps after one dose when the damage requires repeated dosing). These biomarker characteristics would provide support and confidence to the clinician during dose escalation in early clinical trials or during the transition from single to multiple doses. During later clinical development, a predictive biomarker could monitor for onset in a patient who is beginning to develop toxicity perhaps after several months of drug use. In the absence of evidence that a biomarker can predict onset, an ideal biomarker of drug-induced toxicity would characteristically show detectable changes prior to any serious or irreversible tissue injury and would also show reversibility of the biomarker signal upon removal of the toxicological insult.

In Vitro Toxicity Biomarkers

Having considered some general points on the development and validation of toxicity biomarkers, we now present several working examples that explore the concept of *in vitro* toxicity biomarkers and their translation to the clinic. Opportunities and caveats are explored with an emphasis on assisting clinical progression and decision-making. Toxicity biomarkers in the *in vitro* context are defined as quantitative measurable characteristics that serve as indicators of a pathologic process or related biochemical or molecular events.

Endocrine and Exocrine Pancreatic Toxicity

The potential for development of clinical drug-induced diabetes is well recognized, and molecules such as cyproheptadine and structurally related compounds such as cyclizine and chlorcyclizine have been shown to cause pancreatic islet toxicity. However, this is not unique to this class of structurally related compounds because a similar finding has been observed following oral administration of some quinolone antimicrobial agents. *In vivo*, this effect is characterized by cytoplasmic islet vacuolation, decreased insulin secretory granules, decreased circulating insulin levels, intolerance to glucose challenge, and increased glycosylated albumin. The working hypothesis was that these compounds were associated with functional impairment of insulin synthesis and/or secretion, and an *in vitro* screen would be invaluable in identifying this liability early in the drug development process.

The clonal insulin-producing cell line RINm5F is widely used as a model to detect compound-induced loss and/or reduction of insulin *in vitro* that is comparable to *in vivo* findings. Using this *in vitro* model, it was determined that cyproheptadine and selected analogs caused inhibition of insulin biosynthesis without any effects on pre-pro-insulin mRNA. These data suggested that inhibition of transcriptional regulation of insulin was an unlikely mode of action responsible for cyproheptadine-induced pancreatic islet cell toxicity. The limitation of the RINm5F cells is that they are not glucose-sensitive and as such cannot be used to study glucose-dependent insulin secretion. When this effect is suspected, the use of Min 6 (mouse) and HT15 (hamster) cells is recommended.

Isolated mouse pancreatic acini provide an *ex vivo* tool used to assess potential exocrine pancreatic toxicity. This is achieved by evaluating digestive enzyme secretion through an assessment of basal amylase and amylase release in response to carbachol and cholecystokinin octapeptide (CCK8). *In vitro*, cyproheptadine at high doses causes marked release of amylase without any breaks in structural integrity, thus ruling out cytotoxicity as the likely cause of the marked amylase release. Furthermore, cyproheptadine can abolish carbachol-stimulated amylase release, and this may be related to its anti-muscarinic pharmacological property. This approach can serve as an *in vitro* toxicity screen to eliminate and/or reduce undesirable effects on the exocrine as well as endocrine pancreas, and, additionally, the pathophysiology of these potential effects can be investigated.

Bone Marrow Toxicity

Within the first decade of the twenty-first century, clinical hematopoietic toxicity has accounted for 10–20% of approved drug withdrawals and was a major cause of early candidate drug attrition. Identification of potential hematopoietic toxicity as a hazard is done only when *in vivo* studies are initiated even though there is strong evidence suggesting that *in vitro* assays could have a very good predictability. A three-tiered approach that enables a large number of compounds to be evaluated at the lead optimization phase was developed to improve the assessment of potential hematotoxicity. A key question in this approach is whether there is clinical translation and predictability of this combined *in vitro, in vivo* preclinical evaluation.

This novel approach uses multiple assays, cell lines representative of broad bone marrow lineages (myeloid, erythroid, lymphoid, and stromal), and a non-hematopoietic cell line (tier 1). In general, there is a good correlation between the mouse myeloid cell line and the mouse colony-forming unit – granulocyte, macrophage (CFU-GM) assays, and compounds with lineage-specific *in vivo* bone marrow toxicity showed similar lineage specificity in the tier 1 assays. An erythrotoxicant such as chloramphenicol has an IC_{50} value of 8 μM in the erythroid cell line assay and >100 μM for the other cell lines. CFU assays (tier 2) are then run to qualify and confirm the findings of tier 1. Tier 2 consists of mouse, rat, dog, and human CFU-GM assays to predict species sensitivity, but these assays are tedious and time consuming, enabling only a limited number of compounds to be evaluated. Increased throughput was achieved when a 96-well plate CFU-GM assay was developed that yielded results that correlated well with the traditional CFU-GM assay. The data generated from the tiers 1 and 2 can trigger frontloading of *in vivo* studies to evaluate a more comprehensive hematology profile and include use of additional technologies such as flow cytometry (tier 3). In summary, a three-tiered approach using *in vitro, ex vivo,* and *in vivo* studies allows rapid, high-throughput assessment of compounds with potential chemistry-related bone marrow toxicity to be eliminated in the early preclinical stages of drug development.

Liver Toxicity

The overall goal of developing an *in vitro* toxicity testing system is for inclusion in a toolbox as an element of investigative toxicology. This should contribute to hypothesis-based research aimed at identification of mode-of-action-based toxicity biomarkers, which can extend to preclinical and clinical phases of drug development. The current preclinical and clinical diagnostic panel used to monitor hepatocellular toxicity has been quite reliable except in cases of human drug-induced liver injury, which occurs with a low incidence and prevalence, ranging from 1 in 10 000 to 1 in 100 000. This is also referred to as *idiosyncratic hepatotoxicity*, meaning that the adverse events are a consequence of individual and unpredictable patient responses. It is well recognized that there is an urgent need to identify and develop a more successful prediction model coupled with diagnostic and prognostic biomarkers for drug-induced idiosyncratic liver injury in humans. This type of human liver toxicity is multifactorial, including individual-specific responses, and because of this, it is highly unlikely that any preclinical and/or clinical studies can be powered appropriately to unmask this risk. Furthermore, it is difficult to assess the value of our preclinical toxicology studies and their relevance to predicting drug-induced idiosyncratic liver injury in humans, because we lack understanding of the mode of action and pathophysiology of this unique type of toxicity. However, there is extensive literature related to the predictability of hepatotoxicity, and there is good concordance with *in vitro* and *in vivo* assessments.

Several different strategies have been employed to detect various forms of hepatotoxicity, including, necrosis, apoptosis, and phospholipidosis (PLD). The *in vitro* hepatotoxicity predictive strategy uses hepatic microsomes; immortalized cell lines; primary hepatocyte cell cultures from humans, rodents, and non-rodents; and liver slices. The latter, in particular, enables a better approximation and evaluation of metabolites that may be formed and an assessment of their hepatotoxicity potential determined and *in vitro–in vivo* comparisons made. For example, using this approach, studies were conducted to characterize the metabolic profile of troglitazone, an approved drug that has now been withdrawn from the market because of hepatotoxicity and hepatic failure requiring liver transplants. Data from a series of studies using rat and human hepatic microsomes as well as *in vivo* rat studies suggested that troglitazone hepatotoxicity was caused by several reactive intermediates covalently binding to hepatic proteins that may undergo redox cycling and induce oxidative stress, causing cell damage. This hypothesis was strongly supported by the data suggesting that not only is troglitazone an inducer of P4503A, but it is also responsible for metabolism of the thiazolidinedione ring, a key structural element of many of the peroxisome proliferator–activated receptor (PPAR) agonists. Taken together, troglitazone acts as an inducer of enzymes that catalyzes its biotransformation to chemically reactive intermediates; this autoinduction of its own metabolism is ultimately detrimental to the cells. Additionally, this combined approach was able to show that in humans these intermediates could be formed, thus identifying a potential etiology for troglitazone-induced hepatotoxicity in humans. Utilization of liver slice technology, co-cultures, and whole liver perfusion can be quite useful, but these are low-throughput assays that are not always suitable for screening. The advantage is that toxicity to the biliary and other hepatic cellular constituents can be assessed. Thus, predicting the hepatotoxicity potential using *in vitro* cell culture systems can also employ the well-recognized preclinical and clinical biomarkers of hepatocellular damage, such as ALT (alanine aminotransferase), AST (aspartate aminotransferase), GLDH (glutamate dehydrogenase), MDH (malate dehydrogenase), PNP (purine nucleoside phosphorylase), and PON-1 (paraoxonase-1). Other enzymes, such as ALP (alkaline phosphatase), GGT (γ-glutamyl transferase), and 5′NT (5′-nucleotidase), can also be used when toxicity to the biliary tree is suspected.

Cytotoxicity (necrosis and apoptosis) *in vitro* is the primary endpoint, and, as such, biomarkers for potential hepatotoxicity clinical monitoring should include necrosis as well as apoptosis. Evidence suggests that this may be possible by measuring the soluble pool of cytokeratin 18 (CK18) and the caspase-cleaved CK18 fragments in conjunction with the traditional markers. The *in vitro* surrogate system can also provide valuable information on the suitability of the appropriate biomarker and potential mechanism of action related to toxicity, such as mitochondrial dysfunction, metabolic pathways, and/or CYP450 induction.

In summary, *in vitro* hepatocyte cell culture systems and tissues derived from *in vivo* studies will identify the appropriate biomarker for monitoring preclinical and clinical hepatotoxicity, but these data can also contribute to hypothesis-driven mode-of-action investigative toxicity studies. This will include molecular profiling, metabonomics, transcriptomics, and proteomics, all of which are useful tools that could aid in the identification of novel biomarkers of hepatotoxicity.

Phospholipidosis

PLD is characterized by concentric-layered multilamellar intracellular lysosomal inclusion bodies that are often composed of complex phospholipids, parent drug, and/or metabolites. Accumulation will often occur in several different cell types of the hepatobiliary, immune, and nervous systems. Typically, hepatocytes, biliary epithelial cells and macrophages of lymph nodes, and pulmonary alveoli and ganglia and nonganglia neuronal cell bodies of the central nervous system can be affected. The evidence to date suggests that PLD is a structural-related toxicity of cationic amphiphilic compounds irrespective of pharmacologic action. The finding of PLD may also be associated with inflammation, severe organ damage, and possibly impairment of immune function. Although drugs are marketed that cause PLD preclinically and clinically, this is an undesirable profile for potentially new candidate drugs, and, as such, this liability should be identified and avoided early in the drug discovery process. Therefore, high-throughput *in vitro* predictive screens can add value, particularly if PLD potency can be ranked and supported by *in vivo* data.

Several methodologies have been evaluated to assess neutral and phospholipid content as an index of PLD in cells growing in culture. However, accumulation of phospholipid probe *N*-(7-nitrobenz-2-oxa-1,3-diazol-4-yl)-1,2-dihexadecanoyl-sn-glycero-3-phosphoethanolamine (NBD-PE) as a result of cytotoxicity induces false-positive results, particularly at high concentrations. Recently, a high-throughput, validated, predictive, sensitive, and selective multichannel fluorescence-based *in vitro* PLD assay was developed to reduce the false-positive limitation of cytotoxicity. This assay uses I-13.35 adherent mouse spleen macrophages cultured in 96-well plates with fluorescent-tagged phospholipids. Cells with an intact nucleus were differentiated from dead cells using ethidium staining and cell gating that rejects dead cells. Using this improved technique, 26 of 28 positive phospholipidogenic compounds were identified. These findings aided application of this methodology to other techniques, such as flow cytometry, which may be used in preclinical toxicology studies and clinical trials. For example, flow cytometric analysis coupled with Nile Red staining was used to detect neutral and phospholipids in a monocyte cell line, U397. Application of this methodology was utilized in (*in vivo*) toxicology studies, and this raised the possibility that preclinical toxicology and clinical assessment of PLD could be done using peripheral blood cells and flow cytometry.

Conclusions

Overall, there is a compelling need for *in vitro* toxicity biomarkers for clinical endpoints. Toxicity biomarkers in the *in vitro* context are defined as quantitative measurable characteristics that serve as indicators of a pathologic process or related biochemical or molecular events. Conceptually, bridging biomarkers of toxicity should include not only the traditional parameters of biofluid and physiological measurements but also measurable endpoints that could serve as indicators of potential adverse events preclinically and clinically even though they have been derived from *in vitro, ex vivo*, and *in vivo* studies. The use and application of this combination will undoubtedly improve the toxicologist's ability to identify human hazards so that the appropriate risk assessment and management strategy can be developed. In developing *in vitro* assays to identify biomarkers with potential clinical application and utility, a clear understanding and determination of what that biomarker will assess must be defined. For example, exaggerated pharmacologic action of a molecular target may be associated with an undesirable effect resulting in toxicity. In such cases, single/multiple markers and/or assays may be required for utilization in screens during the early drug discovery phase, with continued assessment in preclinical and clinical development. If successful, these data can be invaluable in deriving the therapeutic index of a drug. In the case of a specific target organ toxicity, the ultimate goal is to identify biomarkers of toxicity that can be used as a general screen that reflects cellular damage, regardless of mode of action in a high-throughput manner. Therefore, it is imperative that *in vitro* assays and the appropriate platforms be developed to identify relevant toxicity biomarkers that will be useful during preclinical and clinical development.

Reference

1 Atkinson, A.J. Jr., Colburn, W.A., DeGruttola, V.G. et al. (2001). Biomarkers and surrogate endpoints: preferred definitions and conceptual framework. *Clin. Pharmacol. Ther.* 69: 89–95.

Part VI

Biomarkers in Clinical Trials

20

Opportunities and Pitfalls Associated with Early Utilization of Biomarkers: A Case Study in Anticoagulant Development

Kay A. Criswell

Westbrook Biomarker & Pharmaceutical Consulting, LLC, Westbrook, CT, USA

Introduction

Recent indicators suggest a potential upturn in pharmaceutical R&D productivity. In 2014, the U.S. Food and Drug Administration (FDA) approved 41 new NDA (new drug application) and BLA (biologics license application), the highest number of approvals in a decade [1]. An overall decline in development pipelines suggested that the pharmaceutical industry is improving in their ability to "fail fast." Additionally, the number of terminated projects in Phase III also declined, indicating selection of better candidates with improved safety profiles and efficacy [1]. Looking further into the 41 successful new applications in 2014, 11 new molecular entities (NMEs) were for oncology indications and seven received orphan drug status [1]. This indicates that the pharmaceutical industry is also getting better at selecting the right patient population for individual drugs. Biomarker development and utilization is a critical component in target selection, optimization of safety signals, and proof of efficacy. Understanding the opportunities and pitfalls associated with early biomarker utilization in preclinical testing and clinical trials will help sustain forward progress in pharmaceutical R&D.

Studying the effects of markers on disease processes is not new. It has existed for centuries, as evidenced by such obsolete practices as tasting urine to determine the diabetic status of a patient. Utilization of biomarkers for disease diagnosis and prognosis has progressed dramatically during this century with the improved diagnostic power, reliability, and speed of routine clinical pathology, biochemical, and genomic testing. However, when studying the specific effects of a drug on a specific indicator of activity, utilization of biomarkers has a more recent history. One of the early documented cases is the trials of lamotrigine, which effectively utilized the electroencephalograph (EEG) as a

Biomarkers in Drug Discovery and Development: A Handbook of Practice, Application, and Strategy, Second Edition. Edited by Ramin Rahbari, Jonathan Van Niewaal, and Michael R. Bleavins.

biomarker and demonstrated decreased epileptiform activity [2]. Thirty years after the lamotrigine study, the urgency for relevant and early biomarkers of safety and efficacy during drug development has been recognized, and it is not uncommon that each new drug development program is accompanied by a request for biomarker development.

Not only has the frequency of biomarker development exploded, but the timing of biomarker utilization has been pushed earlier in development. There is an ever-increasing demand to test and screen potential candidates before committing to studies conducted under good laboratory practices (GLPs) and progression to human clinical trials. Two problems inherently complicate the use of early biomarkers to preclinical outcomes: (i) truly understanding the disease and therapeutic intervention processes and (ii) recognizing species-to-species differences that may alter the translation and application of the biomarker.

Understanding the disease process appears obvious, but understanding the target is still a key area of focus for decreasing candidate molecule attrition. Regardless of the expanding body of knowledge surrounding disease processes, infectious diseases may be one example where the disease process is best understood. Successful therapeutic intervention and drug sensitivity can be predicted accurately with well-characterized biomarkers of causal organism growth and survival [3]. For biomarkers to be truly effective in drug development, they need to reflect a highly specific biochemical or molecular change that occurs in the disease process, is altered by the therapeutic intervention, and occurs prior to any downstream effects on clinical endpoints. This is a lofty goal, as most disease processes affect multiple pathways, and feedback mechanisms further complicate the overall biology. The added impact of non-translatable, species-specific characteristics makes daunting the task of developing biomarkers early in development. There are, however, diseases and biological pathways that have a fairly broad acceptance regarding the essential components. Additionally, certain areas of drug development target activation or inhibition of a highly specific molecule or receptor within those pathways. These conditions facilitate biomarker success when conducted early in the drug discovery process.

The coagulation pathway is an example of a well-characterized model, and specific inhibitors of this pathway are being explored actively as new candidates for anticoagulant therapy. One of the documented areas is the utilization and challenge of early implementation of biomarkers to predict the clinical outcomes of factor Xa (FXa) inhibitor compounds. Nearly one million people are affected by deep vein thrombosis or pulmonary embolism in the United States alone, and 10–30% of these individuals will die within one month of diagnosis [4]. Although these conditions have been treated for many years with the vitamin K antagonist warfarin (coumadin), or with intravenous or subcutaneous heparin administration, the opportunity to provide therapeutic intervention with an improved safety profile, fewer side effects, less intra- and intersubject variability, a better route of administration, and less need for continuous coagulation monitoring is attractive. Novel coagulation therapeutics have generated tremendous interest due to the medical need and opportunity to improve on current treatment strategies. FXa occupies a pivotal position within the coagulation cascade and is a highly attractive candidate as a target for novel coagulation intervention. This enzyme links the intrinsic and extrinsic coagulation pathways and is the rate-limiting step in thrombin formation [5].

Changes within the intrinsic coagulation pathway are routinely monitored by a clotting assay called the *activated partial thromboplastin time* (aPTT), whereas changes in the extrinsic pathway are assessed by increases or decreases in another clotting assay called the *prothrombin time* (PT) (5, Figure 20.1). Both assays are assessed spectrophotometrically or mechanically and are expressed as the time to clot formation after addition of an exogenous clot-activating agent. The aPTT is used routinely to monitor the safety and level of anticoagulation of heparin therapy, while coumadin treatment is monitored via the PT assay. Safety profiles for these compounds have been established following years of use so that fold increases over the predose aPTT or PT can be used to predict therapeutically effective doses of these drugs versus levels associated with inadequate or excessive coagulation. Additionally, aPTT and PT are readily available as routine and highly standardized clinical pathology assays. Testing reagents are standardized, and instrumentation has

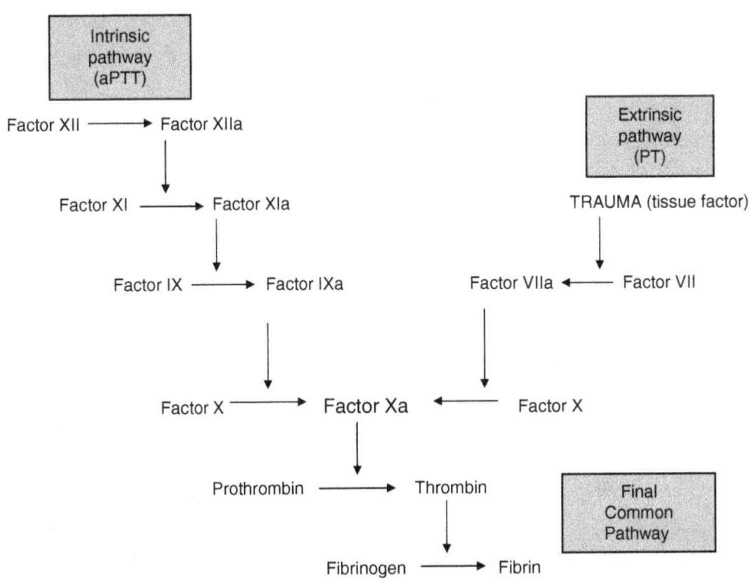

Figure 20.1 Intrinsic and extrinsic coagulation pathways.

undergone rigorous scrutiny to pass FDA requirements because of its routine use in clinical diagnostics. The reliability of aPTT and PT as biomarkers of anticoagulant therapy safety is also well established. Since activation of factor X is required for completion of both the intrinsic and extrinsic pathways, it is logical that inhibition of factor X activation should prolong both PT and aPTT.

It would be anticipated that early assessment of novel anticoagulants such as FXa inhibitors could readily be assessed with the aPTT and PT assays. Furthermore, if the new compound does not require metabolic activation, *ex vivo* incubation of human plasma with the compound of interest, followed by aPTT and PT assessment, may provide a reliable method to rapidly assess efficacy. Despite their reliability and acceptability, PT and aPTT do not always reflect the anticoagulant activity of novel compounds. Direct factor Xa inhibitors are being used clinically as substitutes for vitamin K antagonists or low molecular weight heparin [6, 7]. Rivaroxaban is an oral direct inhibitor of activated factor X [8]. In a rat venous model of the inferior vena cava, rivaroxaban produced dose-dependent inhibition of FXa and an increase in PT [8]. An inhibition of 32% was associated with a 1.8-fold increase in PT, and nearly 100% inhibition produced a 3.2-fold increase in PT. However, in a rabbit model, 92% inhibition of FX was associated with only a modest 1.2-fold increase in PT. This demonstrates that species-specific sensitivity is one of the problems associated with developing and monitoring the newer anticoagulants with standard coagulation parameters. Even coumadin provides a good example of species-specific effects that allows it to be a useful therapeutic for humans while also being an effective pesticide for rodents. Despite this challenge, utilization of coagulation testing for *in vivo* and *ex vivo* screening is well documented in the development of FXa inhibitors [6–11].

Rarely is a single biomarker considered definitive for evaluation of safety or efficacy. The drug development approach for Otamixaban and DU-176b incorporated a series of clotting parameter assays and assays to measure the effects on thrombus formation [9–11]. Clotting parameters would be the preferred biomarkers based on the ability to monitor in plasma, availability of relatively simple and reproducible assays, predictable cost, and the ability to easily monitor a clinical population. The evidence, however, is not established showing that clotting assays and thrombus formation assays are interchangeable.

For Otamixaban, *in vitro* coagulation parameters were assessed for their ability to produce a doubling of PT and aPTT. This allowed ranking of anticoagulant effects by species (rabbit > human > monkey > rat > dog) [10]. Additionally, aPTT was a more sensitive biomarker in all species, with aPTT doubling occurring at drug concentrations that were less than half the concentration required for PT doubling. Multiple pharmacological models of thrombosis in rats, dogs, and pigs were also conducted with Otamixaban. In rats, thrombus mass was markedly reduced by nearly 95%, with a corresponding increase in aPTT of 2.5-fold and in PT of 1.6-fold [10]. In contrast, intravenous administration of 1, 5, or 15 μg/ml Otamixaban in the pig model effectively eliminated coronary flow reserves related to this stenosis model at the middle and high dose. PT was also prolonged at the middle and high dose, but aPTT was prolonged only at the high dose. Although pigs were not listed as assessed in the species-specificity model, it suggests that the clotting parameter of choice may vary per species and may not correlate well with thrombosis assays. Furthermore, clinical trial outcomes showed that at anticipated antithrombotic and therapeutic concentrations of 100 ng/ml Otamixaban, neither PT nor aPTT changed appreciably. In contrast, alternative clotting parameters such as the HepTest clotting time and the Russell viper venom clotting time showed substantial prolongation, again suggesting that alternatives to standard PT and aPTT may be preferable [10].

Further work with the oral FXa inhibitor DU-176b provides additional evidence that selection of the right biomarker and appropriate correlation to functional assays is critical. This study was conducted in 12 healthy male volunteers [11]. The antithrombotic effect of DU-176b was assessed by measuring the difference in size of acutely formed, platelet-rich thrombus, pre- and post-drug administration using a Badimon perfusion chamber model under low and high shear force. Subjects received a single 60-mg dose of DU-176b, and pharmacokinetic and pharmacodynamic assessments were conducted at 1.5, 5, and 12 hours post-dosing. Pharmacodynamic assessments included PT, international normalized ratio (INR), aPTT, thrombin generation, and anti-factor Xa activity. Drug levels were also assessed. Badimon chamber results demonstrated a strong antithrombotic effect at 1.5 hours with a progressive return toward baseline by 12 hours. All pharmacokinetic endpoints showed significant change from pre-treatment, suggesting that any of the parameters might be an effective biomarker of DU-176b safety and/or efficacy. However, a close statistical look at these data raised some questions. A comparison of drug concentration level to anti-factor Xa activity and clotting parameters showed the strongest correlation with anti-factor Xa activity ($r^2 = 0.85$), a similar correlation with PT and INR ($r^2 = 0.795$ and 0.78, respectively), but a fairly weak correlation with aPTT ($r^2 = 0.40$). This suggests that although Otamixaban and DU-176b are both FXa inhibitors, arbitrary selection of PT or aPTT as a better predictor of drug concentration is problematic. Furthermore, when the antithrombotic effects of DU-176 assessed by Badimon chamber were compared to those obtained by clotting parameters, the correlation was even more challenging. PT showed a correlation of $r^2 = 0.51$ at both high and low shear stress, and the correlation with aPTT was only $r^2 = 0.39$ and 0.24 [11]. This suggests that although aPTT is used for monitoring of heparin therapy and PT is utilized for clinical safety of coumadin, their routine use as factor Xa inhibitors by themselves is insufficient.

A Case Study Data with an Experimental FXa Inhibitor

Beyond the published literature that is available, data and personal observations collected during the development of another FXa inhibitor at Pfizer Global Research & Development are now provided. This case study approach to biomarker utilization during anticoagulant development demonstrates the testing and decision paths used for a novel molecule. Development of this particular FXa inhibitor ultimately was discontinued even though it had good activity, as the compound required intravenous administration and as such lacked marketability compared to oral FXa inhibitors. However, the lessons learned provide further documentation of species-specific and interpretational complications that arise with the utilization of accepted coagulation biomarkers to monitor anticoagulant efficacy and safety for FXa inhibitors.

Dose Selection for *Ex vivo* Experiments

In designing *ex vivo* experiments to evaluate potential biomarkers, selection of the appropriate drug concentration is critical. Furthermore, when experiments are conducted in multiple species, selection of the same drug concentration for all species is typically not ideal, due to species-specific drug sensitivity and pharmacokinetics. Factor X concentrations vary by species, and the level of FXa inhibition is also variable. Therefore, the concentrations of FXa inhibitor utilized in this *ex vivo* evaluation were selected to achieve a range of FXa inhibition that was modest to nearly complete in all species examined. Pharmacology studies predicted that this FXa inhibitor would result in species-specific factor Xa sensitivity in the order human > dog > rat. Interestingly, this species specificity was not the same as that observed with Otamixaban [9, 10], demonstrating that extrapolating biomarker data even between compounds in the same class may be misleading. For this developmental FXa inhibitor, human plasma was spiked to obtain final drug concentrations of 0, 0.2, 0.6, 1.2, and 6.0 μg/ml. Drug concentrations of 0, 0.4, 2.0, 8.0, and 15.0 μg/ml were selected for dog assessments, and 0, 1.0, 4.0, 12.0, and 24.0 μg/ml were used for *ex vivo* assessments in rats to achieve a comparable range of FXa inhibition compared to that observed in human samples.

Thromboplastin is the reagent that induces clot formation in the PT assay. There is ample documentation that the type and sensitivity of thromboplastin is a critical factor in the effective and safe monitoring of coumadin administration [12–15]. To minimize this variability in PT assays, a calibration system was adopted by the World Health Organization (WHO) in 1982. This system converts the PT ratio observed with any thromboplastin into an INR. This value was calculated as follows: INR = observed PT ratio^c, where the PT ratio is subject PT/control PT and ^{c} is the power value representing the international sensitivity index (ISI) of the particular thromboplastin [16]. This system has proven to be an effective means of monitoring human oral anticoagulant therapy with coumadin and has been implemented almost universally. It allows individuals to be monitored at multiple clinics using varying reagents and instrumentation while still achieving an accurate assessment of true anticoagulation. However, there is little or no information regarding selection of thromboplastin reagents or use of the INR for monitoring of FXa inhibitors.

Typically, the higher the ISI value, the less sensitive the reagent and the longer the PT time produced. The most commonly used thromboplastin reagents for PT evaluation are either rabbit brain thromboplastin (of variable ISI values, depending on manufacturer and product) or human recombinant thromboplastin, typically with an ISI of approximately 1.0. Use of the INR is accepted as a more relevant biomarker of anticoagulant efficacy than are absolute increases in PT alone, at least for coumadin therapy [15]. To more fully evaluate the effect of this FXa inhibitor on INR, PT was evaluated using rabbit brain thromboplastin, with ISI values of 1.24, 1.55, and 2.21 and a human recombinant thromboplastin (0.98 ISI). Although either human recombinant thromboplastin or rabbit thromboplastin are considered acceptable reagents for the conduct of PT testing, it was unclear whether these reagents would produce similar results in the presence of an FXa inhibitor or whether the sensitivity of the thromboplastin itself would affect results.

Effect on Absolute Prothrombin Time

PT data obtained using rabbit brain thromboplastin with the three increasing ISI values during these *ex vivo* studies are presented in Table 20.1. The source and sensitivity of thromboplastin used in the assay affected the absolute PT value in all species, clearly demonstrating the need to standardize this reagent in preclinical assessment and to be cognizant of this impact in clinical trials or postmarketing, when reagents are less likely to be standardized. As anticipated, addition of the FXa inhibitor to plasma under *ex vivo* conditions increased the PT in a dose-dependent manner. This increase in PT time length was observed regardless of the ISI value (sensitivity) of the thromboplastin used and occurred in all species (Table 20.1). Although the absolute time for clot formation generally increased with increasing ISI, this was not true for all assessments.

Table 20.2 summarizes the maximum change in PT and the range of variability when rabbit brain thromboplastin of varying ISI values was compared to human recombinant thromboplastin. Again, in general, the higher the PT value, the larger the deviation between reagent types. For example, although there was a 2.1-second difference between human and rabbit thromboplastin in untreated human plasma, the difference increased from 5.5, 9.2, 12.0, 13.6, and 38.2 seconds in samples containing 0.2, 0.6, 1.2, 1.8, or 6.0 μg/ml FXa inhibitor, respectively. Dogs were substantially less sensitive to the type of thromboplastin used and showed smaller maximum changes in PT values. In contrast, rat PT values were highly dependent on the source of thromboplastin, and samples tested with rabbit brain thromboplastin were markedly longer than those with

Table 20.1 *In vitro* effect of an experimental factor Xa inhibitor on absolute prothrombin time using a rabbit brain thromboplastin.

Concentration of factor Xa inhibitor (μg/ml)	International sensitivity index[a)]			
	0.98	1.24	1.55	2.21
PT (s) – human plasma				
0	11.4 ± 0.11	13.4 ± 0.17[a)]	12.9 ± 0.13[a)]	10.9 ± 0.14
0.2	18.3 ± 0.37	23.8 ± 0.50[a)]	23.4 ± 0.44[a)]	16.8 ± 0.43[a)]
0.6	30.8 ± 0.76	37.3 ± 0.86[a)]	39.9 ± 0.70[a)]	27.0 ± 0.93[a)]
1.2	47.0 ± 1.98	51.4 ± 1.51[a)]	58.9 ± 1.08[a)]	38.2 ± 1.39[a)]
1.8	61.2 ± 2.04	61.8 ± 1.58	74.8 ± 1.54[a)]	48.3 ± 2.08[a)]
6.0	133.0 ± 3.82	111.4 ± 3.17[a)]	152.3 ± 3.08[a)]	94.8 ± 4.12[a)]
PT (s) – dog plasma				
0	7.8 ± 0.14	8.4 ± 0.07[a)]	7.1 ± 0.07	6.7 ± 0.06[a)]
0.4	11.0 ± 0.26	11.0 ± 0.13	10.5 ± 0.19	9.2 ± 0.15[a)]
2.0	17.6 ± 0.42	16.6 ± 0.23[a)]	17.9 ± 0.41	15.0 ± 0.31[a)]
8.0	32.0 ± 0.98	27.6 ± 0.49[a)]	33.6 ± 0.92	27.1 ± 0.63[a)]
15.0	45.4 ± 1.52	36.5 ± 0.73[a)]	46.7 ± 1.35	36.8 ± 0.88[a)]
PT (s) – rat plasma				
0	9.1 ± 0.05	15.1 ± 0.07[a)]	17.1 ± 0.15[a)]	13.1 ± 0.07[a)]
1.0	13.3 ± 0.06	20.7 ± 0.13[a)]	31.3 ± 0.27[a)]	23.2 ± 0.19[a)]
4.0	20.1 ± 0.29	30.8 ± 0.23[a)]	51.8 ± 0.52[a)]	36.9 ± 0.52[a)]
12.0	31.1 ± 0.72	46.2 ± 0.49[a)]	83.6 ± 0.84[a)]	56.2 ± 0.84[a)]
24.0	42.7 ± 1.09	60.6 ± 0.73[a)]	109.4 ± 0.79[a)]	75.4 ± 1.60[a)]

a) Mean value ± S.E.M. for 10 individual subjects significantly different from 0.98 ISI thromboplastin means at 5% level by *t*-test, separately by increasing ISI value for individual rabbit thromboplastins.

human recombinant thromboplastin. Rats showed this high level of thromboplastin dependence even in untreated control samples. Variability of PT in FXa inhibitor–treated human and dog plasma was similar to that observed in controls and did not change appreciably with increasing concentration of drug (Table 20.2). FXa inhibitor–treated rat plasma showed an approximately twofold increase in variability compared to control.

Effect on PT/Control Ratio and INR

Generating a PT/control ratio by dividing the number of absolute seconds in the treated sample by the number in the control (untreated) sample provides

Table 20.2 Comparison of human recombinant thromboplastin and rabbit brain thromboplastin on prothrombin time in plasma samples containing increasing concentrations of factor Xa inhibitor.[a)]

Species	Intended drug concentration (μg/ml)	Maximum change in PT[b)]	Range of variability[c)] (%)
Human	0	2.1	−4 to +18
	0.20	5.5	−8 to +30
	0.60	9.2	−12 to +30
	1.20	12.0	−19 to +25
	1.80	13.6	−21 to +22
	6.00	38.2	−29 to +14
Dog	0	1.1	−15 to +7
	0.40	1.8	−7 to −5
	2.00	2.5	−14 to +2
	8.00	4.9	−15 to +5
	15.00	8.9	−20 to +3
Rat	0	8.0	+43 to +88
	1.00	18.0	+56 to +136
	4.00	31.7	+53 to +158
	12.00	52.5	+48 to +168
	24.00	66.7	+42 to +156

a) Samples spiked with a factor Xa inhibitor *in vitro*.
b) Maximum change in prothrombin time compared to 0.98 ISI human recombinant thromboplastin.
c) Variability of three increasing ISI levels of rabbit brain thromboplastin compared to human recombinant.

a second method of assessing PT. If the ISI of the thromboplastin used is close to 1.0, the INR should be similar to the PT/control ratio (Table 20.3). The PT/control ratio could be used effectively to normalize thromboplastin differences in untreated human, dog, or rat samples. At predicted efficacious concentrations of FXa inhibitor, the PT/control ratio effectively normalized reagent differences. However, at high concentrations of FXa inhibitor, particularly in the rat, this method lacked the ability to normalize results effectively.

Table 20.4 shows the corresponding INR values obtained in human, dog, and rat plasma when assessed with rabbit brain thromboplastins of increasing ISI. As anticipated, the PT/control ratio and INR were similar when the ISI was approximately 1. In contrast to the modest differences in PT when expressed as either absolute seconds or as a ratio compared to control value, the INR showed dramatic increases (Table 20.3). The magnitude of the INR value rose consistently with increasing ISI value and was marked. At the

Table 20.3 *In vitro* effect of an experimental factor Xa inhibitor on prothrombin time/control ratio using a rabbit brain thromboplastin.

Concentration of factor Xa inhibitor (μg/ml)	International sensitivity index[a]			
	0.98	1.24	1.55	2.21
PT/control ratio (:1) – human plasma				
0	1.0 ± 0.00	1.0 ± 0.00	1.0 ± 0.00	1.0 ± 0.00
0.2	1.6 ± 0.02	1.8 ± 0.12	1.8 ± 0.21	1.5 ± 0.02
0.6	2.7 ± 0.05	2.8 ± 0.04	2.9 ± 0.03	2.5 ± 0.06
1.2	4.1 ± 0.15	3.8 ± 0.08	4.6 ± 0.06[a]	3.5 ± 0.09
1.8	5.4 ± 015	5.6 ± 0.07	5.8 ± 0.08	5.4 ± 0.15
6.0	11.7 ± 0.26	8.3 ± 0.17[a]	11.9 ± 0.16	8.7 ± 0.29[a]
PT/control ratio (:1) – dog plasma				
0	1.0 ± 0.00	1.0 ± 0.00	1.0 ± 0.00	1.0 ± 0.00
0.4	1.4 ± 0.01	1.3 ± 0.01	1.5 ± 0.02	1.4 ± 0.02
2.0	2.2 ± 0.02	2.0 ± 0.02	2.3 ± 0.04	2.3 ± 0.04
8.0	4.1 ± 0.07	3.3 ± 0.05[a]	4.3 ± 0.10	4.1 ± 0.08
15.0	5.8 ± 0.14	4.4 ± 0.08[a]	6.5 ± 0.16[a]	5.5 ± 0.11
PT/control ratio (:1) – rat plasma				
0	1.0 ± 0.00	1.0 ± 0.00	1.0 ± 0.00	1.0 ± 0.00
1.0	1.5 ± 0.02	1.4 ± 0.02	1.6 ± 0.02	1.6 ± 0.01
4.0	2.2 ± 0.03	2.0 ± 0.04	3.0 ± 0.02[a]	2.8 ± 0.03[a]
12.0	3.4 ± 0.08	3.1 ± 0.07	4.9 ± 0.06[a]	4.3 ± 0.05[a]
24.0	5.5 ± 0.17	7.3 ± 0.24[a]	15.3 ± 0.21[a]	11.3 ± 0.28[a]

a) Mean value ± S.E.M. for 10 individual subjects significantly different from 0.98 ISI thromboplastin means at 5% level by *t*-test, separately by increasingly ISI value for individual rabbit thromboplastins.

highest dose tested, the INR ranged from 11.1 with the 0.98 ISI reagent to 121.9 with the 2.21 ISI reagent in human samples, 5.6–43.4 in dogs, and 4.6–48.1 in rats. Assessment of PT in human, dog, or rat plasma containing this developmental FXa inhibitor was affected by the ISI of the thromboplastin selected for the assay. However, it was not affected to the same degree as was coumadin. Consequently, using the correction calculation designed for coumadin fluctuations to obtain an INR with the Pfizer Factor Xa inhibitor molecule grossly exaggerated the INR value. Although INR has been used clinically to monitor anticoagulant status during coumadin therapy, it probably should not be used with FXa inhibitor administration. Coumadin therapy typically produces INR values of 2, 4, and 6 as therapeutic, above therapeutic, and critical levels, respectively. INR values of 10–15 may be observed in acute

Table 20.4 *In vitro* effect of an experimental factor Xa inhibitor on international normalization ratio using a rabbit brain thromboplastin.

Concentration of factor Xa inhibitor (μg/ml)	International sensitivity index[a)] 0.98	1.24	1.55	2.21
International normalized ratio (:1) – human plasma				
0	1.0 ± 0.10	1.0 ± 0.02	1.0 ± 0.02	1.0 ± 0.03
0.2	1.6 ± 0.04	2.0 ± 0.06[a)]	2.6 ± 0.07[a)]	2.6 ± 0.16[a)]
0.6	2.7 ± 0.07	3.6 ± 0.10[a)]	5.7 ± 0.16[a)]	7.5 ± 0.57[a)]
1.2	4.0 ± 0.16	5.3 ± 0.19[a)]	10.5 ± 0.29[a)]	16.2 ± 1.31[a)]
1.8	5.2 ± 0.17	6.7 ± 0.21[a)]	15.3 ± 0.47[a)]	27.4 ± 2.618
6.0	11.1 ± 0.31	13.8 ± 0.49[a)]	46.0 ± 1.41[a)]	121.9 ± 11.54[a)]
International normalized ratio (:1) – dog plasma				
0	1.0 ± 0.03	1.0 ± 0.01	0.9 ± 0.02	1.0 ± 0.0
0.4	1.4 ± 0.04	1.4 ± 0.01	1.7 ± 0.05[a)]	2.0 ± 0.08[a)]
2.0	2.2 ± 0.05	2.3 ± 0.04	3.8 ± 0.14[a)]	6.0 ± 0.28[a)]
8.0	4.0 ± 0.12	4.4 ± 0.10[a)]	10.1 ± 0.43[a)]	22.1 ± 1.19[a)]
15.0	5.6 ± 0.18	6.2 ± 0.16[a)]	16.7 ± 0.74[a)]	43.4 ± 2.31[a)]
International normalized ratio (:1) – rat plasma				
0	1.0 ± 0.03	1.0 ± 0.00	1.0 ± 0.02	1.0 ± 0.01
1.0	1.5 ± 0.02	1.5 ± 0.02	2.6 ± 0.03[a)]	3.5 ± 0.07[a)]
4.0	2.2 ± 0.03	2.4 ± 0.06[a)]	5.6 ± 0.11[a)]	9.9 ± 0.31[a)]
12.0	3.3 ± 0.07	4.0 ± 0.13[a)]	11.7 ± 0.34[a)]	25.1 ± 0.82[a)]
24.0	4.6 ± 0.12	5.6 ± 0.22[a)]	17.8 ± 0.20[a)]	48.1 ± 2.27[a)]

a) Mean value ± S.E.M. for 10 individual subjects significantly different from 0.98 ISI thromboplastin means at 5% level by t-test, separately by increasingly ISI value for individual rabbit thromboplastins.

coumadin poisoning, but INR values higher than 15 rarely occur [17]. Clearly, the magnitude of the INR obtained in this experiment (>120 in humans), combined with the incremental increase that occurred with increasing ISI value, shows that INR values in these FXa inhibitor–treated samples were an artifact of the calculation and not associated with the true anticoagulant effects of the FXa inhibitor itself. This suggests that when INR is used in clinical trials, it is important to select a thromboplastin with an ISI value close to 1.0. In this manner, the INR will closely approximate the PT/control ratio and give a true estimate of the anticoagulated status of the individual.

Table 20.5 indicates the maximum change in PT/control ratio and INR using thromboplastins with increasing ISI values (1.24–2.21). Changes in the

Table 20.5 Comparison of PT/control ratio and international normalization ratio in plasma samples containing increasing concentrations of factor Xa inhibitor.[a)]

Species	Intended drug concentration (μg/ml)	PT/control ratio (0.98 ISI)	INR (0.98 ISI)	Maximum change in PT/control ratio[b)]	Maximum change in INR[b)]
Human	0	1.00	0.99	0	0.02
	0.2	1.61	1.58	0.22	1.03
	0.6	2.70	2.65	0.38	4.88
	1.2	4.13	4.00	0.65	12.24
	1.8	5.37	5.19	0.97	22.21
	6.0	11.67	11.10	3.39	110.84
Dog	0	1.00	1.00	0	0.08
	0.4	1.42	1.41	0.10	0.61
	2.0	2.24	2.22	0.27	3.77
	8.0	4.11	4.00	0.81	18.06
	15.0	5.82	5.60	1.46	37.76
Rat	0	1.00	1.00	0	0.01
	1.0	1.45	1.45	0.38	2.08
	4.0	2.20	2.19	0.82	7.7
	12.0	3.42	3.34	1.48	21.74
	24.0	5.51	4.55	9.82	43.54

a) Samples spiked with a factor Xa inhibitor *in vitro*.
b) Results obtained by selecting the maximum result obtained with rabbit brain thromboplastin and subtracting from result obtained with human recombinant thromboplastin.

PT/control ratio were modest at drug concentrations that produced increases of fourfold or less, the maximum targeted therapeutic PT value for clinical trials. The mean PT/control ratio in human samples increased maximally from 2.7 to 3.1 at twice the therapeutic dose (0.6 μg/ml). Absolute PT and PT ratios compared to baseline values were only modestly different using thromboplastin from various manufacturers, sources (human recombinant versus rabbit), and ISI. This finding indicates that absolute PT or PT/control ratios were more effective biomarkers of FXa inhibitor concentration than was INR.

Pursuing Biomarkers Beyond PT, INR, and aPTT

Values obtained with aPTT under *ex vivo* conditions were less sensitive than PT to FXa inhibitor–induced elevations and often underestimated drug concentration (data not shown). Beyond PT, INR, and aPTT, the most commonly used assay to evaluate FXa inhibitors is probably the anti-factor Xa assay (anti-FXa). It seems logical that a parameter named anti-factor Xa assay should be the ideal biomarker for an FXa inhibitor. Additionally, this assay is used routinely in clinical settings to monitor the safety of heparin, a substance that also inhibits FXa production [17, 18]. However, this assay is little more than a surrogate marker for drug concentration. A standard curve is prepared using the administered heparin (or other FXa inhibitor), and the chromogenic assay allows determination of the drug concentration in the plasma samples via production of FXa [17–19]. For heparin, the anti-FXa assay appears relevant. Years of use has allowed the development of a strong correlation between the number of international units of heparin determined via the assay and clinical safety. Reference ranges have been defined for the assay and provide a rapid estimation of under, over, or therapeutic levels of heparin administration [18]. Variability in the anti-FXa assay has been reported and is attributable to a number of factors, including instrumentation, assay technique, specificity of the commercially available kits, heparin preparations used in generating the standard curve, and approaches to data fitting [18]. In contrast, this experience does not exist for anti-Xa values obtained during FXa inhibitor administration. Just as the PT and INR may not be as beneficial for predicting FXa inhibitor effects as they are for coumadin, it should not be assumed that the anti-FXa assays have equivalent predictivity for heparin and other FXa inhibitors. For the Pfizer developmental FXa inhibitor, the anti-FXa assay offered little more than the PT as a monitor of drug concentration. Laboratory assessment of rivaroxaban was consistent with this finding, showing that the anti-factor Xa method was a good method of assessing drug concentration but did not predict the intensity of the drug's anticoagulant activity [20].

An additional assay called the factor X clotting (FX:C) assay was also evaluated in the Pfizer trial. This assay was conducted using genetically engineered factor-deficient plasma containing serial dilutions of purified human factor X. Concentrations of factor X in plasma were then determined by extrapolation from the standard curve. Since factor X must be converted to factor Xa for clot formation to occur, a functional clotting assay for factor X can also be used to assess the effects of factor Xa inhibitors. The FX:C assay provides several unique features that may make it a valuable biomarker for monitoring factor Xa inhibitor therapy: (i) the assay provides a rapid, reliable assessment of drug concentration and the percent inhibition of FXa achieved during drug inhibitor administration; (ii) the assay can be performed on a high-throughput automated platform that is available in most hospital-based coagulation laboratories; and (iii) individual factor X concentrations range from 60% to 150% between subjects [21]. This fairly high level of baseline intersubject variability suggests that a standard dose of drug may have a substantially different impact on total factor X inhibition. The FX:C assay defines baseline factor X activity and thereby allows continued dosing to achieve a targeted factor X concentration. Literature is available concerning factor X concentrations and bleeding history in patients with either inherited or acquired factor X deficiency, so minimally there is some understanding that correlates the impact of reductions in FX:C evaluations and bleeding potential [21–23]. By determining the actual concentration of functional factor X remaining, physicians may have increased confidence in the administration of factor Xa inhibitors.

As with all the other coagulation biomarkers used for monitoring FXa inhibition, it was not immediately clear whether the FX:C assay was applicable in multiple species. *Ex vivo* experiments allowed this evaluation. To provide effective anticoagulant activity, a 30% reduction in FX:C activity was predicted to be the minimal requirement for this compound. The FXa inhibitor concentrations in the *ex vivo* experiments were selected to bracket a range of factor X inhibition predicted to range from approximately 30% to 100%. Table 20.6 shows the intended concentrations of this FXa inhibitor in each species, the

Table 20.6 Factor X activity and percent inhibition in plasma samples containing increasing concentrations of a factor Xa inhibitor.[a)]

Species	Intended drug concentration (μg/ml)	FX:C activity[b)](%)	Percent inhibition[c)]
Human	0	106.1 ± 1.9	NA
	0.2	64.3 ± 1.6	39.4
	0.6	32.2 ± 1.0	69.7
	1.2	16.5 ± 0.7	84.4
	1.8	10.0 ± 0.5	90.6
	6.0	2.3 ± 0.2	97.8
Dog	0	143.0 ± 4.5	NA
	0.4	112.9 ± 8.6	21.0
	2.0	42.4 ± 4.3	70.3
	8.0	11.1 ± 1.4	92.2
	15.0	5.4 ± 0.8	96.2
Rat	0	84.8 ± 2.8	NA
	1.0	52.6 ± 1.8	38.0
	4.0	26.2 ± 0.9	69.1
	12.0	11.8 ± 0.6	86.0
	24.0	6.7 ± 0.3	92.1

a) Samples spiked with a Factor Xa inhibitor *in vitro*.
b) Mean ± SD of 10 samples/concentration.
c) Calculated from species-specific control value. NA, not applicable.

resulting FX:C activity, and the percent inhibition achieved. Assessment of these drug concentrations induced factor Xa inhibition of approximately 20% to >90%, showing that the targeted range could be predicted and achieved in all species. These *ex vivo* experiments demonstrated that the predicted efficacious dose of 0.2–0.3 μg/ml achieved the required 30–40% inhibition of FXa, providing early confidence in the dose selection process for phase I human trials. Additionally, these early *ex vivo* studies confirmed species-specific differences. The drug concentrations required to produce similar levels of FXa inhibition across species were markedly different. The FX:C assay was used effectively in preclinical rat and dog studies with this developmental FXa inhibitor. Knowledge of the species-specific concentration of drug required to induce the required 30% inhibition of FXa drove the selection of the low dose, whereas nearly complete inhibition of FXa drove the selection of the high dose. The FX:C assay helped determine the drug concentration required for complete inhibition of factor Xa in these species and the relative bleeding risk associated with a range of factor X concentrations. Prior knowledge of the impact of this drug on FXa inhibition through fairly simple clotting assessments helped eliminate undue risks of over-anticoagulation in preclinical studies, and there was no loss of animals due to excessive hemorrhage. The data also addressed questions of whether dosing had been pushed to high enough levels when only minimal bleeding was observed at the highest dose. Because nearly 100% inhibition was achieved during the study, using higher doses was not indicated and the lack of bleeding under conditions of complete FXa inhibition in rats and dogs suggested a good safety profile. Inclusion of these biomarkers in preclinical studies provided greater confidence for selection of target stopping criteria for the first-in-human trial. The FX:C assay was translated and used as part of the first-in-human clinical trial with this compound. The FX:C assay provided data consistent with *in vitro* modeling, suggesting that it is predictive of drug concentration.

Conclusions

One of the goals for new anticoagulant therapies is a superior safety profile compared to marketed anticoagulants, thereby minimizing or eliminating the need for clinical monitoring. Although clinical monitoring with standardized coagulation assays may appear to be a simple solution to monitoring the safety and efficacy of anticoagulants, there are inherent issues that make the elimination of clinical monitoring highly desirable. The obvious factors of cost and labor are minor in comparison to the problems associated with lack of patient compliance, delayed time to achieve therapeutic benefit, and the high degree of variability in the assay itself due to instrumentation, reagents, technique, and the inherent variability among subjects. Phase I studies using biomarkers are generally more cost-effective than phase II clinical endpoint studies. Additionally, new anticoagulants pose a relatively undetermined and possible safety risk, due to the possibility of excessive bleeding. Therefore, biomarkers will continue to be essential until safety profiles can be established for this newer generation of anticoagulants. Although PT and INR are an effective reflection safety and efficacy for coumadins, they are less than ideal as biomarkers of new FXa inhibitor drugs. Assessing anti-FXa activity has been similar to drug concentration analysis for some inhibitors but fail to predict the true intensity of the factor Xa inhibitor anticoagulant activity ([20], current case study). FX:C clotting activity may provide another alternative but requires additional exploration. Regardless of the hope for the development of safer anticoagulants that are monitoring-free, the reality is that development of these drugs requires extensive patient monitoring to ensure safety. Compared to heparin and coumadin, which are monitored fairly effectively with aPTT and PT, respectively, development of the new FXa inhibitors is typically accompanied by a long list of probable biomarkers. This process is likely to continue until safety is firmly established through prolonged use and clinical experience with these agents.

It seems that most of these coagulation assays could be used as a bioassay of drug concentration and as an indicator of pharmacologic response. In Stern's evaluation of biomarkers of an antithrombin agent, he concluded that "not all biomarkers are created equal" [24]. He suggested that "if a proposed biomarker measurement requires a drug and its molecular target to be combined in the same assay, it may be more a pharmacokinetic than a pharmacodynamic assessment. Also, such assays should not be assumed to demonstrate an in vivo effect" [24]. As such, these biomarkers face a significant hurdle to replace such pharmacodynamic endpoints as the Badimon chamber. So, what does this mean for the early use of biomarkers in the development of new anticoagulants? It suggests that the greatest benefit for early utilization of coagulation biomarkers remains in allowing optimal selection of compounds, attrition of compounds without appropriate characteristics, and the opportunity to provide an early *ex vivo* assessment against marketed competitors. It also demonstrates that efforts expended in understanding species-specific and reagent differences are critical in performing those early experiments.

References

1 CMR International (2015). Pharmaceutical R&D Factbook CMR International. https://web.archive.org/web/20160618154941/http://cmr.thomsonreuters.com/pdf/Executive_Summary_Final.pdf (accessed 05 May 2019).

2 Jawad, S., Oxley, J., Yuen, W.C., and Richens, A. (1986). The effect of lamotrigine, a novel anticonvulsant, on interictal spikes in patients with epilepsy. *Br. J. Clin. Pharmacol.* 22: 191–193.

3 Smith, M.B. and Woods, G.L. (2001). In vitro testing of antimicrobial agents. In: *Henry's Clinical Diagnosis and Management by Laboratory Methods*, 20e (eds. F.R. Davey, C.J. Herman, R.A. McPherson, et al.), 1119–1143. Philadelphia, PA: W.B. Saunders.

4 Data and Statistics | DVT/PE | NCBDDD | CDC (2016). http://www.cdc.gov/ncbddd/dvt/data.html (accessed 05 May 2019).

5 Colman, R.W., Clowes, A.W., George, J.N. et al. (2001). Overview of hemostasis. In: *Hemostasis and Thrombosis: Basic Principles and Clinical Practice* (eds. R.W. Colman, J. Hirsh, V.J. Marder, et al.), 3–16. Philadelphia, PA: Lippincott Williams & Wilkins.

6 Cabral, K.P. and Ansell, J.E. (2015). The role of factor Xa inhibitors in venous thromboembolism treatment. *Vasc. Health Risk Manag.* 11: 117–123.

7 Yeh, C.H., Fredenburgh, J.C., and Weitz, J.I. (2012). Oral direct factor Xa inhibitors. *Circ. Res.* 111: 1069–1078.

8 Kakar, P., Watson, T., and Gregory-Lip, Y.H. (2007). Drug evaluation: rivaroxaban, an oral, direct inhibitor of activated factor X. *Curr. Opin. Invest. Drugs* 8 (3): 256–265.

9 Guertin, K.R. and Choi, Y.M. (2007). The discovery of the factor Xa inhibitor otamixaban: from lead identification to clinical development. *Curr. Med. Chem.* 14: 2471–2481.

10 Hylek, E.M. (2007). Drug evaluation: DU-176b, an oral, direct factor Xa antagonist. *Curr. Opin. Invest. Drugs* 8 (9): 778–783.

11 Zafar, M.U., Vorchheimer, D.A., Gaztanaga, J. et al. (2007). Antithrombotic effect of factor Xa inhibition with DU-176b: phase-1 study of an oral, direct factor Xa inhibitor using an ex-vivo flow chamber. *Thromb. Haemost.* 98: 883–888.

12 Becker, D.M., Humphries, J.E., FB, W. et al. (1993). Standardizing the prothrombin time. Calibrating coagulation instruments as well as thromboplastin. *Arch. Pathol Lab. Med.* 117: 602–605.

13 Poller, L. (1987). Progress in standardization in anticoagulant control. *Hematol. Rev.* 1: 225–228.

14 Kirkwood, T.B. (1983). Calibration of reference thromboplastin and standardization of the prothrombin time ratio. *Thromb. Haemost.* 49: 238–244.

15 Jeske, W., Messmore, H.L., and Fareed, J. (1998). Pharmacology of heparin and oral anticoagulants. In: *Thrombosis and Hemorrhage*, 2e (eds. J. Loscalzo and A.I. Schafer), 257–283. Baltimore, MD: Williams & Wilkins.

16 Crowther, M.A., Ginsberg, J.S., and Hirsh, J. (2001). Practical aspects of anticoagulant therapy. In: *Hemostasis and Thrombosis: Basic Principles*

and Clinical Practice (eds. R.W. Colman, J. Hirsh, V.J. Marder, et al.), 1497–1516. Philadephia, PA: Lippincott Williams & Wilkins.

17 Kitchn, S., Theaker, J., and Fe, P. (2000). Monitoring unfractionated heparin therapy: relationship between eight anti-Xa assays and a protamine titration assay. *Blood Coagul Fibrin.* 11: 55–60.

18 Bauer, K.A., Kass, B.L., and ten Cate, H. (1989). Detection of factor X activation in humans. *Blood* 74: 2007–2015.

19 (1998). Fifth ACCP Consensus Conference on Antithrombotic Therapy. *Chest* 119 (Suppl): 1S–769S.

20 Semana, M.M., Contant, G., Spiro, T.E. et al. (2013). Laboratory assessment of rivaroxaban: a review. *Thromb. J.* 11: 11–17.

21 Herrmann, F.H., Auerswald, G., Ruiz-Saez, A. et al. (2006). Factor X deficiency: clinical manifestation of 102 subjects from Europe and Latin American with mutations in factor 10 gene. *Haemophilia* 12: 479–489.

22 Choufani, E.B., Sanchlorawala, V., Ernst, T. et al. (2001). Acquired factor X deficiency in patients with amyloid light-chain amyloidosis: incidence, bleeding manifestations, and response to high-dose chemotherapy. *Blood* 97: 1885–1887.

23 Mumford, A.D., O'Donnell, J., Gillmore, J.D. et al. (2000). Bleeding symptoms and coagulation abnormalities in 337 patients with AL-amyloidosis. *Br. J. Haematol.* 110: 454–460.

24 Stern, R., Chanoine, F., and Criswell, K. (2003). Are coagulation times biomarkers? Data from a phase I study of the oral thrombin inhibitor LB-30057 (CI-1028). *J. Clin. Pharmacol.* 43: 118–121.

21

Integrating Molecular Testing into Clinical Applications

Anthony A. Killeen

University of Minnesota, Minneapolis, MN, USA

Introduction

The clinical laboratory plays a critical role in modern health care. It is commonly estimated that approximately 70% of all critical clinical diagnoses are to some extent dependent on a laboratory finding. The clinical laboratory has various roles in the diagnosis and treatment of disease, including determining disease risks, screening for disease, establishing a diagnosis, monitoring of disease progression, and monitoring of response to therapy. Not surprisingly, the size of the global *in vitro* diagnostics (IVDs) market is large, being estimated to reach over US$ 67 billion by 2020 [1]. Today, molecular testing is used in many areas of the clinical laboratory, including microbiology and virology, analysis of solid and hematologic tumors, inherited disorders, tissue typing, and identity testing (e.g. paternity testing and forensic testing). The growth has occurred rapidly over the last 25 years. In this chapter, we examine the principal issues surrounding the integration of molecular testing into the clinical laboratory environment.

Clinical Laboratory Regulation

The clinical laboratory environment in the United States is one of the most extensively regulated areas of medical practice and comes under the federal Clinical Laboratory Improvement Amendments Act (CLIA) of 1988 and the corresponding federal regulations (http://www.cms.hhs.gov/clia/). Any implementation of molecular diagnostic is therefore governed by the provisions of the CLIA. The history of the CLIA dates back to the 1980s, when public and congressional concern was raised by reports of serious errors being made in clinical laboratories. In response to these concerns, legislation

Biomarkers in Drug Discovery and Development: A Handbook of Practice, Application, and Strategy, Second Edition. Edited by Ramin Rahbari, Jonathan Van Niewaal, and Michael R. Bleavins.

was introduced with the intention of improving laboratory testing. These regulations cover most aspects of laboratory practice and include monitoring and clinical proficiency guidelines. Any laboratory testing that is performed in the United States for clinical purposes such as diagnosis, monitoring, deciding appropriate treatment, and establishing prognosis must be performed in a CLIA-certified laboratory. These regulations, however, do not apply to purely research studies or to early research and development work for molecular or other testing in a non-CLIA-certified environment. As soon as such testing that has genuine clinical utility is made available, it then must be performed in a certified environment.

The initial application for a CLIA certificate is usually made to the state office of the Centers for Medicare and Medicaid Services (CMS). A successful application will result in a certificate of registration, which allows a laboratory to perform clinical testing pending its first formal inspection. Depending on whether the laboratory is certified by CMS or by an accrediting organization, a successful inspection will result in a grant of either a certificate of compliance or a certificate of accreditation (Figure 21.1). These are essentially equivalent for the purposes of offering clinical testing. Accrediting organizations function as surrogates for CMS in the laboratory accreditation process and must be approved by CMS to accredit clinical laboratories. The major accrediting organizations are the College of American Pathologists (CAP), the Council on Laboratory Accreditation (COLA), the Joint Commission, the American Association of Blood Banks (AABB), the American Society for Histocompatibility and Immunogenetics (ASHI), and the American Association of Bioanalysts (AAB). Some of these, such as the ASHI, accredit laboratories that perform only

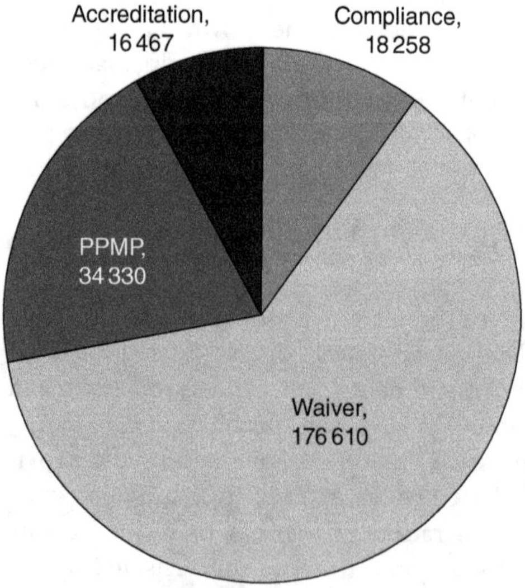

Figure 21.1 Distribution of CLIA certificates by type in non-CLIA-exempt states in 2016. Source: Data from the CLIA database, http://www.cms.hhs.gov/CLIA/.

limited types of testing. Others, such as the CAP, accredit laboratories for all types of clinical testing, including molecular diagnostic testing.

Clinical tests are categorized for the purposes of the CLIA into several levels of complexity. This categorization is the function of the US Food and Drug Administration (FDA). The type of CLIA certificate that a laboratory requires parallels the complexity of its test menu. The lowest level of test complexity is the waived category. Tests in this category are typically simple methods with little likelihood of error or of serious adverse consequences for patients if performed incorrectly. Commonly, such tests are performed in physician office laboratories. It should be noted that the term *waived* applies to a test, not to the need for the laboratory to have a CLIA certificate to perform any clinical testing. The next highest level is the moderate-complexity test, including a category known as provider-performed microscopy. The highest level is the high-complexity test, which is applicable to most molecular tests. Laboratories that perform high-complexity testing must have a certificate to perform this type of testing. When the CLIA was written 25 years ago, there was relatively little molecular testing, and as a result, molecular diagnostics does not have specific requirements in the regulations, unlike most areas of clinical laboratory practice, such as clinical chemistry, microbiology, and hematology. Nevertheless, the general requirements of CLIA can be adapted to molecular testing. Accrediting organizations such as the CAP do have specific requirements for laboratories that perform molecular diagnostic testing. These are available in their laboratory inspection checklists [2]. Whereas the FDA is responsible for categorizing tests, the CMS is responsible for the oversight of the CLIA program, including granting certificates, approving accrediting organizations, approving proficiency testing (PT) programs, inspections, and enforcement actions.

The CLIA is a federal law and applies to all clinical testing performed in the United States and in foreign laboratories that are certified under the CLIA. There are provisions in the CLIA under which individual states can substitute their own laboratory oversight programs if it is determined that such programs are at least as stringent as the federal program. Currently, such programs exist only in New York and Washington states. These are known as "CLIA-exempt" states, although CMS reserves the authority to inspect any aspect of laboratory performance in these states. The CLIA includes the following areas of laboratory testing: PT, pre-analytic testing, analytic testing, and personnel requirements.

Proficiency Testing

PT is one external measure by which a laboratory's performance can be judged. In a PT program, laboratories are sent samples for analysis who return their results to the PT program organizers. The correct result (or range of results) for these programs is determined by the organizers based on a comparison of participant results with results obtained by reference laboratories (accuracy-based grading), or by comparison with other laboratories that use the same analytical methods (peer-group grading). Ideally, all PT programs would use accuracy-based grading, but there are significant practical limitations to this approach. One of the major limitations is the PT material itself. For many analytes, it is not possible to obtain the necessary range of concentrations to test low, normal, and high concentrations using unaltered human samples. This necessitates the use of synthetic samples that have been spiked with the analyte or had the analyte concentration reduced. Such adulterated samples may behave unexpectedly when tested using some analytical equipment and give higher or lower values that would be obtained in a native specimen containing the same concentration or activity of the analyte. This is known as the *matrix effect*. Other limitations may require peer-group grading; for example, recombinant proteins may not be detected equally in different manufacturers' immunoassays, making accuracy-based grading impossible. Enzyme concentrations may be determined by different manufacturers using different concentrations of cofactors, different temperatures, and different substrates, thus giving rise to such inter-method disagreement that accuracy-based grading is impossible.

Molecular testing poses certain challenges to PT programs. It may not be possible to obtain human specimens such as blood from subjects known to carry mutations of interest. This necessitates the use of cell lines, or even DNA aliquots, for PT programs in genetics. Such samples do not test all phases of the analytical process, including extraction of DNA from whole blood (the normal procedure for genetic testing). The same concern applies to molecular testing for infectious diseases such as human immunodeficiency virus-1 (HIV-1). For these reasons, it is not uncommon that PT samples do not fully mimic patient samples.

Under the CLIA, laboratories are required to enroll in PT programs for a group of analytes specified in Subpart I of the regulations. These analytes were chosen based on clinical laboratory testing patterns that existed in 1988, and the list has not been updated since then. As a result, many newer tests, including molecular tests, are not included in this list for mandatory PT participation. For tests not on this list of "regulated" analytes, laboratories must verify the accuracy of their methods by some other methods at least twice a year. This could include comparison of results with those obtained by a different method: sample exchange with another laboratory, or even correlation of results with patients' clinical status. If formal PT programs exist, laboratories should consider enrolling in these. Several of the accrediting organizations do have requirements for participation in PT programs where these exist, including PT programs for molecular testing. In the United States, the CAP offers a wide range of molecular PT programs that cover human inherited and acquired mutations, infectious agents, and next generation sequencing (NGS)

[2]. Laboratories that are accredited by the CAP must participate in these, or other approved PT programs as a condition of their accreditation.

Additionally, published guidance to optimize the quality of NGS laboratory results has been published by the Centers for Disease Control (CDC) and an expert panel [3]. Laboratories that are accredited under the ISO 15189 standard should follow those certification requirements.

Pre-analytic Testing

The CLIA has requirements that cover the pre-analytic phase of testing. These include the use of requisition forms with correct identification of the patient, the patient's age and gender, the test to be performed, the date and time of sample collection, the name of the ordering provider or the person to whom results should be reported, the type of specimen (e.g. blood), and any other additional information needed to produce an accurate result. All of these are critical pieces of information that should be provided to the laboratory. Many so-called "laboratory errors" actually arise at the time of sample collection, and specimen misidentification is one of the most common types of error in the testing process.

In addition to the patient's age and gender, orders for molecular genetic testing should include relevant information about suspected diagnosis, clinical findings, and especially the family history. Many experienced clinical geneticists and genetic counselors will include a pedigree diagram on a requisition form for tests for inherited disorders. This practice is highly desirable and provides useful information to the laboratory. As an example of the importance of this information, current practice guidelines in obstetrics and gynecology in the United States encourage the offering of prenatal testing to expectant Caucasian mothers to determine if they are carriers of mutations for cystic fibrosis. A recommended panel of mutations to be tested by clinical laboratories covers approximately 80–85% of all mutations in this population. In general, a negative screening test for these mutations reduces the risk of being a cystic fibrosis carrier from 1 in 30 to 1 in 141, and the laboratory would report these figures, or, if a mutation were identified, would report the specific mutation. However, these figures are based on the assumption that there is no family history of the disorder in the patient's family. If there is such a history, the risk of being a carrier (both before and after testing) is substantially higher. It is therefore essential that the ordering physician informs the laboratory if there is a family history.

Analytic Testing

The CLIA has detailed requirements for the analytic phase of the testing process. These include the procedure manual, which is a step-by-step set of instructions on how the test should be performed, the process for method calibration, the procedures for preparation of reagents, the use of controls, establishment of the reference range, reporting procedures, and analytical parameters such as sensitivity and specificity. There are no specific CLIA requirements that are unique to molecular testing, and therefore the molecular diagnostics laboratory has to adapt requirements from related areas such as clinical chemistry and microbiology to molecular testing. Some of the accrediting organizations have checklists that include specific requirements for molecular testing. These can provide useful guidance on procedures even for a laboratory that is not accredited by one of these organizations.

Post-analytic Testing

Post-analytic testing refers to steps involved in reporting results to the ordering physician in a timely manner. The patient's name and identification information must be on the report, as should the name and address of the performing laboratory. In addition to the result, the report should include the reference interval and any relevant interpretive comments. The laboratory should be able to provide information on test validation and known interferences on the request of an ordering physician. Results must be released only to authorized persons.

Although certain elements of the post-analytic phase of testing can be controlled by the laboratory, there are also critical elements that are beyond its control, notably the correct interpretation of the result by the ordering physician. Molecular diagnostics (and genetics in general) is an area in which many physicians and other providers never had formal training in medical school. Concern has been expressed about the need to improve genetics education for health care professionals. Where there is a gap in provider knowledge, the laboratory should be able to offer expert consultation on the interpretation of its results to primary care providers [4]. This requires time, patience, and good communication skills on the part of the laboratory director and senior staff. Although such activity may be reimbursable under some health plans, the primary incentives for providing this kind of consultation are good patient care and customer satisfaction.

Personnel Qualifications

Under the CLIA, requirements exist for laboratory personnel qualifications and/or experience. Perhaps the most important qualification requirements apply to the laboratory director. The director of a high-complexity laboratory such as a clinical molecular testing laboratory must hold a license in the state in which he or she works (if the state issues such licenses) and be a physician or osteopathic physician with board certification in pathology. Alternatively, the laboratory director can be a physician with at least 1 year of training in laboratory practice during residency, or a physician with at least 2 years of experience supervising or directing a clinical laboratory. A doctoral scientist holding a degree in a chemical, physical, biological, or clinical laboratory science field with board certification from an approved board may also serve as the laboratory director. There are also provisions that allow for grandfathering of persons who were serving as laboratory directors at the time of implementation of the CLIA. Currently, there are no specific CLIA-required qualifications for the director of a molecular diagnostics laboratory. There are, however, board examinations in this field or similar fields that are offered by the American Board of Pathology, the American Board of Medical Genetics and Genomics, the American Board for Clinical Chemistry, the American Board of Medical Microbiology, and the American Board of Bioanalysts. Individual states may begin to require specific qualifications in molecular diagnostics in the future or even that changes to the CLIA may require such qualifications. Other personnel and their qualifications described in the CLIA for high-complexity laboratories are technical supervisor, clinical consultant, general supervisor, cytology supervisor, cytotechnologist, and testing personnel.

Genetic Testing and Patient Privacy

For many years, there has been concern about the use of genetic information to discriminate against people with genetic diseases or those who are at risk of manifesting genetic disease at some time in the future. Although there are very few reported examples of such discrimination, the possibility of such misuse of genetic information by employers or insurance companies has received considerable attention by both the public and legislative bodies [5]. A comprehensive analysis of applicable laws is beyond the scope of this chapter, but certain principles that apply to the clinical laboratory are worth mentioning. It is generally assumed, of course, that all clinical laboratory testing is performed with the consent of the patient. However, written consent is a legal requirement for genetic testing in some jurisdictions. The laboratory is generally not in a position to collect informed consent from patients, so consent is usually obtained by some other health care worker, such as the ordering physician or genetics counselor. The laboratory director should be aware of applicable laws in this matter and determine, with legal advice if necessary, what testing is covered in his or her jurisdiction and ensure that appropriate consent is obtained. Genetic testing in its broadest meaning can cover more than just nucleic acid testing. For example, some laboratory methods for measuring glycohemoglobin, a test used for following diabetes control, can indicate the presence of genetic variants of hemoglobin, such as sickle-cell hemoglobin. Histopathologic examination of certain tumors can be strongly suggestive of an inherited disorder. Serum protein electrophoresis can reveal α-1 antitrypsin deficiency, an inherited disorder. The laboratory should consider how it reports such findings, which may contain genetic information that is unanticipated by both the ordering physician and the patient.

The most significant federal legislation in this area is the Genetic Information Nondiscrimination Act of 2008. This act offers protection against the use of genetic information as a basis for discrimination in employment and health insurance decisions. Under the provisions of this law, people who are healthy may not be discriminated against on the basis of any genetic predisposition to developing disease in the future. Health care insurers (but not life insurers or long-term care insurers) and employers may not require prospective clients or employees to undergo genetic testing or take any adverse action based on knowledge of a genetic trait. The benefits of this legislation are that some people may feel less trepidation about undergoing genetic testing because of fear that such information could be used by an employer or insurance company to discriminate against them.

Testing in Research Laboratories

As research laboratories report new molecular findings in inherited and acquired diseases, it is not uncommon for clinical laboratories to receive requests to send patient samples to research laboratories for testing. This is an area in which the clinical laboratory must be careful to avoid noncompliance with CLIA regulations. One of the requirements of the CLIA is that certified laboratories must not send samples for patient testing to a non-CLIA-certified laboratory. This rule applies even if the research laboratory is the only one in the world to offer a particular test. Such samples should not be referred by a CLIA-certified laboratory, and the ordering physician should find some other means of arranging for testing if it is considered necessary. For example, it may be possible for testing to be performed under a research protocol. In this case, the local institutional review board may be able to offer useful guidance on the creation and implementation of an appropriate protocol.

There are good reasons to be cautious about performing clinical testing in a research setting. The standards that one expects in a CLIA-certified laboratory are designed to promote quality and improve the accuracy of patient testing. Laboratories that do not follow these extensive requirements may not have all of the necessary protocols and procedures in place to offer the same quality of test result. Research laboratories are often part of academic institutions that may or may not carry malpractice insurance coverage in the event that a reported test result is erroneous.

Molecular Testing from Research to Clinical Application

The usual progression of molecular testing begins with gene and mutation discovery, typically in a research laboratory setting. Publication of these early findings in peer-reviewed literature is the normal means of disseminating new information about a gene of clinical interest and the variations that can cause disease. It is important to document at least the most common disease-causing mutations and benign polymorphisms.

After a disease-causing mutation has been discovered, diagnostic testing on patients (as opposed to research subjects who have consented) requires performance in a laboratory that holds a CLIA certificate. For molecular testing, this almost certainly means a certificate that allows for high-complexity testing. Research laboratories are often not set up to perform clinical testing or to meet the stringent criteria for clinical laboratory operations.

What should a molecular diagnostics laboratory be able to offer to meet clinical needs for molecular testing? First, the quality of the test result must be of a very high standard; that is, the results are reliable. Of course, all laboratories strive for this goal, which is implicit in the numerous regulations that govern laboratory testing. This is achieved by careful attention to the pre-analytic, analytic, and post-analytic factors mentioned above and to the hiring of qualified and skilled personnel. The laboratory should offer turnaround times that are appropriate to the clinical needs of a specific test and which may vary from one test to another. For example, testing for some infectious diseases is likely to require a faster turnaround time than is testing for a genetic predisposition to a chronic disease. Information should be readily available on the requirements for specimen type and the needs for special handling. The laboratory should be able to offer interpretations and consultations to ordering physicians regarding results of patient testing. If the genetic test result is a risk factor for future development of disease or for carrier status (e.g. cystic fibrosis carrier screening in pregnancy), the laboratory should be able to reevalulate such risks if additional family history is provided at a later time. Many laboratories have a formal relationship with a genetic counselor who can interact with both patients and other health care workers and provide a variety of very useful services.

As clinical testing becomes more widespread, there can be significant changes to the knowledge and thinking about the relationship between disease and underlying genetic mutation. An example of this is illustrated by the hereditary hemochromatosis gene, *HFE*. Discovery of this gene and the common mutations, *C282Y* and *H63D*, led to the view that the homozygous states, especially homozygosity for *C282Y*, would lead to chronic iron overload and hemochromatosis [6]. That view is no longer correct in light of more recent population studies of the penetrance of these mutations. Approximately, one-third of patients who are homozygous for *C282Y* do not have elevated ferritin levels and appear not to be at risk of iron overload [7]. The reason for the variability of penetrance is probably related to dietary iron, blood loss, and other genetic factors that have yet to be determined. It is important for the laboratory director to be aware of such changing perspectives in thinking about diseases and to be an educator to others, making them aware of important developments, so that rational ordering patterns are encouraged.

The Role of the FDA in Molecular Testing

As discussed earlier, the FDA has broad regulatory oversight of IVD devices. A list of FDA-cleared or -approved nucleic acid–based tests, and a list of *in vitro* companion diagnostic devices are available online at the FDA website [8, 9]. These include tests for inherited and acquired mutations in humans, the use of certain NGS instruments, and tests for a variety of infectious agents. The companion diagnostic tests concern the recommended use of certain pharmaceutical agents in patients in the presence (or absence) of defined mutations.

The FDA has also announced a proposal to exert its authority over many so-called "laboratory developed tests" (LDTs). These include tests that are developed and used in clinical laboratories, and which have not been formally evaluated or approved by the FDA. Although such tests must meet the validation requirements of the CLIA, there is less emphasis in those regulations on clinical performance (as opposed to analytical performance) than what the FDA requires for approval. At the time of writing, it is unclear how far the FDA's proposal will advance.

Reimbursement for Molecular Testing

In common with all areas of medical practice, reimbursement for molecular testing at the federal level (Medicare) is based on the current Common Procedural Terminology (CPT) coding system. Other state providers such as Medicaid and private insurance companies generally follow the same process. Under CPT coding, a charge and its payment are based on the number of individual items of service provided. Each step in a typical molecular assay ranging from extraction of DNA to performance of a polymerase chain reaction to gel electrophoresis, and final result interpretation has a unique CPT code and an associated reimbursement based (in the case of Medicare) on the published fee schedule. Therefore, the Medicare reimbursement rate is calculable and is based on the individual steps in an assay. Private insurance companies may reimburse at a higher rate than federal payers. The CPT codes are updated annually by the American Medical Association, which retains copyright on the codes. Because of the rapid advances in molecular testing, it is not uncommon for laboratories to use methods that are not listed in the CPT guide. In this case, it may be necessary to seek consultation billing experts on choosing the appropriate fee codes.

Not uncommonly, genetic test prices from commercial laboratories are well above those that can be justified from published fee schedules. Although this may be perfectly legal, it can lead to significant problems for patients whose insurance companies (including Medicare) may not cover the full cost of the testing. In this situation, the patient may have to pay out of pocket for part or all of the cost of the test if it is decided that the testing is essential. This situation can pose a financial risk for hospitals and clinics if they refer a sample for testing to a reference laboratory and thereby possibly incur the charges for a test. One possible option is to notify the patient and ordering physician that such tests are unlikely to be covered by insurance and determine how they propose to pay for testing. For Medicare patients, an advance beneficiary notice (ABN) may be used to formally notify a patient that the test is considered to be a noncovered service [10]. These types of situations should be discussed with hospital management.

Summary

Molecular testing is firmly established in clinical laboratories for a wide variety of disorders. According to published reports, molecular diagnostics is and will continue to remain one of the fastest areas of growth in clinical testing. In the United States, the clinical laboratory operates under the regulations of the Clinical Laboratory Improvement Amendments of 1988, which provide the framework for producing high-quality results. The clinical laboratory differs significantly both from the research laboratory in practice and from a regulatory point of view. Careful attention should be paid to issues such as patient privacy and reimbursement for molecular testing.

References

1 Deloitte 2016. Global life science outlook. Available from https://www2.deloitte.com/global/en/pages/life-sciences-and-healthcare/articles/global-health-care-sector-outlook.html (accessed 27 April 2019).

2 College of American Pathologists. Available from www.cap.org (accessed 27 April 2019).

3 Gargis, A.S., Kalman, L., Berry, M.W. et al. (2012). Assuring the quality of next-generation sequencing in clinical laboratory practice. *Nat. Biotechnol.* 30 (11): 1033–1036.

4 Harvey, E.K., Fogel, C.E., Peyrot, M. et al. (2007). Providers' knowledge of genetics: a survey of 5915 individuals and families with genetic conditions. *Genet. Med.* 9 (5): 259–267.

5 Harmon, A. (2008). Insurance fears lead many to shun DNA tests. *The New York Times.* http://web.archive.org/web/20190108162929/https://www.nytimes.com/2008/02/24/health/24dna.html.

6 Feder, J.N., Gnirke, A., Thomas, W. et al. (1996). A novel MHC class I-like gene is mutated in patients with hereditary haemochromatosis. *Nat. Genet.* 13 (4): 399–408.
7 Olynyk, J.K., Trinder, D., Ramm, G.A. et al. (2008). Hereditary hemochromatosis in the post-HFE era. *Hepatology* 48 (3): 991–1001.
8 FDA (2016). Nucleic acid based tests. Available from http://www.fda.gov/MedicalDevices/ProductsandMedicalProcedures/InVitroDiagnostics/ucm330711.htm (accessed 27 April 2019).
9 FDA (2016). List of cleared or approved companion diagnostic devices (in vitro and imaging tools). Available from http://www.fda.gov/MedicalDevices/ProductsandMedicalProcedures/InVitroDiagnostics/ucm301431.htm (accessed 27 April 2019).
10 Carter, D. (2003). Obtaining advance beneficiary notices for Medicare physician providers. *J. Med. Pract. Manage.* 19 (1): 10–18.

Part VII

Big Data, Data Mining, and Biomarkers

22

IT Supporting Biomarker-Enabled Drug Development

Michael Hehenberger

HM NanoMed, Westport, CT, USA

A Paradigm Shift in Bio-pharmaceutical R&D

The bio-pharmaceutical industry is currently undergoing a transformation driven primarily by the need to move from its proven "Blockbuster model" to a new "Stratified medicine" model [1, 2]. This paradigm shift has been anticipated widely in whitepapers such as IBM's "Pharma 2010" report [3] and is accompanied by serious efforts to streamline operations and to address research and development (R&D) productivity issues as described and quantified by DiMasi [4] and DiMasi et al [5]. In its "Critical Path" initiative (https://web.archive.org/web/20090512144844/www.fda.gov/oc/initiatives/criticalpath/whitepaper.pdf), the US Food and Drug Administration (FDA) has been guiding industry towards use of biomarkers (http://web.archive.org/web/20060318101147/http://www.fda.gov/cder/genomics/PGX_biomarkers.pdf) that address efficacy and safety issues and hold the promise of increased R&D productivity [6]. In the excellent biomarker review paper by Trusheim, Berndt, and Douglas [1], "clinical biomarkers" are defined as biomarkers that associate a medical treatment to a patient subpopulation that has historically exhibited a differential and substantial clinical response. Clinical biomarkers can be based on "*genotypes, proteins, metabonomic patterns, histology, medical imaging, physician clinical observations or even self-reported patient surveys. A clinical biomarker is not defined by its technology or biological basis, but rather by its reliable, predictive correlation to differential patient responses.*" It is generally believed that "biomarker-enabled" drug development will lead to better and earlier decision-making and that clinical biomarkers will pave the way towards targeted therapeutics combining "precision drugs" for stratified patient populations with diagnostic tests designed to identify not

Biomarkers in Drug Discovery and Development: A Handbook of Practice, Application, and Strategy,
Second Edition. Edited by Ramin Rahbari, Jonathan Van Niewaal, and Michael R. Bleavins.

only "responders" (who will benefit) but also patient cohorts that will not respond, as well as those most at risk for adverse side effects.

Biomarker-enabled R&D is evolving into a new discipline with a strong patient focus. Organizations who believe in biomarker-enabled R&D are investing in tools and making the necessary organizational changes to implement the new concepts and associated processes. Among biomarkers, "imaging biomarkers" have received particular attention, because of the noninvasive nature of imaging technologies and the obvious link to diagnostic procedures and clinical care. Imaging technologies are increasingly used as core technologies in bio-pharmaceutical R&D, both in the preclinical and the clinical phases of the R&D process.

Disease areas most affected by this paradigm shift are cardiology, oncology, and neurology. Below we will discuss in more detail how biomarker-related data types and their increasing volumes are challenging existing information technology (IT) infrastructures, and how IT architectures have to be enhanced and modified to integrate genomic, imaging, and other biomarker data. We will also address the new opportunities provided by "Cognitive Computing": the use of advanced text analytics algorithms.

Processes, Workflows, IT Standards, and Architectures

The conventional R&D process – documented extensively by most research based bio-pharmaceutical companies – is sequential (Figure 22.1). After target identification and validation by the biologists, the medicinal chemists take over and screen extensive libraries of hundreds of thousands of chemical compounds against the target to eventually find a suitable candidate for a drug. Before the IND (Initial New Drug) application to the FDA, or an equivalent process for countries outside the United States, the drug candidate is tested in preclinical animal studies. It is then handed over to the clinical development organization for clinical trials that proceed through three phases. If the drug candidate survives through Phase III testing in significant numbers of patients, all the collected supporting information will be submitted to the FDA in form of a "New Drug Application" (NDA) dossier that has to be compiled such that FDA's rules and regulations are respected and followed. After FDA approval, Phase IV trials may be conducted to collect postmarketing surveillance data

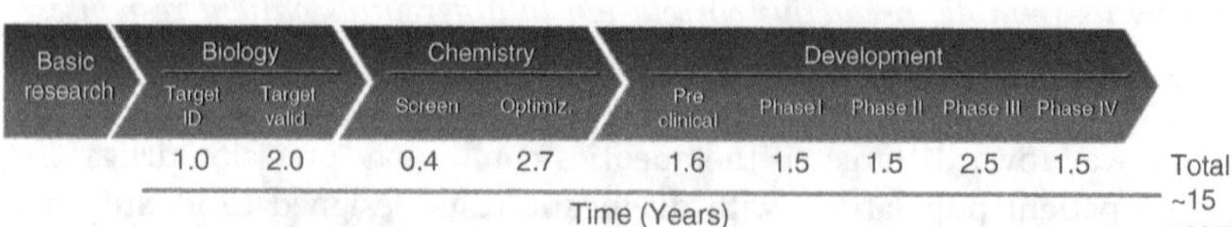

Figure 22.1 Sequential R&D process. Source: From Corr 2007 [7].

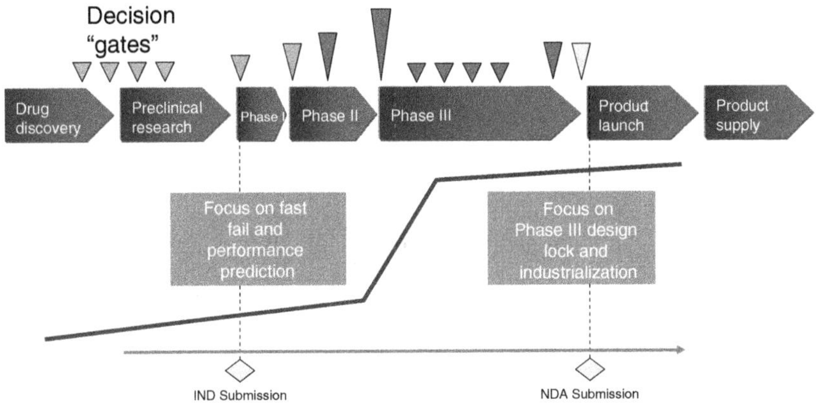

Figure 22.2 Decision gates (milestones) to manage sequential R&D processes. Source: From McCormick [8].

about adverse drug reactions or to position the drug for new indications not yet covered by a given FDA approval.

To manage the process effectively and to terminate moribond projects as early as possible, R&D organizations have created a set of disciplined processes designed to optimize project portfolios and to track the progress of individual projects by "milestones" or "decision gates" (Figure 22.2):

Industry leaders such as Novartis [9] have realized that biomarker-enabled R&D has to be organized differently. In particular, there is an increasing emphasis on parallel processes and accelerated "proof-of-concept" in humans, to "learn" quickly and to "confirm" (i.e. conduct extensive clinical trials) only if the "learning" is yielding promising results (Figure 22.3). A high-level depiction of this new approach can also be found in IBM's above-mentioned Pharma 2010 report [3].

It is the role of IT standards and architectures to support business strategies and enable their implementation. IT standards relevant in this context are as follows:

- Data Standards proposed by the Clinical Data Interchange Standards Consortium (CDISC; www.cdisc.org) such as the following:
  - SEND (Standard for Exchange of Non-Clinical Data) covering animal data
  - SDTM (Study Data Tabulation Model) covering human data
  - ODM (Operational Data Model) covering study data including EDC generated data
  - ADaM (Analysis Data Model) covering analysis data sets
- HL7 (Health Level Seven) Clinical Document Architecture (CDA; www.hl7.org)

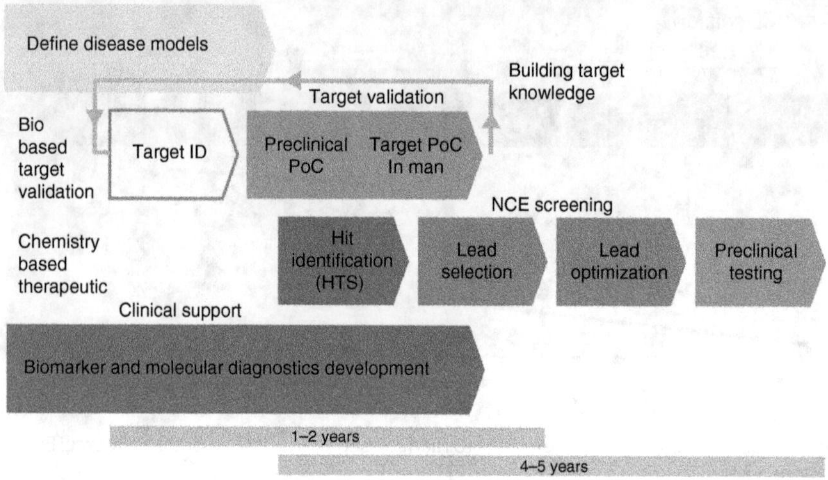

Figure 22.3 Parallel biomarker-enabled processes up to preclinical development. Source: From McCormick [8].

- DICOM (Digital Imaging and Communications in Medicine) to transmit medical images (Medical.nema.org)
- JANUS Data Model (https://www.fda.gov/industry/fda-resources-data-standards) developed by FDA and IBM (in a Collaborative Research and Development Agreement – CRADA) and implemented by NCI (National Cancer Institute) and FDA.

A comprehensive overview of relevant standards along with associated web sites is shown in Figure 22.4.

Based on the technical challenges for integrating a diverse set of data sources for a biomarker-based clinical data submission, IBM has previously proposed an IT architecture (Figure 22.5) that addresses a majority of those requirements.

While this architecture includes software products and assets belonging to IBM, one can logically extend it to fit other vendor products as well. The idea was to present a general purpose platform for managing clinical submissions of patient data, enhanced with genomic and imaging data.

At the bottom data layer, summarized clinical submission data in CDISC's SDTM format receive feeds from a clinical data management system (CDMS) that stores case report forms (CRFs). First, the associated metadata for the SDTM submission (such as the vocabularies used for the adverse event codes, lab codes, etc.) are mapped into the tables of the JANUS Data Model. An Extract–Transform–Load (ETL) tool is then used to improve the scalability and data validation, cleansing, and transformation needed to load the SDTM data into JANUS. One may also need to build a collection of applications and

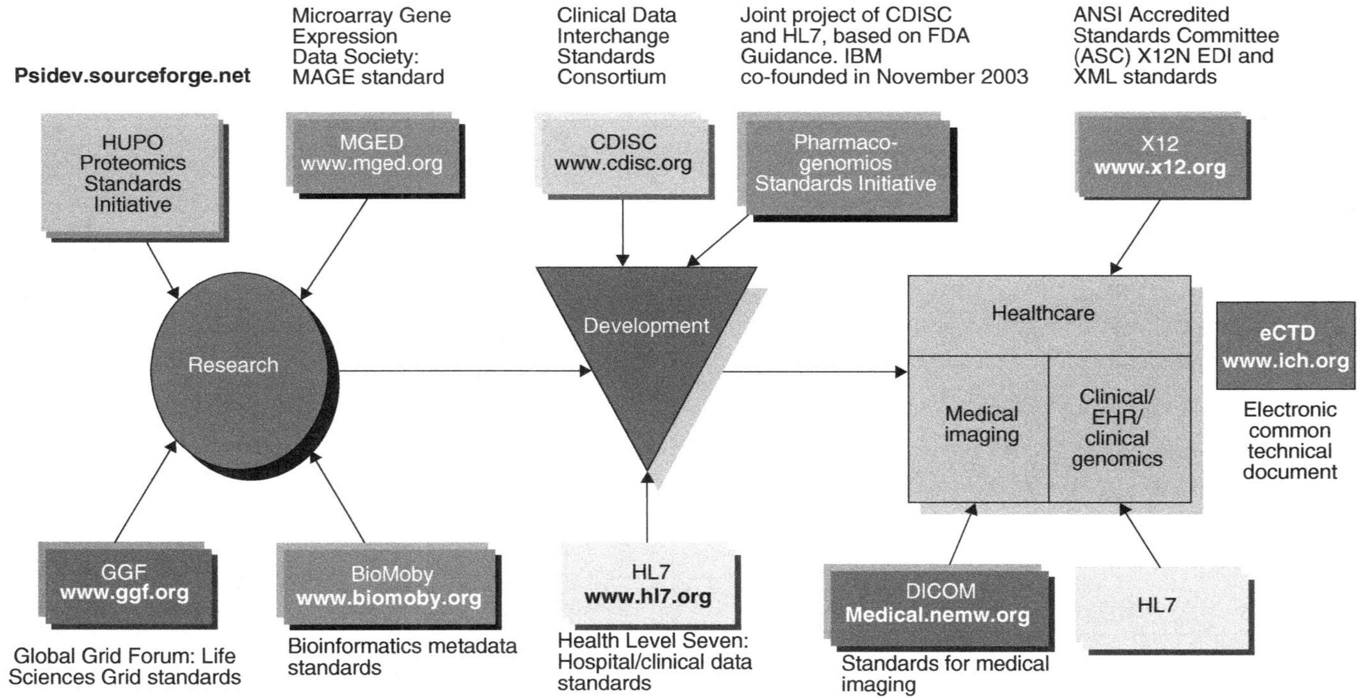

Figure 22.4 Healthcare & Life Sciences Standards Organizations.

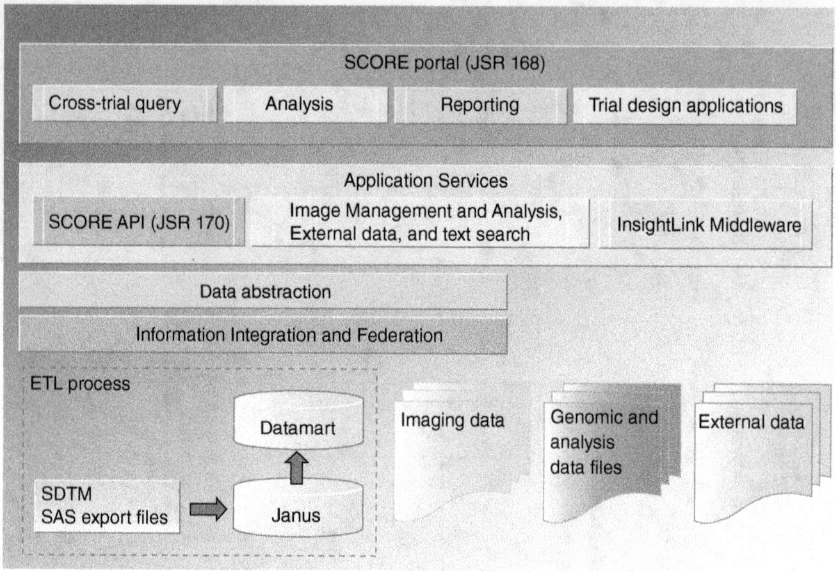

Figure 22.5 Proposed IT architecture for biomarker-based clinical development [10].

use case-specific data marts[1] on top of JANUS. These can be designed using Star-Schema–based dimensional models for optimal query performance. In addition to the clinical submission data in JANUS, one would need to establish links to the imaging data that often reside in picture archiving and communications systems (PACS). After extraction, the images can be centrally managed using a standardized imaging broker service. Finally, external reference databases such as PubMed, GenBank, dbSNP, SwissProt, and so on are integrated using unstructured information management technologies. All these content stores can be searched dynamically using a federated warehouse that uses a "wrapper based"[2] technology for linking diverse data sources. On top of the federation layer, we propose a data abstraction and query layer designed to expose a user-centric logical data model (based on XML) mapped on top of the physical data model.

To support processes and workflows that may include the transfer of clinical biomarker data between pharmaceutical entities, contract research organizations (CROs), imaging core labs, investigator sites (often academic medical

1 A **data mart** is the access layer of the data warehouse environment that is used to get data out to the users. Data marts are small slices of the data warehouse. Whereas data warehouses tend to have an enterprise-wide depth, the information in data marts pertains to a single department.
2 In federated databases, "wrappers" are used to provide access and to allow management of external data.

research centers), and ultimately regulatory agencies, the IT architecture should be designed as a "Service Oriented Architecture" (SOA). IBM's SCORE architecture satisfied this requirement and could therefore serve as a basis for the enablement of "Biomarker-Based R&D."

Clinical Biomarkers and Associated Data Types

Among clinical biomarkers, the genomic and imaging data types are creating the greatest IT challenges. By recommending CDISC's SDTM (see above) as the standard for drug submission, the FDA has taken an important step toward the integration of genomic and imaging data with conventional clinical patient data. Other global regulatory agencies are often following FDA's lead regarding IT standards.

CDISC SDTM is an easily extendable model that incorporates the FDA submission data structures. Based on strong collaboration between the bio-pharmaceutical industry, clinical research organizations, clinical trial investigator sites, IT vendors, and the FDA, SDTM represents the collective thoughts from a broad group of stakeholders.

Conventional clinical data are categorized into four classes of the SDTM data model, namely "Events," "Interventions,, "Findings," and "Other." In SDTM's hierarchy of definitions, classes are subdivided into domains:

- ***Events*** include specific domains covering adverse events, subject disposition, and medical history.
- ***Interventions*** cover exposure (to study drug), concomitant medications, and substance use.
- ***Findings*** contain the assessment information such as echocardiograms, laboratory results, physical exams, vital signs, subject questionnaire data, and so on.
- The ***Other*** class was created to group specialized categories of information such as clinical trial design, supplemental qualifiers, trial summary, and related records, where related records provide linkages across the different files.

To support biomarker data submission, two new SDTM domains have been added:

- The ***pharmacogenomic*** (PG) and ***pharamacogenomics results*** (PR) domain will support submission of summarized genomic (genotypic data).
- A new ***imaging*** (IM) domain includes a mapping of the relevant DICOM metadata fields required to summarize an imaging submission.

The ***PG domain*** belongs to the ***Findings*** class and is designed to store panel ordering information. The detailed test level information (such as genotype/SNP [single-nucleotide polymorphism] summarized results) is reported in the PR domain. Figure 22.6 shows what a typical genotype test might look like in terms of data content and usage of the HUGO (Human Genome Organization; https://www.genenames.org/) nomenclature. The PG domain supports the hierarchical nature of pharmacogenomic results, where for a given genetic test (such as EGFR, CYP2D6, etc.) from a patient sample (listed in the parent domain), multiple genotypes/SNP's can be reported (listed in the child domain).

To support the use of imaging biomarkers, DICOM metadata tags have to be mapped into the fields of the new IM domain. Table 22.1 illustrates this mechanism.

While the FDA has proposed the SDTM data model for submission data, it is clear that this is only an interchange format for sponsors to submit summarized clinical study data in a standardized fashion to the FDA. The FDA identified a need for an additional relational repository model to store the SDTM data sets. The requirement was to design a normalized and extensible relational repository model that would scale up to a huge collection of studies going back into the past and supporting those in the future, for easy cross-referencing.

The purpose of FDA's proposed data standards is to facilitate the exchange of data between researchers and FDA. The human and animal study data will be stored in the Janus Clinical Trials Repository (CTR), a repository that allows users to generate views for analysis with different end-user tools.

The FDA is also working on the development and adoption of standards based on HL7 Reference Information Model (RIM), to support meaningful information representation and exchange between systems in use by clinical researchers, FDA (Janus CTR), and health care providers (electronic health record systems).

Since 2007, FDA has been collaborating with CDISC and other stakeholders within the HL7 Regulated Clinical Research Information Management (RCRIM) Workgroup on the development of study data exchange standards based on HL7 version 3. General RCRIM Workgroup information is available at http://www.hl7.org/Special/committees/rcrim/index.cfm.

Data Integration and Management

As scientific breakthroughs in genomics and proteomics, and new technologies such as biomedical and molecular imaging, are incorporated into R&D processes, the associated experimental activities are producing ever-increasing volumes of data that have to be integrated and managed. There are two major approaches to solving the challenge of enterprise-wide data access. The creation of data warehouses [11] is an effective way of managing large and complex data that have to be queried, analyzed, and mined in order to generate new knowledge. In order to build such warehouses, the various data sources have

Parent domain:

STUDYID	USUBJID	PGSEQ	PGGRPID	PGREFID	PGOBJ	PGTESTCD	PGTEST
NSCLC10	ZBI000-0007	1	EGFR-KD-001	SPEC001		EGFR-KD	EGFR-KD (EGFR Gene, Protein kinase domain assoc
NSCLC10	ZB10000-007	7	CYP1A2-00001	SPEC002		CYP1A2	CYP1A2 Mutation
NSCLC10	ZB10000-008	1	CYP1A2-00003	SPEC001		CYP1A2	CYP1A2 Mutation DNA Analysis (provided
NSCLC10	ZB10000-009	1	CYP2D6-00001	SPEC001		CYP2D6	CYP2D6 test
NSCLC10	ZB10000-009	11	CYP2C19-00001	SPEC001		CYP2C10 (*2,	Cytochrome P450 2C19 Test

PGMETHCD	PGASSAY	PGORRES	PGSTRESC	PGSTRESN
88323, 88380, 83890 (X2), 838	12700056	EGFR	EGFR	
83891, 83892 x2, 83998	50-776	CYP 1A2	CYP 1A2	
x2, 83998	50-777	CYP1A2	CYP1A2	
83891, 83892, 83901 x2	50-574	CYP2D6	CYP2D6	
83891, 83892, 83901 x2	50-575	CYP2C19	CYP2C19	

Child domain:

STUDYID	USUBJID	PGSEQ	PGGRPID	PGREFID	PGOBJ	PGTESTCD	PGTEST
NSCLC10	ZB10000-009	1	CYP2D6-00001	SPEC001	HGNC:2625	CYP2D6	CYP2D6 GENE.g.-1584C>G
NSCLC10	ZB10000-009	2	CYP2D6-00001	SPEC001	HGNC:2625	CYP2D6	CYP2D6 GENE.g.100C>T
NSCLC10	ZB10000-009	3	CYP2D6-00001	SPEC001	HGNC:2625	CYP2D6	CYP2D6 GENE.g.124G>A
NSCLC10	ZB10000-009	4	CYP2D6-00001	SPEC001	HGNC:2625	CYP2D6	CYP2D6 GENE.g.883G>C
NSCLC10	ZB10000-009	5	CYP2D6-00001	SPEC001	HGNC:2625	CYP2D6	CY.P2D6 GENE.g.1023C>T

PGMETHCD	PGASSAY	PGORRES	PGSTRESC	PGSTRESN
MOLGEN	50-574	M33388:g.-1584GG		
MOLGEN	50-574	M33388:g.100TG		
MOLGEN	50-574	M33388:g.124GC		
MOLGEN	50-574	M33388:g.883GC		
MOLGEN	50-574	M33388:g.1023CG		

Figure 22.6 Partial sample of pharmacogenomics (PG) SDTM domain.

Table 22.1 Mapping of DICOM imaging metadata tags into SDTM Imaging (IM) domain.

CDISC SDTM IM domain		DICOM tags		
Variable name	**CDISC notes (for domains) or description (for general classes)**	**Tag**	**Attribute name**	**Attribute description**
Unique subject identifier	Unique subject identifier within submission.	(0012, 0040)	Clinical trial subject ID	The assigned identifier for the clinical trial subject; shall be present if clinical trial subject reading ID is absent; may be present otherwise.
Sequence number	Sequence number given to ensure uniqueness within a data set for a subject. It can be used to join related records.	(0020, 0013)	Instance number	A number that identifies this image. Note: this attribute was named Image Number in earlier versions of this standard.
Imaging reference ID	Internal or external identifier. Example: UUID for external imaging data file.	(0008, 0018)	SOP instance UID	Uniquely identifies the standard operating procedure instance.
Test or examination short name	Short name of the measurement, test or examination. It can be used as a column name when converting to a data set from a vertical to a horizontal format.	(0008, 1030)	Study description	Institution-generated description or classification of the study (component) performed.

to be extracted, transformed, and loaded (E–T–L) into repositories built upon the principles of relational databases [12]. Warehousing effectively addresses the separation of transactional and analysis/reporting databases and provides a data management architecture that can cope with increased data demands over time. The E–T–L mechanism provides a means to "clean" the data extracted from the capture databases and thereby ensures data quality. However, data warehouses require significant efforts for their implementation.

Alternatively, a virtual, federated model can be employed [13]. Under the federated model, operational databases and other repositories remain intact and independent. Data retrieval and other multiple-database transactions take place at query time, through an integration layer of technology that sits above the operational databases and is often referred to as middleware or a meta-layer. Database federation has attractive benefits, an important one being that the individual data sources do not require modification and can continue to function independently. In addition, the architecture of the federated model allows for easy expansion when new data sources become available. Federation requires less effort to implement but may suffer in query performance compared to a centralized data warehouse.

Common to both approaches is the need for sorting, cleaning, and assessing the data, making sure it is valid, relevant, and presented in appropriate and compatible formats. The cleaning and validation process would eliminate repetitive data stores, link data sets, and classify and organize the data to enhance its utility. Both approaches can coexist, suggesting a strategy where stable and mature data types are stored in data warehouses and new, dynamic data sources are kept federated.

Genomic data are a good example of the dynamic data type. Since genomics is a relatively new field in bio-pharmaceutical R&D, individual organizations use and define data their own way. Only as the science behind genomics is better understood, the business definitions are modified to better represent these new discoveries.

The integration of external (partly unstructured) sources such as GenBank (http://www.ncbi.nlm.nih.gov/Genbank/), Swiss-Prot (http://www.ebi.ac.uk/swissprot/), dbSNP (http://www.ncbi.nlm.nih.gov/SNP/), and so on can be complicated especially if the evolving systems use does not match the actual lab use. Standardized vocabularies (i.e. ontologies) will link these data sources for validation and analysis purposes. External data sources tend to represent the frontier of science especially since they store genetic biomarkers associated to diseases and their associated testing methods. Having a reliable link between genetic testing labs, external data sources for innovations in medical science, and clinical data greatly improves the analytical functionality, resulting in more accurate outcome analysis. These links have been designed into the CDISC PG/PR domains to facilitate the analysis and reporting of genetic factors in clinical trial outcomes.

As standards continue to evolve, the need for "semantic interoperability" is becoming increasingly clear. In order to effectively use standards to exchange information, there must be an agreed-upon data structure, and the stakeholders must share a common definition for the data content itself. The true benefit of standards is attained when two different groups can reach the same conclusions based on access to the same data, because there is a shared understanding of the meaning of the data and the context of use of the data.

Imaging Biomarker Data, Regulatory Compliance, and SOA

Under FDA's strict 21 CFR Part 11 (https://www.accessdata.fda.gov/scripts/cdrh/cfdocs/cfcfr/CFRSearch.cfm?CFRPart=11) guidelines, new drug submissions must be supported by documentation that is compliant with all regulations. The FDA requires reproducibility of imaging findings so that an independent reviewer can make the same conclusion or derive the same computed measurements as that of a radiologist included in a submission. As a result, a unified architecture is required for a DICOM-based imaging data management platform that supports heterogeneous image capture environments and modalities and allows web-based access to the independent reviewers. Automated markups and computations are recommended to promote reproducibility, but manual segmentation or annotations are often needed to compute the imaging findings. A common vocabulary is also needed for the radiological reports that spell out the diagnosis and other detailed findings, as well as for the specification of the imaging protocols.

An imaging data management solution should therefore include the following:

- image repository to store the image content and associated metadata
- collaboration layer providing image lifecycle tasks shared across sponsors, CROs, and investigator sites
- image services providing functionality such as security and auditing
- integration layer providing solutions for integration and interoperability with other applications and systems
- image taxonomy definition(s) to develop image data models including naming, attributes, ontologies, values, and relationships
- image storage policy definition to define and help to manage policies and systems for image storage and retention
- regulatory interpretation assisting interpretation of regulations and guidelines for what is required for compliance
- portal providing a role-based and personalized user interface.

In addition, the solution design should incorporate the customized design, implementation, and monitoring of image management processes, and should

Table 22.2 Five levels of SOA maturity.

Discrete	Partial	Enterprise	Partner	Dynamic partner
Hard-coded application	Cross line of business processes	Cross enterprise business processes	"Known"	"Any trusted partner"

be fully based on principles of SOA [14]. SOA is taking "application integration" to a new level. To take full advantage of the principles of SOA, it is important to start with a full understanding of business processes to be supported by an IT solution. Such a solution must be architected to support the documented business processes. The component business modeling (CBM; http://www.ibm.com/services/us/gbs/bus/html/bcs_componentmodeling.html) methodology identifies the basic building blocks of a given business, leading to insights that will help overcome most business challenges. CBM allows analysis from multiple perspectives – and the intersection of those views offers improved insights for decision-making. In the case of biomarker-enabled R&D, CBM will break down the transformed processes and identify the respective roles of in-house bio-pharmaceutical R&D functions and outside partners such as CROs, imaging core labs, investigator sites, genotyping services, and possible academic or biotechnology research collaborators.

After mapping of workflows, it is then possible to define a service-oriented IT architecture that supports the processes and workflows (Table 22.2). In its most advanced form, SOA will support a complex environment with integrated data sources, integrated applications, and a dynamic network of partners.

Biomarkers and Cognitive Computing

By improving algorithm development and data processing, the field of text analytics has entered the era of cognitive computing – defined as computing systems that learn and interact naturally with people to extend what either humans or machine could do on their own, and helping human experts make better decisions by penetrating the complexity of Big Data (http://www.research.ibm.com/cognitive-computing/index.shtml#fbid=pBmAByM5xsF). By the end of 2010, a major milestone of cognitive computing was reached by IBM Research when the Watson computer was able to beat the Jeopardy game show champions Ken Jennings and Brad Rutter. Clearly, the Jeopardy playing computer system Watson was smarter than a search engine. More than 40 years after the publication of Simmons' paper [15] on "Natural Language Question-Answering Systems," the Watson–Jeopardy team was able to beat the human brain at tasks previously thought as too difficult to do for even the most powerful supercomputer. Watson could not only understand complicated sentences in natural language but also perform temporal and geospatial reasoning and "statistical paraphrasing," a way to bridge the gap between, for example, specialized medical terminology and everyday language. As we are anticipating an increasing impact of text analytics/cognitive computing on biomarker based R&D, it may be useful to add an overview of data sources that are expected to be of value to scientists active in this field.

Chemical and Biomedical Literature, Patents, and Drug Safety

Medline/PubMed

Medline started out as a repository of journal citations and abstracts for biomedical literature from around the world, managed by the National Library of Medicine (NLM). It is now part of PubMed: As of 2019, PubMed comprises more than 29 million citations for biomedical literature from Medline, life science journals, and online books. About 500,000 entries are added each year. Citations may include links to full-text content from PubMed Central and publisher web sites. Managed by the National Center of Biotechnology Information (NCBI) of NIH/NLM, MeSH (Medical Subject Headings) is NLM's controlled vocabulary used to index articles for Medline. Unified Medical Language System (UMLS) is a compendium of biomedical vocabularies and mappings between them.

Web site:

http://www.ncbi.nlm.nih.gov/pubmed/

Embase

Embase is a proprietary repository of biomedical abstracts managed by the publishing house Elsevier. It includes over 30 million records, covers 90 countries, and includes journals not covered by Medline.

Web site:

http://www.elsevier.com/online-tools/embase

Beilstein/Reaxys:

In 1881, the first edition of Beilstein's Handbook of Organic Chemistry was published. It has since served the field of organic chemistry. Chemical compounds are uniquely identified by their Beilstein Registry Number. The fourth edition was published in 503 volumes (over 440 000 pages) from 1918 to 1998 and covered the literature on organic chemistry comprehensively from 1771 up to 1959 and then more selectively for heterocyclics up to 1979. Since 2009, the content has been maintained and distributed by Elsevier Information Systems in Frankfurt under the product name "Reaxys."

Web site:

http://www.elsevier.com/online-tools/reaxys/about

CAS by ACS

The American Chemical Society (ACS) is a very large professional organization with a long history of collecting and providing information related to the world's known chemical compounds, related literature and patents, and other relevant data, CAS (Chemical Abstracts Service). SciFinder makes the information available to nonexperts, and STN is the premier single source for the world's disclosed scientific and technical research. CAS STN is the only platform with complete CAS content.

Web site:

http://www.cas.org/index

The World of Patent Data

The United States Patent and Trademark Office (USPTO) cooperates with the European Patent Office (EPO) and the Japan Patent Office (JPO) as one of the Trilateral Patent Offices. The USPTO is an agency in the US Department of Commerce that issues patents to inventors and businesses for their inventions, and trademark registration for product and intellectual property identification. WIPO is the World Intellectual Property Organization, the global forum for intellectual property services, policy, information, and cooperation.

Web sites:

www.uspto.gov
www.epo.org
www.jpo.go.jp
http://www.wipo.int/pct/en/

Thomson-Reuters Derwent World Patent and World Drug Index

Derwent World Patent Index (DWPI) is the world's most comprehensive database of *enhanced* patent documents. Subject experts correct, analyze, abstract, and manually index every patent record.

Web site:

https://clarivate.com/products/derwent-world-patents-index/

World Drug Index is an authoritative index for marketed and development drugs. It includes internationally recognized drug names, synonyms, trade names, trivial names, and trial preparation codes in one source, plus compound structures, and activity data.

WDI and other resources managed by Thomson-Reuters can be found at the following web site:

http://www.daylight.com/products/wdi.html

Drug Safety and Pharmacovigilance

World Health Organization (WHO) and FDA provide information about drug safety.

Web sites:

https://www.who-umc.org/
http://www.fda.gov/default.htm

Genes and Proteins

GenBank

GenBank is the NIH genetic sequence database, an annotated collection of all publicly available DNA sequences [16]. GenBank is part of the International Nucleotide Sequence Database Collaboration, which comprises the DNA DataBank of Japan (DDBJ), the European Molecular Biology Laboratory (EMBL), and GenBank at NCBI. These three organizations exchange data on a daily basis.

Web site:

http://www.ncbi.nlm.nih.gov/genbank/

OMIM

Online Mendelian Inheritance in Man (OMIM) is a comprehensive, authoritative compendium of human genes, genetic disorders, and genetic phenotypes. OMIM is freely available and updated daily. OMIM is authored and edited at the McKusick-Nathans Institute of Genetic Medicine, Johns Hopkins University School of Medicine.

Web site:

http://www.ncbi.nlm.nih.gov/omim

Protein Data Bank (PDB)

Announced in 2003 [17], the worldwide Protein Data Bank (wwPDB) is the repository of over 120 000 three-dimensional structural data of proteins and nucleic acids; wwPDB consists of organizations that act as deposition, data processing, and distribution centers for Protein Data Bank (PDB) data. Members are the United States' Biological Macromolecular Resource (RCSB), PDBe (Europe) and PDBj (Japan), and Biological Magnetic Resonance Data Bank (BMRB).

Web site:

http://www.wwpdb.org/

UniProt: ExPASy, PIR, Swiss-Prot, TrEMBL, and GOA

The mission of UniProt is to provide the scientific community with a comprehensive, high-quality, and freely accessible resource of protein sequence and functional information. The UniProt consortium comprises the European Bioinformatics Institute (EBI), the Swiss Institute of Bioinformatics (SIB), and the Protein Information Resource (PIR). EBI is located at the Wellcome Trust Genome Campus in Hinxton, UK. SIB, located in Geneva, Switzerland, maintains the **ExPASy** (Expert Protein Analysis System) servers that are a central resource for proteomics tools and databases. **PIR**, hosted by the National Biomedical Research Foundation (NBRF) at the Georgetown University Medical Center in Washington, DC, USA, is heir to the oldest protein sequence database, Margaret Dayhoff's *Atlas of Protein Sequence and Structure,* first published in 1965.

The UniProt protein knowledge base includes **Swiss-Prot**, which is manually annotated and reviewed, and **TrEMBL**, which is automatically annotated and is not reviewed.

The UniProt Gene Ontology annotation (GOA) program aims to provide high-quality Gene Ontology (GO) (see below) annotations to proteins in the UniProt Knowledgebase (UniProtKB)

Web sites:

www.uniprot.org
www.expasy.org
http://pir.georgetown.edu
http://www.uniprot.org/help/uniprotkb
http://www.ebi.ac.uk/GOA

ENZYME

ENZYME is a repository of information relative to the nomenclature of enzymes. It is primarily based on the recommendations of the Nomenclature Committee of the International Union of Biochemistry and Molecular Biology (IUBMB), and it describes each type of characterized enzyme for which an EC (Enzyme Commission) number has been provided.

Web site:

http://enzyme.expasy.org/

Ensembl

Established in 1999, Ensembl is a joint project between EBI, an outstation of the EMBL, and the Wellcome Trust Sanger Institute (located next door outside Cambridge, UK) to develop a software system which produces and maintains automatic annotation on selected eukaryotic genomes, including human genomes. The Ensembl portal features extensive cross-references to other databases and provides tools for processing of Ensembl and user data.

Web site:

http://useast.ensembl.org/info/about/index.html

Hugo Gene Nomenclature Committee (HGNC)

Supported by the National Human Genome Research Institute (NHGRI) and by Wellcome Trust, Hugo Gene Nomenclature Committee (HGNC) is managed by the EBI (see above) and is the only worldwide authority that assigns standardized nomenclature to human genes. HGNC is responsible for approving unique symbols and names for human loci, including protein coding genes, ncRNA genes, and pseudogenes, to allow unambiguous scientific communication.

Web site:

www.genenames.org

Mouse Genome Informatics (MGI)

Managed by the Jackson Lab in Bar Harbor, Maine, Mouse Genome Informatics (MGI) is the international database resource for the laboratory mouse, providing integrated genetic, genomic, and biological data to facilitate the study of human health and disease. Projects contributing to this important resource are the Mouse Genome Database (MGD), Gene Expression Database

(GXD), Mouse Tumor Biology (MTB) Database, Gene Ontology (GO), MouseMine, and MouseCyc with focus on *Mus musculus* metabolism, including cell level processes such as biosynthesis, degradation, energy production, and detoxification.

Web site:

www.informatics.jax.org

Gene Ontology (GO)

The Gene Ontology project is a major bioinformatics initiative with the aim of standardizing the representation of gene and gene product attributes across species and databases. The Gene Ontology Consortium is supported by a grant from the NHGRI of NIH.

Web site:

www.geneontology.org

Gene Expression Omnibus (GEO)

Gene Expression Omnibus (GEO) is a public functional genomics data repository supporting MIAME (Minimum Information about a Microarray Experiment)-compliant data submissions, where MIAME is a standard created by the Functional Genomics Data (FGED; fged.org) Society for reporting microarray experiments. Array- and sequence-based data are accepted. Tools are provided to help users query and download experiments and curated gene expression profiles.

Web site:

http://www.ncbi.nlm.nih.gov/geo/

The Cancer Genome Atlas (TCGA)

Since 2006, TCGA's (The Cancer Genome Atlas) goal is to accelerate our understanding of the molecular basis of cancer through the application of genome analysis technologies, including large-scale genome sequencing. TCGA is a joint effort of the NCI and the NHGRI, 2 of the 27 institutes and Centers of NIH. Criteria for Cancers selected by TCGA for study include "Poor prognosis and overall public health impact" and "Availability of human tumor and matched-normal tissue samples that meet TCGA standards for patient consent, quality and quantity." As of May 2014, the cancer selected by TCGA are breast C., central nervous system C. (glioblastoma, glioma), endocrine C., gastrointestinal C., gynecologic C., head and neck C., hematologic C.

(leukemia, lymphoma), skin C. (melanoma), soft tissue C. (sarcoma), thoracic C. (incl. lung, mesothelioma), and urologic C. (including prostate). There are at least 200 forms of cancer and many more subtypes. Each of these is caused by errors in DNA that cause cells to grow uncontrolled. By studying the cancer genome, scientists can discover what letter changes are causing a cell to become a cancer. The genome of a cancer cell can also be used to tell one type of cancer from another. In some cases, studying the genome in a cancer can help identify a subtype of cancer within that type, such as HER2+ breast cancer. Understanding the cancer genome may also help a doctor select the best treatment for each patient.

Web site:

http://cancergenome.nih.gov/

Biochemical Pathways, Protein Interactions, and Drug Targets

KEGG

The Kyoto Encyclopedia of Genes and Genomes (KEGG) is a collection of manually curated databases dealing with genomes, biological pathways, diseases, drugs, and chemical substances. KEGG is utilized for bioinformatics research and education, including data analysis in genomics, metagenomics, metabolomics and other "omics" studies, modeling and simulation in systems biology, and translational research in drug development. KEGG was initiated in 1995 by Minoru Kanehisa, Professor at the Institute for Chemical Research, Kyoto University, under the then ongoing Japanese Human Genome Program. KEGG is categorized into systems, genomic, chemical, and health information, including DISEASE and DRUG databases where diseases are viewed as perturbed states of the molecular system, and drugs as perturbants to the molecular system.

Web site:

http://www.genome.jp/kegg/

BioCarta

BioCarta was founded in April 2000 with headquarters in San Diego, California, to become a developer, supplier, and distributor of data related to biochemical pathways, reagents, and assays for biopharmaceutical and academic research. BioCarta's goal is to create a complete map of how proteins act in human health and disease. BioCarta's web site has a community-focused maintenance policy, enabled by an open forum for information exchange and collaboration between researchers, educators, and students.

Web site:

http://www.biocarta.com/Default.aspx

Ingenuity Pathway Analysis (IPA)

Ingenuity Pathway Analysis (IPA) was introduced in 2003 and is used to help life sciences researchers analyze "omics" data and model biological systems. IPA is a commercial biochemical pathway suite–based expert curated content. In 2013, Ingenuity Systems Inc. (Redwood City, California) was acquired by QIAGEN, a molecular diagnostics company with its headquarters in the Netherlands.

Web site:

http://www.ingenuity.com/products/ipa

MetaCore

MetaCore is an integrated software suite provided by Thomson Reuter for functional analysis of genomic and proteomic data, based on a high-quality, manually curated database of the following:

- transcription factors, receptors, ligands, kinases, drugs, and endogenous metabolites as well as other molecular classes
- species-specific directional interactions between protein–protein, protein–DNA and protein-RNA, drug targeting, and bioactive molecules and their effects
- signaling and metabolic pathways represented on maps and networks
- rich ontologies for diseases and processes with hierarchical or graphic output.

Web site:

http://lsresearch.thomsonreuters.com/pages/solutions/1/metacore

InterPro

InterPro provides functional analysis of proteins by classifying them into families and predicting domains and important sites. To classify proteins in this way, InterPro uses predictive models, known as signatures, provided by several different databases (referred to as member databases) that make up the InterPro consortium.

Web site:

www.ebi.ac.uk/interpro

DrugBank

The DrugBank database is a bioinformatics and cheminformatics resource that combines detailed drug data with comprehensive drug target (i.e. sequence, structure, and pathway) information. By 2016, the database includes 8206 drug entries, including 1991 FDA-approved small molecule drugs, 207 FDA-approved biotech (protein/peptide) drugs, 93 nutraceuticals, and over 6000 experimental drugs.

Web site:

www.drugbank.ca

DGIdb

Managed by Genome Institute at Washington University in St. Louis, the Drug–Gene Interaction database (DGIdb) mines existing resources that generate hypotheses about how mutated genes might be targeted therapeutically or prioritized for drug development. It provides an interface for searching lists of genes against a compendium of drug–gene interactions and potentially "druggable" genes.

Web site:

dgidb.genome.wustl.edu

Conclusions

Biomarkers are key drivers of the ongoing health care transformation towards the new paradigm of stratified/personalized or precision medicine. In this chapter, we focused on the role of IT in supporting the use of biomarkers in bio-pharmaceutical R&D. When doing so, we need to keep in mind that the desired benefits to patients and consumers will only be realized if the new biomedical knowledge is "translated" into stratified/personalized patient care. The bio-pharmaceutical industry will have to participate not only as a provider of drugs and medical treatments but also as a contributor to the emerging biomedical knowledge base and to IT infrastructures needed to enable biomarker-based R&D and clinical care.

It is therefore critical to define the necessary interfaces between the respective IT environments and to agree on standards that enable data interchanges. IT standards and architectures must support the integration of new biomarker data with conventional clinical data types and the management of the integrated data in (centralized or federated) data warehouses that can be queried and analyzed. Analysis and Mining of Biomarker/Healthcare Data is mathematically challenging but needed to support diagnostic and treatment decisions by providers of personalized care. SOAs are required to support the resulting processes and workflows covering the various health care stakeholders.

Finally, a few words about "Cloud Computing" and its role in biomarker-based drug development. Cloud computing provides shared computer processing resources and data to computers and **other devices** on demand. It enables ubiquitous, on-demand access to a shared pool of configurable computing resources such as computer networks, servers, storage, applications, and services. Those resources can be shared over the internet, and they can be rapidly increased and released with minimal management effort. As users demand access to resources outside their office environments, and on various mobile devices, cloud computing is enjoying increasing popularity. However, it is also generating new challenges in areas like data security and privacy.

Acknowledgments

The ideas described above are based on input from and discussions with a number of my former IBM colleagues. In particular, I'd like to acknowledge Avijit Chatterjee, Terrence McCormick, Kathleen Martin, Jill Kaufman, David Martin, Chris Hines, Joyce Hernandez, and Houtan Aghili.

References

1 Trusheim, M.R., Berndt, E.R., and Douglas, F.L. (2007). Stratified medicine: strategic and economic implications of combining drugs and clinical biomarkers. *Nat. Rev.* 6: 287–293.

2 Hehenberger, M. (2015). *Nanomedicine: Science, Business, and Impact*. Pan Stanford Publishing. ISBN: 978-981-4613-76-7.

3 Steve Arlington, Sam Barnett, Simon Hughes, Joe Palo (2002). Pharma 2010: The threshold of innovation. http://www-07.ibm.com/services/pdf/pharma_es.pdf (accessed 30 April 2019)

4 DiMasi, J.A. (2002). The value of improving the productivity of the drug development process: faster times and better decisions. *PharmacoEconomics* 20 (suppl. 3): 1–10.

5 DiMasi, J.A., Hansen, R.W., and Grabowski, H.G. (2003). The price of innovation: new estimates of drug development costs. *J. Health Econ.* 22: 151–185.

6 Lesko, L.J. and Atkinson, A.J. Jr. (2001). Use of biomarkers and surrogate endpoints in drug development and regulatory decision making: criteria, validation, strategies. *Annu. Rev. Pharmacol. Toxicol.* 41: 347–366.

7 Corr, P. (2005). IBM Imaging Biomarker Summit I, Palisades, NY (15–17 December 2005).

8 McCormick, T., Martin, K., and Hehenberger, M.; IBM Institute for Business Value (July 2007). The evolving role of biomarkers: Focusing on patients from research to clinical practice. http://www.ibm.com/industries/healthcare/doc/jsp/resource/insight/ (accessed 30 April 2019).

9 Kroll, W. (2007). IBM Imaging Biomarker Summit III, Nice, France (24–26 January 2007).

10 Hehenberger, M., Chatterjee, A., Reddy, U. et al. (2007). IT solutions for imaging biomarkers in bio-pharmaceutical R&D. *IBM Syst. J.* 46 (1): 183–198.

11 Kimball, R. and Caserta, J. (2004). *The Data Warehouse ETL Toolkit: Practical Techniques for Extracting, Cleaning, Conforming, and Delivering Data.* Wiley. 416 pp.

12 Codd, E.F. (1981). The significance of the SQL/data system announcement. *Computerworld* 15 (7): 27–30. See also: https://dblp.uni-trier.de/pers/hd/c/Codd:E=_F=.

13 Haas, L., Schwarz, P., Kodali, P. et al. (2001). DiscoveryLink: a system for integrated access to Life Sciences Data. *IBM Syst. J.* 40 (2): 489–511.

14 Carter, S. (2007). *The New Language of Business: SOA & Web 2.0.* IBM Press. ISBN-10: 0-13-195654-X; ISBN-13: 978-0-13-195654-4. 320 pp.

15 Simmons, R.F. (1970). Natural language question-answering systems: 1969. *Commun. ACM* 13: 15–30.

16 Benson, D.A., Cavanaugh, M., Clark, K. et al. (2013). *Nucleic Acids Res.* 41 (D1): D36–D42. https://doi.org/10.1093/nar/gks1195.

17 Berman, H.M., Henrick, K., and Nakamura, H. (2003). Announcing the worldwide Protein Data Bank. *Nat. Struct. Biol.* 10: 980.

23

Identifying Biomarker Profiles Through the Epidemiologic Analysis of Big Health Care Data – Implications for Clinical Management and Clinical Trial Design: A Case Study in Anemia of Chronic Kidney Disease

Gregory P. Fusco

Epividian, Inc., Chicago, IL, USA

Big Healthcare Data must be considered simply as a substrate, not a substitute, for critical thinking, scientific discipline and methodological rigor.

Revere the faculty of discernment. On this faculty rests protection from false perceptions – inconsistent with nature and the constitution of rational beings. It promises freedom from hasty judgement, …

Marcus Aurelius Antoninus Augustus

Introduction

Conceptually, a biomarker is generally thought to be any chemical or biological product that can be isolated in the laboratory or measured in the clinic [1]. These products may be, or may become, tools used for development, diagnostic, assessment, treatment, and prognostic purposes [2]. The term "biomarker" often directs its interpreter to imagine such things as molecules, proteins, enzymes, and genetic mutations – things invisible, but yet somehow cognitively tangible. Depending on the disease or condition being evaluated, expansion of the biomarker concept beyond the "tangible" to include the nebulous may prove beneficial, maybe even necessary, for effective drug development, accurate clinical diagnosis, and successful therapeutic management.

The path to identifying a valid biomarker most often travels from the bench to the clinic prior to becoming medically useful at the bedside [2], a process usually involving painstaking "–omic" laboratory research using tools such as mass spectrometry which is then followed by clinical evaluation and validation

Biomarkers in Drug Discovery and Development: A Handbook of Practice, Application, and Strategy, Second Edition. Edited by Ramin Rahbari, Jonathan Van Niewaal, and Michael R. Bleavins.

processes, with the latter two occurring most commonly with small numbers of subjects and/or tissue samples [1, 2]. However, the process of identification and validation of biomarkers does not always begin in the laboratory as a substantial proportion of clinically useful biomarkers have begun this process within the clinic itself [2]; these tend to be patterns of "multi-marker profiles" rather than single, identifiable genes or proteins.

Molecular epidemiology, conceived and nurtured in the 1980s, applies mathematical and molecular techniques to uncover exposure–disease relationships with a primary focus on: (i) the development of hypotheses, (ii) the identification of exposures, and (iii) the quantification of susceptibilities [3]. Examples include viral–disease, chemical–disease, and chromosomal–disease associations [4–6]. This discipline and these techniques will continue to evolve and contribute to the development and identification of biomarkers as both "–omic" tools and "big data" mature.

The concept of big health care data began to evolve in the 1990s with the term "big data" entering the popular lexicon later that decade. Similar to molecular epidemiology's combining of mathematical and molecular techniques to the evaluation of clinical and tissue sample data; clinical epidemiology combines experimental design and biological/behavioral rationale with mathematical/statistical techniques for the analysis of population-level clinical data. The rapidly expanding volume of available-to-be-analyzed data, especially that of health care data, has given rise to a generalized discipline of data science and to the common application of data-mining techniques, all of which may entice the researcher to continually comb through data in search of answers to sometimes, ill-defined hypotheses. Carefully constructed experimental design, combined with scientific knowledge and mathematical rigor, characterizes sound epidemiologic investigations, and these traits are especially important in the analysis of big health care data in order to reduce the likelihood of uncovering spurious associations or dismissing true ones. This is particularly true when searching for medically useful biomarkers and biomarker profiles lurking in the depths of big data sets.

Medically useful biomarkers, as stated previously, are most commonly perceived as being biochemical in nature (e.g. genes, molecules, proteins). Useful biomarkers may also be biological or physiological concepts, especially in circumstances involving a potential physiologic threshold. Here, as with the elucidation of "biomarker profiles" within the clinic, population-level epidemiologic studies can also identify "profiles"; profiles which may facilitate more tailored monitoring and more nuanced therapeutic management. This case study in anemia of chronic kidney disease (CKD) describes the identification of a "physiologic threshold" through population-level analysis [7] and suggests the use of physiologic thresholds combined with laboratory monitoring based on hemodynamic principles to create a "biomarker profile" that may guide not only therapeutic management but the design of interventional clinical trials as well.

Considerations on Epidemiologic Design and the Analysis of Big Health Care Data

The fundamental objective of epidemiology is to identify and describe the relationship between cause and effect, whether there are direct and/or indirect causal pathways, confounding interactions, and perturbations [8]. Developing and synthesizing the evidence for effective medical and public health interventions, in order to meet this fundamental objective, becomes increasingly challenging as these interventions become increasingly complex [9]. Any resulting evidence base should ideally describe whether, when, why, and how interventions heal or harm [9].

In pursuing this evidence base, epidemiological research has focused its attention toward the isolation of causative factors, whether biological or behavioral, of diseases or adverse outcomes [10] with computational and methodological advancements having allowed the use of hierarchical regression models which consider the contribution of multiple, potentially confounding, factors [11]. Regression approaches, however, lack the ability to account for the relational dynamism and reciprocity between factors and events [11] and poorly differentiate between confounding and mediation [12]. Even so, recent advances have facilitated atheoretical, "black box" epidemiological explorations capable of generating hypotheses [13] while simultaneously risking chance associations without necessarily providing a theoretical (biological or sociological) rationale for the (potentially statistically significant) association's existence [14].

Describing causal associations and isolating factors in these relationships requires a focused and clear investigative plan. Examining complex interventions requires acknowledging and addressing many characteristics of complexity such as multiple components, malleability (e.g. dose adjustments) of the intervention, nonlinear interrelationships, positive/negative feedback, mediation and moderation, and so on [9]. Even with complex medical interventions, however, a narrowly targeted investigation related to a confined set of outcomes is a legitimate pursuit; the prerequisite of which is a clear understanding of the biological basis of the disease and the clinical condition, combined with a well-focused question [9].

Targeted epidemiologic investigations should ideally be based upon biological or physiological concepts such as homeostatic mechanisms [15]. Incorporating physiologic elements such as homeostasis, tissue specificity, and feedback mechanisms may improve epidemiologic models meant to advance theory, guide public health efforts, and analyze data [15]. Defining a focused question, based on biological mechanisms, is the initial step in tracing causal pathways and is ideally supported with careful identification of model parameters and an even more careful identification of parameter values. Critical to causal inference are the assumptions about parameters, the values of which should necessarily incorporate tangible (biological or behavioral) evidence [12].

Comparative effectiveness assessments within complex systems can entice the researcher into statistically adjusting an excessive number of variables in order to estimate the total direct and indirect causal effect of an exposure. This may lead to over-, as well as, unnecessary adjustment [16] which may, in turn, obscure a true causal relationship or identify a spurious one [17]. Efforts to identify true causal relationships should, if possible, construct a model which is (i) centered on a focused question, (ii) predicated on biological evidence, (iii) complemented by clinical knowledge, and (iv) devoid of excessive statistical adjustment.

For example, the analysis for this case study [7] addressing the question "is there a safe(r) path for hemoglobin reconstitution" takes into account the clinical condition of the patient (e.g. hemoglobin variability or "oscillations"), but does not include variables for dose or dose escalation. The specific question is about "following a particular path" with the most relevant point being the path actually taken, not how one was initially propelled down that path or how one managed to follow that path. So, to reliably represent the biological process [15], the oscillations variable substituted for a range of clinical variables including dose and dose escalations [18], variables which would be required to assess initial propulsion and ongoing management, but would not be necessary to assess the path having already been traveled.

The interplay between host, disease, treatment, and biology presents wide-ranging complexity requiring considerable attention. Assessing such complex situations does not necessarily require complex research questions or complex models [9]. The process of formulating, constructing, and analyzing a causal model should consider: (i) targeting inferences about the most plausible theories, (ii) incorporating the appropriate characteristics in the mathematical form necessary to faithfully and predictably represent the biological system, and (iii) capturing the essence of a complex system with simple models [15, 19], all of which to adhere to the principle of Occam's razor: analyses should be as complex as necessary, but not more so [19].

In keeping with the principle of Occam's razor while analyzing a complex physiologic system, covariates of interest in this case study [7] were based on: (i) a priori considerations of clinical significance [20], (ii) pertinence to the specific question [17], (iii) considerations of over-adjustment [16], and (iv) considerations of possible mediating effects [21], with the goal of incorporating biologically rational "exposures" [19].

Clinical Background of Anemia of CKD and Study Premises

The connection of CKD to anemia was identified in the nineteenth century [22], while the biochemical link between kidney function and erythropoietic stimulation was uncovered a century later [23]. CKD is defined as kidney damage lasting for ≥3 months, as determined by structural or functional abnormalities, which can lead to a decrease in glomerular filtration rate (GFR). These anatomic or physiologic abnormalities may or may not initially coincide with a decreased GFR and may manifest in pathological or pathophysiological abnormalities (e.g. composition of blood or urine). Additionally, CKD can also be defined as a GFR < 60 ml/min/1.73 m^2 for ≥3 months with or without kidney damage [24–26].

The direction of clinical inquiry and the foundation for all medical interventions have been based on the understanding that normal physiologic ranges confer a survival advantage as compared to abnormal (i.e. above/below) values (e.g. hypertension, polycythemia, anemia, hyperglycemia). Clinical [27] and epidemiologic studies [28] have demonstrated that a chronically anemic state increases the risk of cardiovascular adverse events and have provided the rationale – normalize hemoglobin and therefore normalize cardiovascular risk – for medical intervention and therapeutic development. The development of recombinant erythropoietin, followed by erythropoiesis stimulating agents (ESAs), marked a significant milestone in the management of anemia of CKD.

Anemia, being a common complication of CKD, is associated with an increased cardiovascular morbidity and mortality [29, 30], and ESAs have been used to normalize hemoglobin/hematocrit levels. Early clinical studies in dialysis patients demonstrated various benefits of correcting anemia with ESAs [31–44], and numerous epidemiologic studies demonstrated an increased risk of mortality with the anemic state [28, 45–48], along with other benefits of higher [28, 49–51] as compared to lower [30, 52, 53] hemoglobin concentrations. Data from various randomized clinical trials (RCTs), designed to compare high hemoglobin/hematocrit target ranges to low hemoglobin/hematocrit target ranges, however, suggested just the opposite; returning to normal hemoglobin/hematocrit ranges actually increased risk, or had no benefit whatsoever when compared to maintaining an anemic state [54–58]. Furthermore, the FDA noted in the licensing review for darbepoetin-α that a separate and distinct factor seemed to increase cardiovascular risk: a rate of hemoglobin rise ≥0.5 g/dl/week (≥2 g/dl/month) [59].

With respect to dosing algorithms, studies have, again, yielded conflicting results focusing the associated risk debate on dose escalations [60] vs. a patient's actual ability to respond [61]. Initial responses to ESA dosing are highly variable, and subsequent doses depend upon the initial response along with the overall clinical condition and general patient (hypo) responsiveness [62, 63]. Some of these studies have related ESA responsiveness to clinical factors [64–68], and others have related the dose to clinical factors necessitating the dose [61, 62, 69–71]. Both RCT and epidemiology study designs each have yielded conflicting dose-outcome results. The large RCTs [54, 55, 57, 58] all concluded that risk increased with higher hemoglobin targets, with a dose–outcome relationship being seen in one secondary analysis [72] while not in another [63]. Similarly, epidemiologic studies, with varying levels of nuance, have suggested that dose is both related [73–75] and unrelated [76–78] to adverse outcomes.

Three basic experimental designs (i.e. small clinical studies, large randomized, interventional trials, and observational studies) provided conflicting evidence about the benefit and risk of complete correction of hemoglobin/hematocrit. As a result, various factors such as hemoglobin target range achieved, hemoglobin rate of rise, ESA dose, ESA dose escalations, and off-target effects of ESAs, as well as the overall clinical profile, have all been implicated as culprits with respect to the increased cardiovascular risk. Together, this evidence has resulted in treatment guidelines recommending partial, rather than complete, correction of anemia; treating to a high-anemic, but not normal, hemoglobin level [79–83], even though the link between the anemic state and increased cardiovascular risk is widely understood.

Study Concept, Biological Rationale, and Study Design

Evaluating these three basic study designs (with respect to hemoglobin reconstitution) in aggregate highlights the difficulty in disentangling true risks from spurious ones. Applying first principles to this collection of conflicting data suggests a flaw either in: (i) the algorithm for intervention or (ii) the designs for scientific inquiry to assess the intervention, or (iii) both. With respect to the foundational, first principles of biology, that the normal physiologic state confers a survival benefit, a return to a normal hemoglobin level from the anemic state should, in and of itself, not be harmful as some of the epidemiologic data demonstrates [28, 30, 45–53]. However, various epidemiologic studies suggest otherwise [73–78], and the large interventional trials suggested that a return to normal hemoglobin ranges was harmful [54–58]. The resulting conflict with first principles argues for a reevaluation of the treatment algorithm through a reconceptualized study design.

The large RCTs evaluated anemia therapy with ESAs by evaluating subjects who were reconstituted to normal hemoglobin ranges as compared to those who were maintained in an anemic state, comparing "high" vs. "low" hemoglobin levels as conceptualized in Figure 23.1. All of these studies either demonstrated harm with, or no clinical benefit of, full hemoglobin reconstitution. Figure 23.2 describes this reconceptualized observational epidemiology study (the basis for this chapter) which evaluated subjects achieving the same level of hemoglobin reconstitution; however, subjects were able to achieve the same hemoglobin levels at differing rates: a "fast" vs. "slow" concept. This "fast"

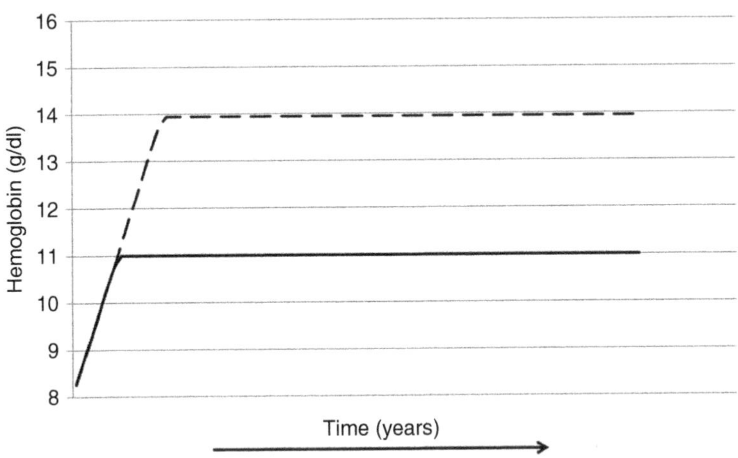

Figure 23.1 Randomized clinical trial concept – high vs. low.

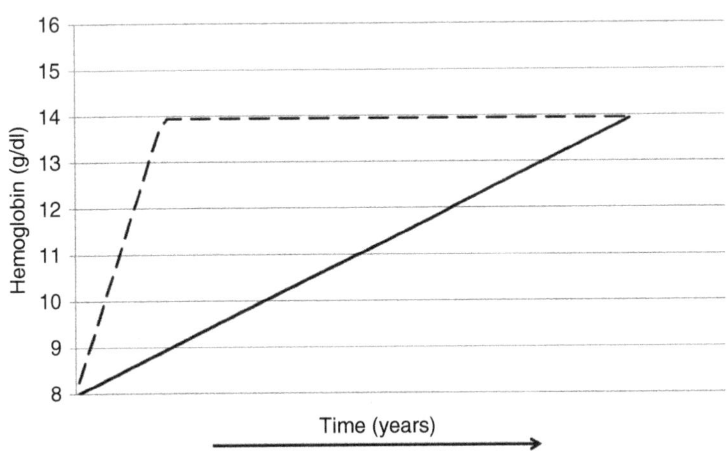

Figure 23.2 Concept for this case study – fast vs. slow.

vs. "slow" study design was based on the hemodynamic principle of viscosity with the hypothesis being that a rapid rate of hemoglobin rise is directly linked to changes in whole blood viscosity (WBV). And it is a rapid change in WBV that results in the adverse effects of stroke and myocardial infarction (MI).

WBV is the major determinant of normal blood flow [84], and hematocrit is the major determinant of WBV [85], having a logarithmically linear relationship to it [85, 86]. Disease states with elevated hemoglobin/hematocrit, such as polycythemia vera, have demonstrated such relationships between hematocrit and WBV, where WBV increases exponentially with a rise in hematocrit [86]. This relationship tended to be log-linear at any given protein concentration [87] and at both high and low rates of shear [85]. Experiments to induce polycythemia in animals showed a marked reduction in cerebral blood flow due to WBV, with approximately 60% of this reduction attributed to the increase in hematocrit [88].

A rapid hemoglobin recovery, as with traditional ESA therapy, leads to a rapidly expanding hemoglobin concentration which then results in an exponential increase in WBV. This exponential increase in WBV leads to macrovascular and microvascular blood flow impairment with resulting complications of tissue ischemia, thrombosis, cerebrovascular strokes, and MIs. Additionally, escalating ESA doses results in an increased production of hyper-reactive platelets, which leads to an elevated thrombotic tendency as well [89–92].

Hemodynamic principles of both normal and pathophysiology suggest the existence of biological thresholds and in order to properly address the viscosity hypothesis, the isolation of a possible hemodynamic threshold was required.

The conceptual basis for the design of this epidemiology study is described graphically in Figure 23.3. A "fast" vs. "slow" study design called for the identification of a "theoretically harmless" rate of hemoglobin rise (slope S in Figure 23.3). It was hypothesized that this slope would necessarily mirror the "normally observed" pathophysiological rate of hemoglobin decline in the setting of CKD. Slope Q represents this hypothesized "normally observed" rate of hemoglobin decline; however, a value for slope Q had not previously been identified and has never been "normally observed" in the clinic. Slope R in Figure 23.3 represents an amalgamation of the various rate(s) of hemoglobin rise seen in the large RCTs, all of which demonstrated a harmful effect of hemoglobin reconstitution. Identifying slope S required solving for slope Q, which involved juxtaposing pathological observation (hypertrophy of the nephron mass), physiologic experimentation (hemoglobin/hematocrit response to parenchymal function), and epidemiologic investigation (a time course of renal functional decline) to yield this hypothesized "normally observed" rate of hemoglobin decline. Merely inverting slope Q provided the "theoretically harmless" trajectory, slope S [7].

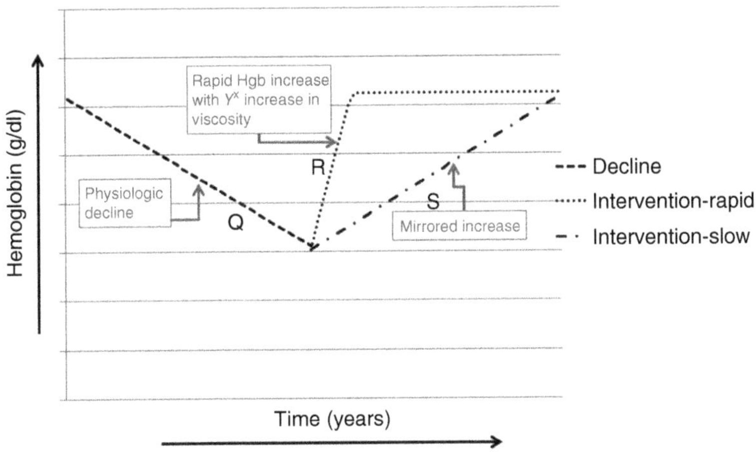

Figure 23.3 Identifying the "slow" trajectory based upon WBV hypothesis.

The process for identifying the variables and cut-points of these variables along with the statistical modeling is described elsewhere [7] with the development of the analytical hypothesis following these steps:

1. *The investigational hypothesis*: hemoglobin rate of rise is directly linked to changes in WBV which is, in turn, linked to increases in the adverse outcomes of MI and stroke.
2. *The specific question*: if change in viscosity is the factor related to the increased risk, then:
   (1) Is there a trajectory (i.e. rate of rise) that is safe(r) than the others and, if so,
   (2) How far/high (i.e. to what hemoglobin milestone or target range) along that trajectory can one go and still maintain a positive benefit/risk balance?
3. *The working hypothesis*: the working hypothesis analyzed was that increasing hemoglobin leads to increasing blood viscosity, and a higher hemoglobin rate of rise magnifies these changes in viscosity, which results in a higher risk of MI and stroke as compared to a lower hemoglobin rate of rise.
4. *The epidemiologic hypothesis*: hemoglobin rate of rise is a mediator between the independent variable (ESA therapy, dose, and dose escalations) and the dependent variable (composite outcome of MI and stroke).
5. *The main analytical hypothesis*: addressing specific question #2.1: a slower hemoglobin rate of rise ($0 < \text{g/dl/month} \leq 0.125$) is associated with a lower incidence of cardiovascular events among ESA users as compared to a faster hemoglobin rate of rise ($0.125 < \text{g/dl/month} \leq 2.0$, and ≥ 2.0 g/dl/month).

6. *The milestone (target range) analytical hypothesis*: addressing specific question #2.2: a slower hemoglobin rate of rise is associated with a lower incidence of cardiovascular events as compared to a faster hemoglobin rate of rise *within* each hemoglobin milestone (i.e. target range) achieved including the normal range (>12.5 g/dl).

With slope S identified (now as group B in Figure 23.4), the incidence rates for cardiovascular events (MI/stroke) were calculated for various rates of hemoglobin rise. As Figure 23.4 describes, slope S (group B) had a markedly lower incidence rate when compared to all of the other rates of rise (groups C–H) and the rates of decline (group A). So, group B represents a threshold trajectory above which the incidence of cardiovascular events rises dramatically. Further statistical analyses were performed to determine if there was an actual statistically significant difference between rates of rise throughout the hemoglobin increase as well as whether there were statistical differences between the rates of rise when subjects achieved the same hemoglobin levels (i.e. target ranges or milestones) by following different trajectories.

In this analysis, the individual groups A through H were not the primary statistical comparison. For the statistical comparison, the groups in Figure 23.4 were combined into three groups: slow (group B), medium (groups C–F; now referred to as group C′), and fast (groups G and H; now referred to as group D′). Figure 23.5 provides a graphical visualization of the proportional hazards modeling. Along the entire rising pathway, crossing all target ranges, the overall slope of the "medium" rate of rise group had more than a 1/3 reduction in risk compared to the "fast" rate of rise group. Most importantly, the "slow" rate

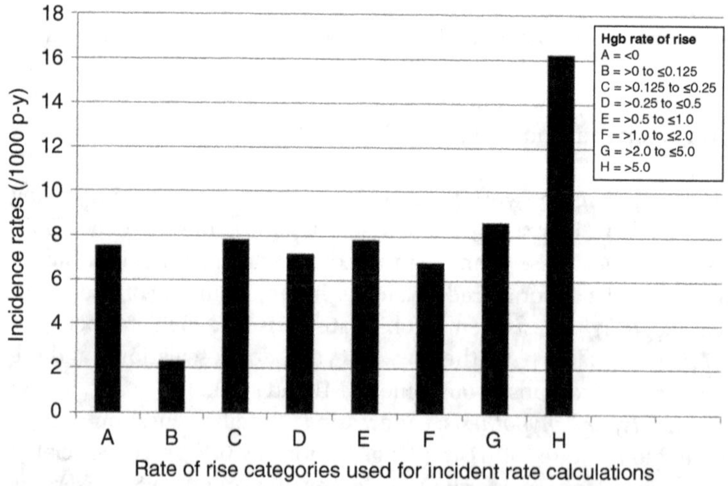

Figure 23.4 Incidence rates per 1000 person-years (*y*-axis) of various hemoglobin rates of rise (*x*-axis).

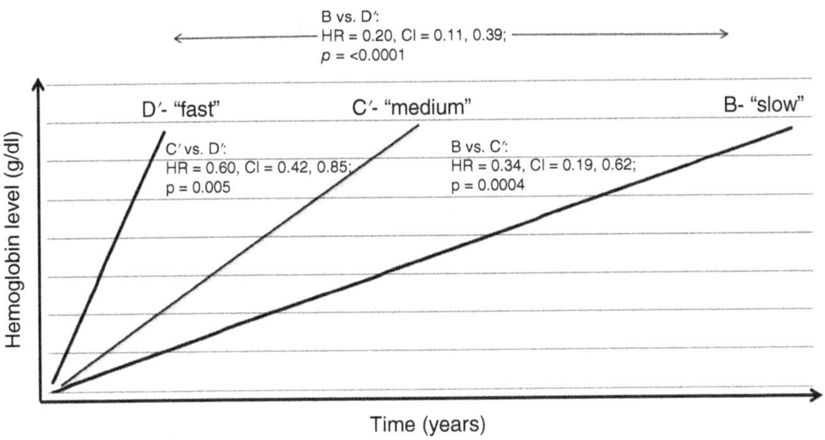

Figure 23.5 Graphical representation of the comparisons of hazard ratios between fast, medium, and slow trajectories across all hemoglobin ranges.

of rise group (group B, slope S) reduced the risk by another 2/3, resulting in an overall 80% reduction in risk between the "fast" and "slow" groups. Similar trends were seen when evaluating patients that started treatment at the same hemoglobin levels (<10 g/dl, 10.1–11.0 g/dl, 11.1–12.5 g/dl).

A more granular evaluation of rising trajectories is visualized in Figure 23.6. Here, the analyses compared the fast (D′), medium (C′), and slow (B) trajectories to one another; however, each comparison was restricted to those patients

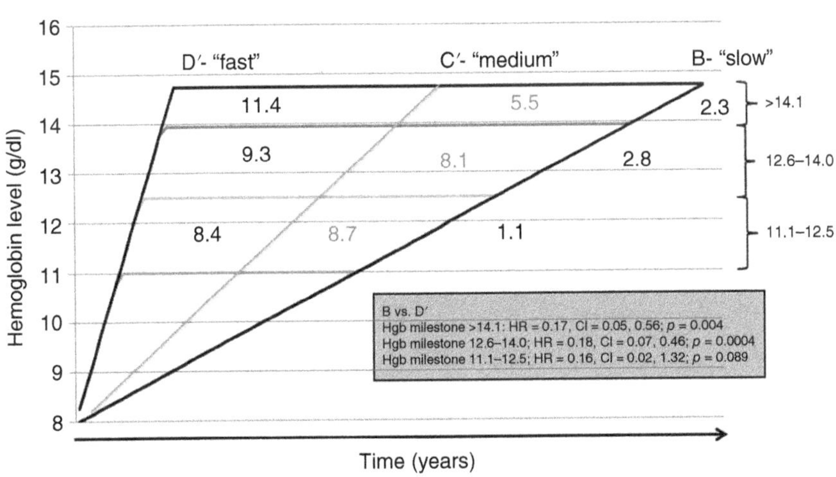

Figure 23.6 Graphical representation of risk for each trajectory within each target range achieved.

that achieved the same level of hemoglobin reconstitution by following different trajectories. In Figure 23.6, incidence rates are shown for each trajectory within each hemoglobin milestone achieved (D′ = 11.4, 9.3, 8.4; C′ = 5.5, 8.1, 8.7; B = 2.3, 2.8, 1.1). Again, at each level, the slow group did best having markedly lower incidence rates within each milestone achieved and having hazard ratios <1/5 that of the "fast" at each hemoglobin target range. This reduction in risk for the "slow" rate of rise group held true even for those achieving "normal" hemoglobin levels (>12.5 g/dl).

The data so far strongly suggest that the increased cardiovascular risk seen with hemoglobin reconstitution may be completely mediated by the trajectory followed, results not seen in any of the clinical trials or in the previous observational studies of outcomes with hemoglobin reconstitution. To complement the rate of rise analyses seen above, additional analyses were performed in this same data set that mimicked the other studies' designs; Figure 23.7 graphically represents the sub-analysis that mimicked the clinical trial designs. By following the fast trajectory only and comparing the outcomes at different target ranges achieved (i.e. a "high" vs. "low" analysis), similar results are demonstrated as those of the clinical trials – an increased risk is seen with complete hemoglobin correction to the normal levels, in this case, a statistically significant increase of over 200%. Figure 23.8 does the same, only this time, mimicking the previously performed epidemiology studies. This sub-analysis concurs with these previous epidemiology studies showing that full hemoglobin reconstitution improves outcomes.

So, in reality, the clinical trials did adhere to the first principles of hemodynamics and viscosity, showing that rapid and full reconstitution resulted

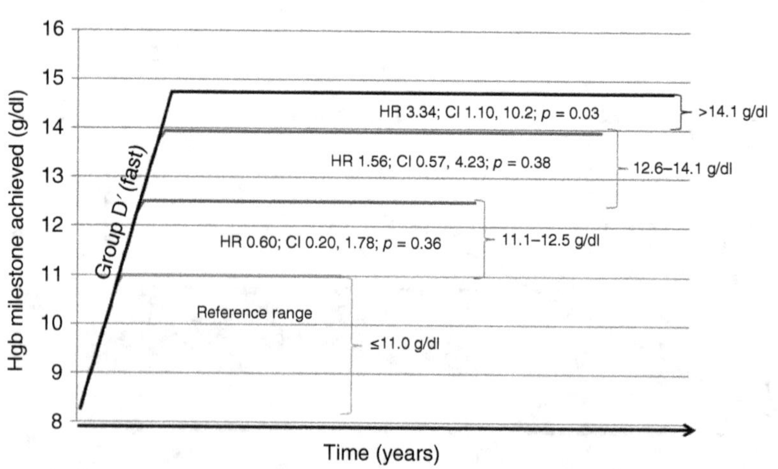

Figure 23.7 Mimicking the clinical trial designs (a graphical representation) – hazard ratios following only the fastest trajectory for various Hgb milestones achieved.

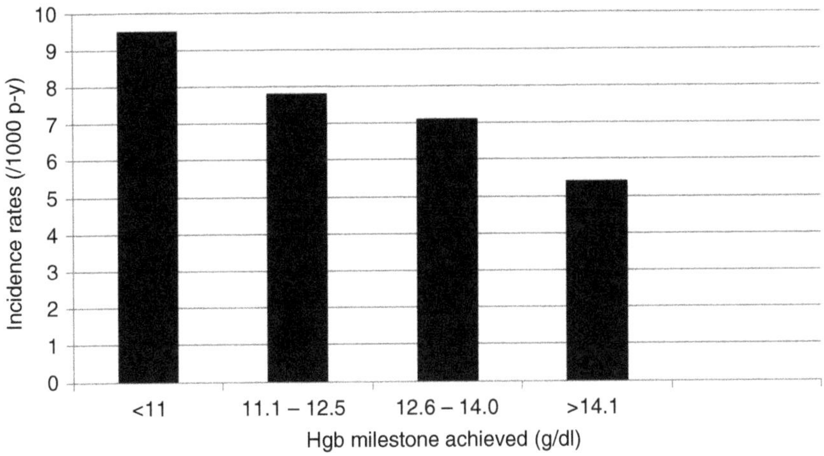

Figure 23.8 Mimicking the epidemiologic designs – incidence rates by Hgb milestone achieved.

in adverse cardiovascular effects. The epidemiology studies also adhered to the first principle that normal physiologic ranges, as in achieving "normal" hemoglobin levels, confer a survival benefit. The conflict between the previous interventional and non-interventional study designs was due to an incomplete and non-simultaneous accounting of both first principles.

The Biomarker Profile: Implications for Clinical Management

The biomarker profile extrapolated from this epidemiology study, which sought to isolate the specific effects of hemoglobin rate of rise, is composed of two measurements to be routinely repeated for monitoring purposes: (i) the trajectory of the rise in hemoglobin and (ii) WBV.

Hemoglobin variability is generally evaluated over a three- to six-month period using all available hemoglobin measurements taken during this time frame [18]. Incorporating this threshold trajectory of 0.125 g/dl/month into the clinical management of anemia of CKD will require repeated hemoglobin measurements, as the current laboratory measurement techniques are insufficient to assess this trajectory with such precision. However, with repeated monthly measures over a three- to six-month period ($0.125 \times 3 = 0.375$), it is quite possible to distinguish it from a rate of 0.150 g/dl/month ($0.150 \times 3 = 0.450$) while still utilizing individual hemoglobin measurements to guide dosing adjustments in order to navigate toward a target trajectory of 0.125 g/dl/month.

Since the relationship between hematocrit and WBV tended to be log-linear at any given protein concentration [87] and at both high and low rates of shear, [85] subsequent recommendations emphasize measuring WBV at low rates of shear to assess the risk of vascular complications [87] as the basis for these complications is erythrocyte aggregation which occurs at low rates of shear. This log-linear relationship between hematocrit and WBV forms the basis for the development of monitoring and therapeutic guidelines.

The rheologic properties of whole blood suggest that routine viscosity measurement may be beneficial in the management of anemia of CKD. The viscosity of whole blood, a non-Newtonian liquid, is dependent on the shear stress applied [93]. Shear stress, the shearing force of adjacent layers of fluid, generally decreases with decreasing flow rate. As blood flow, along with shear stress, decreases, viscosity markedly increases, hence, the recommendations for measurements at low rates of shear [87]. Plasma, unlike whole blood, is a Newtonian fluid with viscosity not depending on flow characteristics, age, or gender and so may be a viable option for measuring viscosity [94]. Reference ranges for the measurement of WBV, plasma, and serum viscosity have been made available in the literature [95, 96].

The Biomarker Profile: Implications for Clinical Trial Design

Applying this biomarker profile to clinical trial designs for the treatment of anemia of CKD highlights potential complications with the "high" vs. "low" trial designs and some design considerations for future comparative, interventional trials. Figure 23.9 graphically illustrates the complications with the "high" vs. "low" designs [54–58]. The subjects randomized to the "high" arm in these large randomized trials reached the higher hemoglobin target range by following trajectories that were all much faster than the "slope S/group B" trajectory identified above. If the initial responses to ESA therapy were inadequate, doses were escalated until desired responses were achieved and the higher target ranges were reached. Whether initially responding or whether responding after dose escalations, these trajectories were adding negative effects to these "higher" treatment arms, namely, exponential increases in viscosity were occurring in these subjects. Additionally, with those receiving dose escalations, another negative effect was added – stimulation of hyper-reactive platelet production which leads to increases in thrombotic events [89–92]. Whether a result of the specifics of the treatment protocols, the limitations of study follow-up or the primary and secondary analysis plans, the data from the low responders (i.e. hypo-responders – those in groups B, C, D, and E in Figure 23.4) would likely be censored, while that of the normal responders (those propelled to higher targets rapidly) would not be censored. This resulting censoring leads to the

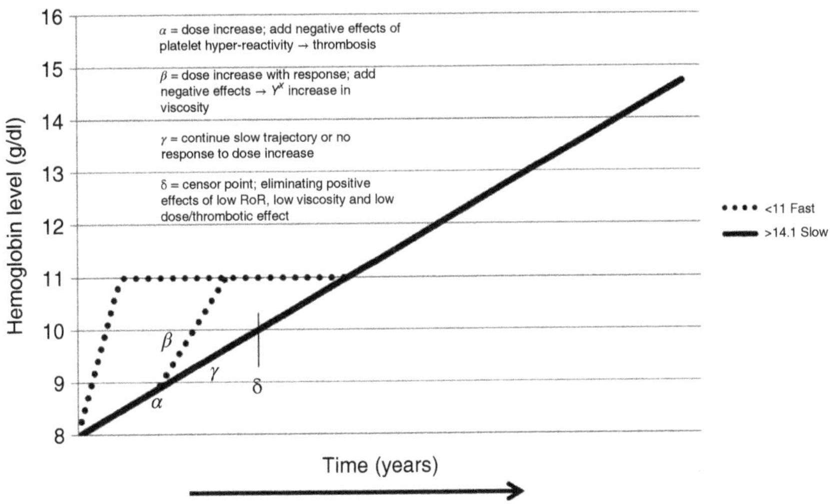

Figure 23.9 Effects of dose escalations and right censoring (a graphical representation).

elimination of the subjects who would have benefited from the positive effects of a slow trajectory (group B) – slow/low changes in viscosity with low thrombotic effects and adverse cardiovascular events.

Clinical studies of investigational medications for the treatment of anemia due to CKD should consider incorporating this biomarker profile into the protocols. By navigating along the 0.125 g/dl/month hemoglobin trajectory in both the main and the comparator arms and monitoring for changes in WBV, the trials can potentially minimize the negative impact of rapid rates of hemoglobin rise which would allow for a more pure comparison of the effects, in terms of both efficacy and safety, due to the interventions themselves and reduce the physiologically mediated adverse cardiovascular effects that result from rapid changes in WBV.

References

1 Mischak, H., Critselis, E., Hanash, S. et al. (2015). Epidemiologic design and analysis of proteomic studies: a primer on – omic technologies. *Am. J. Epidemiol.* https://doi.org/10.1093/aje/kwu462.

2 Ioannidis, J.P.A. (2011). A roadmap for successful applications of clinical proteomics. *Proteomics Clin. Appl.* 5: 241–247.

3 Bonassi, S., Taioli, E., and Vermeulen, R. (2013). Omics in population studies: a molecular epidemiology perspective. *Environ. Mol. Mutagen.* 54: 455–460, 2013.

4 Schiffman, M., Castle, P.E., Jeronimo, J. et al. (2007). Human papillomavirus and cervical cancer. *Lancet* 370: 890–907.
5 Zhang, L., McHale, C.M., Rothman, N. et al. (2010). Systems biology of human benzene exposure. *Chem. Biol. Interact.* 184: 86–93.
6 Hagmar, L., Bonassi, S., Stromberg, U. et al. (1998). Chromosomal aberrations in lymphocytes predict human cancer : a report from the European study group on cytogenetic biomarkers and health (ESCH). *Cancer Res.* 58: 4117–4121.
7 Fusco, G., Hariri, A., Vallarino, C. et al. (2017). A threshold trajectory was revealed by isolating the effects of hemoglobin rate of rise in anemia of chronic kidney disease. *Ther. Adv. Drug Saf.* 8 (10): 305–318. https://doi.org/10.1177/2042098617716819.
8 Diez Roux, A.V. (2007). Integrating social and biologic factors in health research: a systems view. *Ann. Epidemiol.* 17: 569–574.
9 Petticrew, M., Anderson, L., Elder, R. et al. (2013). Complex interventions and their implications for systematic reviews: a pragmatic approach. *J. Clin. Epidemiol.* 66: 1209–1214.
10 Susser, M. (1991). What is a cause and how do we know one? A grammar for pragmatic epidemiology. *Am. J. Epidemiol.* 133: 635–648.
11 Galea, S., Riddle, M., and Kaplan, G. (2010). Causal thinking and complex system approaches in epidemiology. *Int. J. Epidemiol.* 39: 97–106.
12 Galea, S. and Ahern, J. (2006). Considerations about specificity of associations, causal pathways, and heterogeneity in multilevel thinking (invited commentary). *Am. J. Epidemiol.* 163: 1079–1082.
13 Greenland, S., Gago-Dominguez, M., and Castelao, J.E. (2004). The value of risk-factor ("black-box") epidemiology. *Epidemiology* 15: 529–535.
14 Susser, M. and Susser, E. (1996). Choosing a future for epidemiology: II. From black box to Chinese boxes and eco-epidemiology. *Am. J. Public Health* 86: 674–677.
15 Ness, R.B., Koopman, J.S., and Roberts, M.S. (2007). Causal system modeling in chronic disease epidemiology: a proposal. *Ann. Epidemiol.* 17: 564–568.
16 Schisterman, E.F., Cole, S.R., and Platt, R.W. (2009). Overadjustment bias and unnecessary adjustment in epidemiologic studies. *Epidemiology* 20: 488–495.
17 Breslow, N. (1982). Design and analysis of case control studies. *Annu. Rev. Public Health* 3: 29–54.
18 Yee, J., Zasuwa, G., Frinak, S., and Besarab, A. (2009). Hemoglobin variability and hyporesponsiveness: much ado about something or nothing? *Adv. Chronic Kidney Dis.* 16: 83–93.
19 Saracci, R. (2006). Everything should be made as simple as possbile but not simpler. *Int. J. epidemiol.* 35: 513–514.

20 Hernan, M.A., Hernandez-Diaz, S., Werler, M.M., and Mitchell, A.A. (2002). Causal knowledge as a prerequisite for confounding evaluation: an application to birth defects epidemiology. *Am. J. Epidemiol.* 155: 176–184.

21 Greenland, S., Pearl, J., and Robins, J.M. (1999). Causal diagrams for epidemiologic research. *Epidemiology* 10: 37–48.

22 Bright, R. (1836). Cases and observations illustrative of renal disease accompanied with the secretion of albuminous urine. *Med Chir Rev.* I: 388–400.

23 Erslev, A. (1953). Humoral regulation of red cell production. *Blood* 8: 349–357.

24 Levey, A.S., Bosch, J.P., Lewis, J.B. et al. (1999). A more accurate method to estimate glomerular filtration rate from serum creatinine: a new prediction equation. *Ann. Intern. Med.* 130: 461–470.

25 NKF KDOQI Guidelines. Executive Summaries of 2000 Updates Part 4 Guideline 1. http://web.archive.org/web/20081002023324/http://www.kidney.org/Professionals/Kdoqi/guidelines_ckd/p4_class_g1.htm (5 May 2019).

26 Cirillo, M., Lombardi, C., Mele, A.A. et al. (2012). A population-based approach for the definition of chronic kidney disease: the CKD prognosis consortium. *J. Nephrol.* 25 (01): 7–12.

27 Suzuki, M., Hada, Y., Akaishi, M. et al. (2012). Effects of anemia correction by erythropoiesis-stimulating agents on cardiovascular function in non-dialysis patients with chronic kidney disease. *Int. Heart J.* 53: 238–243.

28 Collins, A.J., Li, S., Peter, W.S. et al. (2001). Death, hospitalization, and economic associations among incident hemodialysis patients with hematocrit values of 36 to 39. *J. Am. Soc. Nephrol.* 12: 2465–2473.

29 Weiner, D.E., Tighiouart, H., Vlagopoulos, P.T. et al. (2005). Effects of anemia and left ventricular hypertrophy on cardiovascular disease in patients with chronic kidney disease. *J. Am. Soc. Nephrol.* 16: 1803–1810.

30 Locatelli, F., Pisoni, R.L., Combe, C. et al. (2004). Anaemia in haemodialysis patients of five European countries: association with morbidity and mortality in the dialysis outcomes and practice patterns study (DOPPS). [Erratum appears in Nephrol. Dial. Transplant. 2004, Jun;19(6):1666]. *Nephrol. Dial. Transplant* 19: 121–132.

31 Eschbach, J.W., Abdulhadi, M.H., Browne, J.K. et al. (1989). Recombinant human erythropoietin in anemic patients with end-stage renal disease. Results of a phase III multicenter clinical trial. *Ann. Intern. Med.* 111: 992–1000.

32 Evans, R.W., Rader, B., and Manninen, D.L. (1990). The quality of life of hemodialysis recipients treated with recombinant human erythropoietin. cooperative multicenter EPO clinical trial group. *JAMA* 263: 825–830.

33 Beusterien, K.M., Nissenson, A.R., Port, F.K. et al. (1996). The effects of recombinant human erythropoietin on functional health and well-being in chronic dialysis patients. *J. Am. Soc. Nephrol.* 7: 763–773.

34 Marsh, J.T., Brown, W.S., Wolcott, D. et al. (1991). rHuEPO treatment improves brain and cognitive function of anemic dialysis patients. *Kidney Int.* 39: 155–163.

35 Lundin, A.P., Akerman, M.J., Chesler, R.M. et al. (1991). Exercise in hemodialysis patients after treatment with recombinant human erythropoietin. *Nephron* 58: 315–319.

36 Braumann, K.M., Nonnast-Daniel, B., Boning, D. et al. (1991). Improved physical performance after treatment of renal anemia with recombinant human erythropoietin. *Nephron* 58: 129–134.

37 Veys, N., Vanholder, R., and Ringoir, S. (1992). Correction of deficient phagocytosis during erythropoietin treatment in maintenance hemodialysis patients. *Am. J. Kidney Dis.* 19: 358–363.

38 Sennesael, J.J., Van der Niepen, P., and Verbeelen, D.L. (1991). Treatment with recombinant human erythropoietin increases antibody titers after hepatitis B vaccination in dialysis patients. *Kidney Int.* 40: 121–128.

39 Macdougall, I.C., Lewis, N.P., Saunders, M.J. et al. (1990). Long-term cardiorespiratory effects of amelioration of renal anaemia by erythropoietin. [Erratum appears in Lancet 1990;335:614]. *Lancet* 335: 489–493.

40 Wizemann, V., Kaufmann, J., and Kramer, W. (1992). Effect of erythropoietin on ischemia tolerance in anemic hemodialysis patients with confirmed coronary artery disease. *Nephron.* 62: 161–165.

41 Cannella, G., La, C.G., Sandrini, M. et al. (1991). Reversal of left ventricular hypertrophy following recombinant human erythropoietin treatment of anaemic dialysed uraemic patients. *Nephrol. Dial. Transplant.* 6: 31–37.

42 Pascual, J., Teruel, J.L., Moya, J.L. et al. (1991). Regression of left ventricular hypertrophy after partial correction of anemia with erythropoietin in patients on hemodialysis: a prospective study. *Clin. Nephrol.* 35: 280–287.

43 Goldberg, N., Lundin, A.P., Delano, B. et al. (1992). Changes in left ventricular size, wall thickness, and function in anemic patients treated with recombinant human erythropoietin. *Am. Heart J.* 124: 424–427.

44 Zehnder, C., Zuber, M., Sulzer, M. et al. (1992). Influence of long-term amelioration of anemia and blood pressure control on left ventricular hypertrophy in hemodialyzed patients. *Nephron* 61: 21–25.

45 Izaks, G.J., Westendorp, R.G.J., and Knook, D.L. (1999). The definition of anemia in older persons. *JAMA.* 281: 1714–1717.

46 Wu, W.C., Rathore, S.S., Wang, Y. et al. (2001). Blood transfusion in elderly patients with acute myocardial infarction. *N. Engl. J. Med.* 345: 1230–1236.

47 Groenveld, H.F., Januzzi, J.L., Damman, K. et al. (2008). Anemia and mortality in heart failure patients: a systematic review and meta-analysis. *J. Am. Coll. Cardiol.* 52: 818–827.

48 Denny, S.D., Kuchibhatla, M.N., and Cohen, H.J. (2006). Impact of anemia on mortality, cognition, and function in community-dwelling elderly. *Am. J. Med.* 119: 327–334.

49 Ofsthun, N., Labrecque, J., Lacson, E. et al. (2003). The effects of higher hemoglobin levels on mortality and hospitalization in hemodialysis patients. *Kidney Int.* 63: 1908–1914.

50 Ma, J.Z., Ebben, J., Xia, H., and Collins, A.J. (1999). Hematocrit level and associated mortality in hemodialysis patients. *J. Am. Soc. Nephrol.* 10: 610–619.

51 Xia, H., Ebben, J., Ma, J.Z., and Collins, A.J. (1999). Hematocrit levels and hospitalization risks in hemodialysis patients. *J. Am. Soc. Nephrol.* 10: 1309–1316.

52 Pisoni, R.L., Bragg-Gresham, J.L., Young, E.W. et al. (2004). Anemia management and outcomes from 12 countries in the dialysis outcomes and practice patterns study (DOPPS). *Am. J. Kidney Dis.* 44: 94–111.

53 Sarnak, M.J., Tighiouart, H., Manjunath, G. et al. (2002). Anemia as a risk factor for cardiovascular disease in the atherosclerosis risk in communities (ARIC) study. *J. Am. Coll. Cardiol.* 40: 27–33.

54 Besarab, A., Bolton, W.K., Browne, J.K. et al. (1998). The effects of normal as compared with low hematocrit values in patients with cardiac disease who are receiving hemodialysis and epoetin. *N. Engl. J. Med.* 339: 584–590.

55 Drueke, T.B., Locatelli, F., Clyne, N. et al. (2006). Normalization of hemoglobin level in patients with chronic kidney disease and anemia. *N. Engl. J. Med.* 355: 2071–2084.

56 Parfrey, P.S., Foley, R.N., Wittreich, B.H. et al. (2005). Double-blind comparison of full and partial anemia correction in incident hemodialysis patients without symptomatic heart disease. *J. Am. Soc. Nephrol.* 16: 2180–2189.

57 Pfeffer, M.A., Burdmann, E.A., Chen, C.Y. et al. (2009). A trial of darbepoetin alfa in type 2 diabetes and chronic kidney disease. *N. Engl. J. Med.* 361: 2019–2032.

58 Singh, A.K., Szczech, L., Tang, K.L. et al. (2006). Correction of anemia with epoetin alfa in chronic kidney disease. *N. Engl. J. Med.* 355: 2085–2098.

59 FDA (2001, 2001). *Medical officer clinical review: darbepoetin alfa (Aranesp) for the treatment of anemia associated with chronic renal failure*. Rockville, MD: Office of Therapeutics Research and Review, Center for Biologics Evaluation and Research, Food and Drug Administration https://web.archive.org/web/20100311020055/http://www.fda.gov/downloads/Drugs/DevelopmentApprovalProcess/HowDrugsareDevelopedandApproved/ApprovalApplications/TherapeuticBiologicApplications/ucm086019.pdf (accessed 2 November 2012).

60 Solomon, S.D., Uno, H., Lewis, E.F. et al. (2010). Erythropoietic response and outcomes in kidney disease and type 2 diabetes. *N. Engl. J. Med.* 363: 1146–1155.

61 Bradbury, D.B., Danese, M.D., Gleeson, M., and Critchlow, C.W. (2009). Effect of epoetin alfa dose changes on hemoglobin and mortality in

hemodialysis patients with hemoglobin levels persistently below 11 g/dL. *Clin. J. Am. Soc. Nephrol.* 4: 630–637.

62 Zhang, Y., Thamer, M., Stefanik, K. et al. (2004). Epoetin requirements predict mortality in hemodialysis patients. *Am. J. Kidney Dis.* 44: 866–876.

63 Kilpatrick, R.D., Critchlow, C.W., Fishbane, S. et al. (2008). Greater Epoetin alfa responsiveness is associated with improved survival in hemodialysis patients. *Clin. J. Am. Soc. Nephrol.* 3: 1077–1083.

64 Johnson, D.W., Pollock, C.A., and MacDougall, I.C. (2007). Erythropoiesis-stimulating agent hyporesponsiveness. *Nephrology* 12: 321–330.

65 O'Mara, N.B. (2008). Anemia in patients with chronic kidney disease. *Diabetes Spectr.* 21: 12–19.

66 Nakamoto, H., Kanno, Y., Okada, H., and Suzuki, H. (2004). Erythropoietin resistance in patients on continuous ambulatory peritoneal dialysis. *Adv. Perit. Dial.* 20: 111–116.

67 Sato, Y., Mizuguchi, T., Shigenaga, S. et al. (2012). Shortened red blood cell lifespan is related to the dose of erythropoiesis-stimulating agents' requirement in patients on hemodialysis. *Ther. Apher. Dial.* 16: 522–528.

68 Rice, L., Alfrey, C.P., Driscoll, T. et al. (1999). Neocytolysis contributes to the anemia of renal disease. *Am. J. Kidney Dis.* 33: 59–62.

69 Regidor, D.L., Kopple, J.D., Kovesdy, C.P. et al. (2006). Associations between changes in hemoglobin and administered erythropoiesis-stimulating agent and survival in hemodialysis patients. *J. Am. Soc. Nephrol.* 17: 1181–1191.

70 Coladonato, J.A., Frankenfield, D.L., Reddan, D.N. et al. (2002). Trends in anemia management among US hemodialysis patients. *J. Am. Soc. Nephrol.* 13: 1288–1295.

71 Singh, A.K., Himmelfarb, J., and Szczech, L.A. (2009). Resolved: targeting a higher hemoglobin is associated with greater risk in patients with CKD anemia: pro. *J. Am. Soc. Nephrol.* 20: 1436–1441.

72 Szczech, L.A., Barnhart, H.X., Inrig, J.K. et al. (2008). Secondary analysis of the CHOIR trial epoetin-alpha dose and achieved hemoglobin outcomes. *Kidney Int.* 74: 791–798.

73 Kainz, A., Mayer, B., Kramar, R., and Oberbauer, R. (2010). Association of ESA hypo-responsiveness and haemoglobin variability with mortality in haemodialysis patients. *Nephrol. Dial. Transplant.* 25: 3701–3706.

74 Fukuma, S., Yamaguchi, T., Hashimoto, S. et al. (2011). Erythropoiesis-stimulating agent responsiveness and mortality in hemodialysis patients: results from a cohort study from the dialysis registry in japan. *Am. J. Kidney Dis.* 59: 108–116.

75 Zhang, Y., Thamer, M., Kaufman, J.S. et al. (2011). High doses of epoetin do not lower mortality and cardiovascular risk among elderly hemodialysis patients with diabetes. *Kidney Int.* 80: 663–669.

76 Zhang, Y., Thamer, M., Cotter, D. et al. (2009). Estimated effect of epoetin dosage on survival among elderly hemodialysis patients in the United States. *Clin. J. Am. Soc. Nephrol.* 4: 638–644.

77 Wang, O., Kilpatrick, R.D., Critchlow, C.W. et al. (2010). Relationship between epoetin alfa dose and mortality: findings from a marginal structural model. *Clin. J. Am. Soc. Nephrol.* 5: 182–188.

78 Bradbury, B.D., Do, T.P., Winkelmayer, W.C. et al. (2009). Greater Epoetin alfa (EPO) doses and short-term mortality risk among hemodialysis patients with hemoglobin levels less than 11 g/dL. *Pharmacoepidemiol. Drug Saf.* 18: 932–940.

79 Parashar, A. and Panesar, M. (2006). The 2006 KDOQI anemia guidelines for CKD: key updates. *Dialysis Transplant.* 35: 632–634.

80 National Kidney Foundation (2007). KDOQI clinical practice guideline and clinical practice recommendations for anemia in chronic kidney disease: 2007 update of hemoglobin target. *Am. J. Kidney Dis.* 50: 471–530.

81 Locatelli, F., Covic, A., Eckardt, K.U. et al. (2009). ERA-EDTA ERBP advisory board. Anaemia management in patients with chronic kidney disease: a position statement by the Anaemia Working Group of European Renal Best Practice (ERBP). *Nephrol. Dial. Transplant.* 24: 348–354.

82 National Institute for Health and Clinical Excellence 2007. Anaemia management in chronic kidney disease: national clinical guideline for management in adults and children. https://www.nice.org.uk/guidance/ng8 (accessed 1 October 2014).

83 US Food and Drug Administration (2007). FDA strengthens boxed warnings, approves other safety labeling changes for erythropoiesis-stimulating agents (ESAs). https://web.archive.org/web/20120316114915/http://www.fda.gov/NewsEvents/Newsroom/PressAnnouncements/2007/ucm109024.htm (5 May 2019).

84 Kwaan, H.C. (2010). Role of plasma proteins in whole blood viscosity: a brief clinical review. *Clin. Hemorheol. Microcirc.* 44: 167–176.

85 Begg, T.B. and Hearns, J.B. (1966). Components in blood viscosity: the relative contribution of haematocrit, plasma fibrinogen and other proteins. *Clin. Sci.* 31: 87–93.

86 Kwaan, H.C. and Wang, J. (2003). Hyperviscosity in polycythemia vera and other red cell abnormalities. *Semin. Thromb. Hemost.* 29: 451–458.

87 McGrath, M.A. and Penny, R. (1976). Paraproteinemia: blood hyperviscosity and clinical manifestations. *J. Clin. Invest.* 58: 1155–1162.

88 Massik, J., Tang, Y.L., Hudak, M.L. et al. (1987). Effect of hematocrit on cerebral blood flow with induced polycythemia. *J. Appl. Physiol.* 62: 1090–1096.

89 Vaziri, N.D. and Zhou, X.J. (2009). Potential mechanisms of adverse outcomes in trials of anemia correction with erythropoietin in chronic kidney disease. *Nephrol. Dial. Transplant.* 24: 1082–1088.

90 Stohlawetz, P.J., Dzirlo, L., Hergovich, N. et al. (2000). Effects of erythropoietin on platelet reactivity and thrombopoiesis in humans. *Blood* 95: 2983–2989.

91 Kirkeby, A., Torup, L., Bochsen, L. et al. (2008). High-dose erythropoietin alters platelet reactivity and bleeding time in rodents in contrast to the neuroprotective variant carbamyl-erythropoietin (CEPO). *Thromb. Haemost.* 99: 720–728.

92 Patel, T.V., Mittal, B.V., Keithi-Reddy, S.R. et al. (2008). Endothelial activation markers in anemic nondialysis chronic kidney disease patients. *Nephron. Clin. Pract.* 110: c244–c250.

93 Grotta, J., Ackerman, R., Correia, J. et al. (1982). Whole blood viscosity parameters and cerebral blood flow. *Stroke* 13: 296–301.

94 Kesmarky, G., Kenyeres, P., Rabai, M., and Toth, K. (2008). Plasma viscosity: a forgotten variable. *Clin. Hemorheol. Microcirc.* 39 (1–4): 243–346.

95 Rosenson, R., McCormick, A., and Uretz, E. (1996). Distribution of blood viscosity values and biochemical correlates in healthy adults. *Clin. Chem.* 42 (8): 1189–1195.

96 Nwose, E.U. (2010). Whole blood viscosity assessment issues I: extrapolation chart and reference values. *North Am. J. Med. Sci.* 2: 165–169.

24

Computational Biology Approaches to Support Biomarker Discovery and Development

Bin Li, Hyunjin Shin, William L. Trepicchio, and Andrew Dorner

Takeda Pharmaceuticals International Co., Cambridge, MA, USA

Introduction

Over the past decade, our understanding of disease mechanisms and individual patient biology has been vastly enhanced by multiple technology platforms such as next generation sequencing of DNA and RNA and multiplex protein and metabolite analyses. We have learned, to successfully treat a patient, both disease and patient heterogeneity must guide the choice of therapeutic options. This effort requires the acquisition/analysis of human-derived data and biology-based linkage of therapeutic mechanism of action (MOA) to different subpopulations of patients based on disease heterogeneity, and the in-depth characterization of patients to link their individual disease subtype to an appropriate therapy.

The patient-centric precision approach starts in drug discovery by defining and characterizing patient populations with unmet clinical needs, using genetic/genomic profiling data to better understand disease mechanisms, and characterizing the heterogeneity of disease pathways to identify potential drug targets. Biomarkers based on disease pathways, the effects of target engagement, and patient heterogeneity are now critical components of drug development [1]. The increasing availability of omics data, the digitalization and standardization of clinical exams and medical records, and the wealth of digital information provide the basis for additional options for biomarker identification and precision medicine research. Importantly, use of these data are not a linear flow of information in an easy to utilize format as therapies and patient selection strategies emerging from discovery platforms require

Biomarkers in Drug Discovery and Development: A Handbook of Practice, Application, and Strategy, Second Edition. Edited by Ramin Rahbari, Jonathan Van Niewaal, and Michael R. Bleavins.

validation using biomarkers and patient characterization in the clinic (bench to bedside to bench).

Using oncology biomarker research as an example, it is well known that many cancer drugs are effective only in a subset of cancer patients, and use of drug responsive biomarkers is crucial to find the right drug for the right patient. The preclinical use of such biomarkers based on disease biology in discovery programs can deliver more effective drugs for clinical testing. Early identification of predictive biomarkers is particularly important to test biomarker hypotheses in early clinical trials and further develop and utilize those biomarkers during later clinical development to stratify patients into appropriate treatment arms. An extension of this effort is the development of companion diagnostic tests along with novel treatments. To maximize the opportunity for success, it is ideal to discover and test drug response biomarkers using preclinical data sets derived from *in vitro* and *in vivo* studies as well as aggregated human disease data.

This chapter introduces computational approaches to support translational biomarker research for discovery and development in the following three key areas: (i) text mining–derived diseases/drug maps, (ii) predictive modeling identifying and validating translational biomarkers for patient stratification, and (iii) building translational research platforms to integrate clinical and omics data, with predictive modeling. We will highlight a case study identifying and validating a translational biomarker for patient stratification and disease indication selection.

Mapping Pathways of Diseases and Drugs Through Text Mining

Today, drug discovery efforts can access and utilize the extensive data derived from genetic and genomic studies to inform the selection of targets and specific biomarkers within disease pathways. The establishment of "big data" strategies and data mining tools allows an unprecedented ability to characterize the biological basis of disease and link drug MOA to activity in the investigated disease. These computational methods can be categorized as either "drug based" or "disease based" [2, 3]. Traditional studies mostly have focused on exploring the shared characteristics among drug compounds such as chemical structures [4, 5] and side effects [6]. Current and evolving methods include rescreening the existing pharmacopeia against new targets to uncover novel drug indications [7], looking for similarities of molecular activities across multiple indications [8], or exploring the biological relationships between drugs and diseases based on pathways.

Connectivity Map (CMap, developed and maintained by Broad Institute) [9] uses molecular profiling and drug sensitivity data for both targeting and expanding the use of existing drugs. Library of Integrated Network-Based Cellular Signatures (LINCS) [10] reported large-scale gene expression profiles from human cancer cell lines treated with different drug compounds. CMap aims to construct a detailed map for functional associations among diseases, genetic perturbations, and drug actions. By integrating with other functional genomics databases like Gene Expression Omnibus (GEO) [11] or ArrayExpress [12], CMap has been widely used to explore potential new disease indications. Analysis of biological information in these databases can also reveal new classes of therapeutic targets. For example, recently, noncoding RNAs, especially microRNAs, have been implicated in the regulation of various cell activities, thus becoming promising therapeutic targets [13, 14]. In addition to transcriptomic data, other genomic profiles (e.g. genetic mutations, methylation) can be applied to drug purposing/repurposing [15, 16].

In addition to genome-wide profiling information, phenome-wide association studies (PheWAS, which analyze many phenotypes compared to a single genetic variant or another attribute) have become increasingly popular as a systematic approach to identify important genetic associations with human diseases [17]. Denny et al. performed a large-scale application of the PheWAS using electronic medical records (EMRs), and demonstrated that PheWAS is a useful tool to enhance the analysis of the genomic basis of disease as well as to detect novel associations between genetic markers and human diseases [18]. Besides PheWAS, the biomedical and pharmaceutical knowledge available in the published literature [19, 20] or public databases contains a vast amount of information for drugs and diseases (e.g. a drug or disease associated genes), which can be automatically mined and retrieved through text mining [21, 22]. Clinical side effects are also shown to be capable of profiling drug-related human phenotypic information and can subsequently help discover new therapeutic uses. Yang and Agarwal used drug side effects as features to predict possible clinical indications [23]. This highlights a key advance in computational bioinformatics to link patient phenotypic information obtained through informed consent procedures with genetics and genomic data. Only through the application of knowledge obtained from human data and confirmed in clinical trials can we optimize our drug development to improve efficacy and probability of success.

We developed a text mining–based diseases/drugs mapping approach (Figure 24.1a): identifying disease- or drug-associated genes, checking for significant overlap for pairs of gene sets to define links, then creating interaction networks (maps) (Figure 24.1b). In addition, consensus information can be identified on a pathway level or local network level (Figure 24.1a). These maps can help project teams generate hypotheses on the therapeutic potential of a drug across multiple indications or in combinations with other drugs with known mechanisms, as well as inferring a drug and/or disease's MOAs.

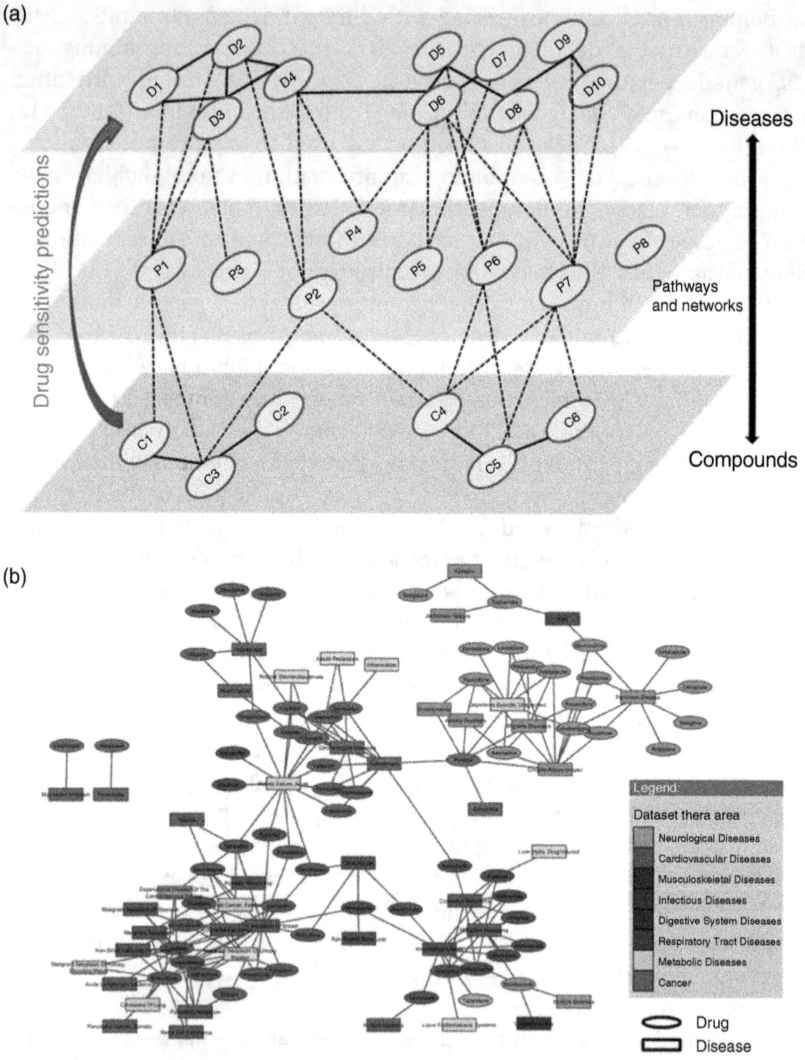

Figure 24.1 Diseases and/or drug maps for drug purposing and repurposing. (a) Information spaces and networks for disease, drugs, and disease drugs. (b) A disease–drug map defined by significant overlapping on pairs of disease/drug associated genes. The associations (i.e. similarities) between diseases or between drugs or between diseases and drugs were measured by taking $-\log_{10}$ of p-values from Fisher's exact test given two gene sets. *(See insert for color representation of this figure.)*

Identification of Biomarkers for Patient Stratification Through Predictive Modeling

Development of drug responsive biomarkers derived from preclinical data is a critical step in drug development for precision medicine as it enables patient stratification and selection of appropriate disease indications. Such translational biomarkers must be evaluated for utility in early clinical trial phases and potentially applied as a patient inclusion parameter in later stage trials. Success may lead to the development of a companion diagnostic to direct drug use to identify an optimally responsive patient subpopulation. Rigorous selection of a translational biomarker for clinical application should take into account the following criteria:

(1) Build a predictive model incorporating cell line and/or *in vivo* animal data, which can be used as a surrogate for patient response to treatment (translational biomarker).
(2) The components of the model-derived signature set should be consistent biologically with the drugs MOA.
(3) The predictive model should be drug or drug class specific.
(4) A model should be evaluated during the training phase, as it is especially important to validate a model in an independent testing set.
(5) In combination with translational data repository efforts, use the predictive model for disease indication selection.

Numerous studies of *in vitro* drug sensitivity screens [24–26] coupled with genomic/genetic profiling data have been conducted on the NCI-60 cell line panel [27–34]. As the NCI-60 panel was of a limited size and tumor-type variety, more comprehensive and diverse cell line panels have been developed. Recently, two very large cell line panel studies were reported with several hundred cancer cell lines tested using dozens of oncology drugs [15, 16]. Powered by the comprehensive molecular characterization of the cancer cell lines in these panels and the drugs' known MOAs, both studies identified important candidate biomarkers for drug sensitivity [15, 16]. The results of these studies require further development by the research community as the studies did not directly generate drug sensitivity predictive models, nor did they validate the biomarkers on independent data sets obtained from treated patients where outcome of treatment was known.

A previous US Food and Drug Administration (FDA)-led initiative [35] evaluated various gene expression modeling methods for predicting clinical endpoints (MAQCII: MicroArray Quality Control II). In the project, 36 independent teams analyzed six microarray data sets to generate predictive models for classifying a sample with 1 of 13 endpoints. Using independent testing data, the study found that the biology of the endpoint was the main performance-associated factor. Thus, 36 independent teams made poor predictions on complex endpoints such as overall cancer survival and chemically induced carcinogenesis [35]. The poor model performance could be improved if more appropriate modeling approaches for the complex clinical endpoints of interest were used. For instance, poor prediction of overall survival for multiple myeloma patients in the MAQCII study could be partly due to applying an arbitrary survival cutoff (24 months) for patients [35]. Both gene expression and overall survival in the multiple myeloma case are continuous variables; therefore, one can build a regression-based prediction model. In fact, when a univariate Cox regression approach is used, it identifies a gene expression signature that significantly predicts a "high-risk" subgroup of patients [36]. This signature was later confirmed in several independent studies and on different regression-based approaches [37–40], highlighting the advantage of a regression approach without predefined class memberships.

Another community effort to evaluate drug sensitivity predictive modeling methods is the NCI-DREAM consortium project: in 2014, NCI-DREAM consortium published results on its drug sensitivity challenge [41], on predicting the sensitivity of breast cancer cell lines to previously untested compounds (https://www.synapse.org/#!Synapse:syn2785778/wiki/70252). Each participating team used their best modeling approaches and optimized their parameter sets on the same training datasets (35 breast cancer cell lines using 31 drugs), then tested their models' performance on the same blinded testing data sets (18 breast cancer cell lines using the same 31 drugs). This provided a fair and consistent comparison to evaluate multiple teams' predictive modeling approaches.

Six types of baseline profiling data were available for generating predictive models – RNA microarray, single-nucleotide polymorphism (SNP) array, RNA sequencing, reverse phase protein array, exome sequencing, and methylation – and each team could choose to use any single or combination of these data to build their drug sensitivity predictive models. Among the six types of profiling data provided by the DREAM challenge, it was found that microarray-based gene expression data provided the most predictive power for any individual profiling data set [41–43].

Ultimate application of patient characterization platforms must also consider sample availability, sample processing, diagnostic accuracy, and cost factors in a clinical setting to ensure optimal data are obtained. We are thus interested in further evaluating whether gene expression alone could present a good source for biomarker identification and provide a parsimonious signature platform linked directly to patient biology. Since NCI-DREAM consortium not only published the results on all 31 drugs [41] but also provided each team's prediction on each drug, it provided an excellent source for method evaluation. We first evaluated drug-screen data (IC_{50} distributions) and excluded 9 compounds from the 31 compounds data set, since these drugs had identical or narrowly

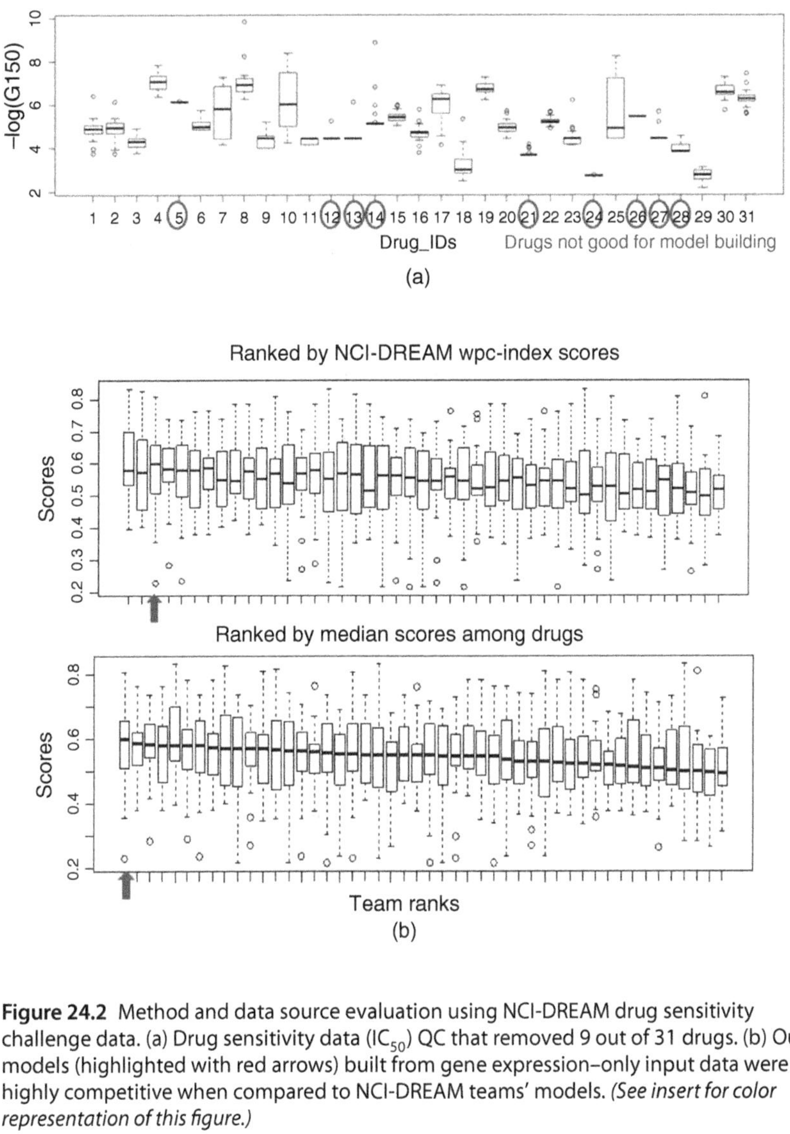

Figure 24.2 Method and data source evaluation using NCI-DREAM drug sensitivity challenge data. (a) Drug sensitivity data (IC_{50}) QC that removed 9 out of 31 drugs. (b) Our models (highlighted with red arrows) built from gene expression–only input data were highly competitive when compared to NCI-DREAM teams' models. *(See insert for color representation of this figure.)*

distributed IC_{50}s among cell lines and are therefore not suitable for building predictive models (Figure 24.2a). For the 22 drugs that passed the internal quality control, our internally developed predictive modeling method was used [44] to generate predictive models only using microarray-based gene expression data. Strikingly, in this follow-up analysis on the 22 drugs, using predictive models, our gene expression–only predictive models performed at a ranking of No. 3 or No. 1 versus the NCI-DREAM participated teams (Figure 24.2b), measured by NCI-DREAM summarized scores or median scores among the 22 drugs, respectively.

The combination of two factors for our gene expression–only models allows it to be highly competitive versus NCI-DREAM challenge participating teams (some teams not only used the six profiling data sets but also incorporated additional public domain data for generating their predictive models). First, gene expression data actually contain key signals related to biology; and second, our predictive modeling method has to be highly effective to identify the signals. For the first point, we reasoned that DNA-level profiling data reflects genetic information, while RNA-level profiling data reflects both genetic and current environmental contributions to patient biology. For the second point, we considered that a combination of a specially designed regression-based modeling framework, a novel splitting strategy to get consensus information, and using pathways to converge biological information made our modeling approach highly competitive [44].

Building Translational Research Platforms Integrating Clinical, Phenotypic, and Genomic Data to Support Predictive Modeling

With the quickly advancing genomics technologies, whole genome sequencing, whole exome sequencing, custom DNA and RNA panels, RNA sequencing, multiplex protein platforms, metabolomics, and biomedical research are transitioning from a hypothesis-driven approach to a data-driven approach [45]. The architecture of translational data repositories will be critical in the management and analysis of combinations of genomic, phenotypic, and clinical data for translational research [46]. Managing and storing large amounts of disparate data types on cloud-based storage facilities is an evolving foundation for this research. In addition, based on informed consent and governing regulations, patient privacy and data security are critical considerations for designing/selecting a translational data repository [47, 48].

Recently, several translational data repository platforms were built. Biology-Related Information Storage Kit (BRISK) provides a cohesive data integration and management platform [49]. BRISK can handle clinical phenotype description and somatic mutation information, and it provides researchers with genome-wide association studies (GWAS) analysis capabilities. The cBio cancer genomics portal [50] was developed at Memorial Sloan-Kettering Cancer Center. It integrates de-identified clinical data, such as phenotype description and survival or disease-free survival data, with large-scale cancer genomics projects like The Cancer Genome Atlas and International Cancer

Genome Consortium. Georgetown Database of Cancer (G-DOC) [51] is a translational informatics infrastructure designed to facilitate translational and systems-based medicine. The associated framework, the Georgetown Clinical and Omics Development Engine (G-CODE, https://icbi.georgetown.edu/g-code), contains a wide array of bioinformatics and systems biology tools dedicated to data analysis and visualization.

TranSMART [52] was initially developed by Johnson & Johnson and later released as an open-source platform. It can handle structured data from clinical trials (demographics, outcomes, laboratory results, and clinical phenotypes) and aligned high-content biomarker data such as gene expression profiles, genotypes, and metabolomics and proteomics data. The tranSMART Foundation (http://transmartfoundation.org) is a precompetitive consortium to provide sustainable code and content development. Pfizer, Roche, Sanofi, and Takeda are pharmaceutical members, and each contributes features/tools to the tranSMART foundation. For example, Takeda contributed the R-interface for tranSMART to enable the use of >2000 R-packages for data analysis, while Pfizer contributed GWAS tools for tranSMART.

A Case Study Identifying Cell Line–Derived Translational Biomarkers for Predicting the Treatment Outcome to Erlotinib or Sorafenib, and Selecting Drug-Sensitive Cancer Indications

Here, we present a case study [44] on building accurate and selective drug sensitivity models from preclinical *in vitro* data, followed by validation of individual models on corresponding treatment arms from patient data generated in clinical trials (Figure 24.3). The case study utilizes real data generated from large erlotinib- and sorafenib-treated cell-line panels, as well as from patient samples collected from the BATTLE clinical trial [53]. A Partial Least Squares Regression (PLSR)–based modeling framework was designed and implemented, using a special splitting strategy and canonical pathways to capture robust information for model building (Figure 24.3a) and baseline gene expression, as well as drug-screen data as inputs (Figure 24.3b). The model-derived signature genes reflect each drug's known MOA (Figure 24.3c). Erlotinib and sorafenib predictive models could be used to identify a subgroup of patients that respond better to the corresponding treatment, and these models are specific to the corresponding drugs (Figure 24.3d). Also, combined with testing using translational data repository efforts (tranSMART), the models predict each drug's potential cancer indications consistent with clinical trial results from a selection of globally normalized GEO data sets (Figure 24.3e).

Identifying translational biomarkers using cell line–derived drug sensitivity models to predict patients' responses are complex and fraught with difficulties

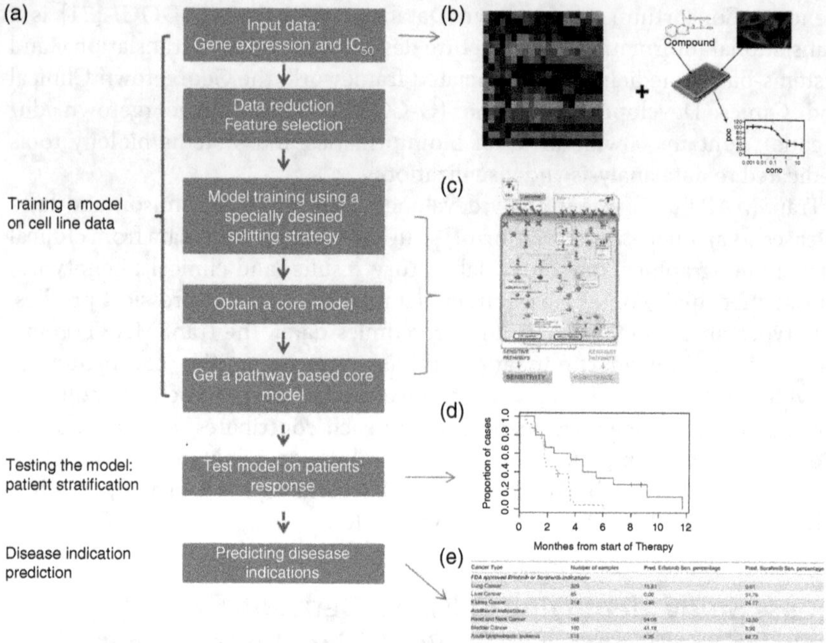

Figure 24.3 A case study on computational supports of translational biomarker research for patient stratification and disease indication selection. (a) Flowchart on the model building, testing, and application. (b) Baseline gene expression and drug sensitivity screen as model inputs. (c) Causal network to depict functional relations between sensitivity-specific and resistance-specific signature genes. (d) Survival analysis on biomarker identified treatment sensitive/resistant subgroups. (e) Combined with translational data repository effort with globally normalized GEO public domain data, the drug sensitivity predictive models were used to predict each drug's sensitive cancer indications. *(See insert for color representation of this figure.)*

related to biology of the research models. Systematic differences between expression patterns in human tumors and *in vitro* tumor cell line models contribute to uncertainty of predictive performance of signatures generated from one source of samples (cell lines) and applied to another source (primary tumors). Specific hazards in the current case study included the following issues. Predictive models were built from an *in vitro* cell line panel with IC_{50}s representing drug sensitivity, and a validation study was conducted on primary tumors from BATTLE clinical trial patients with progression-free survival (PFS) as the clinical endpoint. Moreover, the training data (Oncopanel, cancer cell line panel, Eurofin, http://pharma.eurofins.com) consisted of a mixture of cancer indications, while non-small cell lung cancer (NSCLC) was the only cancer indication in the BATTLE trial. Signatures were generated on the Affymetrix U133plus2 platform and tested on the data generated on

Affymetrix Human Gene 1.0 ST platform. Despite all these differences, the cell line–derived erlotinib and sorafenib sensitivity models predicted BATTLE trial PFS outcomes with accuracies of 84% and 79%, respectively.

After the erlotinib and sorafenib sensitivity models were tested using the BATTLE clinical trial data, they were further used to classify patients' tumor samples as sensitive or resistant from gene expression data in the public domain. An internal data repository was built based on the tranSMART translational medicine platform [54] using data from public GEO data sets. After global normalization, samples in the GEO data sets can be merged into cancer indications. In total, 484 GEO studies with 16 096 samples were normalized on Affymetrix microarray platform U133plus2 and merged into various cancer indications. Also, extensive manual curation and text mining were performed to standardize metadata on patient and clinical features.

Each sample's baseline gene expression profile was used to predict potential erlotinib or sorafenib response, and the predicted percentages of drug sensitive samples were calculated upon each cancer indication. Overall, the differential pattern of drug sensitive indications is generally consistent with clinical trials outcomes for erlotinib and sorafenib, respectively. Lung cancer was predicted to be sensitive to erlotinib, which is consistent with the fact that erlotinib was approved to treat lung cancer patients. On the other hand, kidney cancer was predicted not to be sensitive to erlotinib, which is consistent with a recent erlotinib Phase III clinical trial that showed single agent erlotinib treatment failed to have benefits for kidney cancer patients. For sorafenib, kidney and liver cancer samples were predicted to be sorafenib sensitive, which are consistent with FDA approval of sorafenib on treating kidney and liver cancer patients.

To the best of our knowledge, this erlotinib/sorafenib case study is one clear example of identifying and testing translational biomarkers *in silico* using translational data repositories and of successfully applying the biomarkers for patient stratification and disease indication selection consistent with current therapies.

Conclusions

Biology is changing from an experimental field to a data field, and drug discovery is shifting from compounds screening to patient-driven efforts. Correspondingly, computational biology approaches are becoming increasingly important to the support of drug discovery and development, especially for translational research on patient stratification and disease indication selections. Here, we introduced computational approaches in areas of diseases/drugs maps derived drug purposing/repurposing, drug sensitivity predictions, and translational data repositories. An erlotinib/sorafenib case study was also presented to demonstrate how to design and implement computational approaches to support biomarker and translational medicine research.

References

1 Hood, L. and Flores, M. (2012). A personal view on systems medicine and the emergence of proactive P4 medicine: predictive, preventive, personalized and participatory. *New Biotechnol.* 29: 613–624. https://doi.org/10.1016/j.nbt.2012.03.004.

2 Dudley, J.T., Deshpande, T., and Butte, A.J. (2011). Exploiting drug–disease relationships for computational drug repositioning. *Briefings Bioinf.* 12: 303–311. https://doi.org/10.1093/bib/bbr013.

3 Dudley, J.T., Sirota, M., Shenoy, M. et al. (2011). Computational repositioning of the anticonvulsant topiramate for inflammatory bowel disease. *Sci. Transl. Med.* 3: 96ra76. https://doi.org/10.1126/scitranslmed.3002648.

4 Keiser, M.J., Setola, V., Irwin, J.J. et al. (2009). Predicting new molecular targets for known drugs. *Nature* 462: 175–181. https://doi.org/10.1038/nature08506.

5 Ha, S., Seo, Y.J., Kwon, M.S. et al. (2008). IDMap: facilitating the detection of potential leads with therapeutic targets. *Bioinformatics* 24: 1413–1415. https://doi.org/10.1093/bioinformatics/btn138.

6 von Eichborn, J., Murgueitio, M.S., Dunkel, M. et al. (2011). PROMISCUOUS: a database for network-based drug-repositioning. *Nucleic Acids Res.* 39: D1060–D1066. https://doi.org/10.1093/nar/gkq1037.

7 Gloeckner, C., Garner, A.L., Mersha, F. et al. (2010). Repositioning of an existing drug for the neglected tropical disease onchocerciasis. *Proc. Natl. Acad. Sci. U.S.A.* 107: 3424–3429. https://doi.org/10.1073/pnas.0915125107.

8 Iorio, F., Bosotti, R., Scacheri, E. et al. (2010). Discovery of drug mode of action and drug repositioning from transcriptional responses. *Proc. Natl. Acad. Sci. U.S.A.* 107: 14621–14626. https://doi.org/10.1073/pnas.1000138107.

9 Lamb, J., Crawford, E.D., Peck, D. et al. (2006). The Connectivity Map: using gene-expression signatures to connect small molecules, genes, and disease. *Science* 313: 1929–1935. https://doi.org/10.1126/science.1132939.

10 Vidovic, D., Koleti, A., and Schurer, S.C. (2014). Large-scale integration of small molecule-induced genome-wide transcriptional responses, kinome-wide binding affinities and cell-growth inhibition profiles reveal global trends characterizing systems-level drug action. *Front. Genet.* 5: 342. https://doi.org/10.3389/fgene.2014.00342.

11 Barrett, T., Suzek, T.O., Troup, D.B. et al. (2005). NCBI GEO: mining millions of expression profiles – database and tools. *Nucleic Acids Res.* 33: D562–D566. https://doi.org/10.1093/nar/gki022.

12 Parkinson, H., Sarkans, U., Shojatalab, M. et al. (2005). ArrayExpress – a public repository for microarray gene expression data at the EBI. *Nucleic Acids Res.* 33: D553–D555. https://doi.org/10.1093/nar/gki056.

13 Ding, X.M. (2014). MicroRNAs: regulators of cancer metastasis and epithelial-mesenchymal transition (EMT). *Chin. J. Cancer* 33: 140–147. https://doi.org/10.5732/cjc.013.10094.

14 Wen, X., Deng, F.M., and Wang, J. (2014). MicroRNAs as predictive biomarkers and therapeutic targets in prostate cancer. *Am. J. Clin. Exp. Urol.* 2: 219–230.

15 Garnett, M.J., Edelman, E.J., Heidorn, S.J. et al. (2012). Systematic identification of genomic markers of drug sensitivity in cancer cells. *Nature* 483: 570–575. https://doi.org/10.1038/nature11005.

16 Barretina, J., Caponigro, G., Stransky, N. et al. (2012). The Cancer Cell Line Encyclopedia enables predictive modelling of anticancer drug sensitivity. *Nature* 483: 603–607. https://doi.org/10.1038/nature11003.

17 Hebbring, S.J. (2014). The challenges, advantages and future of phenome-wide association studies. *Immunology* 141: 157–165. https://doi.org/10.1111/imm.12195.

18 Denny, J.C., Bastarache, L., Ritchie, M.D. et al. (2013). Systematic comparison of phenome-wide association study of electronic medical record data and genome-wide association study data. *Nat. Biotechnol.* 31: 1102–1110. https://doi.org/10.1038/nbt.2749.

19 Lu, Z. (2011). PubMed and beyond: a survey of web tools for searching biomedical literature. *Database* 2011: baq036. https://doi.org/10.1093/database/baq036.

20 Islamaj Dogan, R., Murray, G.C., Neveol, A., and Lu, Z. (2009). Understanding PubMed user search behavior through log analysis. *Database* 2009: bap018. https://doi.org/10.1093/database/bap018.

21 Leaman, R., Islamaj Dogan, R., and Lu, Z. (2013). DNorm: disease name normalization with pairwise learning to rank. *Bioinformatics* 29: 2909–2917. https://doi.org/10.1093/bioinformatics/btt474.

22 Wei, C.H., Kao, H.Y., and Lu, Z. (2013). PubTator: a web-based text mining tool for assisting biocuration. *Nucleic Acids Res.* 41: W518–W522. https://doi.org/10.1093/nar/gkt441.

23 Yang, L. and Agarwal, P. (2011). Systematic drug repositioning based on clinical side-effects. *PLoS One* 6: e28025. https://doi.org/10.1371/journal.pone.0028025.

24 Reinhold, W.C., Sunshine, M., Liu, H. et al. (2012). CellMiner: a web-based suite of genomic and pharmacologic tools to explore transcript and drug patterns in the NCI-60 cell line set. *Cancer Res.* 72: 3499–3511. https://doi.org/10.1158/0008-5472.CAN-12-1370.

25 Bussey, K.J., Chin, K., Lababidi, S. et al. (2006). Integrating data on DNA copy number with gene expression levels and drug sensitivities in the NCI-60 cell line panel. *Mol. Cancer Ther.* 5: 853–867. https://doi.org/10.1158/1535-7163.MCT-05-0155.

26 Lee, J.K., Havaleshko, D.M., Cho, H. et al. (2007). A strategy for predicting the chemosensitivity of human cancers and its application to drug discovery. *Proc. Natl. Acad. Sci. U.S.A.* 104: 13086–13091. https://doi.org/10.1073/pnas.0610292104.

27 Shoemaker, R.H., Monks, A., Alley, M.C. et al. (1988). Development of human tumor cell line panels for use in disease-oriented drug screening. *Prog. Clin. Biol. Res.* 276: 265–286.

28 Weinstein, J.N., Myers, T.G., O'Connor, P.M. et al. (1997). An information-intensive approach to the molecular pharmacology of cancer. *Science* 275: 343–349.

29 Lee, J.K., Bussey, K.J., Gwadry, F.G. et al. (2003). Comparing cDNA and oligonucleotide array data: concordance of gene expression across platforms for the NCI-60 cancer cells. *Genome Biol.* 4: R82.

30 Nishizuka, S., Charboneau, L., Young, L. et al. (2003). Proteomic profiling of the NCI-60 cancer cell lines using new high-density reverse-phase lysate microarrays. *Proc. Natl. Acad. Sci. U.S.A.* 100: 14229–14234.

31 Reinhold, W.C., Reimers, M.A., Maunakea, A.K. et al. (2007). Detailed DNA methylation profiles of the E-cadherin promoter in the NCI-60 cancer cells. *Mol. Cancer Ther.* 6: 391–403. https://doi.org/10.1158/1535-7163.MCT-06-0609.

32 Blower, P.E., Verducci, J.S., Lin, S. et al. (2007). MicroRNA expression profiles for the NCI-60 cancer cell panel. *Mol. Cancer Ther.* 6: 1483–1491. https://doi.org/10.1158/1535-7163.MCT-07-0009.

33 Fagan, A., Culhane, A.C., and Higgins, D.G. (2007). A multivariate analysis approach to the integration of proteomic and gene expression data. *Proteomics* 7: 2162–2171. https://doi.org/10.1002/pmic.200600898.

34 Ikediobi, O.N., Davies, H., Bignell, G. et al. (2006). Mutation analysis of 24 known cancer genes in the NCI-60 cell line set. *Mol. Cancer Ther.* 5: 2606–2612. https://doi.org/10.1158/1535-7163.MCT-06-0433.

35 Shi, L., Campbell, G., Jones, W.D. et al. (2010). The MicroArray Quality Control (MAQC)-II study of common practices for the development and validation of microarray-based predictive models. *Nat. Biotechnol.* 28: 827–838. https://doi.org/10.1038/nbt.1665.

36 Zhan, F., Huang, Y., Colla, S. et al. (2006). The molecular classification of multiple myeloma. *Blood* 108: 2020–2028. https://doi.org/10.1182/blood-2005-11-013458.

37 Shaughnessy, J.D. Jr., Zhan, F., Burington, B.E. et al. (2007). A validated gene expression model of high-risk multiple myeloma is defined by deregulated expression of genes mapping to chromosome 1. *Blood* 109: 2276–2284. https://doi.org/10.1182/blood-2006-07-038430.

38 Zhan, F., Barlogie, B., Mulligan, G. et al. (2008). High-risk myeloma: a gene expression based risk-stratification model for newly diagnosed multiple myeloma treated with high-dose therapy is predictive of outcome in

relapsed disease treated with single-agent bortezomib or high-dose dexamethasone. *Blood* 111: 968–969. https://doi.org/10.1182/blood-2007-10-119321.

39 Decaux, O., Lodé, L., Magrangeas, F. et al. (2008). Prediction of survival in multiple myeloma based on gene expression profiles reveals cell cycle and chromosomal instability signatures in high-risk patients and hyperdiploid signatures in low-risk patients: a study of the Intergroupe Francophone du Myelome. *J. Clin. Oncol.* 26: 4798–4805.

40 Mulligan, G., Mitsiades, C., Bryant, B. et al. (2007). Gene expression profiling and correlation with outcome in clinical trials of the proteasome inhibitor bortezomib. *Blood* 109: 3177–3188.

41 Costello, J.C., Heiser, L.M., Georgii, E. et al. (2014). A community effort to assess and improve drug sensitivity prediction algorithms. *Nat. Biotechnol.* 32: 1202–1212. https://doi.org/10.1038/nbt.2877.

42 Wan, Q. and Pal, R. (2014). An ensemble based top performing approach for NCI-DREAM drug sensitivity prediction challenge. *PLoS One* 9: e101183. https://doi.org/10.1371/journal.pone.0101183.

43 Creighton, C.J. (2013). Widespread molecular patterns associated with drug sensitivity in breast cancer cell lines, with implications for human tumors. *PLoS One* 8: e71158. https://doi.org/10.1371/journal.pone.0071158.

44 Li, B., Shin, H., Gulbekyan, G. et al. (2015). Development of a drug-response modeling framework to identify cell line derived translational biomarkers that can predict treatment outcome to erlotinib or sorafenib. *PLoS One* 10: e0130700. https://doi.org/10.1371/journal.pone.0130700.

45 Merelli, I., Perez-Sanchez, H., Gesing, S., and D'Agostino, D. (2014). Managing, analysing, and integrating big data in medical bioinformatics: open problems and future perspectives. *Biomed Res. Int.* 2014: 134023. https://doi.org/10.1155/2014/134023.

46 Canuel, V., Rance, B., Avillach, P. et al. (2015). Translational research platforms integrating clinical and omics data: a review of publicly available solutions. *Briefings Bioinf.* 16: 280–290. https://doi.org/10.1093/bib/bbu006.

47 Perakslis, E.D. and Stanley, M. (2016). A cybersecurity primer for translational research. *Sci. Transl. Med.* 8: 322ps322. https://doi.org/10.1126/scitranslmed.aaa4493.

48 Perakslis, E.D. (2014). Cybersecurity in health care. *N. Engl. J. Med.* 371: 395–397. https://doi.org/10.1056/NEJMp1404358.

49 Tan, A., Tripp, B., and Daley, D. (2011). BRISK – research-oriented storage kit for biology-related data. *Bioinformatics* 27: 2422–2425. https://doi.org/10.1093/bioinformatics/btr389.

50 Cerami, E., Gao, J., Dogrusoz, U. et al. (2012). The cBio cancer genomics portal: an open platform for exploring multidimensional cancer genomics data. *Cancer Discovery* 2: 401–404. https://doi.org/10.1158/2159-8290.CD-12-0095.

51 Madhavan, S., Gusev, Y., Harris, M. et al. (2011). G-DOC: a systems medicine platform for personalized oncology. *Neoplasia* 13: 771–783.
52 Scheufele, E., Aronzon, D., Coopersmith, R. et al. (2014). tranSMART: an open source knowledge management and high content data analytics platform. *AMIA Jt. Summits Transl. Sci. Proc.* 2014: 96–101.
53 Blumenschein, G.R. Jr., Saintigny, P., Liu, S. et al. (2013). Comprehensive biomarker analysis and final efficacy results of sorafenib in the BATTLE trial. *Clin. Cancer Res.* 19: 6967–6975.
54 Perakslis, E.D., Van Dam, J., and Szalma, S. (2010). How informatics can potentiate precompetitive open-source collaboration to jump-start drug discovery and development. *Clin. Pharmacol. Ther.* 87: 614–616.

Part VIII

Lessons Learned: Practical Aspects of Biomarker Implementation

25

Biomarkers in Pharmaceutical Development: The Essential Role of Project Management and Teamwork

Lena King[1], *Mallé Jurima-Romet*[2], *and Nita Ichhpurani*[3]

[1] *Innovative Scientific Management, Guelph, ON, Canada*
[2] *Celerion, Montreal, QC, Canada*
[3] *Innovative Scientific Management, Toronto, ON, Canada*

Introduction: Pharmaceutical Project Teams

The research-based pharmaceutical industry is one of the most complex industries in the world. Discovery and development teams constitute a well-established model to manage the complexity and integrated activities to guide projects in pharmaceutical development. Organizational models and composition of these teams vary between companies, depending on the size and business strategy of the company, but they are always multidisciplinary in nature. The discovery team is charged with discovering and developing new leads. This team may include scientists with expertise in disease models, target identification, high-throughput screening, molecular biology, combinatorial chemistry, medicinal chemistry, and imaging. The development team is generally formed once a decision has been made to fund development of a new pharmaceutical lead for eventual registration. Development teams include preclinical disciplines (pharmacology, pharmacokinetics, and toxicology), pharmaceutical development (pilot and production chemists and/or biopharmaceutical expertise, formulation), regulatory affairs, clinical development, and commercial and marketing expertise.

The development team often has a formal project management structure with a project team leader and a project manager. In smaller organizations, a project manager may also serve as the team leader. Project management serves a critical role in supporting and driving forward the drug development process for the drug candidate chosen.

Particularly in recent years, the organizational structure of the discovery and development teams has been changing to adapt to internal and external demands and the decreasing productivity and increasing costs associated

Biomarkers in Drug Discovery and Development: A Handbook of Practice, Application, and Strategy, Second Edition. Edited by Ramin Rahbari, Jonathan Van Niewaal, and Michael R. Bleavins.

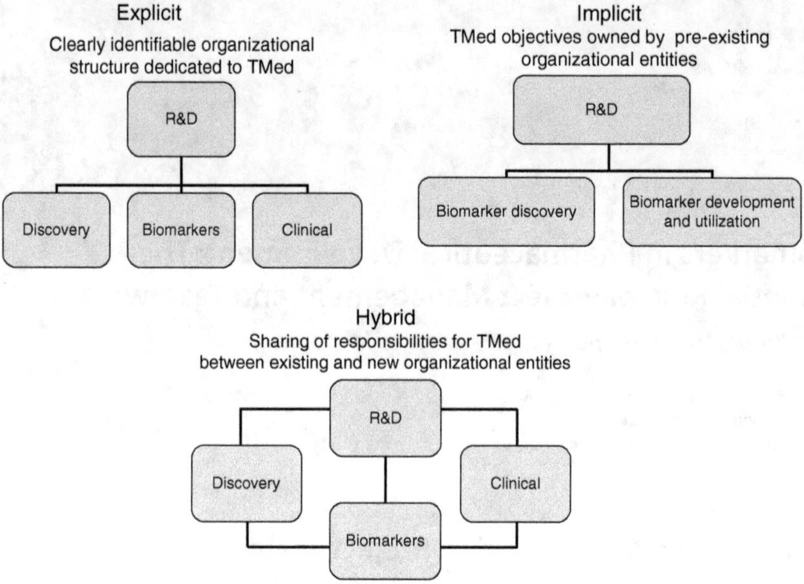

Figure 25.1 Translational research (TMed) organizational models. Source: Adapted from Hurko 2006 [1].

with pharmaceutical development. To meet these challenges, capitalize on new technologies, and improve quality of decision-making, companies are fostering collaborations between discovery and development scientists. The discovery teams increasingly include scientists with experience in DMPK (drug metabolism and pharmacokinetics), toxicology, clinical development, and project management to streamline or translate the research from discovery into development. Translational research is being proposed as the bridge between the perceived discovery–development silos and is emerging as a cross-functional discipline in its own right. As illustrated in Figure 25.1, some organizations have created an explicit biomarker or translational research unit that is represented on the development project team. Other organizations have adopted an implicit model in which biomarkers are part of the function of existing discovery and development units. A third option is a hybrid model that partners biomarker work in discovery and development without the creation and funding of a separate biomarker unit [1].

In addition to internal organizational restructuring, partnering between companies and outsourcing some parts of development (or in the case of virtual companies, all of development) to contract research organizations (CROs) are also becoming more common. These partnerships or alliances cover a wide spectrum of transactions and disciplines. Formalized alliance structures with contracts, governance, and team-specific guidance may not be in place for pharmaceutical development teams. However, even when a drug development team includes only partners from one company, it has been suggested that the project team is an implicit alliance and when including external partners may be an explicit alliance. For small discovery startup companies, CROs may provide not only the conduct of studies necessary to help new candidates to progress through discovery and development but often also the essential development expertise and will act in implicit partnership with the sponsor. Thus, the concepts and processes developed for alliances, and their success stories, are instructive for drug development teams [2]. In research and development (R&D), an alliance provides a venue for access to complementary knowledge, new and different ideas, and processes with shared risk and reward. The following core principles pertain to alliances:

1. Goals and outcomes are shared with equitable sharing of risk and reward.
2. Participants have equal say in decisions, and each participant should have full management support.
3. Decision criteria must be based on what is best for the project rather than for individual participants.
4. The team operates in a culture of open and honest communication.

A survey of formalized R&D alliances evaluated the contribution of alliance design (i.e. the number and type of partners, geographic proximity, R&D knowledge, and capabilities of each partner) and alliance management (governance agreements and processes) to the success of the alliance. The results showed that the alliance could generally be designed with appropriate and complementary expertise. The number of partners and the presence of competitors among the partners had no overall effect on the success of the alliance. Effective contractual provisions and governance had a positive effect on the measures for alliance success. However, the most pronounced positive predictors of success were the frequency of communication and how ambitious a project was. The more ambitious projects were a strong predictor for success [3]. The success factors identified for other R&D alliances apply also to successful project teams involved in pharmaceutical development (Table 25.1).

Managing the project within budget and with appropriate resources is a major responsibility. For the pharmaceutical industry, the need for cost containment is providing compelling arguments for introducing high-value decision gates earlier in the development process. As illustrated in Table 25.2, biomarkers are one of the most important and tangible tools for facilitating translational research, moving data-driven decision-making earlier into development, and for guiding development to the most appropriate indication and patient subpopulation. Although these additional decision gates can be helpful for the team, the inclusion of biomarkers adds complexity to the traditional linear model of drug development with a more reiterative process for the project team to manage.

Table 25.1 Success factors of a Drug Development Project Team.

Predictors of success of R&D alliances	Successful Drug Development Team
Appropriate number, type of partner with complementary R&D knowledge and capabilities	"The right partners."
Effective contractual provisions and governance	"Good plan and good execution" – a well-understood and management-supported plan
Excellent and transparent communication	Consultative team interactions
	Team leader and project manager guide the team for decisions that are "on time, within scope and budget"
	Develop trust between team members and the ability to work efficiently and effectively in a context of imperfect, incomplete, and unexpected information
	Solutions are sought without attribution of blame. Innovative thinking and ideas are encouraged
Ambitious projects	With development time lines spanning decades, these are inherently ambitious projects that require champions to obtain resources and management support

Table 25.2 Biomarkers in the pharmaceutical development cycle.

Discovery/ preclinical stage	Phase I–IIa	Phase IIb–III	Phase IIIb–IV
Defining mechanism of action Compound selection PK/PD modeling Candidate markers for clinical trials Better prediction by animal models through translational research	Demonstrating clinical proof of concept Dose and scheduling optimization Optimization of patient population Applications in new therapeutic indications	Minimize trial sizes through accurate inclusion and exclusion Maximize success rates by early confirmation of efficacy Potential for primary or secondary surrogate endpoints	Differentiation of products in marketplace through superior profiling of response Differentiation in subpopulations (gender, race, genetics) Personalized medicine (co-development of diagnostic)

Team Dynamics: Pharmaceutical Project Teams

The development team has members with complementary technical skills. The management of the complex process of pharmaceutical development requires that these highly skilled knowledge workers engage, relate, and commit to a shared goal with defined milestones. These team interactions have to occur in a dynamic environment where (i) studies and experiments continually generate results that may fundamentally change the course of development, (ii) management support and priority may be low compared to other projects, (iii) team members may be geographically dispersed, and (iv) resources for conduct of studies and other activities often are not controlled directly by the team. At its best, the pharmaceutical development team provides an environment that is mutually supportive, is respectful, and enables discussion on controversial issues. It is open to new ideas, agile and constructive in addressing new issues, and has goals and strategy supported and understood by management. The project leader and the project manager should strive to generate an environment that is as conducive as possible to provide this ideal. These are not features specific to biomarker development, but as mentioned below, including novel biomarkers will add to the complexity of the development project and require additional attention and management due to the increased number of communication channels. Following are some of the general principles of a productive team environment:

1. Include and plan for a project kickoff meeting and face-to-face meetings.
2. Define the roles and responsibilities of each team member.
3. Operate in a spirit of collaboration with a shared vision.
4. Practice active listening.
5. Practice transparent decision-making; determine how decisions will be made and the role of each team member in the process.
6. Encourage all team members to engage in debate about strategic issues.
7. Spend time and energy to define objectives.
8. Engage and communicate actively with management.
9. Decide but revisit which communication tools are optimal.
10. Recognize and respect differences.
11. Plan for adversity.
12. Plan for the expected as well as the unexpected.

There are a number of excellent books that discuss team dynamics and team management [4–6]. Pharmaceutical scientific organizations are beginning to offer continuing education courses in program management, and dedicated pharmaceutical training courses are available [7]. However, effective drug development project leaders and managers do not come out of university programs or training centers. The understanding of how all the complex pieces of drug development come together can best be learned through hands-on experience as a team member, team leader, or project manager. Typically, it takes many years of working within the industry to gain sufficient knowledge of the drug development process to be an effective project team leader or manager.

Consequences of Biomarkers in Pharmaceutical Development Strategies

Biomarkers are not new in pharmaceutical development. The interpretation, clinical significance, and normal variation of established biomarkers are generally well understood and widely accepted (discussed elsewhere in this book). Their utility, normal variation, and significance have been evaluated and corroborated in many different research and clinical studies. However, novel biomarkers that are now emerging may be available at only a single or a few vendors or laboratories. The assays may be technically and scientifically complex, results dependent on platform, and limited data may be available on their normal variation and biological significance. Modern computational techniques allow for powerful multiplex analysis, binning of multiple parameters, and analysis of multiple biomarker on an individual animal or patient basis. These capabilities provide exciting opportunities for advancing the science; however, there are few published or marketed tools for choosing, planning, implementing, and evaluating the risk–cost benefit of biomarkers in pharmaceutical development. The risk–cost benefit for the biomarker may also be dependent on the size of the company, their portfolio, and the financing model. Large pharmaceutical companies' investment decisions for including novel biomarker strategy may be different from those of startup companies. A larger company may be able to offset the costs of biomarker development and implementation by applying the biomarkers to multiple projects and compounds. A startup company may include a biomarker throughout development despite uncertainty as to its ultimate utility; the company accepts the risk associated with new information emerging during the development process. By contrast, a larger company may require prior assessment of the value of including the biomarker in expediting development and improving development decisions. The integral role of biomarkers in decision-making was discussed in Chapter 3 of this book, but this aspect of biomarkers also has implications for project management and teamwork within a drug development team. Following are some of the consequences of employing novel biomarkers or a unique biomarker strategy in pharmaceutical development:

- high levels of investment in infrastructure
- multiple technological platforms and specialized expertise
- high demands on data management
- increased complexity in study designs, sample logistics, and study data interpretation

- uncertainty and ambiguity for strategic decision-making
  - confidence and acceptance in translation and interpretation of results may be low
  - lack of precedence for using biomarkers in novel regulatory alternatives such as exploratory investigational new drug (IND)
- ethical issues, for example, tissue banking, privacy, and data integrity
- evolving regulatory environment with changing requirements and expectations.

Project Management

The following systematic tools and processes approach available for project management [8, 9] can be applied to management of biomarker programs:

Gantt charts	Decision trees
Contracts, scope documents	Risk analysis logs
Meeting minutes	PERT (performance, evaluation, review, tracking) charts
Communication plans	Work breakdown structures budget tracking
RACI (responsible, accountable, consulted, informed) charts	Lean-sigma tools
Lessons-learned tools	
Milestone charts	

Process mapping with the team is useful to ensure that all aspects of the biomarker project management are well understood. The level of detail can range from Gantt charts (Figure 25.2) designed principally to track time lines to program-wide integrated biomarker strategies. An example of the latter for development of an oncology candidate is illustrated in Figure 25.3. The development strategy has to include open discussion about the advantages and disadvantages of including the biomarker, recognizing that this often has to occur in the absence of clear and straightforward knowledge of the value of the biomarker across species and in a specific disease or subset of patients. Increasingly, project teams are expected to analyze risk associated with different activities and develop contingency plans far in advance of the actual activity occurring. Risks associated with various biomarkers (e.g. timely assay qualification, sampling logistics, patient recruitment, regulatory acceptance) have to be part of this analysis and contingency plan development.

There are also numerous stakeholders beyond the development team who influence and may guide the development of the pharmaceutical:

- sponsor (may include different departments with diverse goals/interests)
- business analysts
- investors (small companies) or shareholders

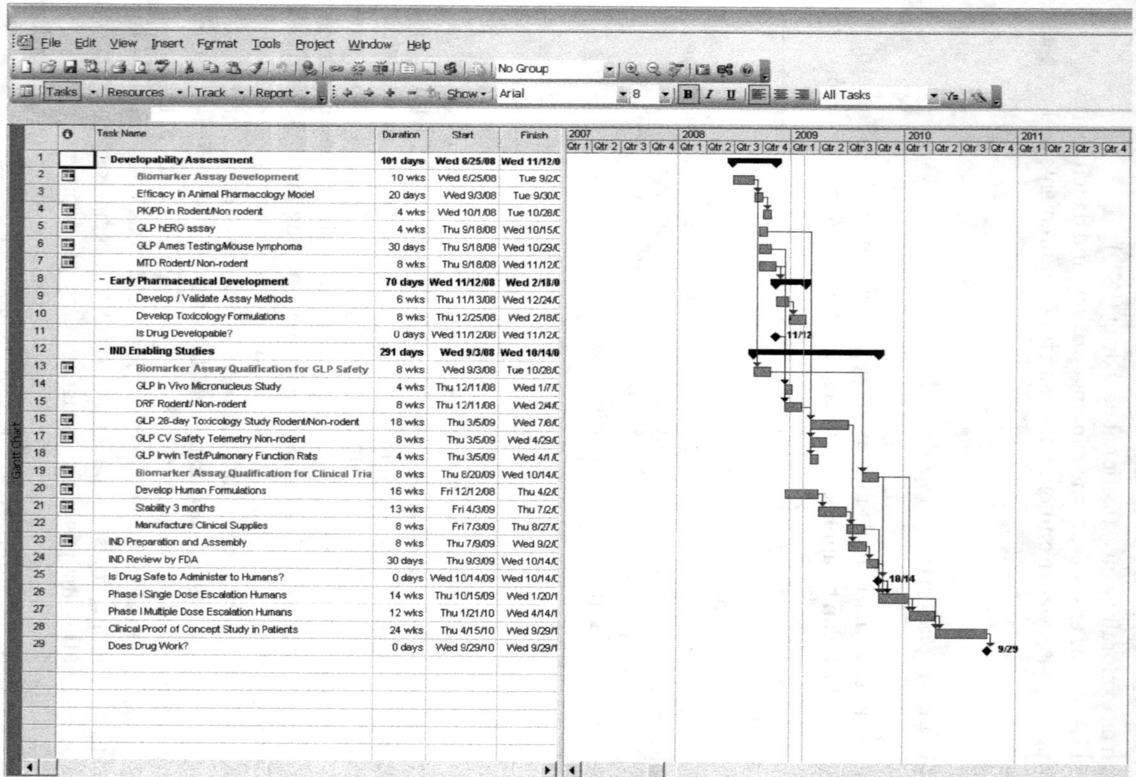

Figure 25.2 Sample Gantt chart of a drug development plan incorporating biomarkers. ADME, absorption, distribution, metabolism, and elimination.

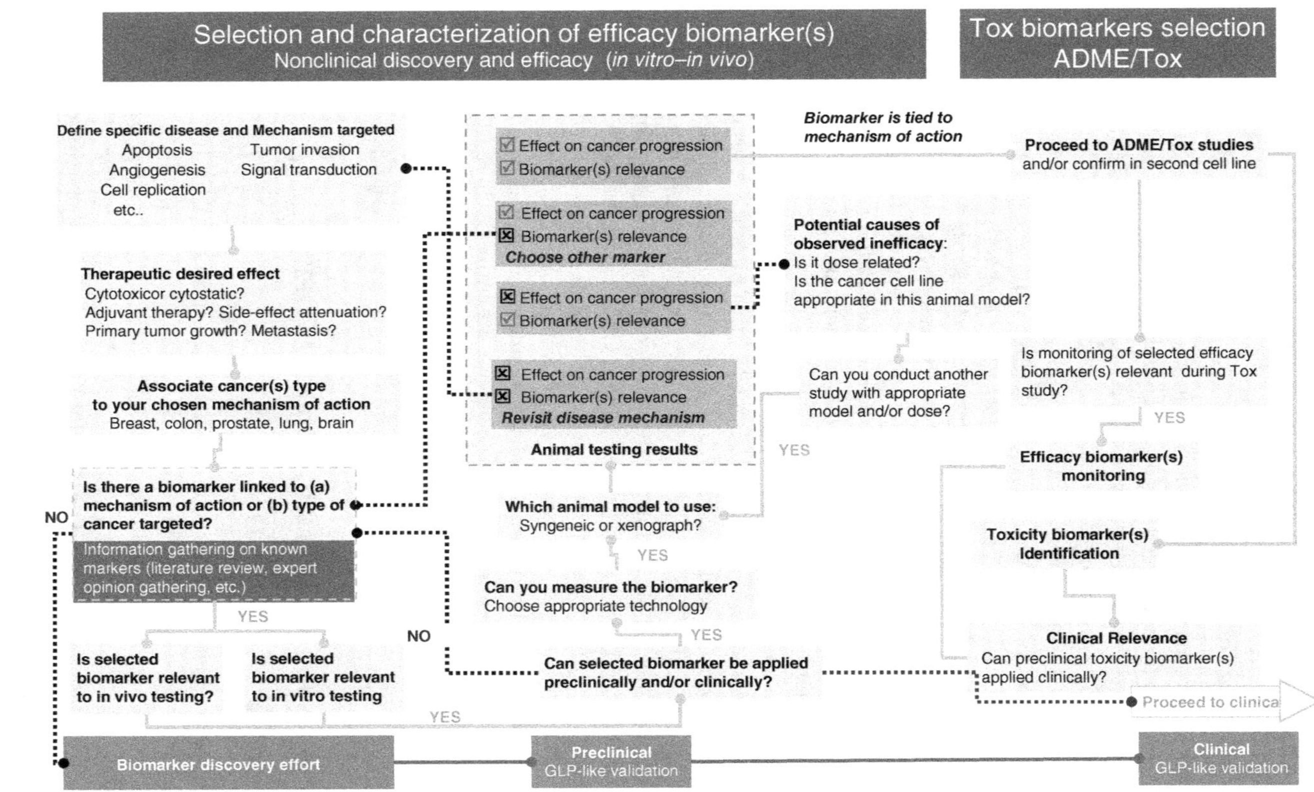

Figure 25.3 Oncology biomarker map.

- regulators
- CROs and/or biomarker labs
- investigators
- patients
- patient support/interest groups.

For the project management team, it is important to identify the stakeholders and evaluate their diverse and potentially competing priorities, particularly their perspectives on the benefit–risk effects of the new pharmaceutical. Novel targets with a drug producing an effect on a pharmacodynamic (PD) biomarker may be the last ray of hope for patients with serious or life-threatening diseases. Patients and sometimes physicians dealing with these diseases may have a different and often highly personal benefit–risk perspective compared to regulators, investors, and sponsors. Concerns about statistical significance, translation of the effect of the PD biomarker to clinical efficacy, and market share may carry little weight for patient advocacy groups under certain situations. Even concerns for safety biomarkers can be viewed as too restrictive: at best, perceived to delay access to potentially valuable medicines, and at worst, to stop their development.

Sponsors and investors, eager to see hints of efficacy as early as possible, can sometimes become overly confident about positive biomarker results before statistical analysis, normal variability, or relationships to other clinical efficacy markers are available. This may be more common in small emerging companies that rely on venture capital to finance their drug development programs than in larger established pharmaceutical companies. CROs or laboratories performing the assays and/or statistical analyses may be more cautious in their interpretations of biomarker data, sometimes seemingly unnecessarily so, but are motivated by the need to maintain quality standards as well as a neutral position. Consensus and communication problems are more likely to occur when these perspectives are widely disparate. Although it may be difficult at times, it is essential to achieve a common ground between stakeholders for effective communication.

Challenges Associated with Different Types of Biomarkers

The definition and characteristic of *biomarkers* proposed by the National Institutes of Health working group, "a characteristic that is objectively measured and evaluated as an indicator of normal biological processes, pathogenic processes, or pharmacological responses to a therapeutic intervention" [10], provide a categorization of biomarkers into efficacy, patient stratification, and safety biomarkers. There are different but overlapping challenges for the team, depending on the category of biomarkers.

Efficacy Biomarkers

Efficacy biomarkers range from PD biomarkers, markers quantifying drug–target interaction, and markers reflecting the underlying pathology of the disease to those with established links to clinical outcomes accepted as surrogate endpoints for regulatory approval. Effect of a pharmaceutical in development on a biomarker associated with efficacy is guaranteed to generate enthusiasm and momentum in the team.

PD biomarkers have a long history in pharmaceutical development and form one of the cornerstones of hypothesis-driven approaches to drug discovery and development. These biomarkers are commonly generated as part of the discovery process. The biomarker may fulfill multiple key criteria in *in vitro* or animal models: (i) it may be used to characterize the pharmacology models; (ii) it may be used in knockout or knockin genetic models to further validate the target; and (iii) it may demonstrate a characteristic PD/pharmacokinetic (PK) relationship with the drug under development. The changes reflecting underlying pathology range from largely unknown to those clearly indicative of potential market impact. An example of the latter is atorvastatin administration, resulting in decreases in serum triglycerides in normolipidemic subjects in clinical pharmacology studies [11, 12]. This is more an exception than the norm. Typically, interpretation of the clinical significance and potential market impact of a biomarker is less certain, particularly if the pharmaceutical is (i) acting by a novel mechanism of action and (ii) targeting a chronic progressive disease where disease modification rather than cure is the outcome anticipated.

The rationale for including PD biomarkers is generally easy to articulate to management, and particularly for smaller companies, these biomarkers may be essential for attracting investment. While enthusiasm and willingness to include these types of markers is generally not the issue, they are not without significant challenges in implementation and interpretation in the pharmaceutical development paradigm:

- technical aspects
  - stability of the biomarkers
  - technical complexity of the assay
  - assay robustness, sensitivity, and specificity
  - throughput of the assay
- biological samples
  - access to matrices that can or should be assayed
  - sample collection volume or amount and timing in relation to dosing
  - feasibility, cost, and resolution capabilities for imaging modalities for interactions with targets in the central nervous system, testis, poorly vascularized tumors, and so on.
- data interpretation
  - normal values, inter- and intraindividual variability

- values in disease vs. healthy conditions
- diurnal and environmental effects in animals
- effects of diet, lifestyle, concomitant medications in humans
- impact on development of no change or unexpected changes in biomarkers in the continuum from discovery to clinical.

Patient Stratification Biomarkers

The use of patient stratification biomarkers in pharmaceutical development and medical practice forms the foundation of what has been called *personalized, individualized,* or *stratified* therapy. Patient stratification biomarkers focus on patients and/or underlying pathology rather than on the effect of the pharmaceutical on the target. For small-molecule drugs, genotyping for polymorphic drug-metabolizing enzymes responsible for elimination or activation/inactivation of the compound is now an established practice in clinical trials. Results about potential effects attributed to certain genotypes may be reflected in labeling recommendations for dose adjustments and/or precautions about drug–drug interactions [13]. For example, determination of genotype for polymorphic metabolizing enzymes is included on the labels for irinotecan [14] and warfarin [15] as considerations to guide selection of dosing regimen.

Targeted therapy in oncology is the best established application of patient stratification biomarkers. The development of Herceptin, the monoclonal antibody trastuzumab, with an indication restricted to breast tumors overexpressing HER2/neu protein [16], is a clinical and commercial success story for this approach. Oncology indications also include examples of the potential of using serum proteomics to classify patients according to the highest potential for clinical benefit. For example, Taguchi et al. [17] used matrix-assisted laser desorption ionization (MALDI) mass spectroscopy (MS) analysis to generate an eight-peak MALDI MS algorithm of unidentified proteins to aid in the pretreatment selection of appropriate subgroups of non-small cell lung carcinoma patients for treatment with epidermal growth factor receptor inhibitors (erlotinib or gefitinib).

As illustrated by the examples above, patient stratification biomarkers encompass a wide range of technologies, including algorithms of unknown proteins. Challenges for the development team are to understand and identify the potential for including patient stratification biomarkers either as part of or as the major thrust in the development process. This is often a major challenge, since the technologies may lie outside the core knowledge areas of the team members, making it difficult to articulate and discuss their value within the team and to communicate effectively to management. These challenges can be particularly pertinent for some of the "omics" technologies, which can be highly platform dependent and rely on complex statistical methodologies to analyze large sets of data to principal components. The results often have little intuitive inference in the underlying targeted disease pathology and may be one of the reasons that these powerful methodologies are not used more commonly. Some considerations for including patient stratification biomarkers are summarized as follows:

- Strategic issues
  - What is the purpose of including the patient stratification biomarker?
  - Will it be helpful in reaching go/no go decisions?
  - Is it required for registration purposes?
  - What will the implications be for marketing and prescribing practices?
- Practical considerations
  - Is the biomarker commercially available and accessible?
  - If a diagnostic biomarker is essential to the development of the pharmaceutical, should co-development be considered?
  - Are there IP and marketing restrictions?
  - What are the implications of the biomarker technology on the conduct of the clinical trial?

Safety Biomarkers

The considerations for safety during development are paramount; not surprisingly, it is one of the most regulated aspects of pharmaceutical development. Safety biomarkers have spurred on interesting and innovative regulatory and industry initiative and collaborations to develop and qualify novel biomarkers. Examples are the guidance of the US Food and Drug Administration (FDA) for voluntary submissions of genomic data [18] and partnerships among government, academia, and industry for qualification of safety biomarkers [19]. Data qualifying the interpretation and significance of changes in safety biomarkers are needed to guide pharmaceutical development as well as evaluation of risk to patients or healthy volunteers in clinical trials. The purpose of safety biomarkers in clinical trials can be (i) to exclude patients at risk of developing adverse effects, (ii) to increase sensitivity of the adverse event monitoring, and (iii) to evaluate the clinical relevance of toxicity observed in the preclinical studies. Introducing novel or more uncommon biomarkers into a development project to address any of these aspects will not be embraced universally. There may be concerns not only about the added testing burden but also about the sensitivity and specificity of the biomarker, its relevance, and its relationship to well-established biomarkers. Nevertheless, including novel or uncommon biomarkers may be a condition for the conduct of a clinical trial as mandated by either regulatory bodies or institutional review boards. For example, there may be requirements to include sperm analysis in healthy volunteers and adapting experimental genotoxicity assays to humans to address effects observed in preclinical safety studies on the male reproductive tract and in genotoxicity evaluation, respectively. These will directly affect the conduct of the trials, the investigator, his or her comfort level with the assay, and the ability to communicate the significance of baseline values and any changes in the biomarkers to the clinical trial participant. However, novel and uncommon biomarkers will also have strategic and practical implications for the overall development program:

- Strategic issues
  - Will including the safety biomarker be a requirement for the entire pharmaceutical development program?
  - Are the efficacy and/or PK properties sufficiently promising to warrant continued development?
  - Can the identified safety concern be managed after approval?
- Practical considerations
  - What are the implications for the clinical trial program, locations of trials, linking to testing laboratories?
  - Will additional qualification of the biomarker assay be required as the development advances and for regulatory approval?

Management of Logistics, Processes, and Expectations

The logistical aspects of biomarker management are often a major undertaking, particularly if these include multisite global clinical trials (Figure 25.4). In clinical trials, specialized or esoteric assays, sometimes including very

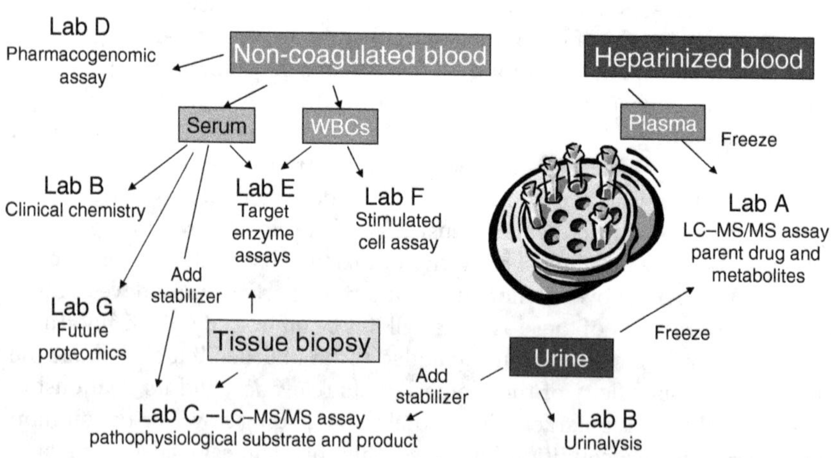

Figure 25.4 Sample logistics. LC–MS/MS, liquid chromatography–mass spectrometry/mass spectrometry.

few samples, may require processes and systems that are not commonly in place for high-throughput analytes with standard operating procedures (SOPs) and established contract service providers. In addition, managing the logistics requires recognition and integration of the different expertise, experience, expectations, culture, and mindset in each discipline within the team. Regulations and guidelines governing the different disciplines as well as generally accepted practices have a major impact on the culture and mindset. Transitioning from the less regulated discovery process into development more accustomed to good laboratory practices (GLPs), good manufacturing practices (GMPs), and good clinical practices (GCPs) can be a major cultural shift. The regulatory requirements will vary depending on the purpose of the biomarker. Safety biomarkers will require a high degree of formalized regulatory compliance, whereas there are no requirements for GLP compliance for PD biomarker assays. The question of whether or not to conduct the assay under GLP regulations will need to be considered for all types of biomarkers.

In the extensive project management coordination required to include particularly novel biomarkers in clinical trials, the long-term vision for program direction can become lost. The long-term view of the impact of the biomarker results on the pharmaceutical product under development as well as guidance for further discovery efforts should be considered. Questions about the impact of different outcomes have to be considered from both a strategic and a scientific perspective. For example, in which preclinical and clinical studies should the biomarker be included if normal values and variations are largely unknown? What will be the impact on future development of no change or unexpected effects in toxicology studies or in a first-in-human study? If there are changes to the assay or new technologies become available, should they be included to provide additional functional information about the target? Particularly when limited information is available about normal values and variation, adding additional parameters may not be of value for the decision-making process. It may be tempting to include a large number of biomarkers simply because they are available. Moreover, the increase in cost, complexity of the studies, and risk for erroneous results should be weighed carefully against the value added at each step of the development process.

The evaluation of whether to include a biomarker in a drug development program may not be straightforward. There is no doubt that biomarkers have proven valuable in pharmaceutical development to provide guidance for dose selection in early clinical studies, to enhance understanding of disease mechanisms and pathobiology, and to support decision-making and strategic portfolio considerations. Success stories for the use of biomarkers in translating from discovery concept to clinical development have been published. One example is the first proteasome inhibitor bortezomib approved for treatment of multiple myeloma. The proteasome is a key component in the ubiquitin–proteasome pathway involved in catabolism of protein and peptides as well as cellular signaling [20]. *Ex vivo* determination of proteasome inhibition was used in discovery and continued through toxicology and early- and late-stage clinical studies. Although not a clinical endpoint, proteasome inhibition provided valuable information that the drug interacted with the intended target [21].

However, in contrast to well-publicized success stories such as the example above, it is more difficult to obtain information and find examples of decisions taken when the PD biomarker did not yield the results expected. Why and where in the continuum of development did the PD marker fail, and what were the consequences of its failure? Strong confidence in the assay and the mode of action of the biomarker, as well as expectations about enhanced effects in patients compared to healthy volunteers, may result in progression of the biomarker despite lack of apparent interaction with the target. Decisions based on biomarkers require making a judgment call taking into account all the data available. For the development team, the consequences of no effect or an unexpected effect of the drug on the PD marker should be considered and debated openly before the relevant studies are initiated.

Questions that should be discussed and understood by the team in the inclusion of biomarkers are as follows:

- Cost and logistics
  - What are the costs and logistics associated with including efficacy biomarkers?
  - What are the costs associated with not including a biomarker (i.e. progressing a compound without use of a biomarker or panel of biomarkers)?
- Confidence in the biomarker
  - Will the team/management accept go/no go decisions on the basis of the results of the efficacy biomarker?
  - How many patients are required to obtain meaningful results and/or to demonstrate response?
  - What degree of change or lack of progression of disease is considered acceptable?

These may appear to be relatively simple to answer, but it will take courage and conviction from the team to make decisions to discontinue development based, or at least partially based, on results of unproven efficacy biomarkers. There may be pressures from patient groups or specific patients for access or continuation of clinical development if the drug is perceived to be beneficial. Management may be reluctant to accept the decision if significant resources have been spent in development.

Summary

The successful launch of a novel pharmaceutical product represents the culmination of years of discovery and development work driven by knowledgeable people passionate about their project and the pharmaceutical. The development process will be challenging, require perseverance, and cannot be successful without coordination and teamwork. Novel biomarkers, organizational structures with multiple stakeholders, and a need to bring data-driven decision-making strategies earlier in development make the paradigm more complex and place higher demands on team communication and project coordination. Effective program leadership together with formalized program management and communication tools and processes facilitate this endeavor. As biomarkers in discovery and development are here to stay, more attention will be paid to best practices for project management and teamwork, as these roles are recognized increasingly to be essential for successful pharmaceutical development.

References

1 Hurko, O. (2006). Understanding the strategic importance of biomarkers for the discovery and early development phases. *Drug Discovery World* 7: 63–74.

2 Bamford, J.D., Gomes-Casseres, B., and Robinson, M. (2002). *Mastering Alliance Strategy: A Comprehensive Guide to Design, Management, and Organization*. New York, NY: Jossey-Bass.

3 Dyer, J.H., Powell, B.C., Sakakibara, M., and Wang, A.J. (2006). Determinants of success in R&D alliances. Advanced Technology Program NISTIR 7323. https://nvlpubs.nist.gov/nistpubs/Legacy/IR/nistir7323.pdf (accessed 3 February 2019).

4 Means, J.A. and Adams, T. (2005). *Facilitating the Project Lifecycle: The Skills and Tools to Accelerate Progress for Project Managers, Facilitators, and Six Sigma Project Teams*. Hoboken, NJ: Wiley.

5 Parker, G.M. (2002). *Cross-Functional Teams: Working with Allies, Enemies, and Other Strangers*. Hoboken, NJ: Wiley.

6 Wong, Z. (2007). *Human Factors in Project Management: Concepts, Tools, and Techniques for Inspiring Teamwork and Motivation*. Hoboken, NJ: Wiley.

7 Tufts Center for the Study of Drug Development. http://csdd.tufts.edu program.

8 Atkinson, A.J., Daniels, C.E., Dedrick, R.L. et al. (2001). *Principles of Clinical Pharmacology*, 351–364. San Diego, CA: Academic Press.

9 PMI Standards Committee (2004). *A Guide to the Project Management Body of Knowledge (PMBOK Guide)*, 3e. Newtown Square, PA: Project Management Institute, Inc.
10 Biomarkers Definitions Working Group (2001). Biomarkers and surrogate endpoints: preferred definitions and conceptual framework. *Clin. Pharmacol. Ther.* 69: 89–95.
11 Cilla, D.D., Gibson, D.M., Whitfield, L.R., and Sedman, A.J. (1996). Pharmacodynamic effects and pharmacokinetics of atorvastatin after administration to normocholesterolemic subjects in the morning and evening. *J. Clin. Pharmacol.* 36: 604–609.
12 Posvar, E.L., Radulovic, L.L., Cilla, D.D. et al. (1996). Tolerance and pharmacokinetics of a single-dose atorvastatin, a potent inhibitor of HMG-CoA reductase, in healthy subjects. *J. Clin. Pharmacol.* 36: 728–731.
13 Huang, S.-M., Goodsaid, F., Rahman, A. et al. (2006). Application of pharmacogenomics in clinical pharmacology. *Toxicol. Mech. Methods* 16: 89–99.
14 Pfizer Inc. (2018). Camptosar (irinotecan) label. http://www.pfizer.com/pfizer/download/uspi_camptosar.pdf (accessed 3 February 2019).
15 Bristol-Myers Squibb Company (2017). Coumadin (warfarin) label. https://packageinserts.bms.com/pi/pi_coumadin.pdf (accessed 3 February 2019).
16 Genentech (2018). Herceptin (trastuzumab). https://www.gene.com/download/pdf/herceptin_prescribing.pdf (accessed 3 February 2019).
17 Taguchi, F., Solomon, B., Gregorc, V. et al. (2007). Mass spectrometry to classify non–small-cell lung cancer patients for clinical outcome after treatment with epidermal growth factor receptor tyrosine kinase inhibitors: a multicohort cross-institutional study. *J. Natl. Cancer Inst.* 99: 838–846.
18 Goodsaid, F. and Frueh, F. (2006). Process map for the validation of genomic biomarkers. *Pharmacogenomics* 7: 773–782.
19 Predictive Safety Testing Consortium. www.c-path.org (accessed 26 April 2019).
20 Glickman, M.H. and Cienchanover, A. (2002). The ubiquitin–proteasome proteolytic pathway: destruction for the sake of construction. *Physiol. Rev.* 82: 373–428.
21 EPAR (2004). Velcade (bortezomib). https://www.ema.europa.eu/documents/product-information/velcade-epar-product-information_en.pdf (accessed 3 February 2019).

26

Novel and Traditional Nonclinical Biomarker Utilization in the Estimation of Pharmaceutical Therapeutic Indices

Bruce D. Car, Brian Gemzik, and William R. Foster

Bristol-Myers Squibb Co., Princeton, NJ, USA

Introduction

Accurate projection of the safety margins of pharmaceutical agents from late discovery and early nonclinical development phase studies, including *in vitro* and animal studies, to humans is fundamental to the first decision to move compounds forward into the clinic. The robustness of those estimates over time is also central to the ability to conduct proof-of-concept studies in humans in phases IIa and IIb. After sufficient clinical experience, direct human information renders the nonclinical projections redundant. When nonclinical projections are discrepant with safe clinical exposures, discovery strategies for backup compound selection should be adjusted appropriately.

The essential elements of this work are well-defined no-effect levels (NOELs) or no-adverse effect levels (NOAELs) and lowest observed effect levels (LOELs) or IC_{50}s/EC_{50}s, if a molecular off-target is known, both as unbound drug and plasma protein–bound drug concentrations, expressed as area under the curve (AUC), C_{max}, or concentration at a defined time point. A ratio is calculated from the relevant parameter (C_{max}, AUC, time above a certain concentration, IC_{50}) between the NOEL or NOAEL values and that same parameter at the efficacious human dose projected. The science of the projection of the clinical dose has become more refined [1], allowing preclinically determined therapeutic indices to have greater predictive value.

This ratio is the safety margin. A safety margin is considered a therapeutic index if a relevant efficacy or pharmacodynamic endpoint is included in the animal study. The two terms are frequently used interchangeably, although therapeutic indices by nature of their derivation should generally be considered better predictors of drug safety. Considerable skill is required for accurate human pharmacokinetic and pharmacodynamic prediction to facilitate accurate estimation of therapeutic index. Superimposed on this

Biomarkers in Drug Discovery and Development: A Handbook of Practice, Application, and Strategy, Second Edition. Edited by Ramin Rahbari, Jonathan Van Niewaal, and Michael R. Bleavins.

numerical calculation is the toxicologist's understanding of the therapeutic index that incorporates the severity of the LOELs or other higher-dose effects, its reversibility, how cumulative exposure influences the safety margin estimate over time, and the ability to monitor in clinical evaluation. A therapeutic index of greater than 1 in at least two nonclinical species indicates that a compound can generally be given to humans safely up to the efficacious concentrations projected. Therapeutic indices below 1, which frequently occur with oncologics, alert the clinician that toxicities should be expected and monitored for at exposures below those projected to have benefit for patients.

Drug discovery toxicology and development nonclinical safety groups, together with ADME (absorption, distribution, metabolism, and elimination) and discovery pharmacology groups, influence the progression of compounds through the progressive refinement of estimates of their therapeutic indices. Traditional therapeutic indices calculated from findings in toxicology studies are broad in nature, including such diverse endpoints as liver necrosis, seizure, or prolongation of the electrocardiographic QT interval (interval between Q and T waves of electrocardiogram). Such types of endpoints are unambiguous, and may be refined further with an additional tier of biomarker information, such as clinical chemistry, electroencephalography, or advanced electrocardiography (e.g. instrumented animals monitored by telemetry). When novel biomarkers prove to be more sensitive in detection and time relative to traditional biomarkers, they may supplant or be used to supplement traditional approaches. Examples of these differing tiers of biomarkers are provided in Table 26.1.

When pharmacodynamic or efficacy endpoints are available for a nonclinical species and form the denominator of the therapeutic index equation, the most predictive safety margins may be calculated, assuming that the particular species has exposure, metabolite profiles, and other ADME characteristics similar to those of humans. Creative research in applications or technologies to validate such endpoints in nonclinical species greatly enhances the predictive power of animal models of toxicity. Several examples of these are included in Table 26.2.

In Vitro Therapeutic Indices

When pharmacology targets and homologous or nonhomologous secondary off-targets are molecularly well defined, the temptation to calculate ratios of IC_{50}s at desired efficacy endpoints to safety endpoints leads to the creation of *in vitro therapeutic indices*. Typically, these are large numbers that lull many a discovery working group into a false sense of security. Two examples are provided based on real outcomes, for which large *in vitro* ratios would potentially create an illusion of greater safety. For example, an oncology compound had a

Table 26.1 Tiers of biomarkers.

Toxicity endpoint	Traditional biomarkers	Novel or secondary endpoint
Hepatocellular necrosis	Histopathology, ALT, AST, SDH, LDH	Markers of apoptosis, gene signature, circulating RNA [2]
Renal tubular injury	BUN	Cystatin C, urinary α-GST, urinary GGT, urine sediment
Renal glomerular injury	Creatinine	Urine protein electrophoresis
Seizure	Observational finding	Electroencephalographic seizure, repetitive sharp waves
Retinal degeneration	Histopathology	Electroretinography
Systemic phospholipidosis	Histopathology and electron microscopy	Evidence of organ dysfunction or PL storages in leukocytes, metabonomic [3] or transcriptomic [4] profiles
Small intestinal mucous metaplasia of γ-secretase inhibitors	Histopathology and clinical signs of diarrhea	Peripheral blood genomic biomarkers of Notch signaling inhibition [5]
Na channel inhibition	QRS prolongation	Decreased dP/dT determined telemetrically
hERG channel inhibition	QT interval prolongation	Integration of ion-channel effects in Purkinje fiber assay
Mutagenicity	Ames assay positivity	Genomic biomarker of DNA repair
Carcinogenesis	Malignant tumors in chronic toxicity studies or two-year bioassays	Gene signatures predictive for carcinogenicity [6–10]
Teratogenicity in rats or rabbits	Positive findings in development and reproductive segment II studies	*In vitro* whole embryo culture recapitulating effects at same concentrations

ALT, alanine aminotransferase; AST, aspartate aminotransferase; LDH, lactic acid dehydrogenase; BUN, blood urea nitrogen; GST, glutathione S-transferase; GGT, γ-glutamyl transferase; PL, phospholipid; SDH, sorbitol dehydrogenase.

hERG (Ikr; repolarizing K^+ current) IC_{50} value of 35 μM and a target receptor IC_{50} value of 10 nM, ostensibly providing a 3500-fold safety window.

Caveats for the use of these simple formulas are as follows:

- Plasma protein binding if pharmacologic or toxicologic activity relates to the free fraction.
- Relative concentration in tissue may exceed plasma by many fold.

Table 26.2 Efficacy endpoints of traditional and novel pharmacodynamic biomarkers.

Efficacy endpoint	Traditional pharmacodynamic biomarker	Novel pharmacodynamic biomarker
Decreased anxiety, depression	Neurobehavioral change	CNS receptor occupancy
Cognition improvements	Learning tasks	Altered phosphorylation of proteins in key pathways
Improving Alzheimer dementia	Peripheral blood exposure and CNS Aβ (*ex vivo*)	CSF Aβ concentrations
Cancer xenograft regression	Size and weight of xenograft	Altered phosphorylation of proteins in key pathways
Immune modulation of rheumatoid arthritis	Arthritis scores (ACS 20, 50, 70)	Plasma cytokines, FACS, leukocyte transcriptomics

ACS, acute coronary syndrome; CNS, central nervous system; CSF, cerebral spinal fluid; FACS, fluorescence-activated cell sorter.

- Tissue-bound drug, and thus tissue concentration at efficacy and toxicity targets can be difficult to determine and may influence the expression of efficacy or toxicity.
- C_{max}/trough ratio.
- Typically, one uses IC_{50} values for ion channels, although inhibition of *hERG* at an IC_{10} may still produce clinically important prolongation in the QT interval.
- The efficacy and toxicity of metabolites.

After integrating these various considerations to the calculation of a safety margin, and considering the simple unknown, which is the biological counterpart to *in vitro* activity when measured *in vivo*, the safety margins for QT prolongation due to *hERG* inhibition were in the 5- to 10-fold range.

In a second example, a compound with IC_{50} for phosphodiesterase 4 (PDE4) intravenously of 2 μM was considered safe for a central nervous system efficacy target with EC_{50} of 1000-fold less (2 nM). At the lowest dosage tested in animals, projected to provide a threefold safety multiple, portal vasculitis was observed in rodents, considered likely to be secondary to PDE4 inhibition at a plasma C_{max} of only 70 nM. This toxicity is generally driven by C_{max}, and as peak concentrations occur in the portal vasculature during absorption of drug from the small intestine, the potential toxicity can be markedly exaggerated relative to *in vitro* determined safety windows.

Frequently, the exposure–response relationship for *in vitro* surrogates of *in vivo* toxicity can be quite poor. Although this makes prediction of valid safety windows difficult, *in vitro*–determined numbers are still valuable in permitting the rank ordering of compounds within a chemical series for selection and for refining *in vivo* assessments.

Novel Metabonomic Biomarkers of Toxicity

The application of systems biology technologies, including metabonomics, proteomics, and transcriptomics to biomarker development, is a nascent science, with relatively few examples of the impactful prospective uses of these technologies [11–13]. The following example describes how understanding the mechanism of rat urinary bladder carcinogenesis combined with metabonomic profiling of urine yielded a mechanism-specific biomarker that could be evaluated in studies with patients.

Muraglitazar-Induced Urinary Bladder Transitional Cell Carcinoma

Peroxisome proliferator–actuated receptors (PPARs) are nuclear hormone receptors targeted for therapeutic modulation in diabetes. Specifically, PPARα agonism will control dyslipidemia, while PPARγ agonism affords improved glucose homeostasis. Nonclinical and clinical safety issues have prevented PPARαγ agonists from becoming drugs [14, 15]. The results of two-year rodent carcinogenicity studies, including hemangiosarcoma, liposarcoma, and urinary bladder transitional cell carcinoma, have generally clouded a clear human risk assessment. With widespread distribution of PPARα and PPARγ receptors in tissues, including those transformed in carcinogenesis, a clear separation of the potentially beneficial role of receptor agonism from the potentially adverse contribution to tumor development is complex to research and understand.

The investigative approaches directed toward establishing a cogent human risk assessment for dual PPAR agonist–induced urinary bladder transitional cell carcinoma in rodents are described here for the PPARα/γ agonist muraglitazar [16–19]. An increased incidence of ventral bladder wall transitional cell papillomas and carcinomas of the urinary bladder was noted in rats at doses as low as eight times the projected human exposure at 5 mg/kg [18]. Histopathology and scanning electron microscopy revealed early microscopic injury associated with the presence of calcium phosphate crystals. Crystalluria was confirmed in studies designed to document the fragile and sometimes transient crystals in male rats dosed with muraglitazar. The crystal-induced epithelial injury was hypothesized as initiating the increased turnover as confirmed in bromodeoxyuridine (BrdU)-labeling experiments of the ventral bladder urothelium, a proliferative response strongly suspected in the genesis of tumor development. To determine the potential role of crystalluria in injury, and carcinogenesis, crystals were solubilized in rats through urinary acidification with 1% dietary ammonium chloride. Urinary acidification of male rats dosed with muraglitazar abrogated crystalluria, early urothelial injury, and cell proliferation (urothelial hyperplasia), and ultimately, urinary bladder carcinogenesis. This mode of action is recognized as a nongenotoxic mechanism of urinary bladder carcinogenesis in rats [20].

To evaluate a potential role for pharmacology, the regulation of genes downstream of PPARα and PPARγ in the rat bladder urothelium were evaluated in the presence of PPARαγ agonist–treated crystalluric and acidified diet, noncrystalluric rats. No changes in gene expression or traditional endpoints were observed, suggesting that PPAR-mediated changes were not directly causative in urothelial proliferation or carcinogenesis [17].

To investigate further the mechanism of muraglitazar-induced crystalluria, urine samples were collected from treated rats for metabonomic analysis. Nuclear magnetic resonance (NMR) spectroscopic evaluation of urine from treated compared to control rats revealed a striking reduction in divalent acids, including citrate and 2-oxoglutarate. Subsequent analytical-grade analyses of urinary citrate to creatinine concentrations confirmed and extended these metabonomic findings. It was hypothesized that male rat–specific decreased urinary excretion of divalent acids, and in particular citrate, contributed to a milieu highly permissive of calcium phosphate crystal formation.

Based on the results of studies conducted in rats, a final set of experiments examined the absolute excretion of citrate in urine from humans treated with muraglitazar. No reductions in citrate concentrations were observed across many patients compared to placebo and pretest populations. Therefore, a research strategy based on determining the role of urinary crystallogenesis in rats suggested that muraglitazar was unlikely to pose any risk to humans in inducing the early procarcinogenic change observed in rats.

The muraglitazar example demonstrates how preclinical metabonomics evaluations may identify biomarkers with potential clinical impact; however, the potential for this technology to yield specific and sensitive individual or multiple-entity biomarkers is also largely unrealized [11, 12, 21].

Novel Transcriptomic Biomarkers

Disease- and toxicity-specific transcriptional and metabonomic biomarkers are an as yet largely untapped reservoir; however, publications investigating such biomarkers have become increasingly visible in the literature [21–23]. In a retrospective review of several years of toxicogenomic analyses of drug safety studies, biomarkers of pharmacology were readily identified in 21% of studies (40% of drug targets) [23]. An unvalidated version of such an mRNA signature is that of proliferation inhibition observed consistently in the liver of rats given oncologics. A set of such genes is illustrated in Table 26.3.

Table 26.3 Genes commonly changed by diverse oncologic agents in rat liver.

Gene	Transcriptional change	Gene function	Gene name
Rrm2	Repression	Proliferation	Ribonucleotide reductase M2
Cdc2a	Repression	Proliferation	Cell division cycle 2 homolog A
Cdkn1a	Repression	Proliferation	Cyclin-dependent kinase inhibitor 1A
Ccnb1	Repression	Proliferation	Cyclin B1
Dutp	Repression	Proliferation	Deoxyuridine triphosphatase
Csnk1a1	Induced	Cell survival	Casein kinase 1, α 1

These observations can readily be adapted to transcriptomic signatures and used to determine which doses, for example in a toxicology study, demonstrate compound efficacy. When combined with traditional endpoint data, a therapeutic index may then be derived. This approach is particularly useful when the pharmacology of a compound has not been evaluated in the test species. Although early transcriptomic signatures consistent with previously identified pathology are frequently observed (in approximately 50% of studies of target tissues profiled at times preceding pathology), the target tissues involved are rarely analyzed transcriptionally such that finding valid predictive signatures will continue to be problematic [22].

Transcriptomic signatures may also provide insight toward pharmacologic effect in distinct patient groups. The demonstration of increased expression of wild-type and mutant Kras by both immunohistochemistry and transcriptional profiling or real-time polymerase chain reaction (RT-PCR) led to the hypothesis that the selection of certain patients dosed with epidermal growth factor receptor (EGFR) inhibitors based on the expression and presence of mutation could markedly increase the responder rate of patients [24, 25]. The ability to triage patient groups and eliminate patients for whom potentially toxic medicines offer no benefit is clearly a huge advance in the practice of oncology.

Conclusions

Accurate determination of therapeutic indices from nonclinical studies across multiple species, overlain with an understanding of the human risk associated with nonclinically identified liabilities, provides an invaluable tool for advancing compounds with reduced potential for harm and reduced likelihood of attrition for safety concerns. Novel approaches for identifying and validating biomarkers combined with highly refined clinical dose projections will allow toxicologists to predict and therefore avoid clinically adverse outcomes with increasing accuracy.

References

1 Huang, C., Zheng, M., Yang, Z. et al. (2007). Projection of exposure and efficacious dose prior to first-in-human studies: how successful have we been? *Pharm. Res.* 25 (4): 713–726.

2 Miyamoto, M., Yanai, M., Ookubo, S. et al. (2008). Detection of cell-free, liver-specific mRNAs in peripheral blood from rats with hepatotoxicity: a potential toxicological biomarker for safety evaluation. *Toxicol. Sci.* 106 (2): 538–545.

3 Delaney, J., Neville, W.A., Swain, A. et al. (2004). Phenylacetylglycine, a putative biomarker of phospholipidosis: its origins and relevance to phospholipid accumulation using amiodarone treated rats as a model. *Biomarkers* 3: 271–290.

4 Sawada, H., Takami, K., and Asahi, S. (2005). A toxicogenomic approach to drug-induced phospholipidosis: analysis of its induction mechanism and establishment of a novel in vitro screening system. *Toxicol. Sci.* 83: 282–292.

5 Milano, J., McKay, J., Dagenais, C. et al. (2004). Modulation of notch processing by gamma-secretase inhibitors causes intestinal goblet cell metaplasia and induction of genes known to specify gut secretory lineage differentiation. *Toxicol. Sci.* 82: 341–358.

6 Ellinger-Ziegelbauer, H., Gmuender, H., Bandenburg, A., and Ahr, H.J. (2008). Prediction of a carcinogenic potential of rat hepatocarcinogens using toxicogenomics analysis of short-term in vivo studies. *Mutat. Res.* 637 (1–2): 23–39.

7 Fielden, M.R., Brennan, R., and Gollub, J. (2007). A gene expression biomarker provides early prediction and mechanistic assessment of hepatic tumor induction by nongenotoxic chemicals. *Toxicol. Sci.* 99: 90–100.

8 Fielden, M.R., Nie, A., McMillian, M. et al. (2008). Interlaboratory evaluation of genomic signatures for predicting carcinogenicity in the rat: Predictive Safety Testing Consortium; Carcinogenicity Working Group. *Toxicol. Sci.* 103 (1): 28–34.

9 Nie, A.Y., McMillian, M., Parker, J.B. et al. (2006). Predictive toxicogenomics approaches reveal underlying molecular mechanisms of nongenotoxic carcinogenicity. *Mol. Carcinog.* 45 (12): 914–933.

10 Andersen, M.E., Clewell, H., Bermudez, E. et al. (2008). Genomic signatures and dose-dependent transitions in nasal epithelial response to inhaled formaldehyde in the rat. *Toxicol. Sci.* 105: 368–383.

11 Lindon, J.C., Holmes, E., and Nicholson, J.K. (2004). Metabonomics: systems biology in pharmaceutical research and development. *Curr. Opin. Mol. Ther.* 6: 265–722.

12 Robertson, D.G. (2005). Metabonomics in toxicology: a review. *Toxicol. Sci.* 85: 809–822.

13 Car, B.D. (2006). Enabling technologies in reducing drug attrition due to safety failures. *Am. Drug Discovery* 1: 53–56.

14 Balakumar, P., Rose, M., Ganti, S.S. et al. (2007). PPAR dual agonists: are they opening Pandora's box? *Pharmacol. Res.* 2: 91–98.

15 Rubenstrunk, A., Hanf, R., Hum, D.W. et al. (2007). Safety issues and prospects for future generations of PPAR modulators. *Biochim. Biophys. Acta* 1771: 1065–1081.

16 Dominick, M.A., White, M.R., Sanderson, T.P. et al. (2006). Urothelial carcinogenesis in the urinary bladder of male rats treated with muraglitazar, a PPAR alpha/gamma agonist: evidence for urolithiasis as the inciting event in the mode of action. *Toxicol. Pathol.* 34: 903–920.

17 Achanzar, W.E., Moyer, C.F., Marthaler, L.T. et al. (2007). Urine acidification has no effect on peroxisome proliferator–activated receptor (PPAR) signaling or epidermal growth factor (EGF) expression in rat urinary bladder urothelium. *Toxicol. Appl. Pharmacol.* 223: 246–256.

18 Tannehill-Gregg, S.H., Sanderson, T.P., Minnema, D. et al. (2007). Rodent carcinogenicity profile of the antidiabetic dual PPAR alpha and gamma agonist muraglitazar. *Toxicol. Sci.* 98: 258–270.

19 Waites, C.R., Dominick, M.A., Sanderson, T.P., and Schilling, B.E. (2007). Nonclinical safety evaluation of muraglitazar, a novel PPARalpha/gamma agonist. *Toxicol. Sci.* 100: 248–258.

20 Cohen, S.M. (1999). Calcium phosphate-containing urinary precipitate in rat urinary bladder carcinogenesis. *IARC Sci. Publ.* 147: 175–189.

21 Robertson, D.G., Reily, M.D., and Baker, J.D. (2007). Metabonomics in pharmaceutical discovery and development. *J. Proteome Res.* 6: 526–539.

22 Fielden, M.R. and Kolaja, K.L. (2006). The state-of-the-art in predictive toxicogenomics. *Curr. Opin. Drug Discovery Dev.* 9: 84–91.

23 Foster, W.R., Chen, S.J., He, A. et al. (2006). A retrospective analysis of toxicogenomics in the safety assessment of drug candidates. *Toxicol. Pathol.* 35: 621–635.

24 Lièvre, A., Bachet, J.B., Le Corre, D. et al. (2006). KRAS mutation status is predictive of response to cetuximab therapy in colorectal cancer. *Cancer Res.* 66: 3992–3995.

25 Di Nicolantonio, F., Martini, M., Molinari, F. et al. (2008). Wild-type BRAF is required for response to panitumumab or cetuximab in metastatic colorectal cancer. *J. Clin. Oncol.* 26: 5705–5712.

Part IX

Where Are We Heading and What Do We Really Need?

27

Ethics of Biomarkers: The Borders of Investigative Research, Informed Consent, and Patient Protection

Sara Assadian[1,2], *Michael Burgess*[2], *Breanne Crouch*[1,2], *Karen Lam*[1,2], *and Bruce McManus*[1,2]

[1] *PROOF Centre of Excellence, Vancouver, British Columbia, Canada*
[2] *University of British Columbia, Vancouver, British Columbia, Canada*

Introduction

In 2000, the Icelandic Parliament (Althingi) authorized an Iceland-based subsidiary of deCODE genetics to construct a biobank of genetic samples from the Icelandic population [1–4]. The Althingi also granted deCODE (which had a 5-year commercial agreement with the Swiss pharmaceutical company Roche Holdings) a 12-year exclusive commercial license to use the country's medical records, in return for an annual 70 million kronur (approximately US$ 1 million in 2000). These records were to be gathered, together with lifestyle and extensive genealogical data, into the Icelandic Health Sector Database. The resulting public outcry and academic critique have been well documented [3, 5, 6]. Several hundred articles appeared in newspapers [7], many of them referring to the sale of the "genetic heritage" of the nation (see http://www.mannvernd.is/english/home.html for a list of media articles). A grass-roots lobby group, Mannvernd, emerged to fight the project, complaining principally about the use of "presumed consent" and the commercial aspects of the agreement [4]. Despite these critiques, Iceland was one of the first countries to discuss how to structure a biobank at the political level [8].

When a population geneticist from Stanford University announced plans for a Human Genome Diversity Project (HGDP), he received a similar reception. This project aims to challenge the ethnocentrism of the Human Genome Project by studying 722 diverse "anthropologically unique" human populations [9]. Laboratories across the world have contributed 1064 lymphoblastoid cell lines from 52 populations to the HGDP collection [10]. Although there are large gaps in the collection strategy, it was important for HGDP to begin collecting cell lines to understand if the project was worth expanding. Indigenous

Biomarkers in Drug Discovery and Development: A Handbook of Practice, Application, and Strategy, Second Edition. Edited by Ramin Rahbari, Jonathan Van Niewaal, and Michael R. Bleavins.

activists were, however, unconvinced. Debra Harry, a Paiute spokesperson from Nevada, worried that "these new 'scientific findings' concerning our origins can be used to challenge aboriginal rights to territory, resources, and self-determination" [11]. The Canada-based Rural Advancement Foundation International (RAFI), now the ETC Group (Action Group on Erosion, Technology and Concentration), characterized the list of 722 as a list of peoples who had suffered most at the hands of Western "progress" and campaigned against this "bio-colonial Vampire Project." The project has since stimulated productive dialogue about the importance of race and ethnicity to health and genetic research.

The UK Biobank opened its doors to donors across the country and recruited 500 000 participants aged 40–69 years from 2006 to 2010 [12]. This biobank is a prospective cohort study hoping to contribute to disease risk prediction through the identification of biomarkers [13]. The UK Biobank has recognized the need to build public trust and knowledge. This has led to public engagement, although some critics suggest that public acceptance of this project has been carefully cultivated, with varying success, in a context of controversy and distrust [14, 15]. The United Kingdom is no stranger to human tissue scandals. In 2001, it became known that the organs of deceased children were routinely kept for research purposes at Alder Hey Hospital in Liverpool and Bristol Royal Infirmary without their parents' knowledge or consent [16]. Public outrage led to a near moratorium on tissue banking and research. An expensive system of accreditation of specimen collections by the newly formed Human Tissues Authority eventually followed [17].

These examples illustrate the increasingly visible role of large biobanking projects within biomedical research. They publicly announce the complexity of international collaborations, commercial involvement, and public–private partnerships that have become the norm in biomedical research. They also reveal major public concerns with the social and ethical implications of these projects: for privacy, indigenous identity and self-determination, ownership and control over body parts, and medical data for individuals and their families.

Traditionally, the interests of patient protection and investigative research have been served jointly by research ethics boards and the guiding principles of biomedical ethics: respect for autonomy, beneficence, nonmaleficence, and justice. These have been enacted through the process of obtaining informed consent, alongside measures to protect privacy and confidentiality of research participants and guard against discrimination. They have ensured, to a reasonable degree, the ethical enactment, legitimacy, and public acceptance of research projects. Today, however, the demands of biomedical research, of the informed consent process and of patient protection, especially privacy, are beginning to jostle against each other uncomfortably. They are engaged in an increasingly public struggle and there appears to be ever-decreasing space in which to maneuver. If biomarker research is to proceed without unnecessary constraint toward improving patient care in a manner that individuals and society at large deem ethical, radical intervention is needed.

This chapter begins by outlining the diversity of social and ethical issues surrounding biomarker-related research and its applications. Focusing in on the ever-more central process of banking of human biological materials and data, it then traces a recent trend toward large-scale population biobanks. Advances in genomics and computational biology have brought a whole raft of new questions and concerns to the domain of biomedical ethics. The peculiarities of these large biobanks, in the context of divergent legislative frameworks and increasing demands for international networking and collaboration, make such challenges ever starker. Privacy advocates argue that studies using DNA can never promise anonymity to their donors [18, 19]. Prospective collections of human DNA and tissues seem doomed either to fail the demands of fully informed consent, or face the crippling financial and administrative burden of seeking repeated consent. Population biobanks are increasingly conceived as national resources [20]. Indigenous populations and wider publics are now vocal in their concerns about ownership, commercialization, and privacy: essentially, about who uses their DNA, and how.

We do not set out here to design new governance frameworks for biobanking, or suggest the best ethical protocols for biomarker research, although these are sorely needed. The aim of this chapter is to suggest legitimate processes for so doing. In our search, we veer outside the realm of ethics as traditionally conceived, into the domain of political science. New theories of deliberative democracy facilitate public participation in policy decision-making; they aim for deliberation and communicative actions rather than strategic action; they have much to offer. Our conclusion is that ethics must embrace politics. Those involved in biomarker-related research are essential – as informers and participants in democratic public deliberation.

Biomarkers, Ethics, and Investigative Research

What are the ethics of biomarkers? The application of biomarkers to assess the risks of disease, adverse effects of drugs, and organ rejection and for the development of targeted drugs and treatments is essential. Yet the search for biomarkers of exposure, effect and susceptibility to disease, toxic chemicals, or pharmaceutical drugs raises many diverse ethical questions. Some of the most common debates surround the impact of developing predictive genetic tests as biomarkers for disease and their use in pharmacogenomics.

Neuroimaging, for example, promises much for the identification of biomarkers of diseases such as Alzheimer's, offering earlier prediction capability than is currently available. But this technology may have unintended social or ethical consequences [21]. It could lead to reduced autonomy for patients at an earlier age if they are not allowed to work or drive. New tests may not be distributed equitably if certain health insurance plans refuse to include the test. Physicians may not be adequately prepared to counsel patients and interpret biomarker test results. Most importantly, the value of early prediction is questionable for a disease that as yet has no effective treatment.

Ethical concerns surrounding the use of biomonitoring in the workplace or by insurers have also been voiced within the health sciences literature and the wider media [22–24]. Biomarkers offer some hope for monitoring exposure to toxic chemicals in the workplace and protecting the health of employees. In an environment of high exposure to carcinogens, for example, a test could be developed to identify persons with an increased genetic risk of developing cancer from a specific dose, who could, for example, be excluded from the workplace. This would probably reduce the number of workers at risk of developing cancer.

There are, however, concerns about discrimination, as well as the reliability of such tests for measuring risk [23]. Is it right for an employer to exclude people from an occupation or workplace on genetic grounds rather than reducing carcinogen exposure for all employees? Some high-risk individuals could spend a lifetime working under high exposure to carcinogens and never develop cancer, whereas some low-risk co-workers might. There are also fears that insurance companies could use biomonitoring methods to exclude people from insurance opportunities on the basis of genetic risk[1] [22, 25, 26]. Confidentiality, interpretation of biomarker data, and the problem of obtaining genuinely informed consent emerge as the key *ethical tension zones* identified by occupational health stakeholders involved in one research project in Quebec [22].

The promise of pharmacogenomics and the ethical issues it raises have also been the subject of lengthy debate. The Human Genome Organization (HUGO) Ethics Committee released a statement in 2007 recognizing that "pharmacogenomics has the potential to maximize therapeutic outcomes and minimize adverse reactions to therapy, and that it is consistent with the traditional goals of public health and medical care to relieve human suffering and save lives" but noting many ethical concerns. These include the implications for developing countries and for those seeking access to therapy for neglected diseases, the impact on health care costs and on research priorities, and the fear that pharmacogenomics could reinforce genetic determinism and lead to stigmatization of individuals and groups [27].

---

1 In the United Kingdom, such fears were voiced by a coalition of 46 organizations in a Joint Statement of Concern presented to a House of Commons Cross Party Group on 14 February 2006. The issue has also been the subject of much debate and policy analysis in the United States, given its system of private health insurance. The Genetic Information Nondiscrimination Act was passed in the U.S. House of Representatives on 25 April 2007. See U.S. National Institutes of Health fact sheet at https://www.genome.gov/about-genomics/policy-issues/Genetic-Discrimination.

Perhaps the widest range of social and ethical issues emerging from biomarker research, however, surrounds the process of collection, storage and use of human biological samples, and associated data for research purposes: to identify new biomarkers of exposure, effect, and susceptibility to disease and pharmacogenomic products. Many genetic and epidemiological studies require access to samples of annotated human blood, tissue, urine, or DNA and associated medical and lifestyle data. Often, they need large numbers of samples and repeated sampling over many months or years. Often, the outcomes of the research are uncertain, technological advances in research methodologies are unpredictable, and neither can be anticipated.

This discussion focuses on ethical issues relating to the biobanking process. The development of large-scale population databases has rendered the ethics of this technology complex, controversial, and publicly visible. Debates about biobanking also reveal the increasing inadequacy of the old ethics guidelines, frameworks, and protocols that have served us for the last 50 years.

Population Biobanks and the Challenge of Harmonization

The "banking" of human biological samples for research is not a twenty-first century phenomenon. Human tissue has been gathered and collected for at least 100 years. According to the U.S. National Bioethics Advisory Committee, by 1999, a total of 282 million unique tissue specimens were being held in the United States [28]. The term *biobank*, however, is relatively new. It appeared in *PubMed* for the first time in 1996 [29] and was not common nomenclature until the end of the decade. The sequencing of the human genome, advances in computational biology, and the emergence of new disciplines such as biomarker discovery, pharmacogenomics, and nutrigenomics have sparked unprecedented demand for samples of human blood, tissue, urine, DNA, and associated data. Three-fourths of the clinical trials that drug companies submit to the US Food and Drug Administration for approval now include a provision for sampling and storing human tissue for future genetic analysis [3]. Over 1000 biobanking publications in 600 distinct journals are now being published per year [30]. Biobanking has become deserving of its own name and has gained a dedicated society, the International Society for Biological and Environmental Repositories (ISBER), the European, Middle Eastern and African Society for Biopreservation and Biobanking, as well as two worldwide congresses: The WorldWide BioBank Summits (organized by IBM Healthcare and Life Sciences) and Biobanking and Biorepositories (organized by Informa Life Sciences).

The collection of human samples and data for research has not just accelerated; it has evolved. Four features differentiate biobanks today from those of 25 years ago: the emergence of large population-level biobanks, increased levels of commercial involvement, the desire for international collaborations requiring samples and data to be shared beyond national borders, and finally, the prospective nature of many emerging collections. The increased speed and scale of biobanking has contributed to the increasing public and academic concern with the ethical and social implications of this technology.

The rules and practices of research and research ethics developed prior to the consolidation of these trends now inhibit the ability to construct biobanks and related research efficiently. They also provide ineffective protection for individuals and populations.

Small genetic databases containing a limited number of samples, attached to one research project, focused on a specific disease were once standard. Such collections still exist: clinical collections within hospital pathology departments, and case- or family-based repositories for genetic studies of disease. Larger provincial, national, and international repositories are now increasingly common, as is the networking of existing collections. Provincial examples include the CARTaGENE project in Quebec (Canada). National disease-based biobanks and networks include the Alzheimer's Genebank, sponsored jointly by the U.S. National Institute on Ageing and the Alzheimer's Association. Examples of national or regional population-level biobanks include the Estonian Genome Project (Estonia), Biobank Japan (Japan), Icelandic Health Sector Database (Iceland), UK Biobank (UK), Medical Biobank (Sweden), the Singapore Tissue Network (Singapore), and the Canadian Partnership for Tomorrow Project (Canada). International collaborations include the European GenomEUtwin Project, a study of twins from Denmark, Finland, Italy, the Netherlands, Sweden, the United Kingdom, France, Australia, Germany, Lithuania, Poland, and the Russian Federation (http://www.genomeutwin.org/) and the Biobanking and BioMolecular Resources Research Infrastructure – European Research Infrastructure Consortium which is a distributed infrastructure of biobanks and biomolecular resources to facilitate collaboration and support research (http://www.bbmri-eric.eu/).

Many of these large population databases are designed as research infrastructures. They do not focus on one specific disease or genetic characteristic, but contain samples from sick and healthy persons, often across several generations. DNA, blood, or other tissues are stored together with health and lifestyle data from medical records, examinations, and questionnaires. These large population databases support research into complex gene interactions involved in multifactorial diseases and gene–gene and gene–environment interactions at the population level. There are few clinical benefits to individual donors. Benefits are expected to be of long term and often cannot be specified at the time of data and tissue collection. It is a major challenge to the requirement of informed consent that persons donating biological and data samples cannot know the specific future research purposes for which their donations will be used.

This proliferation of biobanks, along with the advent of population-wide and transnational biobanking endeavors, has triggered a variety of regulatory responses. Some national biobanks have been created in association with new legislation. Estonia and Lithuania enacted the Human Genes Research Act (2000) and the Human Genome Research Law (2002), respectively, possibly motivated by the inadequacy of existing norms, a belief that genetic data and research require different regulation than traditional medicine, as well as by the need for democratic legitimacy [20]. Finland's Biobank Act (2013) [31], Iceland's Act on Biobanks (2000) [32], the UK Human Tissue Act (2004), Sweden's Act on Biobanks (2002), and the Norwegian Act on Biobanks (2003) all pertain to the storage of biological samples [33]. Other national initiatives do not treat genetic data as exceptional. They remain dependent on a network of existing laws.

A series of national and international guidelines have also been produced, such as the World Medical Association's Declaration on Ethical Considerations Regarding Health Databases (2002), guidelines from the U.S. National Bioethics Advisory Commission (1999), the Council of Europe Committee of Ministers (2006), and two guidelines by the Economic Cooperation and Development: Human Biobanks and Genetic Research Databases and Biological Resource Centers (2009). As with national regulation, however, the norms, systems, and recommendations for collection and processing of samples, informed consent procedures, ownership rights, and even the terminology for degrees of anonymization of data differ substantially between guidelines.

Ownership laws related to biological samples vary between different countries or remain unclear. In the United Kingdom, donors have some ownership rights over samples in biobanks, but they do not retain rights if samples were collected for research purposes [34]. Similarly, in the United States, donors do not hold ownership of samples donated for research. In Canada, although the courts have not made a decision regarding the ownership issue, donors have a common law right to access their health information. In Estonia, the samples are owned by the institution overseeing the biobank and in Portugal, donors own their donated samples. Evidently, clarification and harmonization of ownership rights are required, especially with regard to international biobanks.

Anonymization terminology illustrates the confusion that can result from such diversity. European documents distinguish five levels of anonymization of samples [35]. Within European documents, *anonymized* describes samples used without identifiers but that are sometimes coded to enable reestablishing the identity of the donor. In most English, Canadian, and US texts, however, *anonymized* means that the sample is irreversibly de-identified. Quebec follows the French system, distinguishing between reversibly and irreversibly anonymized samples. In European documents, *coded* usually refers to instances where researchers have access to the linking code. But the U.S. Office for Human Research Protection (OHRP) uses the word to refer to situations where the researcher does not have access to the linking code [35]. To add to the confusion, UNESCO has been criticized for creating new terms, such as *proportional* or *reasonable anonymity*, that do not correspond to existing categories [20].

Such confusion has led to repeated calls for harmonization of biobank regulations. The Public Population Project in Genomics consortium (P3G) is one attempt, a nonprofit consortium aiming to promote international collaboration and knowledge transfer between researchers in population genomics. With 14 institutional members and over 530 individual members from over 40 countries, P3G declares itself to have "achieved a critical mass to form the principal international body for the harmonization of public population projects in genomics" (http://www.p3g.org).

In 2010, the first biobank education and training center, the Office of Biobank Education and Research (OBER), was created in Canada. In partnership with the UBC Department of Pathology, the BC Cancer Agency Tumor Repository, and the Canadian Tissue Repository Network, the overall objective is to establish an international center of excellence in biobanking education, best practices, standards, and policies to further translational health research. Through OBER, biobanks are registered and certified in accordance with provincial, national, and international biobanking organizations. Their strategy is to address the challenges biobanks face including lack of common standards, limited ability and capacity to ensure quality, and disconnection from donors resulting in reduced accrual capacity and jeopardizing public confidence [36].

Standardization also has its critics, notably among smaller biobanking initiatives. In 2006, the U.S. National Cancer Institute (NCI) launched guidelines spelling out best practices for the collection, storage, and dissemination of human cancer tissues and related biological specimens. These high-level guidelines are a move toward standardization of practice, following revelations in a 2004 survey of the negative impact of diverse laboratory practices on resource sharing and collaboration [37]. The intention is that NCI funding will eventually depend on compliance. The guidelines were applauded in *The Lancet* by the directors of major tissue banks such as Peter Geary of the Canadian Tumor Repository Network. They generated vocal concerns from other researchers and directors of smaller banks, many of which are already financially unsustainable. Burdensome informed consent protocols and the financial costs of infrastructural adjustments required were the key sources of concern. This is a central problem for biobanking and biomedical ethics: the centrality, the heavy moral weight, and the inadequacy of individual and voluntary informed consent.

Informed Consent: Centrality and Inadequacy of the Ideal

Informed consent is one of the most important doctrines of bioethics. It was introduced in the 1947 Nuremberg Code, following revelations during the Nuremberg trials of Nazi medical experimentation in concentration camps. It developed through inclusion in the United Nations' Universal Declaration of Human Rights in 1948 and the World Medical Association's Declaration of Helsinki in 1964. Informed consent is incorporated in all prominent medical, research, and institutional ethics codes, and is protected by laws worldwide.

The purpose of informed consent in research can be viewed as twofold: to minimize harm to research subjects and to protect their autonomous choice. Informed consent requires researchers to ensure that research participants consent voluntarily to participation in research and that they be fully informed of the risks and benefits. The focus of informed consent has slowly shifted: from disclosure by health professionals toward the voluntary consent of the individual based on the person's understanding of the research and expression of their own values and assessments [38]. Simultaneously, health research has shifted from predominantly individual investigator-designed protocols with specific research questions to multiple investigator and institution projects that gather many forms of data and samples to understand complex phenomena and test emerging hypotheses. Further, informed consent as a protection for autonomy has become important in arguments about reproductive autonomy. Informed consent has been described as representing the dividing line between "good" genetics and "sinful" eugenics [38].

Unprecedented computational power now makes it possible to network and analyze large amounts of information, making large-scale population biobanks, and genetic epidemiology studies more promising than ever before. This research context raises the stakes of research ethics, making it more difficult to achieve individual consent and protect privacy while serving as the basis for strong claims of individualized and population health benefits. Large-scale biobanks and cohorts by their very nature cannot predict the exact uses to which samples will be put ahead of time. There can be a high level of uncertainty at the time of consent regarding the type of research projects that will use the stored samples, the risks to donors that may arise, the ethical aspects of project objectives or procedures, and the potential benefits to donors [39]. Additionally, informed consent requires the disclosure of commercial involvement as this knowledge has been shown to affect a donor's decision to participate [34]. However, financial support from private entities may be introduced after consent has been obtained. The ideal of voluntary participation based on knowledge of the research appears to require new informed consent for every emergent hypothesis that was not part of the original informed consent. The practicality of such an ideal approach is less than clear.

Genetic testing can also use samples that were not originally collected for genetic studies. Tissue biopsies collected for clinical diagnosis are now providing information for gene expression studies [40]. The precise nature of future technologies that will extract new information from existing samples cannot be predicted. The rapidly changing, and at times, uncertain direction of genetic developments prevents the explicit disclosure of future research to donors at the time the biobank is created. On the other hand, seeking repeated consent from biobank donors is a costly and cumbersome process for researchers that can impede or even undermine research. Response rates for data collection (e.g. questionnaires) in any large population may vary between 50% and over 90%. The need for renewed consent could therefore reduce participation in a project and introduce selection bias [41]. Repeat consent may also be unnecessarily intrusive to the lives of donors or their next of kin.

Other forms of consent have been suggested and implemented for biobanking purposes. These include consent with several options for research use: presumed consent, broad consent, and blanket consent. The rise in the number of international biobanks that have open-ended research purposes necessitates a broader type of consent. Many European guidelines, including a memorandum from the Council of Europe Steering Committee on Bioethics, laws in Sweden, Iceland, and Estonia, and the European Society for Human Genetics guidelines, consider broad consent for unknown future uses to be acceptable as long as such future projects gain approval from Research Ethics Boards and people retain the right to withdraw samples at any time [35]. The Canadian Tri-Council Policy Statement on the ethical conduct for research involving humans also considers consent for future unspecified uses acceptable provided this is made explicit in the consent form, that donation is optional, and any known plans to consent for future projects are disclosed [42]. The U.S. Office for Human Research Protection went one step further in 2004, proposing to broaden the definition of nonidentifiable samples, upon which research is allowed under US federal regulations without the requirement of informed consent.

Broad consent appears to be the emerging consent model as it attempts to respect the rights and interests of participants while also enabling research to be conducted efficiently [43]. However, critics argue that it lacks the element of being truly "informed." A participant's autonomy and dignity may not be addressed as in the traditional informed consent. Some international organizations, including UNESCO's International Bioethics Committee and the Council for International Organizations of Medical Sciences, do not expressly accept broad consent for unspecified future use [39].

The problem is that no informed consent mechanism – narrow or broad – can address all ethical concerns surrounding the biobanking of human DNA and data [44]. Such concerns include the aggregate effects of individual consent upon society as a whole and upon family and community members given the inherently "shared" nature of genetic material. If people are given full choice as to which diseases their samples can be used to research, and they choose only to donate for well-known diseases such as cancer, rare disease may be neglected. The discovery that Ashkenazi Jews may have particular mutations predisposing them to breast, ovarian, and colon cancer has generated fears that they could become the target of discrimination [45].

Concerns include irreconcilable trade-offs between donor desires for privacy (best achieved by unlinking samples), control over the manner in which their body parts and personal information are used (samples can be withdrawn from a biobank only if a link exists), and access to clinically relevant information discovered in the course of research. For some individuals and communities, cultural or religious beliefs dictate or restrict the research purposes for which their samples can be used. The Nuu-chah-nulth nations of Vancouver Island became angry in 2000 after discovering that their samples, collected years before for arthritis research, had been used for the entirely different purpose of migration research [46, 47]. In some cases, a history of colonialism and abusive research makes a group demand that their samples be used for research that benefits their community directly.

Complete anonymization of samples containing human DNA is technically impossible, given both the unique nature of a person's DNA and its shared characteristics. Consequently, in 2003, the Icelandic Supreme Court ruled that the transfer of 18-year-old student Ragnhildur Gudmundsdottir's dead father's health data infringed her privacy rights: "The court said that including the records in the database might allow her to be identified as an individual at risk of any heritable disease her father might be found to have had—even though the data would be made anonymous and encrypted" [48]. Reasonable privacy protection in a biobanking context is tough to achieve, extending to the technological capacity to protect privacy through linked or unlinked anonymized samples without risk of error.

Informed consent cannot provide a basis for participants to evaluate the likelihood of benefit arising from their participation in a biobank when these merits are contested by experts. Critics of UK Biobank, for example, have little faith in the value and power of such prospective cohort studies, compared to traditional case–control studies, for isolating biomarkers and determining genetic risk factors. Supporters argue that the biobank will be a resource from which researchers can compile nested case–control studies. Critics claim that it will only be useful for study of the most common cancers, those that occur with enough frequency among donors. Others claim that even UK Biobank's intended 500 000 participants cannot provide reliable information about the genetic causes of a disease without a study of familial correlations [13].

Informed consent is inadequate as a solution for ensuring that the impacts of biobanking and related research will be beneficial to individuals and society, will uphold the autonomy of the individual, or will facilitate justice. Given its historical importance and bureaucratic and legal dependence [49], it is not surprising that informed consent remains central to contemporary discussions of ethical and social implications of biobanking, biomarkers, and biomedical research. Unfortunately, the substance of such debates centers upon the inadequacy of both ideal and current procedures. As Hoeyer points out in reference to Medical Biobank run by UmanGenomics in northern Sweden, informed consent offers an illusion of choice without real consideration of the implications of such choices, "by constructing a diffuse arrangement of donors who can only be semi-accountable agents" [50].

Other Areas that Warrant Consideration: Commercialization of Biobanks

Commercialization of biobanks can entail commercialization of biobank resources, such as biological samples or data, or the resulting research results or products from those resources. It can also refer to the introduction of private funding provided by industry to an existing publicly funded biobank. Biobanks are increasingly seeking support from the private sector to secure funding and scientists are under significant pressure to work with industry and commercialize their research. A recent survey of 456 biobanks revealed that funding shortage was a concern for 71% of biobanks and it was the greatest challenge for 37% of biobanks [34].

Levels of commercial involvement can vary among biobanks. The Icelandic Biobank was founded as a public–private partnership between the Icelandic government and deCODE genetics. UmanGenomics was given exclusive rights to commercialize information derived from Sweden's Medical Biobank. The Singapore Tissue Network, by contrast, is publicly funded and will not be involved in commercialization. Biotechnology companies involved in biobanking include Orig3n, which has established the world's largest cell repository in Boston to accelerate regenerative medicine, and CTIBIOTECH, which is a European company that is also licensed as a biobank (CTIBIOBANK) to store cells and tissue.

The introduction of private funding to a previously publicly funded biobank may increase ethical and privacy issues including the requirement for additional oversight, consent challenges, and uncertainties regarding ownership and control of biobank resources. Furthermore, commercial involvement has a considerable influence on the public's trust in biobanks [34]. The public may fear losing control over how their donated samples and associated data will be used and shared and the potential for use in research that may be stigmatizing or discriminatory.

Since biobanks, including those with commercial involvement, are expensive to maintain, insufficient funding can greatly impact sustainability. The Icelandic Biobank declared bankruptcy in 2009, then sold the biobank to Saga Investments LCC [34]. Following little success, the biobank was then acquired by Amgen for US$ 415 million. The biobank's management and objectives have been maintained, and it remains located in Iceland. However, Amgen has not provided a written guarantee that this arrangement will persist over time. This case highlights the ethical and policy issues that can arise with respect to the outcome of participant samples and data. Whether they are sold, transferred to another country, or destroyed would have great implications for donors' privacy, autonomy, and dignity. The biobank's policy, in addition to the informed consent form, would serve an essential role in determining the outcome of the participants' samples and data. As such, clear policies are crucial in protecting participants and preserving the public's trust.

Science, Ethics, and the Changing Role of the Public

Novel and innovative norms and models for biobank management have been proposed by bioethics, social science, and legal practitioners and theorists in recent years, in an attempt to deal with some of these issues. Alternative ethical frameworks based on social solidarity, equity, and altruism have been suggested [51, 52]. These formed the basis for the Human Genome Organisation's Ethics Committee statement on pharmacogenomics [27]. Onora O'Neil has argued for a two-tiered consent process in which public consent for projects is solicited prior to individual consent for donation of samples [53]. The charitable trust model has also been proposed for biobanking, as a way of recognizing DNA both as a common heritage of humanity and as uniquely individual, with implications for family members. "All information would be placed in a trust for perpetuity and the trustees overseeing the information would act on behalf of the people who had altruistically provided information to the population collection. They would be accountable to individuals but could also act as representatives for the community as a whole" [54].

It is not clear, however, whether such models could ever gain widespread public endorsement and legitimacy without direct public involvement in their design. Appeals to the need for community consultation [54] and scientific citizenship [55] may be more suited to the current mood. There is growing awareness globally, among government, policymakers, regulators, and advocacy groups alike, of the importance of public engagement, particularly in relation to emerging technologies.

In the United Kingdom, crises over bovine spongiform encephalopathy (BSE), otherwise known as "mad cow disease," and genetically modified (GM) crops have forced the government to proclaim the value of early public participation in decision-making [56, 57]. A statement by the UK House of Lords Select Committee in 2000 concluded that "today's public expects not merely to know what is going on, but to be consulted; science is beginning to see the wisdom of this and to move out of the laboratory and into the community to engage in dialogue aimed at mutual understanding" [58]. In Canada, the provincial government of British Columbia pioneered a citizens' assembly in 2003, charging 160 citizens with the task of evaluating the existing electoral system. A BC Conversations on Health project aims to improve the health system by engaging in "genuine conversation with British Columbians" during 2007.

Indeed, public consultations have become the norm for soliciting public support for new technologies. In the United Kingdom, these have included Weekends Away for a Bigger Voice, funded by the National Consumer Council in 2001 and the highly publicized government-funded GM Nation consultation in 2002. In Canada, notable examples include the 1999 Canadian Citizen's Conference on Biotechnology and the 2001 Canadian Public Consultation on Xenotransplantation. In Denmark, more than 20 consensus conferences have been run by the Danish Board of Technology since 1989, on topics as diverse as GM foods, electronic surveillance, and genetic testing [59]. In New Zealand, the government convened a series of public meetings in 2000 as part of its Royal Commission on genetic modification. UK Biobank marketing is careful to assert that the project has "undergone rigorous review and consultation at all levels" (http://www.ukbiobank.ac.uk/about-biobank-uk/).

Traditional public consultations have their limitations, however. Past examples of consultations have either been unpublicized or restricted to stakeholder involvement, undermining the claim to be representative of the full range of public interests [8]. Some critics suspect consultations of being a front to placate the public, a means of researching market strategy, and speeding product development [60] or as a mechanism for engineering consent [14]. GM Nation is one example of a consultation that has been criticized for "capture" by organized stakeholder groups and as misrepresentative of the public it aimed to consult [61].

Public Consultation and Deliberative Democracy

The use of theories and practices of deliberative democracy within such public consultations is a more recent and innovative trend. Deliberation stands in opposition to the aggregative market model of representational democracy and the strategic behavior associated with voting. It offers a model of democracy in which free and equal citizens exchange reasons through dialogue, and shape and alter their preferences collectively, and it is rapidly gaining in popularity, as evidenced by the growth of nonprofit organizations such as the Everyday Democracy (http://www.everyday-democracy.org), AmericaSpeaks (http://www.americaspeaks.org/), and National Issues Forums (http://www.nifi.org/) throughout the United States.

Origin stories of this broad deliberative democracy "movement" are as varied as its incarnations, and practice is not always as closely linked to theory as it could be. But most theorists will acknowledge a debt to the work of either (or both) Habermas or Rawls. Habermas's wider program of discourse ethics provides an overarching rationale for public deliberation [62]. This asserts that publicly binding norms can make a legitimate claim to rationality – and thus legitimacy – only if they emerge from free argument between all parties affected. Claims about what "any reasonable person" would accept as right can only be justified by putting them to the test. This is then a far cry from the heavily critiqued [14, 63] model of public consultation as a tool for engendering public trust or engineering acceptance of a new technology.

By asking participants to consider the perspectives of everyone, deliberation orients individuals away from consideration of self-interest and toward consideration of the common good. Pellizzoni characterizes this *governance virtue* as one of three key virtues of deliberative democracy [64]. The second is *civic virtue*, whereby the process of deliberation produces more informed, active, responsible, cooperative, and fair citizens. The third is *cognitive virtue*, the notion that discussion oriented to understanding rather than success enhances the quality of decisions, gives rise to new or unarticulated points of view, and allows common understanding of a complex problem that no single person could understand in its entirety.

Deliberative democracy is not devoid of challenges when applied to complex issues of science and technology, rich as they can be in future uncertainties and potential societal impact. But it offers much promise as a contribution to biobanking policy that can provide legitimate challenges to rigidly structured research ethics.

Conclusions

In 2009, *Time* Magazine declared biobanking as one of the top 10 ideas changing the world. Biomarker research is greatly advanced by good-quality annotated collections of tissues, or biobanks. Biobanks raise issues that stretch from evaluation of the benefits and risks of research through to the complexity of informed consent for collections for which the research purposes and methods cannot be described in advance. This range of ethical and organizational challenges is not managed adequately by the rules, guidelines, and bureaucracies of research ethics. Part of the problem is that current research ethics leaves too much for the individual participant to assess before the relevant information is available. But many other aspects of biobanks have to do with how benefits and risks are defined, achieved, and shared, particularly those that are likely to apply to groups of individuals with inherited risks, or those classified as having risks or as being more amenable to treatment than others. These challenges raise important issues of equity and justice. They also highlight trade-offs between research efficiency and benefits, biobank sustainability and ownership rights, privacy and individual control over personal information, and tissue samples. These issues are not resolvable by appeal to an existing set of rules or ethical framework to which all reasonable people agree. Inevitably, governance decisions related to biobanks will need to find a way to create legitimate policy and institutions. The political approach of deliberative democracy may hold the most promise for well-informed and representative input into trustworthy governance of biobanks and related research into biomarkers.

Acknowledgments

The authors thank Genome Canada, Genome British Columbia, and the Michael Smith Foundation for Health Research for their essential support. They also appreciate the support and mutual commitment of the University of British Columbia, the British Columbia Transplant Society, Providence Health Care, and Vancouver Coastal Health, and all participants in the Biomarkers in Transplantation program. The contributions of authors of a previous edition of this chapter, Dr. Richard Hegele, Jacquelyn Brinkman, and Janet Wilson-McManus, are deeply appreciated.

References

1 Sigurdsson, S. (2001). Ying-yang genetics, or the HSD deCODE controversy. *New Genet. Soc.* 20 (2): 103–117.
2 Sigurdsson, S. (2003). Decoding broken promises. Open Democracy. www.opendemocracy.net/theme-9-genes/article_1024.jsp (accessed 1 June 2004).
3 Abbott, A. (2003). With your genes? Take one of these, three times a day. *Nature* 425: 760.
4 Mannvernd, Icelanders for Ethics in Science and Medicine (2004). A landmark decision by the Icelandic Supreme Court: the Icelandic Health Sector Database Act stricken down as unconstitutional.
5 Merz, J.F., McGee, G.E., and Sankar, P. (2004). "Iceland Inc."?: On the ethics of commercial population genomics. *Soc. Sci. Med.* 58: 1201–1209.

6 Potts, J. (2002). At least give the natives glass beads: an examination of the bargain made between Iceland and deCODE genetics with implications for global bio-prospecting. *Va. J. Law Technol.* 8: 40.

7 Pálsson, G. and Rabinow, P. (2001). The Icelandic genome debate. *Trends Biotechnol.* 19: 166–171.

8 Burgess, M. and Tansey, J. (2009). Democratic deficit and the politics of "informed and inclusive" consultation. In: *Hindsight and Foresight on Emerging Technologies* (eds. E. Einseidel and R. Parker). Vancouver, British Columbia: UBC Press.

9 Morrison Institute for Population and Resource Studies (1999). Human Genome Diversity Project: Alghero Summary Report. http://web.archive.org/web/20070819020932/http://www.stanford.edu/group/morrinst/hgdp/summary93.html (accessed 2 August 2007).

10 Cavalli-Sforza, L.L. (2005). The Human Genome Project: past, present and future. *Nat. Rev. Genet.* 6: 333–340.

11 Harry, D., Howard, S., and Shelton, B.L. (2000). Indigenous people, genes and genetics: what indigenous people should know about biocolonialism. Indigenous Peoples Council on Biocolonialism. http://www.ipcb.org/pdf_files/ipgg.pdf.

12 Sudlow, C., Gallacher, J., Allen, N. et al. (2015). UK Biobank: an open access resource for identifying the causes of a wide range of complex diseases of middle and old age. *PLoS Med.* 12 (3): e1001779.

13 Barbour, V. (2003). UK Biobank: a project in search of a protocol? *Lancet* 361: 1734–1738.

14 Peterson, A. (2007). Biobanks "engagements": engendering trust or engineering consent? *Genet. Soc. Policy* 3: 31–43.

15 Peterson, A. (2005). Securing our genetic health: engendering trust in UK Biobank. *Sociol. Health Illness* 27: 271–292.

16 Redfern, M., Keeling, J., and Powell, M. (2001). *The Royal Liverpool Children's Inquiry Report*. London: House of Commons.

17 Royal College of Pathologists' Human Tissue Advisory Group (2005). *Comments on the Draft Human Tissue Authority Codes of Practice 1 to 5*. London: The Royal College of Pathologists.

18 Lin, Z., Owen, A., and Altman, R. (2004). Genomic research and human subject privacy. *Science* 305: 183.

19 Roche, P. and Annas, G. (2001). Protecting genetic privacy. *Nat. Rev. Gene.* 2: 392–396.

20 Cambon-Thomsen, A., Sallée, C., Rial-Sebbag, E., and Knoppers, B.M. (2005). Population genetic databases: Is a specific ethical and legal framework necessary? *GenEdit* 3: 1–13.

21 Illes, J., Rosen, A., Greicius, M., and Racine, E. (2007). Prospects for prediction: ethics analysis of neuroimaging in Alzheimer's disease. *Ann. N.Y. Acad. Sci.* 1097: 278–295.

22 Caux, C., Roy, D.J., Guilbert, L., and Viau, C. (2007). Anticipating ethical aspects of the use of biomarkers in the workplace: a tool for stakeholders. *Soc. Sci. Med.* 65: 344–354.

23 Viau, C. (2005). Biomonitoring in occupational health: scientific, socio-ethical and regulatory issues. *Toxicol. Appl. Pharmacol.* 207: S347–S353.

24 The Economist (2007). Genetics, medicine and insurance: Do not ask or do not answer? http://www.economist.com/science/displaystory.cfm?story_id=9679893 (accessed 31 August 2007).

25 Genewatch UK (2006). Genetic discrimination by insurers and employers: still looming on the horizon. Genewatch UK Report. http://www.genewatch.org/uploads/f03c6d66a9b354535738483c1c3d49e4/GeneticTestingUpdate2006.pdf (accessed 31 August 2007).

26 Rothenberg, K., Fuller, B., Rothstein, M. et al. (1997). Genetic information and the workplace: legislative approaches and policy challenges. *Science* 275: 1755–1757.

27 Human Genome Organisation Ethics Committee (2007). HUGO statement on pharmacogenomics (PGx): solidarity, equity and governance. *Genom. Soc. Policy* 3: 44–47.

28 Lewis, G. (2004). Tissue collection and the pharmaceutical industry: investigating corporate biobanks. In: *Genetic Databases: Socio-ethical Issues in the Collection and Use of DNA* (eds. R. Tutton and O. Corrigan). London: Routledge.

29 Loft, S. and Poulsen, H.E. (1996). Cancer risk and oxidative DNA damage in man. *J. Mol. Med.* 74: 297–312.

30 Scott, C., Caulfield, T., Borgelt, E., and Illes, J. (2012). Personal medicine – the new banking crisis. *Nat. Biotechnol.* 30: 141–147.

31 Tupasela, A., Snell, K., and Canada, J. (2015). Constructing populations in biobanking. *Life Sci. Soc. Policy* 11: 1–18.

32 Astrin, J. and Betsou, F. (2016). Trend in biobanking: a bibliometric overview. *Biopreserv. Biobanking* 14: 65–74.

33 Maschke, K.J. (2005). Navigating an ethical patchwork: human gene banks. *Nat. Biotechnol.* 23: 539–545.

34 Caulfield, T., Burningham, S., Joly, Y. et al. (2014). A review of key issues associated with the commercialization of biobanks. *J. Law Biosci.* 1: 94–110.

35 Elger, B.S. and Caplan, A.L. (2006). Consent and anonymization in research involving biobanks. *Eur. Mol. Biol. Rep.* 7: 661–666.

36 UBC Department of Pathology and Laboratory Medicine (2017). Office of Biobank Education and Research (OBER): who we are. http://pathology.ubc.ca/education-resource/ober/who-we-are/ (accessed 14 February 2017).

37 Hede, K. (2006). New biorepository guidelines raise concerns. *J. Nat. Cancer Inst.* 98: 952–954.

38 Brekke, O.A. and Thorvald, S. (2006). Population biobanks: the ethical gravity of informed consent. *BioSocieties* 1: 385–398.

39 Moreno, P. and Joly, Y. (2015). Informed consent in international normative texts and biobanking policies; seeking the boundaries of broad consent. *Med. Law Int.* 15: 216–245.

40 Cambon-Thomsen, A. (2004). The social and ethical issues of post-genomic human biobanks. *Nat. Rev. Genet.* 5: 866–873.

41 Hansson, M.G., Dillner, J., Bartram, C.R. et al. (2006). Should donors be allowed to give broad consent to future biobank research? *Lancet Oncol.* 7: 266–269.

42 Canadian Institutes of Health Research, Natural Sciences and Engineering Research Council of Canada, and Social Sciences and Humanities Research Council of Canada (2014). Tri-Council Policy Statement: Ethical Conduct for Research Involving Humans. http://www.pre.ethics.gc.ca/pdf/eng/tcps2-2014/TCPS_2_FINAL_Web.pdf (accessed 14 February 2017).

43 Caulfield, T. and Knoppers, B. (2010). Policy brief No. 1. consent, privacy & research biobanks. In: *Proceedings from the Genomics, Public Policy, and Society Meet Series*, 1–10. Genome Canada https://www.genomecanada.ca/sites/default/files/pdf/en/GPS-Policy-Directions-Brief.pdf.

44 Burgess, M.M. (2001). Beyond consent: ethical and social issues in genetic testing. *Nat. Rev. Genet.* 2: 147–151.

45 Weijer, C. and Emanuel, E. (2000). Protecting communities in biomedical research. *Science* 289: 1142–1144.

46 Baird, L. and Henderson, H. (2001). Nuu-Chah-Nulth Case History. In: *Continuing the Dialogue: Genetic Research with Aboriginal Individuals and Communities, pp. 30–43. Proceedings of a workshop sponsored by the Canadian Commission for the United Nations Educational, Scientific, and Cultural Organization (UNESCO), Health Canada, and the National Council on Ethics in Human Research, pp. 26–27* (eds. K.C. Glass and J.M. Kaufert), Vancouver, British Columbia, Canada.

47 Tymchuk, M. (2000). *Bad blood: Management and Function*. Canadian Broadcasting Company, National Radio.

48 Abbott, A. (2004). Icelandic database shelved as court judges privacy in peril. *Nature* 429: 118.

49 Faden, R.R. and Beauchamp, T.L. (1986). *A History and Theory of Informed Consent*. New York, NY: Oxford University Press.

50 Hoeyer, K. (2004). Ambiguous gifts: public anxiety, informed consent and biobanks. In: *Genetic Databases: Socio-ethical Issues in the Collection and Use of DNA* (eds. R. Tutton and O. Corrigan). London: Routledge.

51 Chadwick, R. and Berg, K. (2001). Solidarity and equity: new ethical frameworks for genetic databases. *Nat. Rev. Genet.* 2: 318–321.

52 Lowrance, W. (2002). *Learning from Experience: Privacy and Secondary Use of Data in Health Research*. London: Nuffield Trust.

53 O'Neil, O. (2001). Informed consent and genetic information. *Stud. History Philos. Biol. Biomed. Sci.* 32: 689–704.

54 Kaye, J. (2004). Abandoning informed consent: the case for genetic research in population collections. In: *Genetic Databases: Socio-ethical Issues in the Collection and Use of DNA* (eds. R. Tutton and O. Corrigan). London: Routledge.

55 Weldon, S. (2004). "Public consent" or "scientific citizenship"? What counts as public participation in population-based DNA collections? In: *Genetic Databases: Socio-ethical Issues in the Collection and Use of DNA* (eds. R. Tutton and O. Corrigan). London: Routledge.

56 Bauer, M.W. (2002). Arenas, platforms and the biotechnology movement. *Sci. Commun* 24: 144–161.

57 Irwin, A. (2001). Constructing the scientific citizen: science and democracy in the biosciences. *Public Underst. Sci.* 10: 1–18.

58 House of Lords Select Committee on Science and Technology (2000). *Science and Society, 3rd Report*. London: HMSO.

59 Anderson, J. (2002). Danish Participatory Models: Scenario Workshops and Consensus Conferences, Towards More Democratic Decision-Making. *Pantaneto Forum* 6. http://pantaneto.co.uk/danish-participatory-models-ida-elisabeth-andersen-and-birgit-jaeger/ (accessed 22 October 2007).

60 Myshkja, B. (2007). Lay expertise: Why involve the public in biobank governance? *Genet. Soc. Policy* 3: 1–16.

61 Rowe, G., Horlick-Jones, T., Walls, J., and Pidgeon, N. (2005). Difficulties in evaluating public engagement activities: reflections on an evaluation of the UK GM Nation public debate about transgenic crops. *Public Underst. Sci.* 14: 331–352.

62 Habermas, J. (1996). *Between Facts and Norms: Contributions to a Discourse Theory of Law and Democracy*. Cambridge, MA: MIT Press.

63 Wynne, B. (2006). Public engagement as a means of restoring public trust in science: Hitting the notes, but missing the music? *Commun. Genet.* 9: 211–220.

64 Pellizzoni, L. (2001). The myth of the best argument: power, deliberation and reason. *Br. J. Sociol.* 52: 59–86.

28

Anti-Unicorn Principle: Appropriate Biomarkers Don't Need to Be Rare or Hard to Find

Michael R. Bleavins[1] *and Ramin Rahbari*[2]

[1] *White Crow Innovation, Dexter, MI, USA*
[2] *Innovative Scientific Management, New York, NY, USA*

Introduction

Biomarkers have entered the drug development process with great fanfare, with the impression that this is the first time for a new approach. Projections of solving virtually all major issues faced in drug development, and the potential of new technologies revolutionizing medicine, have been promised by many specialists. In reality, biomarker assays represent the logical progression and integration of laboratory techniques already available within clinical pathology and biochemistry, with additional methods arising from new technologies such as molecular biology, genomics, proteomics, and metabonomics. The emphasis on biomarkers as a new approach often has led to expectations that the ideal biomarker must be novel, be exotic, and employ only the newest technologies possible. To answer the pertinent questions during drug development, teams often embark on the quest for the "unicorn" of biomarkers, sometimes resulting in the best and most practical test being overlooked as the search progresses for elusive cutting-edge methods.

When working with drug development teams, especially those with limited hands-on experience in biomarker applications, it is imperative that the business need and scientific rationale for including a biomarker in the development plan be elaborated prospectively. By clearly defining both the question(s) to be answered and how a biomarker will advance the new drug's development, the likelihood of successful implementation to advance the compound and save time and laboratory resources can be dramatic. In most instances, the most expedient biomarker approach will be identified by focusing on what is necessary to advance the compound at its particular stage of development, independent of how that parameter will be measured. Establishing precisely what

Biomarkers in Drug Discovery and Development: A Handbook of Practice, Application, and Strategy, Second Edition. Edited by Ramin Rahbari, Jonathan Van Niewaal, and Michael R. Bleavins.

is needed, versus what would be nice to have, may reveal opportunities not initially obvious to all participants.

Building a new biomarker on an emerging technology generally should be a last resort. Acceptance, reproducibility, sample handling considerations, quality control, standards, automation, and assessing laboratory-to-laboratory differences become exponentially more complex to characterize with the novelty of technology and lower numbers of groups using those approaches. For translation of a potential biomarker from the bench to the bedside, simpler in all aspects is preferable. Simpler tests to administer and evaluate do not mean less scientifically sound or relevant, and can provide a more solid foundation for acceptance in clinical and regulatory environments. This is not to say that the new technologies aren't braving new territory and having a significant impact on drug safety and development [1]. The emphasis of this chapter is to show how casting a wide net for ideas and approaches can expedite a compound's progress and improve decision-making, keeping in mind that even less novel technologies will often be the best choice.

Under ideal conditions, and in the instances where the quest is for a decision-making biomarker, there should be solid prospective agreement to actually affect the drug's progress based on the biomarker results. To that end, scientists should keep in mind that the primary purpose of biomarkers is to enable better decisions. A better decision is one that can be made more confidently, earlier, less invasively, more efficiently, or the test is transferable to reference laboratories as the drug enters Phase II or later. Therefore, it is essential that people supporting preclinical and clinical teams focus on the test(s) that aid in definitively graduating a compound to its next stage of development or to prove nonviability of the development efforts and compound attrition. This needs to be done without concern as to whether the biomarker derives from exciting new technology or is a new application of an established method using conventional science. In reality, if a team will not expedite, realign, or terminate a compound's development based on the biomarker results, inclusion of a non-decision-enabling biomarker is unlikely to serve a purpose other than to increase costs or complicate study designs.

Sometimes, the appropriate biomarker is a "unicorn" (exciting, rare, and exotic), but more often it is a "horse" (understood, generally accepted, and available) or a "mule" (proven performance, hybrid, and versatile). In biomarker selection, the primary consideration must be whether the test(s) is the best solution for the situation at hand. As a part of the biomarker development package, the biological rationale and significance (as understood at that stage), in addition to confidence in the platform, have to be evaluated. Another important factor is time; the development of a novel biomarker can be as complicated and time-consuming as developing a new chemical entity. This can be an essential consideration in circumstances where multiple companies are working in the same areas or patent timelines are not ideal. Identification, development, testing, characterization, and scaling of a new biomarker, for any one stage of the drug development process, generally require at least six to nine months. By looking beyond the experience base or capabilities of one group or laboratory, it may be that your search for unicorns ends most appropriately by locating a horse residing in the stable of someone else. This chapter highlights examples of unicorn, horse, and mule biomarker approaches.

Unicorn Biomarkers

Advances in technology that enable new applications can be valuable tools in addressing difficult questions of activity, efficacy, and safety for drug developers. The novelty of these approaches and the high-tech equipment create excitement and interest. Several new medicines actually owe their development and successful clinical utilization to tools that did not exist 25 years ago.

Gleevec (imatinib) was developed by Novartis to treat chronic myeloid leukemia (CML). This molecule, along with two subsequent tyrosine kinase inhibitors for CML (Sprycel, Tasigna), was designed using imaging and molecular modeling specifically to target cells expressing the mutant BCR-ABL protein underlying the disease origin. The altered chromosome size arising from the balanced translocation between chromosomes 9 and 20 was a key feature in identifying genetic involvement in leukemia [2, 3]. This formation of the Philadelphia chromosome, as well as the discovery that translocation resulted in the chimeric *BCR-ABL* gene, created new opportunities to target the resulting BCR-ABL oncoprotein and its role in hematologic cell proliferation [4, 5]. The role of BCR-ABL and several other proteins in cell cycle progression is reviewed by Steelman et al. [6]. Cloning, mutational analysis, sequencing, and animal models [4, 7–10] have proven to be valuable tools in designing effective treatments to CML. These techniques have also been important in developing the next generation medicines for this disease since mutations in *BCR-ABL* arise in 50–90% of patients and result in resistance to imantinib [11–13]. Genetic testing has identified specific mutations that can be targeted by new tyrosine kinase inhibitors [12–14] and provides better clinical options for patients with these mutations. The use of specific mutational genotyping is an important consideration in developing new drugs, particularly for the T315I mutation that was resistant to all approved tyrosine kinase inhibitor compounds for CML. In addition to *BCR-ABL* mutational analysis, monitoring the phosphorylation status of Crkl and Stat5 using flow cytometry, Western blotting, and/or enzyme-linked immunosorbent assay (ELISA) techniques provides biomarkers of compound activity [15, 16].

Pharmacogenetics also was an important aspect in the target selection and development of the CCR5 antagonist class of anti-HIV (human immunodeficiency virus) drugs, including compound maraviroc developed by Pfizer,

Inc. (Selzentry). The observation that persons homozygous for the CCR5-Δ32 mutation were resistant to the development of acquired immune deficiency syndrome (AIDS) was shown mechanistically to result from inhibiting the HIV binding to the mutated receptor and entering T-helper lymphocytes [17–20]. Maraviroc binds selectively to the human chemokine receptor CCR5 present on the cell membrane, preventing the interaction of HIV-1 gp120 and CCR5 necessary for CCR5-tropic HIV-1 to enter cells [21]. CXCR4-tropic and dual-tropic HIV-1 are not inhibited by maraviroc.

Genetic testing for the CRR5-Δ32 deletion to stratify clinical trial subjects was useful in determining whether small molecules showed differential activity in these groups, for establishing inclusion and exclusion criteria, and the characterization of safety [22]. These data were also key components in the successful registration of maraviroc in 2007. As HIV treatment advanced, genetic assays for CCR5 mutations, as well as tropism for CXCR4 or CCR5 co-receptor, proved useful in optimal use of these co-receptor antagonists. In fact, the Selzentry label [21] recommends tropism testing to identify appropriate candidates and states that "use of Selzentry is not recommended in patients with dual/mixed or CCR4-tropic HIV-1, as efficacy was not demonstrated in a Phase II study of this patient group." The Trofile assay was used extensively in Pfizer's maraviroc Phase III trials and measures the ability of the patient's specific virus envelope gene to effect entry into cells. This biomarker uses amplified RNA to establish a patient's HIV genome, followed by an assessment of that genome to infect CCR5- and CXCR4-expressing cell lines.

Horse Biomarkers

Factor Xa inhibitors are of therapeutic interest as antithrombotic agents because direct-acting antithrombin drugs often induce bleeding or deficits in fibrin [23–25]. By targeting a specific enzyme in the coagulation cascade, it is generally assumed that toxicity can be reduced and efficacy retained. Having an accurate indicator of bleeding risk is essential in this class of molecules since the major rate-limiting toxicities are associated directly or indirectly with inhibited blood coagulation. During the development of factor Xa inhibitors, a logical biomarker of both safety and efficacy was available by measurement of Xa activity.

In fact, development and decision-making with this class of drug were expedited by being able to monitor factor Xa activity in both preclinical efficacy experiments and toxicology studies, as well as in early clinical trials. By adapting the automated human technique available in most clinical pathology coagulation laboratories to rodent and non-rodent species, direct comparisons were possible between the relative doses and plasma concentrations associated with therapeutic inhibition of clotting and exposures likely to cause unacceptably long coagulation times. At the discovery, preclinical safety, and Phase I clinical stages, drug development teams had a tool that allowed rapid prioritization of molecules and decisions on dosing. Additionally, because reagents were available commercially at reasonable cost, the blood volumes required were small, and instrumentation already existed for the method in clinical pathology laboratories, a reproducible means of determining a factor Xa inhibiting compound's pharmacodynamic characteristics was readily performed in a variety of research and hospital groups. This biomarker has been a useful tool for developing factor Xa inhibitors, although caution must be exercised to assess compound activity beyond the results being a reflection of plasma drug concentration.

Cholesterol analysis isn't often considered very exotic or cutting edge, but has undergone significant evolution during its application in predicting cardiovascular risk. Total cholesterol has typically been measured using enzymatic, immunochemical, chemical, precipitation, ultracentrifugation, and column chromatography methods [26, 27]. Since there can be significant differences in the values obtained for each of the lipoprotein classes using the various techniques, and to provide a standard for comparison, the Centers for Disease Control maintains reference methods for cholesterol, triglycerides, and high-density lipoproteins (HDLs) [28]. Reference methods considered the gold standards for cholesterol fractions have also been developed, validated, and credentialed [29, 30]. As the relative roles and significance of very low-density lipoprotein (VLDL), HDL, and low-density lipoprotein (LDL) major lipid subgroups became better established in clinical practice, techniques for determining each cholesterol subcategory were developed and integrated into standard clinical laboratory use. Automation of these tests has made lipid analysis easily conducted in routine clinical practice. For many years, enzymatic and chemical methods have comprised the primary approaches to monitoring total cholesterol, triglycerides, and phospholipids. Precipitation is generally used for HDL and LDL, with ultracentrifugation considered the gold standard for assessment of new techniques [27, 30]. As the fibrate and statin classes of drugs were integrated into standard clinical practice among cardiologists and general practitioners, cholesterol monitoring became commonplace and continues as a leading indicator of heart disease risk.

Otvos et al. [31] published a nuclear magnetic resonance (NMR) approach to cholesterol monitoring that could be applied to clinically practical samples, as well as correlating the various lipoprotein subclasses to coronary artery disease [32–34], diabetes [35, 36], and genetic polymorphisms having relevance in heart disease [37–39]. NMR has been an analytical technique since the 1950s, although its primary applications had been in research chemistry laboratories, and more recently for metabonomic investigations. As cholesterol profiles and cardiovascular risk assessment have developed, many physicians have asked for increasingly detailed lipid evaluations. NMR profiling of serum lipids has demonstrated differences in the relative effects on various lipoprotein classes with available fibrate [40–42] and statin drugs [43–47].

A better understanding of particle-size distribution within lipid classes using NMR allows tailoring of patient therapeutic approaches beyond the broad goals of raising HDL and lowering LDL. Even among persons taking statin drugs, the best choice for one patient may be atorvastatin, whereas another person may achieve a more desirable lipid profile with rosuvastatin or simvastatin. This approach has been particularly useful in assessing changes in LDL particle numbers as well as total LDL content. The ability to select the optimal drug for each person provides an aspect of personalized medicine using well-accepted clinical pathology parameters determined by a technology with proven reliability in an arena outside the standard hospital laboratory, but readily accessible through specialty reference laboratories. The use of NMR to determine cholesterol and lipoprotein categories also has been useful in studying exercise [48, 49], arthritis [50], and vascular responses [51].

Mule Biomarkers

The anticancer drug trastuzumab (Herceptin) is a case study in stratification to identify patients most likely to respond to a drug using both conventional approaches and newer technologies. In 2007, trastuzumab achieved sales of $1.3 billion in the United States despite serious toxicity considerations and being efficacious in a limited subset of patients. Trastuzumab has primary efficacy in breast cancer patients who are overexpressing the human epidermal growth factor protein (HER2; [52–54]). The drug is a therapeutically engineered monoclonal antibody that targets the HER2 receptor protein [54]. The response rate to trastuzumab is as high as 35% in patients with overexpression of HER2, while the drug lacks a target and is ineffective at tolerated doses in patients who do not have increased HER2 levels. Pretreatment characterization of breast cancer patients to identify HER2-positive cancers and determine a patient's suitability for trastuzumab therapy is now commonly practiced. The serious adverse events that can develop following trastuzumab use (cardiac failure, pulmonary toxicity, infusion reactions) make it highly desirable to minimize exposure of those populations unlikely to benefit from the drug's therapeutic effects. The targeting of persons most likely to respond to the drug also has been a factor in reimbursement by third-party payers and acceptance by governmental health programs.

The primary and US Food and Drug Administration (FDA)–approved approach for determining HER2 expression is immunohistochemistry for HER2 [55]. This "hybrid mule biomarker" combines the well-established but low-specificity technique of histology with the more recent development of probe antibodies directed against specific proteins. Although neither of these modalities is particularly high tech or exotic, they provided the basis for developing a biomarker indicating HER2 over expression as molecular biology and proteomic techniques became available. These approaches quantified the protein resulting from the underlying gene amplification, leading to overexpression and identification and purification of the specific protein involved. As the reason for the progression to cancer became clearer, components of the overall process became suitable biomarker targets. This led, in turn, to the development of specific antibodies suitable for the identification of HER2 in tissue sections. Direct gene amplification and fluorescent in situ hybridization techniques for measuring HER2 expression also exist, but acceptance into clinical practice has been slower, due to their greater complexity and the limited number of laboratories capable of performing them in a diagnostic setting. Although not a perfect biomarker, HER2 expression is a valuable tool for the oncologist, and additional research is being conducted to refine the predictivity of this measure, either by alternative tests or by better characterization of key variables.

Osteoarthritis presents a difficult therapeutic area for pharmaceutical intervention. The progressive nature of the disease, generally late onset in life, and difficulty in reversing existing damage have made development of effective therapies challenging. Additionally, determining clinical efficacy often requires long-term clinical trial designs even in early drug development assessments where little is known about efficacy or side effects. These characteristics act as significant impediments for safe and rapid screening of new molecules. Clearly, this is an ideal place for a predictive biomarker that would allow a rapid assessment of whether a new drug had activity in the disease processes. Underlying the disease is destruction of joint cartilage by matrix metalloproteinases (MMPs; [56, 57]), so a mechanistic approach to identifying a biomarker was a logical approach to improving drug development in this area. As the role of collagen degradation and MMP activity became clearer in arthritis [58–61], interest focused on biomarkers that could be applied to animal models for both compound advancement and clinical monitoring as the ultimate endpoint.

Nemirovskiy et al. [62] reported development of a translatable biomarker of MMP activity with the goal of its application for MMP inhibitor compound selection and improved diagnosis of osteoarthritis. Similar approaches were under way by other groups within the pharmaceutical industry, both specific pharmaceutical companies and external contract research organizations. The approach taken by Nemirovskiy et al. [62] was to identify specific cleavage products using liquid chromatography–tandem mass spectrometry. By studying these MMP-derived peptides from human articular cartilage, they were able to show a 45-mer peptide fragment of collagen type II correlated with the pathology of human osteoarthritis and was present in urine and synovial fluid. An immunoaffinity liquid chromatography–mass spectrometry/mass spectrometry (LC–MS/MS) assay was developed to quantify collagen type

II neoepitope (TIINE) peptides as biomarkers of collagenase modulation. The resulting assay was capable of detecting TIINE peptides in the urine of healthy and afflicted human subjects and preclinical species (e.g. rat, rabbit, guinea pig, dog). This LC–MS/MS assay had excellent sensitivity, high throughput, reasonable costs, and robustness. By including immunoaffinity in the technique, a substantial improvement in assay sensitivity over traditional LC–MS/MS was achieved by eliminating much of the background noise associated with the sample matrix. ELISA methods also were developed to measure TIINE concentrations in urine, but proved to have higher intrasample variability, greater matrix sample effects, and lower specificity for cleavage products, and be more difficult to outsource to external laboratories. Although the ELISA technique remains useful as a research tool, the LC–MS/MS assay had significant advantages for clinical translation and implementation.

The TIINE biomarker was also applied to better characterize osteoarthritis in a surgically induced system using the Lewis rat [63]. This preclinical model has proven valuable to study progression and therapeutic intervention in degenerative joint disease. Using immunohistochemical staining of the joints, the authors were able to compare TIINE expression with proteoglycan loss and histological changes. This study showed that TIINE levels increased in intensity and area in lesions that co-localized with the loss of proteoglycan. From these data, Wancket et al. [63] were able to better define the medial meniscectomy surgical model of osteoarthritis demonstrate a progressive pattern of cartilage damage similar to those seen in human lesions, and further characterize TIINE as a useful biomarker for monitoring cartilage degradation.

Clinically, TIINE has been used to evaluate the mechanism by which doxycycline slows joint space narrowing in patients with knee osteoarthritis. Urinary TIINE and radiographic determinations were conducted over a 30-month period [64]. Although the TIINE measurements were highly reproducible, the authors concluded that high visit-to-visit variability limits the sensitivity of the TIINE assay for detecting changes in clinical monitoring of osteoarthritis, and that increases in urinary TIINE concentration are unlikely to account for doxycycline reductions in joint space narrowing. The value of the TIINE biomarker for other collagen-based diseases remains to be determined. This biomarker does, however, highlight how even with mechanistic approaches, good preclinical correlations, and solid technology, a new technique may not yield a biomarker with strong clinical utility.

Conclusions

The best biomarker for any given situation can come from a wide range of sources, so it is critical that no promising option be excluded. Matching the application and question to be answered are far more important that the platform used for analysis or if a technique is resident within a particular laboratory or company. It should be recognized, however, that it is difficult to consider approaches that one has no idea exist. Nevertheless, those people providing biomarker support to drug discovery and development teams must keep in mind as wide a range of options as possible. The Internet has made specialized testing by reference laboratories, teaching hospitals, and research groups significantly more accessible. In addition, directories of tests and the groups that perform them are available [65], as well as characterized genomic biomarkers being described on the FDA web site [66].

Genomic, proteomic, and metabonomic technologies can provide essential information when identifying new biomarkers, but have been slow to be implemented into clinical applications. Although often critical to identifying new targets or biomarker options, the extensive data sets produced, variability in sample and platform conditions, challenges of validating multiplexed measurements and algorithms, and lack of experience have limited their usefulness in clinical trials to a few diseases. The fields are rapidly progressing and hold great promise, especially when specific focused questions are defined prior to conducting the tests. To paraphrase Helmut Sterz, a former member of the Pfizer Toxicology Senior Leadership Team, "use of a little grey matter at the beginning can save a lot of white powder, chips, instrumentation, and time." All too often, the quantity of information obtained from many "omics" experiments cannot be realized effectively due to limits on data mining tools and the realities of clinical trial conduct. People cannot be subjected to the same degree of environmental and genetic control possible with animal studies, and many diseases represent a constellation of effects rather than changes induced by a single cause or gene.

Our experience in developing, validating, translating, and implementing new biomarkers has emphasized repeatedly that the question to be answered must drive the technology used. It is also vital that the solution be "fit for purpose" with respect to the parameter being measured, platform selected, and level of assay definition or validation [67]. Sometimes the biomarker must utilize cutting-edge technology and novel approaches, but more commonly the question can be answered without an exotic assay, often with a test that already exists in someone else's laboratory.

References

1 FDA (2008). Pharmacogenomics and its role in drug safety. *FDA Drug Saf. Newsl.* 1 (2). http://web.archive.org/web/20090525082220/http://www.fda.gov/cder/dsn/2008_winter/pharmacogenomics.htm.

2 Nowell, P.C. and Hungerford, D.A.A. (1960). Chromosome studies on normal and leukemic leukocytes. *J. Nat. Cancer Inst.* 25: 85–109.

3 Rowley, J.D. (1973). A new consistent chromosomal abnormality in chronic myelogenous leukemia identified by quinacrine fluorescence and Giemsa staining. *Nature* 243: 290–293.

4 Daley, G.Q., Van Etten, R.A., and Baltimore, D. (1990). Induction of chronic myelogenous leukemia in mice by the *P210*$^{bcr/abl}$ gene of the Philadelphia chromosome. *Science* 247: 824–830.

5 Lugo, T.G., Pendergast, A.M., Juller, A.J., and Witte, O.N. (1990). Tyrosine kinase activity and transformation potency of *bcr-abl* oncogenes products. *Science* 247: 1079–1082.

6 Steelman, L.S., Pohnert, S.C., Shelton, J.G. et al. (2004). JAK/STAT, Raf/MEK/ERK, PI3K/Akt and BCR-ABL in cell cycle progression and leukemogenesis. *Leukemia* 18: 189–218.

7 Hariharan, I.K., Harris, A.W., Crawford, M. et al. (1989). A *bcr-abl* oncogene induces lymphomas in transgenic mice. *Mol. Cell. Biol.* 9: 2798–2805.

8 Pfumio, F., Izac, B., Katz, A. et al. (1996). Phenotype and function of human hematopoietic cells engrafting immune-deficient CB 17 severe combined immunodeficiency mice and nonobese diabetic-severe combined immunodeficient mice after transplantation of human cod blood mononuclear cells. *Blood* 88: 3731–3740.

9 Honda, H., Oda, H., Suzuki, T. et al. (1998). Development of acute lymphoblastic leukemia and myeloproliferative disorder in transgenic mice expressing *P210*$^{bcr/abl}$: a novel transgenic model for human Ph^1-positive leukemias. *Blood* 91: 2067–2075.

10 Li, S., Ilaria, R.J., Milton, R.P. et al. (1999). The P190, P210, and P230 forms of the *BCR/ABL* oncogene induce a similar chronic myeloid leukemia-like syndrome in mice but have different lymphoid leukemogenic activity. *J. Exp. Med.* 189: 1399–1412.

11 Gorre, M.E., Mohammed, M., Ellwood, K. et al. (2001). Clinical resistance to STI-571 cancer therapy caused by *BCR-ABL* gene mutation or amplification. *Science* 293: 876–880.

12 Shah, N.P., Nicoll, J.M., Nagar, B. et al. (2002). Multiple *BCR-ABL* kinase domain mutations confer polyclonal resistance to the tyrosine kinase inhibitor imatinib (STI571) in chronic phase and blast crisis chronic myeloid leukemia. *Cancer Cell* 2: 117–125.

13 Branford, S., Rudzki, Z., Walsh, S. et al. (2003). Detection of *BCR-ABL* mutations in mutations in patients with CML treated with imantinib is virtually always accompanied by clinical resistance, and mutations in the ATP phosphate-binding loop (P-loop) are associated with poor prognosis. *Blood* 102: 276–283.

14 Shah, N.P., Tran, C., Lee, F.Y. et al. (2004). Overriding imatinib resistance with a novel *ABL* kinase inhibitor. *Science* 305 (5682): 399–401.

15 Klejman, A., Schreiner, S.J., Nieborowska-Skorska, M. et al. (2002). The Src family kinase Hck couples *BCR/ABL* to STAT5 activation in myeloid leukemia cells. *EMBO J.* 21 (21): 5766–5774.
16 Hamilton, A., Elrick, L., Myssina, S. et al. (2006). *BCR-ABL* activity and its response to drugs can be determined in CD34$^+$ CML stem cells by CrkL phosphorylation status using flow cytometry. *Leukemia* 20 (6): 1035–1039.
17 Dean, M., Carrington, M., Winkler, C. et al. (1996). Hemophilia Growth and Development Study, Multicenter AIDS Cohort Study, Multicenter Hemophilia Cohort Study, San Francisco City Cohort, ALIVE Study.). Genetic restriction of HIV-1 infection and progression to AIDS by a deletion allele of the CKR5 structural gene. *Science* 273: 1856–1862.
18 Carrington, M., Dean, M., Martin, M.P., and O'Brien, S.J. (1999). Genetics of HIV-1 infection: chemokine receptor CCR5 polymorphism and its consequences. *Hum. Mol. Genet.* 8 (10): 1939–1945.
19 Agrawal, L., Lu, X., Quingwen, J. et al. (2004). Role for CCR5Δ32 protein in resistance to R5, R5X4, and X4 human immunodeficiency virus type 1 in primary CD4+ cells. *J. Virol.* 78 (5): 2277–2287.
20 Saita, Y., Kodama, E., Orita, M. et al. (2006). Structural basis for the interaction of CCR5 with a small molecule, functionally selective CR5 agonist. *J. Immunol.* 117: 3116–3122.
21 Pfizer, Inc. (2008). Selzentry label. http://web.archive.org/web/20090118232145/http://www.fda.gov/cder/foi/label/2007/022128lbl.pdf.
22 Vandekerckhove, L., Verhofstede, C., and Vogelaers, D. (2008). Maraviroc: integration of a new antiretroviral dug class into clinical practice. *J. Antimicrob. Chemother.* 61 (6): 1187–1190.
23 Rai, R., Sprengeler, P.A., Elrod, K.C., and Young, W.B. (2001). Perspectives on factor Xa inhibition. *Curr. Med. Chem.* 8 (2): 101–119.
24 Ikeo, M., Tarumi, T., Nakabayashi, T. et al. (2006). Factor Xa inhibitors: new anti-thrombotic agents and their characteristics. *Front. Biosci.* 11: 232–248.
25 Crowther, M.A. and Warkentin, T.E. (2008). Bleeding risk and the management of bleeding complications in patients undergoing anticoagulant therapy: focus on new anticoagulant agents. *Blood* 111 (10): 4871–4879.
26 Rifai, N. (1986). Lipoproteins and apolipoproteins: composition, metabolism, and association with coronary heart disease. *Arch. Pathol. Lab. Med.* 110: 694–701.
27 Bachorik, P.S., Denke, M.A., Stein, E.A., and Rifkind, B.M. (2001). Lipids and dyslipoproteinemia. In: *Clinical Diagnosis and Management by Laboratory Methods*, 20e (ed. J.B. Henry), 224–248. Philadelphia, PA: W.B. Saunders.
28 Myers, G.L., Cooper, G.R., Winn, C.L., and Smith, S.J. (1989). The Centers for Disease Control–National Heart, Lung and Blood Institute Lipid Standardization Program: an approach to accurate and precise lipid measurements. *Clin. Lab. Med.* 9: 105–135.

29 Myers, G.L., Cooper, G.R., Hassemer, D.J., and Kimberly, M.M. (2000). Standardization of lipid and lipoprotein measurement. In: *Handbook of Lipoprotein Testing* (eds. N. Rifai, G.R. Warnick and M. Dominiczak), 717–748. Washington, DC: AACC Press.

30 Rafai, N. and Warnick, G.R. (2006). Lipids, lipoproteins, apolipoproteins, and other cardiovascular risk factors. In: *Tietz Textbook of Clinical Chemistry and Molecular Diagnostics*, 4e (eds. C.A. Burtis, E.R. Ashwood and D.E. Bruns), 903–981. St. Louis, MO: Elsevier Saunders.

31 Otvos, J.D., Jeyarajah, E.J., Bennett, D.W., and Krauss, R.M. (1992). Development of a proton NMR spectroscopic method for determining plasma lipoprotein concentrations and subspecies distribution from a single, rapid measure. *Clin. Chem.* 38: 1632–1638.

32 Freedman, D.S., Otvos, J.D., Jeyarajah, E.J. et al. (1998). Relation of lipoprotein subclasses as measured by proton nuclear magnetic resonance spectroscopy to coronary artery disease. *Arterioscler. Thromb. Vasc. Biol.* 18: 1046–1053.

33 Kuller, L.H., Grandits, G., Cohen, J.D. et al. (2007). Lipoprotein particles, insulin, adiponectin, C-reactive protein and risk of coronary heart disease among men with metabolic syndrome. *Atherosclerosis* 195: 122–128.

34 van der Steeg, W.A., Holme, I., Boekholdt, S.M. et al. (2008). High-density lipoprotein cholesterol, high-density lipoprotein particle size, and apolipoprotein A-1: significance for cardiovascular risk. *J. Am. Coll. Cardiol.* 51: 634–642.

35 MacLean, P.S., Bower, J.F., Vadlamudi, S. et al. (2000). Lipoprotein subpopulation distribution in lean, obese, and type 2 diabetic women: a comparison of African and White Americans. *Obes. Res.* 8: 62–70.

36 Berhanu, P., Kipnes, M.S., Khan, M. et al. (2006). Effects of pioglitazone on lipid and lipoprotein profiles in patients with type 2 diabetes and dyslipidaemia after treatment conversion from rosiglitazone while continuing stable statin therapy. *Diabetic Vasc. Dis. Res.* 3: 39–44.

37 Couture, P., Otvos, J.D., Cupples, L.A. et al. (2000). Association of the C-514T polymorphism in the hepatic lipase gene with variations in lipoprotein subclass profiles: the Framingham Offspring Study. *Arterioscler. Thromb. Vasc. Biol.* 20: 815–822.

38 Russo, G.T., Meigs, J.B., Cupples, L.A. et al. (2001). Association of the Sst-I polymorphism at the *APOC3* gene locus with variations in lipid levels, lipoprotein subclass profiles and coronary heart disease risk: the Framingham Offspring Study. *Atherosclerosis* 158: 173–181.

39 Humphries, S.E., Berglund, L., Isasi, C.R. et al. (2002). Loci for CETP, LPL, LIPC, and APOC3 affect plasma lipoprotein size and sub-population distribution in Hispanic and non-Hispanic white subjects: the Columbia University BioMarkers Study. *Nutr. Metab. Cardiovasc. Dis.* 12: 163–172.

40 Ikewaki, K., Tohyama, J., Nakata, Y. et al. (2004). Fenofibrate effectively reduces remnants, and small dense LDL, and increases HDL particle number in hypertriglyceridemic men: a nuclear magnetic resonance study. *J. Atheroscler. Thromb.* 11: 278–285.

41 Ikewaki, K., Noma, K., Tohyama, J. et al. (2005). Effects of bezafibrate on lipoprotein subclasses and inflammatory markers in patients with hypertriglyceridemia: a nuclear magnetic resonance study. *Int. J. Cardiol.* 101: 441–447.

42 Otvos, J.D., Collins, D., Freedman, D.S. et al. (2006). LDL and HDL particle subclasses predict coronary events and are changed favorably by gemfibrozil therapy in the Veterans Affairs HDL Intervention Trial (VA-HIT). *Circulation* 113: 1556–1563.

43 Rosenson, R.S., Shalaurova, I., Freedman, D.S., and Otvos, J.D. (2002). Effects of pravastatin treatment on lipoprotein subclass profiles and particle size in the PLACI trial. *Atherosclerosis* 160: 41–48.

44 Schaefer, E.J., McNamara, J.R., Taylor, T. et al. (2002). Effects of atorvastatin on fasting and postprandial lipoprotein subclasses in coronary heart disease patients versus control subjects. *Am. J. Cardiol.* 90: 689–696.

45 Blake, G.J., Albert, M.A., Rifai, N., and Ridker, P.M. (2003). Effect of pravastatin on LDL particle concentration as determined by NMR spectroscopy: a substudy of a randomized placebo controlled trial. *Eur. Heart J.* 24: 1843–1847.

46 Soedamah, S.S., Colhoun, H.M., Thomason, M.J. et al. (2003). The effect of atorvastatin on serum lipids, lipoproteins and NMR spectroscopy defined lipoprotein subclasses in type 2 diabetic patients with ischemic heart disease. *Atherosclerosis* 167: 243–255.

47 Schaefer, E.J., McNamara, J.R., Tayler, T. et al. (2004). Comparisons of effects of statins (atorvastatin, fluvastatin, lovastatin, pravastatin, and simvastatin) on fasting and postprandial lipoproteins in patients with coronary heart disease versus control subjects. *Am. J. Cardiol.* 93: 31–39.

48 Nicklas, B.J., Ryan, A.S., and Katzel, L.I. (1999). Lipoprotein subfractions in women athletes: effects of age, visceral obesity and aerobic fitness. *Int. J. Obes. Rel. Metab. Disord.* 23: 41–47.

49 Yu, H.H., Ginsburg, G.S., O'Toole, M.L. et al. (1999). Acute changes in serum lipids and lipoprotein subclasses in triathletes as assessed by proton nuclear magnetic resonance spectroscopy. *Arterioscler. Thromb. Vasc. Biol.* 19: 1945–1949.

50 Hurt-Camejo, E., Paredes, S., Masana, L. et al. (2001). Elevated levels of small, low-density lipoprotein with high affinity for arterial matrix components in patients with rheumatoid arthritis: possible contribution of phospholipase A2 to this atherogenic profile. *Arthritis Rheum.* 44: 2761–2767.

51 Stein, J.H., Merwood, M.A., Bellehumeur, J.L. et al. (2004). Effects of pravastatin on lipoproteins and endothelial function in patients receiving human immunodeficiency virus protease inhibitors. *Am. Heart J.* 147: E18.

52 Hudelist, G., Kostler, W., Gschwantler-Kaulich, D. et al. (2003). Serum EGFR levels and efficacy of trastuzumab-based therapy in patients with metastatic breast cancer. *Eur. J. Cancer* 42 (2): 186–192.

53 Smith, B.L., Chin, D., Maltzman, W. et al. (2004). The efficacy of Herceptin therapies is influenced by the expression of other *erbB* receptors, their ligands and the activation of downstream signalling proteins. *Br. J. Cancer* 91: 1190–1194.

54 Burstein, H.J. (2005). The distinctive nature of *HER2*-positive breast cancer. *N. Engl. J. Med.* 353: 1652–1654.

55 Genentech (2008). Herceptin label.

56 Shinmei, M., Masuda, K., Kikuchi, T. et al. (1991). Production of cytokines by chondrocytes and its role in proteoglycan degradation. *J. Rheumatol. Suppl.* 27: 89–91.

57 Okada, Y., Shinmei, M., Tanaka, O. et al. (1992). Localization of matrix metalloproteinase 3 (stromelysin) in osteoarthritic cartilage and synovium. *Lab. Invest.* 66 (6): 680–690.

58 Billinghurst, R.C., Dahlberg, L., Ionescu, M. et al. (1997). Enhanced cleavage of type II collagen by collagenases in osteoarthritic articular cartilage. *J. Clin. Invest.* 99: 1534–1545.

59 Huebner, J.L., Otterness, I.G., Freund, E.M. et al. (1998). Collagenase 1 and collagenase 3 expression in a guinea pig model of osteoarthritis. *Arthritis Rheum.* 41: 877–890.

60 Dahlberg, L., Billinghurst, R.C., Manner, P. et al. (2000). Selective enhancement of collagenase-mediated cleavage of resident type II collagen in cultured osteoarthritic cartilage and arrest with a synthetic inhibitor that spares collagenase 1 (matrix metalloproteinase 1). *Arthritis Rheum.* 43: 673–682.

61 Wu, W., Billinghurst, R.C., Pidoux, I. et al. (2002). Sites of collagenase cleavage and denaturation of type II collagen in aging and osteoarthritic articular cartilage and their relationship to the distribution of matrix metalloproteinase 1 and matrix metalloproteinase 13. *Arthritis Rheum.* 46: 2087–2094.

62 Nemirovskiy, O.V., Dufield, D.R., Sunyer, T. et al. (2007). Discovery and development of a type II collagen neoepitope (TIINE) biomarker for matrix metalloproteinase activity: from in vitro to in vivo. *Anal. Biochem.* 361 (1): 93–101.

63 Wancket, L.M., Baragi, V., Bove, S. et al. (2005). Anatomical localization of cartilage degradation markers in a surgically induced rat osteoarthritis model. *Toxicol. Pathol.* 33 (4): 484–489.

64 Otterness, I.G., Brandt, K.D., Le Graverand, M.P., and Mazzuca, S.A. (2007). Urinary TIINE concentrations in a randomized controlled trial of doxycycline in knee arthritis: implications of the lack of association between TIINE levels and joint space narrowing. *Arthritis Rheum.* 56 (11): 3644–3649.

65 Hicks, J.M. and Young, D.S. (eds.) (2005). *DORA 05-07: Directory of Rare Analyses.* Washington, DC: AACC Press.

66 FDA (2008). Table of validated genomic biomarkers in the context of approved drug labels. Center for Drug Evaluation and Research. https://web.archive.org/web/20090521181940/http://www.fda.gov/cder/genomics/genomic_biomarkers_table.htm (accessed 10 September 2008).

67 Lee, J.W., Devanarayan, V., Barret, Y.C. et al. (2006). Fit-for-purpose method development and validation of biomarker measurement. *Pharm. Res.* 23 (2): 312–328.

29

Translational Biomarker Imaging: Applications, Trends, and Successes Today and Tomorrow

Patrick McConville and Deanne Lister

Invicro, a Konica Minolta Company, San Diego, CA and Department of Radiology, University of California San Diego, Molecular Imaging Center, Sanford Consortium for Regenerative Medicine

Introduction

The rapid shift to targeted and personalized therapies (precision medicine) has led to an increasing need for specific and predictive biomarkers for diagnosis, patient stratification, and determination of therapeutic response [1]. Imaging methods including magnetic resonance imaging (MRI), positron emission tomography (PET), single photon emission computed tomography (SPECT), computed tomography (CT), and ultrasound provide many approaches to the use of biomarkers that are importantly noninvasive, translational, and spatially resolved. Many imaging biomarkers have been and continue to be used by the academic, medical, and pharmaceutical sectors [2, 3], though use and related success in terms of widespread clinical implementation and impact have been modest so far. Nevertheless, there have been several examples of successes that have driven imaging biomarker applications to become relatively well standardized and routinely used in nonclinical and clinical practice, particularly MRI-based volumetrics and tumor burden assessment [4], and nuclear medicine–based (PET, SPECT) imaging of receptor occupancy and molecular biodistribution using radioisotope labeled agents [5, 6]. Today, a number of new molecular imaging applications, many associated with newer imaging agents [7, 8], highlight further potential for translational imaging biomarkers in facilitating progress with state-of-the-art molecular targets and therapeutic paradigms currently in discovery and soon to be in clinical trials.

This chapter will outline the following:

- the integration and use of imaging biomarkers in the recent past, including existing successes and clinical prevalence of use
- today's needs and gaps that imaging biomarkers have the potential to fill

Biomarkers in Drug Discovery and Development: A Handbook of Practice, Application, and Strategy, Second Edition. Edited by Ramin Rahbari, Jonathan Van Niewaal, and Michael R. Bleavins.

- new trends and directions (in terms of science, infrastructure, and policy) that are addressing, or could address today's unmet needs and facilitate future impact of imaging biomarkers in the clinic and the pharmaceutical industry.

In discussing the above, today's scientific concepts and standards will be highlighted, focusing on major diseases and respective treatment targets. New major therapeutic strategies and paradigms will be described. These biomedical concepts will be coupled with technical advances, regarding hardware and supporting tools and devices, that will pave the way for increasing benefits from quantitative imaging biomarkers for precision medicine.

Yesterday's Imaging Biomarkers: Prototypes for Tomorrow

The major translational imaging modalities including MRI, PET, SPECT, CT, and ultrasound that are household names today were disruptive, breakthrough technologies upon their discovery and commercialization in the 1970s and 1980s [9–14]. With the establishment of specialized preclinical counterparts in the 1990s and the early new millennium, came the birth of the term "translational imaging," and the discovery and use of translational imaging biomarkers exploded in the academic world [15]. Many of these biomarkers consequently began to see routine clinical and pharmaceutical industry use. One of the first imaging biomarkers was tissue volume, which saw rapid adoption in oncology (i.e. tumor volume) and is now used as a widespread foundational standard in cancer diagnosis and patient management, as well as by the pharmaceutical industry in testing new oncology therapies. Tumor volume is one of the few examples of an imaging biomarker that progressed to become a true *surrogate* or *surrogate marker* for a clinical endpoint (lifespan), in a number of indications.

Today, however, the term "imaging biomarker" is more commonly used to describe parameters that measure tissue or molecular properties *beyond* the level of anatomy and volume [16, 17]. The narrative in this chapter will conform to this convention too, though it should be kept in mind that imaging modalities like MRI, CT, and ultrasound were founded as predominantly *diagnostic* and *volumetric* tools, which interrogated tissue structure noninvasively on a macro scale, without depth limitations. Stated simply, structural images provide definition of meaningful boundaries between tissue, tissue substructures, and lesions. These structural imaging applications paved the way for today's myriad of imaging biomarker applications, the future evolution of which is the subject of this chapter. Each imaging biomarker application is based on the premise that while tissue boundaries can be identified in images, the signal intensity at any spatial location in the image or tissue in question is affected by other meaningful properties of the tissue and its underlying molecular and cellular properties. Additionally, the development of *imaging agents* (including contrast agents) whose local concentration or signal modulating properties depend on relevant properties of the tissue or tissue target density (in the case of *targeted* imaging agents) has revolutionized imaging. There is an increasingly large and impressive body of imaging agents and corresponding translational and quantitative imaging biomarkers that have seen clinical use and passed certain aspects of clinical validation.

While anatomical imaging applications originally used for diagnosis (e.g. MRI, CT) have formed a platform for other physiological biomarkers, molecular biomarker applications have also driven expanded applications beyond diagnosis, for example, in measuring disease progression and response to therapies. For example, in nuclear medicine imaging, or namely, PET and SPECT, tracer molecule properties that make them suitable for *diagnosis* have led to their expanded use in measuring *disease progression* and *therapeutic response*. The most notable example of this is 18F-fluorodeoxyglucose (^{18}F-FDG) PET-based detection and diagnosis of cancer [18, 19]. This occurs through administration of an exogenous tracer (e.g. ^{18}F-FDG) and detection of its radioactive signal, the location of which is used to infer the existence and location of diseased tissue (e.g. a solid tumor). In an analogous fashion, the signal intensity (not just location) can be used as a translational biomarker for a relevant tissue property (e.g. glycolytic metabolic rate, using ^{18}F-FDG). Similarly, future nuclear medicine imaging–based biomarkers will leverage the dynamic range of PET- and SPECT-based tracer signal intensities as disease progression and response biomarkers, extending the foundational *location*, or diagnostic functionality.

However, these more sophisticated translational biomarker applications for therapeutic response remain relatively immature. In considering their potential, and the overall goal for imaging biomarkers, it is useful to consider more mature, widely applied, and more extensively described biomarker technologies, such as histopathology [20, 21], and its extension through immunohistochemistry (IHC) [22–24]. In this context, hematoxylin & eosin (H&E) stains can be thought of as the anatomical or *location* platform for more specific biomarkers accessed through antibody-based stains that correlate with other meaningful tissue properties that provide information beyond structure and location of cells. Due to the high sensitivity and specificity of today's IHC biomarkers, IHC has become a gold standard despite the fact that it is invasive and involves tissue destruction. Just as an IHC stain is almost always performed in parallel with a structurally focused H&E stain, translational imaging biomarkers can extend anatomical or diagnostic imaging, with both accomplished in parallel using the same imaging instrument in the same subject. It is particularly noteworthy that as translational imaging evolves to incorporate biomarker breadth and versatility to parallel that already standard with IHC, IHC is evolving to become quantitative, leveraging the strengths of imaging agents and analytics that are already standard in translational imaging.

Recent Evolution

Through the first decade of the new millennium, translational imaging saw unprecedented adoption and growth rates through the use of translational imaging biomarkers in the academic, pharmaceutical, and clinical spheres. Translational imaging was poised to grow exponentially, and the imaging hardware manufacturing space was hectic and crowded. Most of the large pharmaceutical companies had invested heavily in internal preclinical imaging capabilities, as well as clinical imaging expertise. However, the predicted inflection did not occur and the rate of growth declined. The most recent 5–10 years has seen a flattening in the use and adoption of imaging biomarkers for translational imaging.

The cause of the downturn was multifaceted, and poor economic drivers were no doubt influential [25]. Additional important issues also contributed to slower than projected growth, forming a basis for learnings that may promote future growth. These include a general lack of depth of characterization and understanding of imaging biomarkers, including their specificity and relevance, leading to a failure to appropriately limit the breadth and volume of applications in many cases. This resulted in many unjustified uses and failures clinically, leading to large, non-recoverable financial losses. Additionally, there was a general underestimate of fixed and variable costs associated with imaging – by both manufacturers and users. This increased the severity of the cost of failed applications.

Examples of the above issues are many and have led to several outcomes and trends that were not anticipated based on the rate of adoption seen 15 years ago. These include the following:

- major manufacturers of preclinical imaging instrumentation pulling back, or completely out of, the space
- many of the large pharmaceutical companies removing, in part or completely, their internal preclinical imaging infrastructure and turning toward an outsourcing model
- few PET tracers approved clinically for many years after the approval of ^{18}F-FDG
- no non-volumetric MRI biomarkers seeing standardized, broad, routine use, and many that were touted as "game changing" (e.g. ADC, Ktrans) seeing a substantial decline in use
- multiple examples of MRI contrast agents being given black box warnings (e.g. gadolinium contrast agents) and/or being pulled from the market,

including the general clinical failure to this point, of the much-heralded iron oxide contrast agents.

The remainder of this section will examine more deeply the historical trends and related causes associated with translational imaging biomarkers, constructing a platform for future needs and advances.

Latest Trends Evidenced in Peer-Reviewed Literature

Figure 29.1 is a summary of searches using PubMed (2018) that highlights trends in imaging-based pharmaceutical research, based on searches by recent five-year duration time periods for all reviews with a pharmaceutical/drug or cell therapy focus, and the portion of these that reference *in vivo* or translational imaging.

While significant growth in the percent of pharmaceutical/drug reviews that referenced imaging occurred from 1990 to 2005, there was a decrease in the period 2010–2014 and there has been no significant growth in this proportion since then. This may be related to many factors, both economic and political. Nonetheless, it shows that any kind of "explosion" in the use and reliance on imaging for pharmaceutical discovery and development has not occurred. In general, the data supports the notion that imaging currently plays a relatively small, yet significant role in pharmaceutical and drug research, and begs

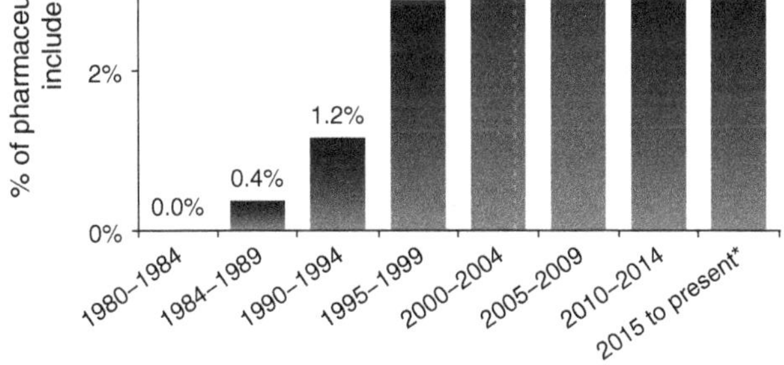

Figure 29.1 Proportion (as %) of pharmaceutical and cell therapy scientific reviews that reference *in vivo* or translational imaging, by 5 year period over the last 30 years. The results show an increase in the use of imaging modalities in pharmaceutical research up to 2010 followed by reduced use from 2010 to present. Source: Data from PubMed.com. (*Data extrapolated to a five-year period beginning in 2015.)

the question of how learnings to date might drive future scientific and clinical returns.

Current Clinical Prevalence

Further evidence of the current state of imaging biomarker use is highlighted by examining clinical reliance on imaging through registered, active clinical trials using the ClinicalTrials.gov database [26]. Figure 29.2 summarizes clinical trial imaging prevalence across several major imaging modalities, expressed as percent of trials incorporating imaging to total number of reported clinical trials. The data suggests reasonable prevalence of translational imaging in today's clinical trials, though it should be noted that Phase 0 (method development studies in healthy volunteers) trials may not be reported in the clinicaltrials.gov database. Later stage trials, where reporting is mandatory, the patient population studies are larger, and the results more relevant in terms of interpretation of efficacy, show the largest use of biomarkers.

On further examination, the data also show that today's clinical trial imaging in the United States includes heavy prevalence of biomarkers that were first used or "discovered" 10–20 years ago. For example, excluding MRI-based volume endpoints, MRI in current trials shows significant usage of biomarkers such as tumor Ktrans, apparent diffusion coefficient, and proton magnetic

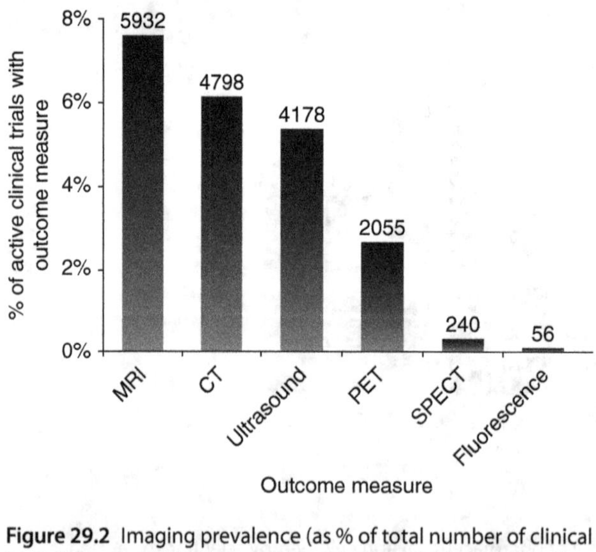

Figure 29.2 Imaging prevalence (as % of total number of clinical trials) in current clinical trials (including Actively recruiting, Not yet recruiting, Active/not recruiting, and Enrollment by invitation). The total number of trials employing each imaging modality is shown at the top of each bar in the graph). Over 17,000 patient imaging trials are ongoing, with MRI being the most prevalently used imaging modality, followed closely by CT, ultrasound, then PET. Source: Adapted from ClinicalTrials.gov.

resonance spectroscopy (^{1}H MRS)-based metabolite biomarkers. For PET, predominance is in the use of ^{18}F-FDG. These biomarkers in clinical use today are relatively nonspecific in terms of association with molecular targets and pathways. They are more appropriately termed *physiological* or *functional* biomarkers that limit the precision and specificity with which pharmaceutical target modulation can be quantified. The realization of true precision medicine will require biomarkers that are more precisely coupled to molecular pathways and targets, and imaging technologies are poised to play a key role. While there has been substantial progress with more precise, specific, and better targeted imaging biomarkers, there is a need for increased translation and clinical optimization of such agents. As our knowledge of, and ability to screen, potential imaging targets improves, so too does our ability to design, optimize, and translate precise, "high fidelity" imaging agents.

Learnings Driving Today's Needs

Other than volumetric biomarkers and FDG, there are few examples of imaging biomarkers that have seen standardized use in routine clinical practice. The clinical trial prevalence of decade-old imaging biomarkers like Ktrans [27, 28], ADC [29, 30], ^{18}F-FLT [31–33], and ^{18}F-FMISO [34, 35], demonstrates success in terms of justifying the cost of using them for the information they can provide. But at the same time, the modest success in translating them into routine clinical practice, for diagnosis, let alone prognosis or treatment response, defines the challenge for tomorrow's imaging biomarker development. Imaging agents developed over the last 10–20 years have rarely reached desired or hoped potential in terms of clinical translation and routine use. Learnings from the past have led to improved knowledge and precision in terms of imaging agent size, PK, mass dose, and clearance properties in animals and humans, as well as choice of imaging label and its window/duration (e.g. based on metabolism and degradation *in vivo*, or radioactive half-life).

The limitations of the older imaging biomarkers and agents still prevalent in today's clinical trials demonstrate the need for improved specificity, and moreover, the need for biomarkers that address the current state of knowledge of diseases in which they are intended to be applied. To date, imaging biomarkers have tended to be developed *after* advances in our knowledge of disease targets for which biomarkers would be an asset. An unfortunate consequence of this sequential approach is that our understanding of major diseases like those under the umbrella of "cancer" has moved so rapidly that the current suite of imaging biomarkers have too quickly become limiting, or even irrelevant for today's applications. Dynamic contrast enhanced magnetic resonance imaging (DCE MRI) and its use for biomarkers targeted at angiogenesis is a good example of this issue. While anti-angiogenic and vascular disrupting therapies were once immensely popular, they lost favor due to inability to obtain significant and durable responses with treatments that targeted this aspect of various solid tumors. This is one example of a set of biomarkers becoming less relevant than originally hoped (due to a therapeutic strategy failure) and ultimately not seeing impactful clinical use.

Another limitation that has plagued even the most successful imaging biomarkers has been standardization. The complexity of the imaging tools and underlying protocols seemingly drives an unmanageable diversity in terms of how the biomarkers are implemented, measured, and analyzed in practice. The poster child biomarkers obtained through FDG, DCE MRI, diffusion MRI, and the like, certainly demonstrate this problem. Despite efforts of consortia and multi-institutional teams, such as the Radiological Society of North America (RSNA) Quantitative Imaging Biomarker Alliance (QIBA) [36], that strive to obtain wide, expert-driven consensus and policies, standardization has remained elusive. The recent, rapid developments in information technology (IT) and software have in some ways amplified this problem by providing widespread and inexpensive access to highly sophisticated software tools, which add many variables, and contrasting and contradicting opinions to an already complicated set of methods that can be used in applying each biomarker [37]. Today, image analysis conversations are as commonly themed around making analysis as simple and minimally dimensionalized as possible, versus around leveraging sophistication to dissect data in increasing ways. Standardization and automation are increasingly high priorities that are simplifying, enhancing versatility, and improving return on investment in imaging biomarker technologies. Recent progress in machine learning and artificial intelligence provides great potential for less subjective, image analysis and imaging biomarkers [38, 39]. Successes with these new endeavors will lead to less human intervention and variability between operators and facilities, increasing information per unit scanning and analysis time.

A recent, promising move to improve imaging biomarker standardization is the QIBA consortium [40–44]. QIBA, which was launched by the RSNA, brings together the major translational imaging societies and prominent translational imaging researchers across different imaging modalities, with the intent of developing standards including scanning and analysis protocols, and related quality control (QC) for major, clinically used biomarkers. To its credit, QIBA has acknowledged the lack of manufacturer attention to ensuring imaging hardware, and generated parameters are truly "quantitative" due to their initial use being dominated by research, and not necessarily advanced clinical standard use. As such, the most common imaging biomarkers (e.g. positron emission tomography standardized uptake value [PET SUV]) in clinical use today have been shown to vary significantly between different instruments. The inherent flexibility built in to these research tools drives other levels of complexity and variability in terms of the choice of scanning and analysis protocol. Today's "standard" remains that the "same" protocol varies widely from site to site in terms of the set of optional parameters that the scanning and analysis professionals choose for their application of that protocol. Efforts by bodies like QIBA hold promise for evolving the imaging biomarker field to adhere to commonly accepted standards, and acceptance criteria, which will better facilitate successful translation of imaging biomarkers. Further, this will drive more highly performing future imaging biomarkers, through improved precision. Along with this will come a better ability to interpret results across the necessary multicenter preclinical validation and development studies and later multicenter clinical trials.

Today's Imaging Biomarkers: Successes and Failures to Leverage Future Needs

Despite the limitations and problems that have led to less successful translation of promising biomarkers and more generally that have tempered the rate of translation of imaging biomarkers, there are numerous examples of imaging biomarkers in routine clinical use. This section will focus on examples of several that have reached the highest levels of usage and are playing a key role in driving clinical decisions both in terms of patient diagnosis and management and pharmaceutical trials.

18F-FDG PET Imaging of Tissue Metabolism

The inherent characteristic of altered metabolism in cancer cells has made ^{18}F-FDG PET a broadly valuable and high impact tool for cancer diagnosis and treatment response assessment. ^{18}F-FDG, a glucose analog, enables whole body, three-dimensional imaging of relative tissue metabolic activity. Cancer lesions are identified by regions of abnormally high ^{18}F-FDG uptake due to increased glucose uptake in tumors relative to normal tissue. Relatively short uptake times and quick clearance from the blood allow for reasonable procedure times for patients and studies, and the short half-life (110 minutes) makes working with ^{18}F-FDG more convenient than other, longer-lived isotopes [45].

While ^{18}F-FDG was discovered as a glucose metabolic reporter, its uptake significantly depends on other general tissue properties including cell viability and blood supply/blood flow. ^{18}F-FDG–based biomarkers are therefore notoriously nonspecific in many cases. However, since glucose metabolism, cell viability, and blood supply are all major indicators of solid tumor "health," the relative non-specificity of ^{18}F-FDG has contributed to its success, as well as its increasingly standard use in staging and treatment monitoring, through PET imaging in many cancer types including lung, colorectal, and lymphoma [46–49]. However, there are known issues with detection of other cancer types, most notably breast and prostate cancer, primarily due to generally lower tissue metabolic rates. Although ^{18}F-FDG–based detection of primary breast and prostate cancers is problematic, the high sensitivity of ^{18}F-FDG PET has facilitated detection of metastatic disease in these and other cancer types due to the commonly increased metabolism of distant metastases versus primary tumors [50–52].

Additionally, ^{18}F-FDG-PET has demonstrated success as a tool for assessing *early* treatment response. Periodic imaging during treatment regimens enables intervention in therapeutic regimens, including early termination of treatment in highly successful cases. In any case, ^{18}F-FDG PET imaging allows researchers and clinicians in many cases to visualize treatment effects earlier than changes in lesion volume that may be detectable by anatomical imaging methods, such as MRI.

18F PET Tracer Imaging in Alzheimer's Disease: Amyloid-β and Tau

For decades, amyloid-β dysfunction has been a known hallmark in Alzheimer's disease (AD). In more recent years, researchers have also found significant links to the tau protein that suggest an active feedback loop between the two may drive disease progression [53]. Amyloid-β and tau are both naturally occurring proteins in the brain. However, in the early AD brain, clearance of amyloid-β proteins becomes dysregulated, leading to buildup of insoluble neuron plaques. These plaques cause decreased neuronal functionality and ultimately lead to neuron death, giving rise to the classical degenerative cognitive signs as disease progresses. Tau proteins, while studied for many decades, have only more recently been linked to a key role in AD progression. The aberrant buildup of amyloid-β leads to hyper-phosphorylation of tau, its subsequent release from cells and change in localization, leading to the second hallmark of AD, termed "neurofibrillary tangles." The results of studies comparing tau knockout and wild-type mice have shown that amyloid-β toxicity is dependent, at least in part, on tau [54].

The important and interconnected roles of amyloid-β and tau activity in AD pathogenesis make them ideal targets for molecular imaging applications. To date, three amyloid-β PET radiopharmaceuticals have been approved by the U.S. Food and Drug Administration (FDA) for use as AD diagnostic agents: ^{18}F-fluorbetapir (2012) [55, 56], ^{18}F-flutemetamol (2013) [57], and ^{18}F-florbetaben (2014) [58]. These agents have each shown success in diagnosing AD-related neurological issues through increased presence of amyloid-β in the brain compared to normal brain tissue [59], which corresponds to the characteristic buildup of plaques. While this technology is promising, there are known limitations of amyloid-β PET imaging [60], including an inherent increased presence of amyloid-β in the normal aging brain. Additionally, other medical conditions, such as Lewy body dementia, show high amyloid-β presence [61]. Although the location of amyloid-β expression is different in these two diseases, PET imaging is not able to distinguish between the two, leading to the potential for misdiagnosis.

Realization of the amyloid-β-tau dependency in the progression of AD has led researchers to consider a move toward targeting tau radiotracers as a diagnostic tool to improve the accuracy of AD diagnosis. Although no tau PET radiotracers are yet approved for use, many are in development and are actively being used at various stages both nonclinically and clinically [62, 63]. Binding of these radiotracers with multiple isoforms of tau characteristic to AD, in combination with the needed specificity to this set of isoforms versus normal brain tau or amyloid-β, has proven to be a challenge but is critical in determining the outlook for tau radiotracers in AD patient management.

PET Tracer Imaging of Neuroreceptor Occupancy

After tumor volumetrics, one of the most broadly used and standardized imaging biomarker applications is PET- and SPECT-based imaging of neuroreceptor occupancy. The concept is based on the radiolabeling of ligands that are proven to be specific to a central nervous system (CNS) target of interest and that are confirmed to provide a reliable measure of receptor occupancy. This companion biomarker concept allows direct measurement of pharmacologic/therapeutic agent occupancy of the target. This provides a PD biomarker and determination of required dose can be achieved on an individual patient basis. There have been numerous successes in this endeavor with PET imaging agents created for many relevant CNS targets. This approach is today being adapted in oncology, for oncology targets (see further discussion below) primarily leveraging the precision targeting of biologics. Additionally, certain CNS targets where imaging agents already exist, have been shown to be targets in other diseases – for example, adenosine 2A (A2A) as a target in oncology. A2A imaging agents previously developed in CNS applications are today being used in new oncology applications.

Magnetic Resonance Spectroscopy

Nuclear magnetic resonance enables differentiation of various, MR-compatible nuclei (e.g. ^{1}H, ^{13}C, ^{19}F, ^{31}P) and changes in the resonant frequencies of these nuclei in the presence of different chemical environments. For example, ^{1}H protons in water resonate at a slightly different frequency from protons bound within various macromolecules, and this is resolved as separation in resonant peaks in spectroscopic data. The term to describe this phenomenon is chemical shift. ^{1}H MRS is the most common application of this technology and has been successful in routine diagnosis of brain cancer, in particular, but has also been extended to other cancers, such as prostate and breast [64].

^{1}H MRS provides information about metabolites that may be tumor lesions, such as choline, N-acetylaspartate (NAA), and creatine. These metabolites are also present in normal brain, but tumor lesion presence, or tumor response to treatment, may cause modulation in the total amount of some or all these. Recently, increasing focus has turned to identifying new metabolites of interest as better predictors of disease and tools for patient stratification. An example of this is the detection of glutamate (Glu) and glutamine (Gln) amino acid levels in the brain. Elevations of Glu and Gln have been shown to correlate with a number of diseases, including brain tumors, various psychiatric disorders, and neurodegenerative disorders [65]. In brain tumors, the degree of Glu/Gln concentration elevation has been shown to correlate with the grade, or severity, of the disease. Similarly, 2-hydroxyglutarate (2-HG) has been correlated with isocitrate dehydrogenase (IDH) mutations in low-grade gliomas [66].

Functional and Pharmacological MRI

Functional magnetic resonance imaging (fMRI) and the related pharmacological magnetic resonance imaging (pHMRI) detect and quantify neurovascular coupling effects related to brain metabolism and modulation of such via interventions or pharmacologic treatments. There are several detection modes that have seen prevalence including T2*-based detection of the blood-oxygenation-level-dependent (BOLD) effect, super paramagnetic iron oxide (SPIO) particle-based relative cerebral blood volume (rCBV) measurement, and arterial spin-labeling (ASL). These methods comprise relatively mature MRI techniques that are seeing continued interest for pharmacodynamic measurement of brain activation/blood flow/metabolism in response to CNS drug treatments [67, 68]. By detecting regions of brain metabolism modulation, fMRI can be used for target validation as well as for enabling reliable biomarkers for drugs such as cognitive enhancers, for example, in Alzheimer's, schizophrenia, and depression. However, fMRI has suffered from challenges associated with data quality, stability, and difficulty in establishing and controlling interventions that are needed in order to measure the BOLD response. These challenges have resulted in limited routine clinical use for fMRI but have at the same time paved the way for the promising newer area of "resting state" fMRI, which will be discussed below.

The Future Imaging Biomarker

What Do We Need?

The advent of precision medicine and molecular targeted approaches to therapeutic interventions requires increasing prevalence of precision biomarkers. Such precision ultimately defines a need for biomarkers to couple tightly and specifically to a relevant cellular, molecular, or gene level target. Latest trends in the pharmaceutical industry and in clinical practice highlight moves

to integrate radiological biomarkers with pathology (increasingly a digital imaging science) and genetic biomarkers. Successes in integrating big data originating from medical imaging, pathology, proteomics, and genomics hold groundbreaking potential for precision medicine and translation of disruptive new clinical approaches.

So far, as discussed above, the most prevalent imaging biomarkers and their respective successes to date have also highlighted inherent limitations, providing templates for what we need in future imaging biomarkers. Arguably, the concept of "specific" and "sensitive" biomarker is somewhat of a misnomer. No imaging biomarker is perfectly specific, and no imaging biomarker has unlimited sensitivity. It is therefore more scientifically accurate (and realistic) to talk in terms of imaging biomarkers that are sensitive and specific *enough* to provide a significant gain versus the current state of the art in at least one indication where there exists a significant unmet need. This necessitates an understanding of the limitations of each biomarker, or said a different way, understanding the specific uses in which the biomarker will suffice. It could be the narrowness of the applications in which this holds true that establishes an undesirable cap on the return on the investment, or inability to sell the biomarker to funding bodies. But the concept of narrow success vs. broad failure is critical in terms of increasing the translational success of future imaging biomarkers.

This concept of "right fit" imaging biomarkers does not overly weight effort in validation and development toward proving maximal or "ideal" specificity or sensitivity, but rather uses these critical concepts in a subtly but importantly different way. That is, answering the question "Is the biomarker sensitive and specific enough to enable a significant clinical impact in one or more defined disease/patient populations?" vs. "Is it sensitive and specific?". Additionally, our academic enthusiasm often has us setting unrealistic goals such as providing a response biomarker for changing clinical patient treatment paradigms, when stratification of patient populations is more appropriate and would still create significant impact, and quite probably in a shorter period of time.

Nevertheless, part of the key to improving the performance of imaging biomarkers (e.g. through optimizing sensitivity and specificity), is tight interfacing of imaging knowledge to our knowledge of the disease in question, which, as referenced above, is advancing at a hectic pace. The need to better-parallel imaging biomarker discovery and development with disease target discovery is clear. A great example of this is the recent massive advances in our knowledge of, and ability to treat, several major cancers. Less than 10 years ago, there was minimal attention to the use of the immune system in this endeavor and yet today, immunotherapies overwhelmingly dominate state-of-the-art approaches to treating cancer [69, 70]. More highly performing imaging biomarkers can be obtained both by (i) leveraging the improving specificity of disease targets (e.g. immune checkpoints in cancer) and (ii) developing imaging biomarkers in *parallel* with identification of the best and most specific targets, and secondarily, therapies to these targets.

Imaging Agents: Precision Tools for Precision Medicine

Imaging agents are defined as any molecule which through inherent chemical properties or through labeling can be detected through imaging. Under the condition that some measurable property of the imaging agent–based signal is modulated by a relevant property of the tissue or disease, imaging agents directly enable imaging biomarkers. Table 29.1 details the commonly used sources for imaging signal modulation (contrast) by imaging modality, and Table 29.2 summarizes the major classes of imaging agents that utilize these contrast moieties or "labels." The imaging agent approach to provision of imaging biomarkers has been very successful, particularly in PET imaging, but limitations and failures have been prevalent in the past and are driving more precisely tuned and performing agents today and in the future.

Importantly, imaging agents can be applied to the same targets as histology agents, and to the extent that pathology is a gold standard for measuring some of the most relevant aspects of disease, finely tuned imaging agents may be used to extract the same kind of information, but in living subjects, allowing time courses and therapeutic response quantification. Advances in both spaces will more tightly couple information coming from each. The move to digital pathology and related automated image based tissue slice analysis is highly key to potential for integrating *in vivo* imaging information with pathology data. Equally important to this is next generation pathology agents that allow greater sensitivity and a substantially larger dynamic range for detection and therefore digital analysis, quantification, and informatics such as those enabled through phosphor-integrated dots (PIDs) [71, 72] Quantitative pathology has

Table 29.1 Commonly employed contrast moieties or "labels" by imaging modality.

Modality	Contrast agent/reporter
MRI	Gadolinium, (U)SPIOs, ^{19}F (e.g. PFCs), CEST agents, hyperpolarized ^{13}C
PET	Positron-emitting radioisotopes (e.g. ^{18}F, ^{64}Cu, ^{68}Ga, ^{89}Zr)
SPECT	Gamma-emitting radioisotopes (e.g. ^{99m}Tc, ^{111}In, ^{123}I, ^{125}I)
Ultrasound	Microbubbles
CT	Iodine, barium
Fluorescence	Fluorescent proteins and synthetic fluorophores (450–850 nm emission range)

These labels are attached to compounds with specific properties to create precision imaging agents and related quantitative biomarkers (PFC, perfluorocarbon; CEST, chemical exchange saturation transfer).

Table 29.2 Major categories of imaging agents, common examples, and corresponding imaging modalities employed.

Probe type	Examples	Common modalities
"Passive" specificity and blood pooling	• Perfusion-driven extracellular localization (e.g. tumors) • Long half-life vascular agents • Size based localization (e.g. tumor-specific nanoparticles)	MRI Ultrasound Fluorescence
Targeted	Antigen/receptor targeting: • Antibody • Antibody fragments • Peptides • Proteins	PET SPECT Fluorescence
Conditional intracellular "trapping"	• Metabolic cycle (e.g. ^{18}F-FDG, DNP ^{13}C-pyruvate) • Cell cycle (e.g. ^{18}F-FLT) • Hypoxic cells (e.g. ^{18}F-MISO)	PET (11C, 18F) MRI (hyperpolarized ^{13}C)
Conditionally activated	• Protease cleavage (e.g. caspases, cathepsins, MMPs) • pH	Fluorescence MRI (CEST)
Gene reporters	• HSV1-tk • PSMA • Dopamine D2 • Ferritin	PET MRI
Phagocytosis-based cell reporters	• Immune cells, stem cells labeled in culture • Macrophages labeled endogenously	MRI (^{19}F PFCs, SPIOs) PET/SPECT (^{89}Zr/^{111}In-oxine) Fluorescently labeled particles

CEST, chemical exchange saturation transfer; PFC, perfluorocarbon; SPIO, super paramagnetic iron-oxide.

become a reality, and the term "imaging agent" is beginning to encompass both translational imaging and histopathology. The sections below will detail the current areas of intense focus that have potential for high impact imaging biomarkers in near future clinical applications.

Immuno-oncology Imaging Agents

The demonstration of unprecedented, durable responses in major cancer indications using immune checkpoint inhibitors [73, 74] and engineered T cells [75, 76] over the last several years has rapidly changed the face of oncology.

The targets and therapeutic strategies involved rely on precision biologics as the treatment agents. Examples of this include antibodies and their derivatives, and chimeric antigen receptor (CAR) expressing T cells. Fortuitously, the state of the art in translational imaging has developed in parallel to enable image-based approaches to quantifying immuno-oncology target expression and/or target engagement in animal models and patients. More specifically, the labeling of biologic molecules and cells is so routine today that it has led to a rapid attention to immuno-oncology imaging agents that can be used to stratify patient populations and provide more comprehensive information than local biopsy. Some of the most promising approaches incorporate PET isotopes such as ^{18}F, ^{68}Ga, and ^{89}Zr to enable imaging agents targeted to CD8 [77, 78], PD-1/PD-L1 [79], CTLA-4 [80, 81],CD3 [82], Granzyme B [83], arabinosyl guanine (AraG) [84], and OX40 (CD134) [85].

A recent study described a novel PET tracer, ^{68}Ga-NOTA-GZP, which was designed to evaluate Granzyme B activity [83]. Granzyme B is a serine protease that is released by cytotoxic T cells and NK cells during an immune response. Preclinical PET imaging data has shown that ^{68}Ga-NOTA-GZP is able to differentiate between responders and non-responders to immunotherapy, and this differentiation is detectable prior to decreases in tumor volume.

A second study utilized a Granzyme B fluorescent nanoprobe to evaluate immune-mediated myocarditis [86]. Fluorescent Molecular Tomography (FMT) imaging in a CMy-mOva transgenic mouse model of lymphocyte-mediated myocarditis demonstrated differentiation between immune activity in diseased and wild-type mice, as well as mice treated with dexamethasone, an anti-inflammatory agent.

^{18}F-AraG [84] PET imaging, OX40 [85] induction and PET imaging have also been recently demonstrated as potentially viable immuno-oncology imaging approaches. ^{18}F-AraG is targeted toward two salvage kinase pathways, preferentially accumulates in activated primary T cells, and may therefore have potential as a biomarker for T cell activation. An anti-OX40 antibody labeled with 64Cu was used to image upregulated OX40 in an immunoenhancing therapeutic strategy demonstrating the use of PET imaging in this immunotherapy setting.

The immuno-oncology imaging agent concept is analogous to that successfully used in neuroscience, most commonly by ^{11}C or ^{18}F labeling of small molecules targeted to neuroreceptor targets. This approach has created a significant positive impact in CNS drug translation by providing companion molecules that directly readout receptor occupancy as a PD biomarker. Through quantifying receptor occupancy, clinicians can choose and stage treatments and adjust dose in a patient-specific manner, directly facilitating precision medicine.

In immuno-oncology, imaging agents hold great potential for use in determining which patients are likely to respond to specific targeted treatments, when to stage treatments, and when to switch treatments. A recent review out of the Gambhir lab proposes a framework for the development of immune checkpoint imaging agents [87]. The framework includes consideration of imaging molecule size and related PK/PD, as well as choice of isotope, which affects imaging window duration as well as practicality of use/safety. Perhaps unlike previous imaging biomarker translation efforts, this is a timely and "up front" attempt to drive consensus and standards into a rapidly progressing new field, based partly on learnings from imaging agents that did not realize their originally perceived potential.

Imaging of Macrophages

Inflammation is a major feature of a broad spectrum of diseases [88–90], and macrophage activation and infiltration are hallmarks of the inflammatory cascade. This is true in oncology also, where tumor-associated macrophages (TAMs) play an important role in tumor treatment response and the immune response generally [91]. Macrophages are a relatively new target for therapeutic intervention. Macrophages and microglia play highly critical roles in many other diseases including rheumatoid arthritis [92], nonalcoholic steatohepatitis (NASH) [93], and inflammatory bowel disease (IBD) [94]. The ability to detect and quantify activation of macrophages in a spatially resolved manner is therefore very valuable, and MRI biomarker approaches to doing this are under development, including the use of ^{19}F and FeOx containing agents that can be administered systemically and are selectively taken up by activated macrophages [95–97]. Quantitative imaging biomarkers would facilitate management of these diseases by allowing noninvasive patient stratification and treatment staging, potentially aiding the development of therapies, which have been difficult to assess using endpoints largely associated with indirect clinical endpoints and biopsy.

Therapeutic Cell Imaging

New areas of medicine based on cell therapies including stem cell/regenerative medicine and the explosive T cell therapeutics being used in immuno-oncology are driving new interest in, and development of, noninvasive cell detection, such as through MRI [98–102]. Current MRI approaches include the *in vitro* labeling of cells with phagocytic properties (including T cells) using benign ^{19}F and FeOx imaging agents, and subsequent detection using MR methods, after cell inoculation. Nuclear medicine approaches also exist including 111-indium and 89-zirconium oxine–based cell labeling [103–105]. Complimentary techniques that include pre-targeting radiochemistry [106] and PET or SPECT imaging and the use of potentially translatable gene reporters [107–109] are also being tested and developed, particularly in the therapeutic immune cell space.

Hyperpolarized ^{13}C MRI

Hyperpolarized 13-carbon (HP ^{13}C) technology has rapidly risen to prominent use [110–112], driven by the prevalence of metabolic targets in oncology. The widespread use of ^{18}F-FDG PET highlighted success in use of a metabolite tracer and respective biomarker for cell metabolism, but at the same time limitations have defined the need for other imaging agents that drive biomarkers for other metabolic pathways. HP ^{13}C MRI has overcome some of the technological hurdles but remains expensive and relatively inaccessible, though this is gradually improving. Nevertheless, several unique features of HP ^{13}C MRI include high sensitivity, lack of background, and the ability to simultaneously detect multiple molecules, including substrates, intermediates, and products (due to the broad chemical shift of ^{13}C nuclei). Selectively labeled ^{13}C probes are available for various metabolic enzymes such as pyruvate, lactic acid dehydrogenase (LDH)-A, alanine transaminase (ALT), glutaminase (GLU), and others. These agents are enabling specific biomarkers of response for metabolically targeted cancer pharmaceuticals and may pave the way for new advances in the pharma industry [110–113].

CEST Imaging Agents

Chemical exchange saturation transfer (CEST) is a MRI contrast enhancement technique that enables indirect detection of endogenous metabolites with exchangeable protons and today presents a very broad array of biomarkers and applications across all major areas of disease research. These imaging agents include glycosaminoglycans, glycogen, myo-inositol, glutamate, and creatine [114–117]. CEST molecular imaging is seeing use in clinical trials and in the pharmaceutical industry for a variety of purposes related to diagnosis, stratification, and treatment monitoring. One further CEST example has been the developments with CEST-based pH biomarkers [118–120] where there exist few reliable alternative *in vivo* assays. This holds significant promise, for example, in interpreting patient response to pH sensitive therapies, which are an active area of research in the pharmaceutical industry.

Resting State fMRI

One of the major new areas of neuroscience and CNS imaging biomarker applications is resting state functional magnetic resonance imaging or rs-fMRI [121–123], an example of a simplified and improved biomarker strategy based on learnings and quantification challenges associated with traditional fMRI biomarkers. rs-fMRI is based on imaging basal (non-intervention driven) neural activation via spontaneous, low-frequency fluctuations in the BOLD signal, and has now become a major translational imaging application in clinical research and the pharmaceutical industry. rs-fMRI relies on sophisticated, though today standard, processing techniques that model temporal correlations in BOLD effect spatial patterns in a rapid imaging time course over typically 10–20 minutes in the whole brain. These algorithms, which typically rely on either seeding or principle component analysis methods, generate functional connectivity maps and highlight one or more physiologically connected networks, parameters which can be used as biomarkers of disease progression and/or pharmacologic response. The most fundamental and well understood network that has been defined in rodents through non-human primates (NHPs), and in human patients, is the "Default Mode Network."

rs-fMRI can be used as a quantitative measure of pharmacologic response, through highly sensitive detection of brain activity. Based on the current state of the art, rs-fMRI is poised to play an increasingly significant role in the pharmaceutical industry in both discovery and translational work. As of today, there are hundreds of ongoing clinical trials in the United States that incorporate rs-fMRI.

New Modalities and Hardware

Future needs in imaging biomarkers are also being fueled by several advancements that have driven novel imaging modalities, and downstream imaging applications, biomarkers, and imaging agents. Two prominent examples are photoacoustic imaging [124–126] and magnetic particle imaging (MPI) [127]. Photoacoustic imaging is a modality that combines the best features of ultrasound and optical detection to enable high-resolution, non-depth limited images through endogenous and exogenous optical imaging probes. The modality leverages detectable acoustic signals in optical imaging agents and has been translated for clinical use. It has the potential to provide clinical translation of optical probes with unique capabilities, as well as unique biomarkers not accessible through other technologies. Similarly, MPI is a novel imaging technology that leverages a magnetic field to directly detect iron oxide–containing tracers with picomolar sensitivity and micron resolution. MPI is poised to enable the use of unique iron oxide agent–based biomarkers that have shown limited translation with MRI due to sensitivity and specificity challenges.

The increasing prevalence and commercial availability of multimodality imaging instruments (where more than one modality can be used in the same bore/gantry) are additionally facilitating imaging biomarker applications. Examples include combinations of two to three modalities for PET, SPECT, MRI, CT, photoacoustic imaging, and MPI. Recent advances in optimizing performance of the individual modalities in these hybrids have been key. By combining modalities that each have complementary strengths, the precision or utility associated with an imaging biomarker under one modality may be improved by co-acquisition of image data using a secondary modality. A common example is the use the MRI or CT anatomical image data to improve the localization and therefore quantification of lower resolution and more poorly tissue delineating PET or SPECT data.

Next Generation Image Informatics: Radiomics, Machine Learning, Deep Learning, and Artificial Intelligence

Massive advancements in computing power and speed are a critical driver for next generation imaging biomarkers. New approaches to image informatics (read: big data image analysis) such as artificial intelligence (AI) (including machine learning and deep learning) and image feature based "radiomics" are enabling extraction of more information from existing data sets and more precise and/or predictive biomarkers. The volume of image data that is being generated by major imaging institutions and consortia as well as those now available in the public domain are enabling artificial intelligence to play an increasing role in medical imaging.

Radiomics

The assessment of patient images has commonly been qualitative, describing broad features of disease pathology (e.g. "moderately heterogenous tumor") which are subjective in nature and prone to interobserver variability [128]. Radiomics has sought to employ high-throughput screening of quantitative image features to more objectively analyze and assess patient data. In radiomics approaches, mathematical algorithms mine image data with a segmented region of interest for numerous quantitative features including shape/size, histogram or filter-based readouts, and tissue texture. Results for a patient can be compared with clinical and genomics databases for more effective diagnosis and to personalize therapy. Most importantly, radiomics offers an objective, quantitative, and *personalized* way to assess image features that may not be apparent to the naked eye. Radiomics has the potential to improve existing biomarkers, but importantly to define new biomarkers extracted from the same image data.

Artificial Intelligence

A major category of AI, machine learning, involves pattern recognition algorithms that can be trained to parse large numbers of data sets, learn information from those data sets, and use that knowledge to make informed decisions. However, these algorithms still need user input and guidance upon inaccurate predictions in order to increase accuracy. Deep learning is a subset of the machine learning category, in which an algorithm creates a structure of artificial neural networks to aid in decision-making and then independently assess whether that decision is accurate.

In the medical imaging field, machine learning and deep learning are being used in a variety of applications including image segmentation, image data augmentation, and radiomics-based texture analysis and feature extraction to improve the accuracy of patient staging and better predict therapeutic response. For example, lung cancer patient staging requires analysis of multiple factors, including tumor histology and presence or absence of metastasis. Algorithms have been designed to "read" this information from patient records and make decisions regarding patient disease stage at the time of diagnosis. After training such algorithms using large data sets from previously staged patients, the accuracy of decisions was greater than 90% [129]. Similarly, analysis of therapeutic response using parameters such as tumor size has been shown to be more accurately predicted in colorectal and lung cancer patients by deep learning algorithms than standard models such as the Linear Regression and support vector machine (SVM) models [130, 131].

Also, paramount to using medical image data for making decisions about patient management is segmentation of the images themselves. Structural segmentation of brain MR images is routinely performed manually. This is not only tedious and time-consuming but also prone to user bias and error. Automated algorithms trained on millions of existing data sets have been successful in reducing segmentation time to as little as a few minutes using rapid graphics processing unit (GPU) processors, and have performed with accuracy rates greater than 85% [132]. Increasing the number of available training data sets and refining the algorithms will allow developers to continue improving upon these advances.

Facilitation and Outlook

While the landscape for tomorrow's imaging biomarkers is highly promising, the degree of success realized will depend not only on the science described above but also on other external factors including cost, reimbursement, and public and private funding. Access to imaging instrumentation and supporting technologies and related cost is another critical factor. The cost of imaging hardware and supporting infrastructure and personnel remain as major challenges, and have led to much churn within the translational imaging industry.

Notably, the preclinical imaging industry is critical to discovery and development of imaging biomarkers. Major manufacturers in this space have in recent years reduced or eliminated their nonclinical footprints (e.g. GE and Siemens), and other manufacturers have taken their place and introduced innovative new platforms (e.g. Mediso, MR Solutions, Sophie Biosciences, Magnetic Insight, Cubresa). Supporting technologies are also highly critical to imaging biomarker development, most notably imaging agent chemistry/radiochemistry, and also state-of-the-art image analytics platforms and applications that are key to biomarker validation and standardization. Both of these have historically been difficult to access, but today the industry is in a significantly better position with regard to these and other supporting technologies than 10 years ago, and imaging biomarker developers are poised to take advantage.

The pharmaceutical industry R&D spending is a further major driver for imaging biomarker validation and development. Recent evolution in this segment has seen a rapid shift from internalizing of large-scale imaging infrastructure to utilization of contract research organizations (CROs) that are more easily able to provide broad and deep capabilities, expertise, and supporting technologies. Today, leading imaging CROs have broad infrastructure and capabilities to support the latest industry demands. The concept of a CRO with large scale, global resources, including full coverage from *ex vivo* to preclinical to clinical imaging, with supporting chemistry, and with image analytics is now a reality. These developments are enhancing the positive outlook for tomorrow's quantitative imaging biomarkers and their provision, standardization, and potential for enabling precision medicine. The CRO model importantly also allows pharmaceutical company's internal R&D to be focused more completely on therapeutics discovery, with CROs providing biomarker and imaging agent platform technologies. This substantially more efficient and powerful model stands to better support imaging biomarker translation than under the previous standards of internal biomarker discovery and validation.

Notably, the pharmaceutical industry, including CROs, and some of the most prominent clinical institutions are combining departments and teams that specialize in historically separate disciplines including translational imaging (radiology) with those specializing in other major disciplines including pathology and genomics. Such efforts are paving the way for realizing success in combining and integrating big data across different "omics" areas, including radiomics in order to provide more powerful and finely tuned biomarkers and more precise information. These initiatives are being facilitated by artificial intelligence approaches to medical image data analysis and biomarker improvement.

Summary

Imaging biomarkers that have emerged in the last 10–20 years have driven major preclinical and clinical applications, yet at the same time have demonstrated limitations in clinical utility, including sensitivity and specificity in their originally intended applications. Based on learnings and recent advancements in areas such as imaging agents and image analysis/informatics, imaging biomarkers are poised to create unique impact in the future, keeping pace with rapidly advancing medical science. Improved processes and understanding of imaging biomarker discovery and validation are being coupled with greater access to imaging technologies and supporting infrastructure and efforts to combine imaging data sets with those coming from other major biomedical disciplines. These trends are expected to pave the way for higher fidelity next generation imaging biomarkers for specific, critical applications in major diseases including cancer, Alzheimer's, NASH and IBD.

References

1 Farwell, M.D., Clark, A.S., and Mankoff, D. (2015). How imaging biomarkers can inform clinical trials and clinical practice in the era of targeted cancer therapy. *JAMA Oncol.* 1: 421–422.
2 Wang, Y.-X. and Deng, M. (2010). Medical imaging in new drug clinical development. *J. Thorac. Dis.* 2: 245–252.
3 Willmann, J.K., van Bruggen, N., Dinkelborg, L.M., and Gambhir, S.S. (2008). Molecular imaging in drug development. *Nat. Rev. Drug Discovery* 7: 591–607.
4 Lohrke, J., Frenzel, T., Endrikat, J. et al. (2016). 25 years of contrast-enhanced MRI: developments, current challenges and future perspectives. *Adv. Ther.* 33: 1–28.
5 Heiss, W.-D. and Herholz, K. (2006). Brain receptor imaging. *J. Nucl. Med.* 47: 302–312.
6 Ding, H. and Wu, F. (2012). Image guided biodistribution and pharmacokinetic studies of theranostics. *Theranostics* 2: 1040–1053.
7 Chen, Z.-Y., Wang, Y.-X., Lin, Y. et al. (2014). Advance of molecular imaging technology and targeted imaging agent in imaging and therapy. *Biomed Res. Int.* 2014: 1–12.
8 Freise, A.C. and Wu, A.M. (2015). In vivo imaging with antibodies and engineered fragments. *Mol. Immunol.* 67: 142–152.
9 Becker, E.D. (1993). A brief history of nuclear magnetic resonance. *Anal. Chem.* 65: 295A–302A.
10 Bradley, W.G. (2008). History of medical imaging. *Proc. Am. Philos. Soc.* 152: 349–361.
11 Wagner, H.N. (1998). A brief history of positron emission tomography (PET). *Semin. Nucl. Med.* 28: 213–220.
12 Hutton, B. (2014). The origins of SPECT and SPECT/CT. *Eur. J. Nucl. Med. Mol. Imaging* 41: 3–16.
13 Pan, X., Siewerdsen, J., La Riviere, P.J., and Kalender, W.A. (2008). Anniversary paper: development of x-ray computed tomography: the role of Medical Physics and AAPM from the 1970s to present. *Med. Phys.* 35: 3728–3739.

14 Troxclair, L., Smetherman, D., and Bluth, E.I. (2011). Shades of gray: a history of the development of diagnostic ultrasound in a large multispecialty clinic. *Ochsner J.* 11: 151–155.

15 Vanderheyden, J.L. (2009). The use of imaging in preclinical drug development. *Q. J. Nucl. Med. Mol. Imag.* 53: 374–381.

16 Abramson, R.G. et al. (2015). Methods and challenges in quantitative imaging biomarker development. *Acad. Radiol.* 22: 25–32.

17 European Society of Radiology (2010). White paper on imaging biomarkers. *Insights Imaging* 1: 42–45.

18 Almuhaideb, A., Papathanasiou, N., and Bomanji, J. (2011). 18F-FDG PET/CT imaging in oncology. *Ann. Saudi Med.* 31: 3–13.

19 Farwell, M.D., Pryma, D.A., and Mankoff, D.A. (2014). PET/CT imaging in cancer: current applications and future directions. *Cancer* 120: 3433–3445.

20 Titford, M. (2006). A short history of histopathology technique. *J. Histotechnol.* 29: 99–110.

21 Musumeci, G. (2014). Past, present and future: overview on histology and histopathology. *J. Histol. Histopathol.* 1: 5.

22 Carvajal-Hausdorf, D.E., Schalper, K.A., Neumeister, V.M., and Rimm, D.L. (2015). Quantitative measurement of cancer tissue biomarkers in the lab and in the clinic. *Lab. Invest.* 95: 385–396.

23 de Matos, L.L., Trufelli, D.C., de Matos, M.G.L., and da Silva Pinhal, M.A. (2010). Immunohistochemistry as an important tool in biomarkers detection and clinical practice. *Biomarker Insights* 2010: 9–20.

24 Eyzaguirre, E. and Haque, A.K. (2008). Application of immunohistochemistry to infections. *Arch. Pathol. Lab. Med.* 132: 424–431.

25 Schuhmacher, A., Gassmann, O., and Hinder, M. (2016). Changing R&D models in research-based pharmaceutical companies. *J. Transl. Med.* 14: 1–11.

26 U.S. National Institutes of Health (2013). clinicaltrials.gov.

27 O'Connor, J.P.B., Jackson, A., Parker, G.J.M. et al. (2012). Dynamic contrast-enhanced MRI in clinical trials of antivascular therapies. *Nat. Rev. Clin. Oncol.* 9: 167–177.

28 Fennessy, F.M., McKay, R.R., Beard, C.J. et al. (2014). Dynamic contrast-enhanced magnetic resonance imaging in prostate cancer clinical trials: potential roles and possible pitfalls. *Transl. Oncol.* 7: 120–129.

29 Chilla, G.S., Tan, C.H., Xu, C., and Poh, C.L. (2015). Diffusion weighted magnetic resonance imaging and its recent trend-a survey. *Quant. Imaging Med. Surg.* 5: 407–422.

30 Gluskin, J.S., Chegai, F., Monti, S. et al. (2016). Hepatocellular carcinoma and diffusion-weighted MRI: detection and evaluation of treatment response. *J. Cancer* 7: 1565–1570.

31 Sanghera, B. et al. (2014). FLT PET-CT in evaluation of treatment response. *Indian J. Nucl. Med.* 29: 65–73.

32 Bollinenia, V.R., Kramerb, G.M., Jansmac, E.P. et al. (2016). A systematic review on [18F]FLT-PET uptake as a measure of treatment response in cancer patients. *Eur. J. Cancer* 55: 81–97.
33 McKinley, E.T., Ayers, G.D., Smith, R.A. et al. (2013). Limits of [18F]-FLT PET as a biomarker of proliferation in oncology. *PLoS One* 8: e58938.
34 Lee, S.T. and Scott, A.M. (2007). Hypoxia positron emission tomography imaging with 18F-fluoromisonidazole. *Semin. Nucl. Med.* 37: 451–461.
35 Fleming, I.N. et al. (2015). Imaging tumour hypoxia with positron emission tomography. *Br. J. Cancer* 112: 238–250.
36 RSNA. Quantitative Imaging Biomarkers Alliance® (QIBA®). https://www.rsna.org/research/quantitative-imaging-biomarkers-alliance/profiles-and-protocols.
37 El-Gamal, F.E.-Z.A., Elmogy, M., and Atwan, A. (2016). Current trends in medical image registration and fusion. *Egypt. Inf. J.* 17: 99–124.
38 Shen, D., Wu, G., and Suk, H.-I. (2017). Deep learning in medical image analysis. *Annu. Rev. Biomed. Eng.* 19: 221–248.
39 de Bruijne, M. (2016). Machine learning approaches in medical image analysis: from detection to diagnosis. *Med. Image Anal.* 33: 94–97.
40 Sullivan, D.C. et al. (2015). Metrology standards for quantitative imaging biomarkers. *Radiology* 277: 813–825.
41 RSNA. QIBA Profiles and Protocols. https://www.rsna.org/research/quantitative-imaging-biomarkers-alliance/profiles-and-protocols.
42 Kinahan, P., Wahl, R., Shao, L. et al. (2014). The QIBA profile for quantitative FDG-PET/CT oncology imaging. *J. Nucl. Med.* 55: 1520.
43 Peters, J., Leal, J., and Subramaniam, R. (2015). The QIBA profile: are we adhering to recommendations stated regarding blood glucose values and radiotracer uptake times? *J. Nucl. Med.* 56: 2603.
44 DCE MRI Technical Committee (2012). DCE MRI Quantification Profile, Quantitative Imaging Biomarkers Alliance. Rsna.Org/Qiba 46.
45 Nayak, T.K. and Brechbiel, M.W. (2009). Radioimmunoimaging with longer-lived positron-emitting radionuclides: potentials and challenges. *Bioconjugate Chem.* 20: 825–841.
46 Cremonesi, M., Gilardi, L., Ferrari, M.E. et al. (2017). Role of interim 18F-FDG-PET/CT for the early prediction of clinical outcomes of Non-Small Cell Lung Cancer (NSCLC) during radiotherapy or chemo-radiotherapy. A systematic review. *Eur. J. Nucl. Med. Mol. Imaging* 44: 1915–1927.
47 Kitajima, K. et al. (2016). Present and future roles of FDG-PET/CT imaging in the management of lung cancer. *Jpn. J. Radiol.* 34: 387–399.
48 Agarwal, A., Marcus, C., Xiao, J. et al. (2014). FDG PET/CT in the management of colorectal and anal cancers. *Am. J. Roentgenol.* 203: 1109–1119.

49 D'souza, M.M., Jaimini, A., Bansal, A. et al. (2013). FDG-PET/CT in lymphoma. *Indian J. Radiol. Imaging* 23: 354–365.

50 Correa, A.F. and Smaldone, M.C. (2017). PET/CT for prostate cancer: from diagnosis to metastasis. *Renal Urology News* 11–18.

51 Jadvar, H. (2009). FDG PET in prostate cancer. *PET Clin* 4: 155–161.

52 Groheux, D., Espié, M., Giacchetti, S., and Hindié, E. (2013). Performance of FDG PET/CT in the clinical management of breast cancer. *Radiology* 266: 388–405.

53 Bloom, G.S. (2014). Amyloid-β and tau: the trigger and bullet in Alzheimer disease pathogenesis. *JAMA Neurol.* 71: 505–508.

54 Rapoport, M., Dawson, H.N., Binder, L.I. et al. (2002). Tau is essential to beta-amyloid-induced neurotoxicity. *Proc. Natl. Acad. Sci. U.S.A.* 99: 6364–6369.

55 Clark, C.M. et al. (2011). Use of florbetapir-PET for imaging beta-amyloid pathology. *JAMA* 305: 275–283.

56 Wong, D.F. et al. (2010). In vivo imaging of amyloid deposition in Alzheimer's disease using the novel radioligand [18F]AV-45 (Florbetapir F 18). *J. Nucl. Med.* 51: 913–920.

57 Hatashita, S., Yamasaki, H., Suzuki, Y. et al. (2014). [18F]Flutemetamol amyloid-beta PET imaging compared with [11C]PIB across the spectrum of Alzheimer's disease. *Eur. J. Nucl. Med. Mol. Imaging* 41: 290–300.

58 Villemagne, V.L. et al. (2012). Comparison of 11C-PiB and 18F-florbetaben for A-beta imaging in ageing and Alzheimer's disease. *Eur. J. Nucl. Med. Mol. Imaging* 39: 983–989.

59 Marcus, C., Mena, E., and Subramaniam, R.M. (2015). Brain PET in the diagnosis of Alzheimer's disease. *Clin. Nucl. Med.* 39: e413–e426.

60 Johnson, K.A. et al. (2013). Appropriate use criteria for amyloid PET: a report of the Amyloid Imaging Task Force, the Society of Nuclear Medicine and Molecular Imaging, and the Alzheimer's Association. *Alzheimers Dementia* https://doi.org/10.1016/j.jalz.2013.01.002.

61 Edison, P., Rowe, C.C., Rinne, J.O. et al. (2008). Amyloid load in Parkinson's disease dementia and Lewy body dementia measured with [11C]PIB positron emission tomography. *J. Neurol. Neurosurg. Psychiatry* 79: 1331–1338.

62 James, O.G., Doraiswamy, P.M., and Borges-Neto, S. (2015). PET imaging of tau pathology in Alzheimer's disease and tauopathies. *Front. Neurol.* 6: 38.

63 Shah, M. and Catafau, A.M. (2014). Molecular imaging insights into neurodegeneration: focus on tau PET radiotracers. *J. Nucl. Med.* 55: 871–874. https://doi.org/10.2967/jnumed.113.136069.

64 García-Figueiras, R., Baleato-González, S., Padhani, A.R. et al. (2016). Proton magnetic resonance spectroscopy in oncology: the fingerprints of cancer? *Diagn. Interventional Radiol.* 22: 75–89.

65 Ramadan, S., Lin, A., and Stanwell, P. (2013). Glutamate and glutamine: a review of in vivo MRS in the human brain. *NMR Biomed.* 26: 1630–1646.

66 Choi, C. et al. (2012). 2-Hydroxyglutarate detection by magnetic resonance spectroscopy in IDH-mutated patients with gliomas. *Nat. Med.* 18: 624–629.

67 Chin, C.L., Fox, G.B., Hradil, V.P. et al. (2006). Pharmacological MRI in awake rats reveals neural activity in area postrema and nucleus tractus solitarius: relevance as a potential biomarker for detecting drug-induced emesis. *Neuroimage* 33: 1152–1160.

68 Carmichael, O., Schwarz, A.J., Chatham, C.H. et al. (2018). The role of fMRI in drug development. *Drug Discovery Today* 23: 333–348.

69 Farkona, S., Diamandis, E.P., and Blasutig, I.M. (2016). Cancer immunotherapy: the beginning of the end of cancer? *BMC Med.* 14: 73.

70 Almåsbak, H., Aarvak, T., and Vemuri, M.C. (2016). CAR T cell therapy: a game changer in cancer treatment. *J. Immunol. Res.* 2016.

71 Yamaki, S., Yanagimoto, H., Tsuta, K. et al. (2017). PD-L1 expression in pancreatic ductal adenocarcinoma is a poor prognostic factor in patients with high CD8+ tumor-infiltrating lymphocytes: highly sensitive detection using phosphor-integrated dot staining. *Int. J. Clin. Oncol.* https://doi.org/10.1007/s10147-017-1112-3.

72 Gonda, K. et al. (2017). Quantitative diagnostic imaging of cancer tissues by using phosphor-integrated dots with ultra-high brightness. *Sci. Rep.* https://doi.org/10.1038/s41598-017-06534-z.

73 Topalian, S.L., Drake, C.G., and Pardoll, D.M. (2015). Immune checkpoint blockade: a common denominator approach to cancer therapy. *Cancer Cell* 27: 451–461.

74 Sharma, P. (2016). Immune checkpoint therapy and the search for predictive biomarkers. *Cancer J.* 22: 68–72.

75 Sharpe, M. and Mount, N. (2015). Genetically modified T cells in cancer therapy: opportunities and challenges. *Dis. Model. Mech.* 8: 337–350.

76 Wang, Z., Wu, Z., Liu, Y., and Han, W. (2017). New development in CAR-T cell therapy. *J. Hematol. Oncol.* 10: 53.

77 Tavaré, R. et al. (2016). An effective immuno-PET imaging method to monitor CD8-dependent responses to immunotherapy. *Cancer Res.* 76: 73–82.

78 Tavaré, R. et al. (2014). Engineered antibody fragments for immuno-PET imaging of endogenous CD8+ T cells in vivo. *Proc. Natl. Acad. Sci. U.S.A.* 111: 1108–1113.

79 Chatterjee, S. et al. (2016). Rapid PD-L1 detection in tumors with PET using a highly specific peptide. *Biochem. Biophys. Res. Commun.* 483: 258–263.

80 Ehlerding, E.B., England, C.G., Majewski, R.L. et al. (2017). W. ImmunoPET imaging of CTLA-4 expression in mouse models of non-small cell lung cancer. *Mol. Pharmaceutics* 14: 1782–1789.

81 Ehlerding, E.B., England, C.G., McNeel, D.G., and Cai, W. (2016). Molecular imaging of immunotherapy targets in cancer. *J. Nucl. Med.* https://doi.org/10.2967/jnumed.116.177493.

82 Larimer, B.M., Wehrenberg-Klee, E., Caraballo, A., and Mahmood, U. (2016). Quantitative CD3 PET imaging predicts tumor growth response to anti-CTLA-4 therapy. *J. Nucl. Med.* 57: 1607–1611.

83 Larimer, B.M. et al. (2017). Granzyme B PET imaging as a predictive biomarker of immunotherapy response. *Cancer Res.* 77: 2318–2327.

84 Ronald, J.A., Kim, B.-S., Gowrishankar, G. et al. (2017). A PET imaging strategy to visualize activated T cells in acute graft-versus-host disease elicited by allogenic hematopoietic cell transplant. *Cancer Res.* 77: 2893–2902.

85 Alam, I.T., Mayer, A.T., Sagiv-Barfi, I. et al. (2018). Imaging activated T cells predicts response to cancer vaccines. *J. Clin. Invest.* 128: 2569–2580.

86 Konishi, M. et al. (2015). Imaging granzyme B activity assesses immune-mediated myocarditis. *Circ. Res.* 117: 502–512.

87 Aaron T. Mayer and Sanjiv S. Gambhir. (2018). The Immuno-Imaging Toolbox, *J Nucl Med August 1*, vol. 59 no. 8 1174–1182.

88 Hunter, P. (2012). The inflammation theory of disease: the growing realization that chronic inflammation is crucial in many diseases opens new avenues for treatment. *EMBO Rep.* 13: 968–970.

89 Fernandes, J.V., Cobucci, R.N., Jatobá, C.A. et al. (2015). The role of the mediators of inflammation in cancer development. *Pathol. Oncol. Res.* 21: 527–534.

90 Lucas, S.-M., Rothwell, N.J., and Gibson, R.M. (2006). The role of inflammation in CNS injury and disease. *Br. J. Pharmacol.* https://doi.org/10.1038/sj.bjp.0706400.

91 Yang, L. and Zhang, Y. (2017). Tumor-associated macrophages: from basic research to clinical application. *J. Hematol. Oncol.* https://doi.org/10.1186/s13045-017-0430-2.

92 Udalova, I.A., Mantovani, A., and Feldmann, M. (2016). Macrophage heterogeneity in the context of rheumatoid arthritis. *Nat. Rev. Rheumatol.* https://doi.org/10.1038/nrrheum.2016.91.

93 Grunhut, J., Wang, W., Aykut, B. et al. (2018). Macrophages in nonalcoholic steatohepatitis: friend or foe? *Eur. Med. J. Hepatol.* 6: 100–109.

94 Steinbach, E.C. and Plevy, S.E. (2014). The role of macrophages and dendritic cells in the initiation of inflammation in IBD. *Inflamm. Bowel Dis.* 20: 166–175.

95 Korchinski, D.J., Taha, M., Yang, R. et al. (2015). Iron oxide as an MRI contrast agent for cell tracking. *Magn. Reson. Insights* 8: 15–29.

96 Fox, M.S., Gaudet, J.M., and Foster, P.J. (2015). Fluorine-19 MRI contrast agents for cell tracking and lung imaging. *Magn. Reson. Insights* 8: 53–67.

97 Makela, A.V., Gaudet, J.M., and Foster, P.J. (2017). Quantifying tumor associated macrophages in breast cancer: a comparison of iron and fluorine-based MRI cell tracking. *Sci. Rep.* 7: 42109.

98 Ahrens, E.T. and Bulte, J.W.M. (2013). Tracking immune cells in vivo using magnetic resonance imaging. *Nat. Rev. Immunol.* 13: 755–763.

99 Ahrens, E.T. and Zhong, J. (2013). In vivo MRI cell tracking using perfluorocarbon probes and fluorine-19 detection. *NMR Biomed.* 26: 860–871.

100 Ahrens, E.T., Helfer, B.M., O'Hanlon, C.F., and Schirda, C. (2014). Clinical cell therapy imaging using a perfluorocarbon tracer and fluorine-19 MRI. *Magn. Reson. Med.* 72: 1696–1701.

101 Bulte, J.W.M. (2009). In vivo MRI cell tracking: clinical studies. *Am. J. Roentgenol.* 193: 314–325.

102 Liu, Z. and Li, Z. (2014). Molecular imaging in tracking tumor-specific cytotoxic T lymphocytes (CTLs). *Theranostics* 4: 990–1001.

103 Hartimath, S.V., Draghiciu, O., van de Wall, S. et al. (2016). Noninvasive monitoring of cancer therapy induced activated T cells using [18F]FB-IL-2 PET imaging. *Oncoimmunology* 6.

104 Stanton, S.E. et al. (2016). Concurrent SPECT/PET-CT imaging as a method for tracking adoptively transferred T-cells in vivo. *J. ImmunoTher. Cancer* 4: 27.

105 Sato, N., Wu, H., Asiedu, K.O. et al. (2015). 89Zr-oxine complex PET cell imaging in monitoring cell-based therapies. *Radiology* 275: 490–500.

106 Keinänen, O. et al. (2017). Pretargeted PET imaging of trans-cyclooctene-modified porous silicon nanoparticles. *ACS Omega* https://doi.org/10.1021/acsomega.6b00269.

107 Keu, K.V. et al. (2017). Reporter gene imaging of targeted T cell immunotherapy in recurrent glioma. *Sci. Transl. Med.* https://doi.org/10.1126/scitranslmed.aag2196.

108 Gschweng, E.H. et al. (2014). HSV-sr39TK positron emission tomography and suicide gene elimination of human hematopoietic stem cells and their progeny in humanized mice. *Cancer Res.* https://doi.org/10.1158/0008-5472.CAN-14-0376.

109 Minn, I.L., Huss, D., Ahn, H.-H. (2018). PSMA gene reporter CAR-T imaging: P300 PSMA-associated PET imaging of CAR T cells. Abstracts of the 29th Annual Scientific Meeting of the Society for Immunotherapy of Cancer.

110 Miloushev, V.Z., Keshari, K.R., and Holodny, A.I. (2016). Hyperpolarization MRI: preclinical models and potential applications in neuroradiology. *Top. Magn. Reson. Imaging* 25: 31–37.

111 Kurhanewicz, J., Bok, R., Nelson, S.J., and Vigneron, D.B. (2008). Current and potential applications of clinical ^{13}C MR spectroscopy. *J. Nucl. Med.* 49: 341–344.

112 Sinharay, S. and Pagel, M.D. (2016). Advances in magnetic resonance imaging contrast agents for biomarker detection. *Annu. Rev. Anal. Chem.* 9: 95–115.

113 Adamson, E.B., Ludwig, K.D., Mummy, D.G., and Fain, S.B. (2017). Magnetic resonance imaging with hyperpolarized agents: methods and applications. *Phys. Med. Biol.* 62: R81–R123.

114 Kogan, F., Hariharan, H., and Reddy, R. (2013). Chemical exchange saturation transfer (CEST) imaging: description of technique and potential clinical applications. *Curr. Radiol. Rep.* 1: 102–114.

115 Walker-Samuel, S. et al. (2013). In vivo imaging of glucose uptake and metabolism in tumors. *Nat. Med.* 19: 1067–1072.

116 Cai, K. et al. (2017). Creatine CEST MRI for differentiating gliomas with different degrees of aggressiveness. *Mol. Imag. Biol.* 19: 225–232.

117 Cai, K. et al. (2012). Magnetic resonance imaging of glutamate. *Nat. Med.* 18: 302–306.

118 Chen, L.Q. and Pagel, M.D. (2015). Evaluating pH in the extracellular tumor microenvironment using CEST MRI and other imaging methods. *Adv. Radiol.* 2015: 1–25.

119 Longo, D.L. et al. (2016). In vivo imaging of tumor metabolism and acidosis by combining PET and MRI-CEST pH imaging. *Cancer Res.* 76: 6463–6470.

120 Moon, B.F. et al. (2015). A comparison of iopromide and iopamidol, two acidoCEST MRI contrast media that measure tumor extracellular pH. *Contrast Media Mol. Imaging* 10: 446–455.

121 Chen, L.M. et al. (2017). Biophysical and neural basis of resting state functional connectivity: evidence from non-human primates. *Magn. Reson. Imaging* https://doi.org/10.1016/j.mri.2017.01.020.

122 de Vos, F. et al. (2018). A comprehensive analysis of resting state fMRI measures to classify individual patients with Alzheimer's disease. *Neuroimage* https://doi.org/10.1016/j.neuroimage.2017.11.025.

123 Lee, M.H., Smyser, C.D., and Shimony, J.S. (2013). Resting-state fMRI: a review of methods and clinical applications. *Am. J. Neuroradiol.* 34: 1866–1872.

124 Xia, J., Yao, J., and Wang, L.V. (2014). Photoacoustic tomography: principles and advances. *Electromagn. Waves (Camb)* 147: 1–22.

125 Liu, C. et al. (2016). Advances in imaging techniques and genetically encoded probes for photoacoustic imaging. *Theranostics* 6: 2414–2430.

126 Weber, J., Beard, P.C., and Bohndiek, S.E. (2016). Contrast agents for molecular photoacoustic imaging. *Nat. Methods* 13: 639–650.

127 Panagiotopoulos, N. et al. (2015). Magnetic particle imaging: current developments and future directions. *Int. J. Nanomed.* https://doi.org/10.2147/IJN.S70488.

128 Yip, S.S.F. and Aerts, H.J.W.L. (2016). Applications and limitations of radiomics. *Phys. Med. Biol.* https://doi.org/10.1088/0031-9155/61/13/R150.

129 Simon, G., DiNardo, C.D., Takahashi, K. et al. (2018). Applying artificial intelligence to address the knowledge gaps in cancer care. *Oncologist* 23: 1–11.

130 Bibault, J.E. et al. (2018). Deep learning and radiomics predict complete response after neo-adjuvant chemoradiation for locally advanced rectal cancer. *Sci. Rep.* https://doi.org/10.1038/s41598-018-30657-6.

131 Parmar, C., Grossmann, P., Bussink, J. et al. (2015). Machine learning methods for quantitative radiomic biomarkers. *Sci. Rep.* https://doi.org/10.1038/srep13087.

132 Akkus, Z., Galimzianova, A., Hoogi, A. et al. (2017). Deep learning for brain MRI segmentation: state of the art and future directions. *J. Digital Imaging* https://doi.org/10.1007/s10278-017-9983-4.

Index

Biomarkers in Drug Discovery and Development: A Handbook of Practice, Application, and Strategy, Second Edition. Edited by Ramin Rahbari, Jonathan Van Niewaal, and Michael R. Bleavins.

e

f

g

h

e

f

g

h